# DESIGN AND CONTROL OF DISTILLATION SYSTEMS FOR SEPARATING AZEOTROPES

# DESIGN AND CONTROL OF DISTILLATION SYSTEMS FOR SEPARATING AZEOTROPES

**WILLIAM L. LUYBEN**
Lehigh University

**I-LUNG CHIEN**
National Taiwan University of Science and Technology

AIChE®

WILEY

A JOHN WILEY & SONS, INC., PUBLICATION

Published by John Wiley & Sons, Inc., Hoboken, New Jersey
Published simultaneously in Canada



***Library of Congress Cataloging-in-Publication Data:***

Luyben, William L.
Design and control of distillation systems for separating azeotropes / William L Luyben, I-Lung Chien.
p. cm.
Includes index.
ISBN 978-0-470-44862-5 (cloth)
1. Azeotropic distillation- -Design. 2. Azeotropes. I. Chien, I-Lung, 1955- II. Title.
TP156. D5L89 2010
660′.2842- -dc22

2009030816

10 9 8 7 6 5 4 3 2 1

This book is dedicated to the fond memory of
Cheng-Ching "CC" Yu, colleague and friend who made
many significant contributions in the field of process control.

# CONTENTS

# PREFACE

Many mixtures of chemical components form azeotropes. An azeotropic mixture has vapor and liquid phases that have identical compositions. The occurrence of this phenomenon means that simple distillation cannot be used to achieve complete separation because distillation relies on differences in compositions between liquid and vapor phases. Azeotropes occur because of nonideal phase equilibrium resulting from the molecular interaction (either repulsion or attraction) of dissimilar chemical components.

The petroleum industry seldom encounters azeotropes because the hydrocarbon components have similar molecular structures. The chemical industry, on the other hand, frequently deals with components of different types, which often produce azeotropes. Common examples are the production of methyl acetate, tetrahydrofuran, tert-amyl methyl ether, isopropanol, vinyl acetate, and many others.

The emerging biofuel processes typically have fermentation products that form azeotropes with the water, which is present in large excess in the fermentor. The most important example is ethanol, which forms a minimum-boiling homogeneous binary azeotrope with water. Butanol is another biofuel example that forms an azeotrope with water. The nonideality in this system is so large that the azeotrope is heterogeneous, forming two liquid phases.

If the molecules exhibit repulsive forces, the azeotrope will be minimum boiling and can be either homogeneous or heterogeneous. If the molecules exhibit attractive forces, the azeotrope will be maximum boiling. Since there are many types of azeotropes with a variety of physical properties, there will be many ways to achieve separation. Homogeneous binary azeotropes can be separated using several methods.

1. *Pressure-swing distillation*: If the composition of the azeotrope changes significantly with pressure, two columns operating at two different pressures can achieve separation.
2. *Azeotropic distillation*: A third component called a light-entrainer is added, which carries one of the components overhead in a distillation column and forms two liquid phases in a decanter. A two-column system is used with one liquid phase from the decanter fed to one column and the other liquid phase fed to the second column. High-purity products are produced from the bottom of the two columns.

3. *Extractive distillation*: A third component called a solvent (or heavy entrainer) is added near the top of the first extractive distillation column. The solvent carries one of the components out the bottom of the extractive column. A second solvent-recovery column regenerates the solvent, which is recycled back to the extractive column. High-purity products are produced from the top of the extractive column and from the top of the solvent-recovery column.

Heterogeneous binary azeotropes can be separated by taking advantage of the liquid–liquid phase separation in a decanter. No third component is required. A two-column system is used with the one liquid phase from a decanter fed to one column and the other liquid phase fed to the second column. High-purity products are produced from the bottom of the two columns.

Azeotropic distillation and extractive distillation have been used for many decades. There is a rich literature of the subject with many hundreds of papers and several books published on the subject. The recent textbooks that contain discussions of azeotropic and extractive distillation are Doherty and Malone[1] and Stichlmair and Fair[2]. The authors provide an excellent treatment of conceptual design of azeotropic systems. Both of these books give examples of azeotropic separation phase equilibrium and flowsheets in both binary and ternary systems. The chapter by Doherty and Knapp "Azeotropes" in the *Kirk–Othmer Encylcopedia of Chemical Processes* also provides an excellent discussion of vapor–liquid–liquid equilibrium (VLLE) fundamentals and conceptual design of a number of azeotropic systems.

Most of the treatments in the above books are qualitative and conceptual in nature, emphasizing VLLE issues and alternative configurations. Few of these books present in-depth rigorous designs that achieve optimum economic criteria. None of these books deal with the control and operation of azeotropic distillation systems. Detailed discussions of these two areas are the main contribution of this book. Rigorous steady-state and dynamic simulation tools (Aspen Plus and Aspen Dynamics) are used for design calculations and rigorous dynamic simulations.

This book studies a broad spectrum of real azeotropic distillation separation methods for a variety of industrially important chemical systems. Economically optimum rigorous steady-state designs are developed for many of these chemical systems. Then practical control structures are developed that provide effective load rejection in the face of typically large disturbances in throughput and feed composition. Trade-offs between steady-state energy savings and dynamic controllability (product quality variability) are demonstrated.

We hope this book will be useful for undergraduate and graduate students in their design, control, mass transfer, and bioprocessing courses as a supplemental reference. We hope it will also be used by young engineers just starting in their technical careers in industry or by older engineers facing an azeotropic separation problem for the first time.

The design and control of azeotropic systems is a challenging and fascinating subject, which requires application in many chemical engineering principles: thermodynamics of phase equilibrium, mass and energy balances, and dynamics and control. We have enjoyed working on these systems and hope the reader will find it equally intriguing and rewarding.

## REFERENCES

1. Doherty M. F. and M. F. Malone, *Conceptual Design of Distillation Systems*, McGraw-Hill, New York, 2001.
2. Stichlmair J. G. and J. R. Fair, *Distillation Principles and Practices*, Wiley-VCH, Weinheim, 1998, 222–342.

# ACKNOWLEDGEMENTS

I would like to thank my wife Pingching, the most important person in my life, and my two wonderful sons, Andrew and Jonathan, for their enduring support and encouragement. I would also like to express my thanks to Professor Hsiao-Ping Huang, for collaboration and work over many years, particularly on the results of acetic acid dehydration in this book. Thanks are also due to Professor David S. H. Wong for an initial collaborative study on IPA dehydration that inspired me to start a series of studies on various heterogeneous azeotropic systems. Thanks go to my former graduate students at National Taiwan University of Science and Technology and at National Taiwan University: Dr. Hao-Yeh Lee, Saiful Arifin, Y. C. Wu, K. Y. Hsu, L. Gunawan, K. L. Zeng, H. Y. Chao, K. H. Liu, T. K. Gau, C. H. Wang, J. Y. Yao, S. Y. Lin, and H. J. Huang. Without their hard work this book would not have been completed.

I-Lung Chien

# PART 1

# FUNDAMENTALS AND TOOLS

# CHAPTER 1

# INTRODUCTION

An azeotrope is a mixture of chemical components that has identical compositions of the liquid and vapor phases in equilibrium with each other. This azeotropic phenomenon occurs because of molecular interactions between different chemical components. If the components contain similar elemental molecules and structure, the molecular interaction is very slight and azeotropes do not occur. Mixtures of hydrocarbons (propane, butane, pentane, etc.) are important examples. These mixtures have phase equilibrium behavior that is close to ideal and often have relative volatilities that are almost constant. However, if the components are dissimilar, the molecules can exhibit either repulsion or attraction. Repulsion increases the effective vapor pressures of the components at a given temperature and can produce minimum-boiling azeotropes. In terms of liquid-phase activity coefficients, discussed in detail in Chapter 2, repulsion produces activity coefficients greater than unity. If the repulsion is large enough, the repulsive forces can lead to the formation of heterogeneous minimum-boiling azeotropes (two liquid phases). The azeotrope boils at a temperature that is lower than the boiling point of the lighter component.

The methanol–water mixture is an example of a system that exhibits modest nonideality. The OH end of the methanol molecule is similar to the OH end of the water molecule, but the hydrocarbon $CH_3$ end of the methanol molecule is different from water. So the system exhibits a modest amount of nonideality, which can be represented by the change in relative volatility as liquid composition changes.

If we add another $CH_2$ group and move to the ethanol–water system, there is more repulsion because the $CH_3$–$CH_2$ end of the ethanol molecule is quite different from the OH end of the water molecule. The system exhibits more nonideality and a minimum-boiling azeotrope occurs. This azeotrope is homogeneous (only one liquid phase in equilibrium with a vapor phase). If we add two more $CH_2$ groups and move to the *n*-butanol–water system, the

*Design and Control of Distillation Systems for Separating Azeotropes*. By William L. Luyben and I-Lung Chien

repulsion is even more extreme. The result is the formation of a *heterogeneous* minimum-boiling azeotrope with two liquid phases in equilibrium with a vapor phase.

In other chemical systems, the molecules can *attract* instead of repulse. The results can be the formation of *maximum*-boiling azeotropes because the molecular attraction reduces the effective vapor pressures of the components. Examples include nitric acid–water, acetone–chloroform, formic acid–water, and *n,n*-dimethyl acetamide–acetic acid.

Chapter 2 is devoted to a detailed discussion of the vapor–liquid–liquid phase equilibrium of azeotropic systems. In this chapter we provide some historical perspective of the field, discuss several typical and important applications, and provide some journal references in the area of control of azeotropic distillation systems.

Also included is an example of one of the important complexities in trying to deal with azeotropic distillation systems. These systems are highly nonlinear and exhibit the phenomenon of multiple steady states.

## 1.1 HISTORY

The existence of azeotropes has been known for many years, probably as long as man has made ethanol. Designing chemical engineering processes to separate azeotropes has been dealt with in some of the earliest chemical engineering books. The first chemical engineering textbook used by the senior author was the 1937 *Principles of Chemical Engineering*,[1] authored by three of the very early workers in the field of chemical engineering. On page 526 of that pioneering book, the *xy* curve for ethanol–water was presented. Subsequent pages showed curves for typical homogeneous minimum and maximum-boiling azeotropes.

The earliest book that concentrated on distillation as a unit operation *Elements of Fractional Distillation*,[2] was first published in 1922 and was into its fourth edition by 1950. Chapter 10 of this work discussed extractive and azeotropic distillation. In an era before computers, the authors presented how plate-to-plate component and phase equilibrium equations can be solved manually (mechanical calculator or slide rule) to design several extractive distillation systems. One interesting example was the maximum-boiling nitric acid–water system using sulfuric acid as the heavy entrainer (solvent). They also presented a detailed design of azeotropic distillation of ethanol–water using benzene as the light entrainer. Ternary diagrams and composition profiles were presented. The labor-intensive calculations probably took a hard-working graduate student many days or weeks to complete. We are fortunate to now have the tools to perform these calculations in seconds.

The earliest book that concentrated exclusively on extractive and azeotropic distillation was published in 1964 by E. J. Hoffman.[3] Ternary diagrams for many azeotropic systems were discussed, including the concepts of residue curves, and alternative flowsheet configurations were presented.

One of the pioneers in developing a method for separating azeotropic mixtures was Donald F. Othmer, who worked for Eastman Kodak before moving to Brooklyn Polytechnic University. Othmer's many papers and patents in the 1930s and 1940s were major contributions to the field. His early paper,[4] discussed the use of azeotropic distillation to facilitate the separation of acetic acid and water (which do not form an azeotrope but which exhibit a severe pinch in the high-water region) by the addition of a light entrainer (ethylene dichloride). Othmer developed these process designs without the aid of modern tools.

The advent of computer technology greatly enhanced our ability to explore azeotropic systems in a quantitative manner. Two of the early workers in developing and applying modern computational tools were Michael Doherty and Michael Malone at the University of Massachusetts. They demonstrated how ternary diagrams and the concepts of residue curves and distillation boundaries could be effectively used to aid in the conceptual design of azeotropic systems. They developed the Mayflower simulation software for performing azeotropic distillation calculations, which was later commercialized by Hyprotech into the DISTIL program and by AspenTech into the SPLIT program. Some of these tools are discussed and demonstrated in Chapter 2.

The early workers in the area of azeotropic separation concentrated on phase equilibrium issues and steady-state designs. There was much less work in the area of control of azeotropic systems. A pioneering paper by Prokopakis and Seider[5] dealing with the dynamic simulation of azeotropic distillation columns was published in 1983. One of the earliest papers that dealt with both design and control was published in 1985.[6] Several early papers by Morari and co-workers considered both design and control of azeotropic systems.[7–9]

A comprehensive review paper by Widagdo and Seider,[10] summarized the literature as of that date concerning the design, simulation, and control aspects of azeotropic systems. More recent control papers are reviewed in Section 1.3.

## 1.2 APPLICATIONS

There are many important industrial applications of azeotropic separations, which employ a variety of methods. In this book we discuss several of these chemical systems and demonstrate the application of alternative methods of separation. The methods presented include pressure-swing distillation, azeotropic distillation with a light entrainer, extractive distillation with a heavy entrainer (solvent), and pervaporation. The chemical systems used in the numerical case studies included ethanol–water; tetrahydrofuran (THF)–water, isopropanol–water, acetone–methanol, isopentane–methanol, *n*-butanol–water, acetone–chloroform, and acetic acid–water. Economic and dynamic comparisons between alternative methods are presented for some of the chemical systems, for example azeotropic distillation versus extractive distillation for the isopropanol–water system.

The systems discussed in this book are typical of many azeotropic systems. So the techniques demonstrated in subsequent chapters should be applicable to a wide range of systems.

## 1.3 LITERATURE

The literature dealing with azeotropes is vast with hundreds of journal articles. There are several excellent textbooks that deal with the conceptual design of azeotropic systems. Most notable are the books by Doherty and Malone[11] and Stichlmair and Fair.[12] Both of these books give examples of azeotropic separation phase equilibrium and flowsheets in both binary and ternary systems. The chapter by Doherty and Knapp "Azeotropes," in the *Kirk–Othmer Encylcopedia of Chemical Processes* provides an excellent discussion of VLLE fundamentals and conceptual design of a number of azeotropic systems.

The vast majority of the literature deals with phase equilibrium and steady-state design aspects of azeotropic separation. Papers dealing the control of azeotropic systems are far fewer in number. Some more recent journal articles are listed in the references 13–32.

## 1.4 MULTIPLE STEADY STATES

The nonlinear algebraic equations that describe a steady-state distillation column consist of component balances, energy balances, and vapor–liquid phase equilibrium relationships. These equations are nonlinear, particularly those describing the phase equilibrium of azeotropic systems. Unlike a linear set of algebraic equations that have one unique solution, a nonlinear set can give multiple solutions; therefore, the possibility of multiple steady states exists in azeotropic distillation.

With exactly the same input variables fixed (feed flow and composition, reflux flowrate, and distillate flowrate), there may be completely different values for the compositions and temperatures throughout the column. This is called "output multiplicity." If this occurs it adds significant complexity to the design and control of these systems. Problems in converging the steady-state program in Aspen Plus frequently are encountered and can be challenging to overcome.

To provide an example of this phenomenon, let us consider the azeotropic distillation column shown in Figure 1.1. This is the first column in a two-column azeotropic distillation system to separate a binary mixture of ethanol and water. Benzene is used as the light entrainer that carries the water overhead and produces a bottoms product of high-purity ethanol. The column has an organic reflux (OR) stream fed to the top tray, which comes from a downstream decanter. A recycle stream from the second column is also fed to the azeotropic column. The overhead vapor is condensed and goes to a decanter. The organic phase is refluxed back to the first column. The aqueous phase is fed to a recovery column whose bottoms product is high-purity water and whose distillate is recycled back to the azeotropic column. Note that the azeotropic column is a stripper with no condenser. Once the

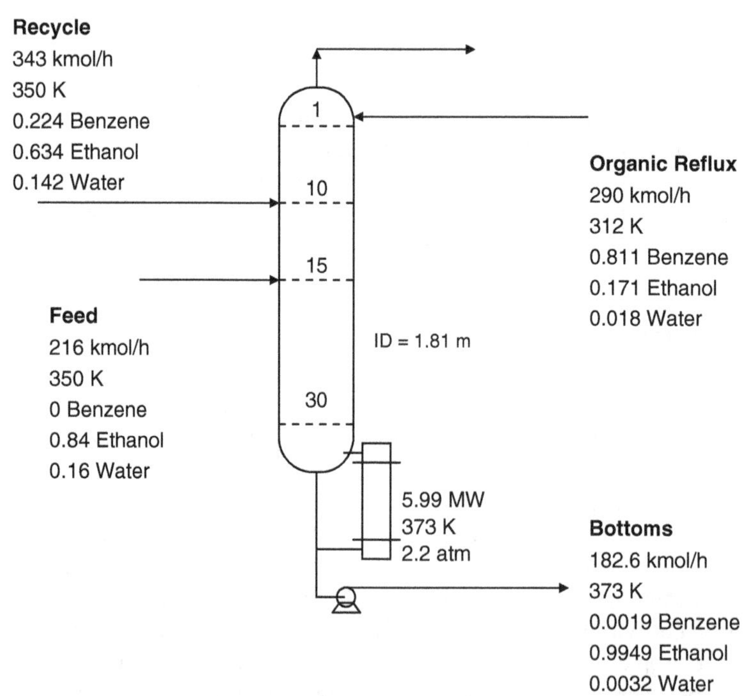


**Figure 1.1** Azeotropic column.

number of stages, the feed locations, the pressure, and the feed conditions (composition, flowrate and temperature) have been fixed, the stripper has only one degree of freedom. This is selected as the bottoms flowrate.

To illustrate what happens when running Aspen Plus with this column, we fix the bottoms flowrate at 182.6 kmol/h and vary the reflux flowrate over a range of values, starting first at low values and then starting at high values. Figure 1.2 gives the results of these positive and negative scans.

If we start at a small value of reflux and gradually increase the flowrate, the resulting trajectories of the benzene and water compositions of the bottoms are shown in the solid lines. For small reflux flowrates, there is not enough benzene being fed, so there is a high concentration of water $x_{B(W)}$ in the bottoms as shown in the lower graph. Of course, there is also very little benzene in the bottoms.

As the reflux is gradually increased, a point is reached at about 315 kmol/h where the water concentration in the bottoms drops abruptly to very small values, while the concentration of benzene begins to gradually increase. Now there is enough benzene in the column to drive the water overhead; however, to have a high-purity ethanol product, there should be very small amounts of both benzene and water.

Once the point of this sharp change has been passed, we can reverse the sequence of cases and begin to reduce the reflux flowrate. These are the dashed lines in Figure 1.2. The benzene concentration gradually decreases; the water concentration remains very low until a reflux flowrate of about 280 kmol/h is reached, where another abrupt change occurs. The column "goes over the cliff" at 315 kmol/h when moving up in reflux, and it again "goes over the cliff" at 280 kmol/h when moving down in reflux. Operating with any reflux flowrate between these two values can yield two possible steady-state conditions.

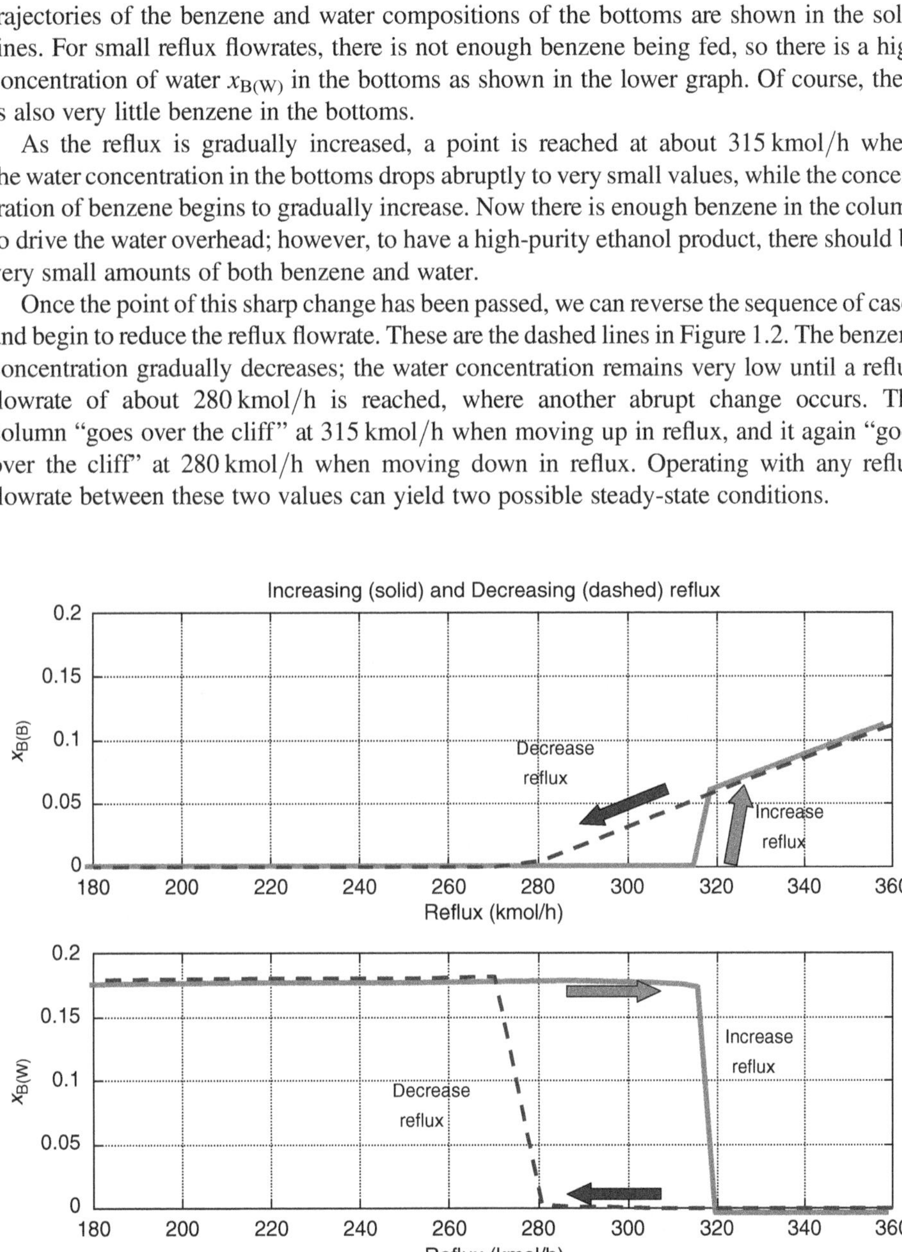


**Figure 1.2** Steady-state convergence.

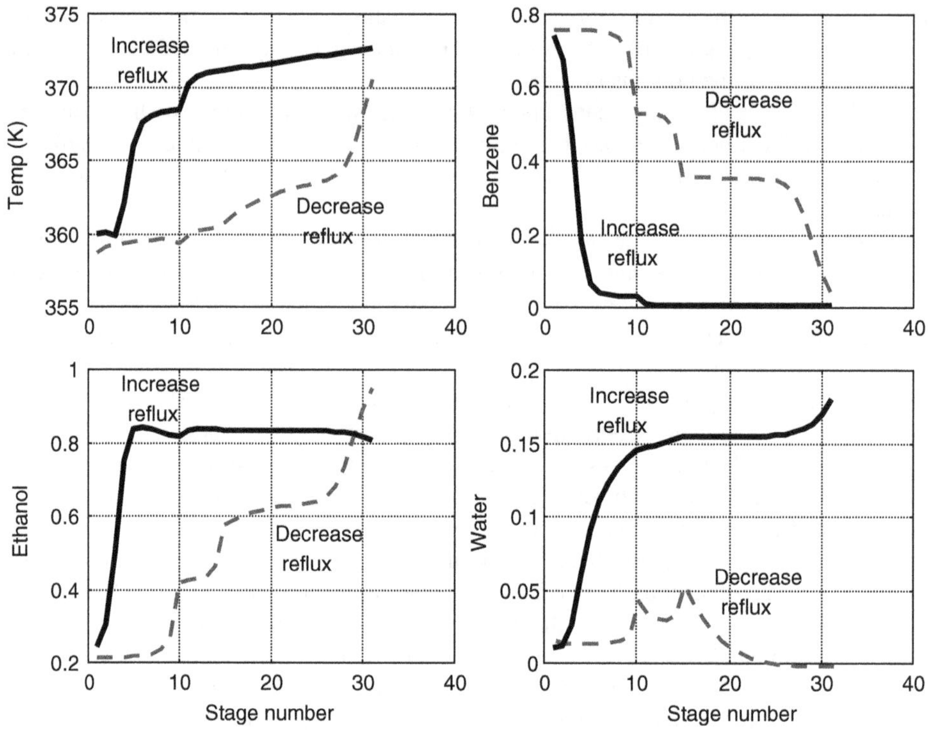


**Figure 1.3** Multiple steady-state profiles; $R = 300$ kmol/h.

Figure 1.3 shows the resulting temperature and compositions profile for two steady-state solutions. Both have exactly the same reflux flowrate (300 kmol/h) and bottoms flowrate (182.6 kmol/h), but the profiles are very different. The solid lines correspond to the case when the reflux has been slowly *increased* to 300 kmol/h from a lower value. There is little benzene in the bottoms but lots of water. Temperatures are high throughout the column.

The dashed lines correspond to the case when the reflux has been slowly *decreased* to 300 kmol/h from a higher value. Now there is little water in the bottoms but increasing values of benzene. Temperatures are low throughout the column.

The desired steady-state design produces a high-purity ethanol bottoms product. Figure 1.1 shows that this is achieved with a reflux flowrate of 290 kmol/h, which is inside the multiple steady-state region (280 to 315 kmol/h). Under these conditions, the bottoms composition is 0.19 mol% benzene, 99.49 mol% ethanol, and 0.32 mol% water.

These results clearly illustrate some of the complexities of dealing with azeotropic distillation. We will encounter these problems and also other issues in dealing with azeotropic systems in subsequent chapters, which make design and control challenging.

## REFERENCES

1. Walker W. H., W. K. Lewis, and E. R. Gilliland, *Principles of Chemical Engineering*, Third Edition, McGraw-Hill, New York, 1937.
2. Robinson C. S. and E. R. Gilliland, *Elements of Fractional Distillation*, McGraw-Hill, New York, 1958.
3. Hoffman E. J., *Azeotropic and Extractive Distillation*, Interscience Publisher, New York, 1964.

4. Othmer D. F., Azeotropic distillation for dehydrating acetic acid. *Chemical and Metallurgical Engineering*, 91–94 (June 1941).
5. Prokopakis G. J. and W. D. Seider, Dynamic simulation of azeotropic distillation towers, *AIChE J.*, **29**, 1017–1029 (1983).
6. Abu-Eishah S. I. and W. L. Luyben, Design and control of a two-column azeotropic system, *Ind. Eng. Chem. Process Des. Dev.*, **24**, 132–140 (1985).
7. Jacobsen E. W., L. Laroche, M. Morari, S. Skogestad, and H. W. Andersen, Robust control of homogeneous azeotropic distillation columns, *AIChE J.*, **37**, 1810–1824 (1991).
8. Andersen H. W., L. Laroche, and M. Morari, Dynamics of homogeneous azeotropic distillation columns, *Ind. Engng. Chem. Res.*, **30**, 1846–1855 (1991).
9. Laroche L., M. Morari, and H. W. Anderson, Effect of design on the operation of homogeneous azeotropic distillation, *Comput. Chem. Engng.*, **19**, 105–122 (1995).
10. Widagdo S. and W. D. Seider, Journal review: azeotropic distillation, *AIChE J.*, **42**, 96–130 (1996).
11. Doherty M. F. and M. F. Malone, *Conceptual Design of Distillation Systems*, McGraw-Hill, New York, 2001.
12. Stichlmair J. G. and J. R. Fair, *Distillation Principles and Practices*, Wiley-VCH, New York, 1998.
13. Kurooka T., Y. Yamahsita, H. Nishitani, Y. Hashimoto, M. Yoshida, and M. Numata, Dynamic simulation and nonlinear control system design of a heterogeneous azeotropic distillation column, *Comput. Chem. Engng.*, **24**, 887–892 (2000).
14. Phimister J. R. and W. D. Seider, Semicontinuous, middle-vessel, extractive distillation, *Comput. Chem. Engng.*, **24**, 879–885 (2000).
15. Phimister J. R. and W. D. Seider, Semicontinuous, pressure-swing distillation, *Ind. Engng. Chem. Res.*, **39**, 122–130 (2000).
16. Chien I. L., C. J. Wang, D. S. H. Wong, C. H. Lee, S. H. Cheng, R. F. Shih, W. T. Liu, and C. S. Tsai, Experimental investigation of conventional control strategies for a heterogeneous azeotropic distillation column, *J. Process Control*, **10**, 333–340 (2000).
17. Chien I. L., W. H. Chen, and T. S. Chang, Operation and decoupling control of a heterogeneous azeotropic distillation column, *Comput. Chem. Engng.*, **24**, 893–899 (2000).
18. Chien I. L., K. L. Zeng, and H. Y. Choa, Design and control of a complete heterogeneous azeotropic distillation column system, *Ind. Engng. Chem. Res.*, **43**, 2160–2174 (2004).
19. Rueda L. M., T. F. Edgar, and R. B. Eldridge, A novel control methodology for a pilot plant azeotropic distillation column, *Ind. Engng. Chem. Res.*, **45**, 8361–8372 (2006).
20. Luyben W. L., Control of a multi-unit heterogeneous azeotropic distillation process, *AIChE J.*, **52**, 623–637 (2006).
21. Luyben W. L., Plantwide control of an isopropyl alcohol dehydration process, *AIChE J.*, **52**, 623–637 (2006).
22. Arifin S. and I. L. Chien, Combined preconcentrator/recovery column design for isopropyl alcohol dehydration process, *Ind. Engng. Chem. Res.*, **46**, 2535–2543 (2007).
23. Arifin S. and I. L. Chien, Design and control of an isopropyl alcohol dehydration process via extractive distillation using dimethyl sulfoxide as an entrainer, *Ind. Engng. Chem. Res.*, **47**, 790–803 (2008).
24. Luyben W. L., Design and control of a fully heat integrated pressure-swing azeotropic distillation system, *Ind. Engng. Chem. Res.*, **47**, 2681–2695 (2008).
25. Luyben W. L., Comparison of extractive distillation and pressure-swing distillation for acetone-methanol separation, *Ind. Engng Chem. Res.*, **47**, 2696–2707 (2008).
26. Luyben W. L., Effect of solvent on controllability of extractive distillation, *Ind. Engng. Chem. Res.*, **47**, 4425–4439 (2008).
27. Luyben W. L., Control of maximum-boiling acetone/chloroform azeotropic distillation system, *Ind. Engng. Chem. Res.*, **47**, 6140–6149 (2008).

28. Luyben W. L., Control of the heterogeneous azeotropic *n*-butanol/water distillation system, *Energy and Fuels*, **22**, 4249–4258 (2008).
29. Luyben W. L., Control of a column/pervaporation process for separating the ethanol/water azeotrope, *Ind. Engng. Chem. Res.*, **48**, 3484–3495 (2009).
30. Wu Y. C., I. L. Chien, and W. L. Luyben, Two-stripper/decanter flowsheet for methanol recovery in the TAME reactive-distillation process, *Ind. Engng. Chem. Res.*, **48**, 10532–10540 (2009).
31. Wu Y. C. and I. L. Chien, Design and control of heterogeneous azeotropic column system for the separation of pyridine and water, *Ind. Engng. Chem. Res.*, **48**, 10564–10576 (2009).
32. Hsu K. Y., Y. C. Hsiao, and I. L. Chien, Design and control dimethyl carbonate-methanol separation via extractive distillation in the dimethyl carbonate reactive-distillation process, *Ind. Engng. Chem. Res.*, **49**, 735–749 (2010).

# CHAPTER 2

# PHASE EQUILIBRIUM

The separation of a mixture of chemical components by distillation is based on the phase equilibrium behavior of the mixture. In Section 2.1 of this chapter, the basis of phase equilibrium will be briefly reviewed by first giving the relationship for each component at equilibrium. Because of the nonideality in the liquid solution, activity coefficient methods will be used to describe the liquid phase behavior. Since the operating pressure is low for all the cases in this book, the vapor phase is assumed to be ideal except for the case involving acetic acid. The equations for calculating vapor pressures and liquid activity coefficients will be given in Section 2.2. With the known thermodynamic model and its parameters, various binary and ternary diagrams of the component system can easily be generated in Aspen Plus. The way to generate *Txy* and *yx* plots with two examples will be shown in Section 2.3. The use of ternary diagrams to display residue curve maps, distillation boundaries, liquid–liquid equilibrium (LLE) envelopes, and isovolatility curves will be demonstrated with examples in Section 2.4. The prediction of the phase equilibrium behavior by a thermodynamic model should be compared to experimental data for verification purposes before the design of the distillation system. The way to do the verification will be described in Section 2.5. If the prediction of the phase equilibrium behavior of a thermodynamic model is not good enough, or if the Aspen Plus built-in parameter set is not available, regression to obtain suitable model parameters should be carried out with known thermodynamic experimental data. The steps that need to be taken for parameter estimation are demonstrated with an example. When no experimental data for the mixture system is available, the UNIFAC group contribution method or its modifications should be used for the prediction of phase equilibrium. This will be discussed in Section 2.6.

*Design and Control of Distillation Systems for Separating Azeotropes*. By William L. Luyben and I-Lung Chien

## 2.1 PHASE EQUILIBRIUM RELATIONSHIPS

### 2.1.1 Vapor–Liquid Equilibrium

The basic relationship for every component $i$ in the vapor and liquid phases of a system at equilibrium is the equality of fugacities in all phases.

$$f_i^{\mathrm{v}} = f_i^{\mathrm{l}} \tag{2.1}$$

where

$f_i^{\mathrm{v}}$ = fugacity of component $i$ in the vapor phase
$f_i^{\mathrm{l}}$ = fugacity of component $i$ in the liquid phase

There are two methods for representing the fugacities in terms of the measurable state variables: the equation-of-state method and the activity-coefficient method. The usual representation in the vapor phase is to use the equation-of-state method.

$$f_i^{\mathrm{v}} = \varphi_i^{\mathrm{v}} y_i P \tag{2.2}$$

where

$\varphi_i^{\mathrm{v}}$ = vapor phase fugacity coefficient of component $i$
$y_i$ = mole fraction of component $i$ in the vapor phase
$P$ = total system pressure

For a vapor at low to moderate pressure, $\varphi_i^{\mathrm{v}}$ is close to unity. Thus, throughout most of the applications in this book, the following vapor fugacity representation is used:

$$f_i^{\mathrm{v}} = y_i P \tag{2.3}$$

For the application system involving acetic acid in Chapter 9, the above representation in Eq. (2.3) is not accurate enough. The reason is that acetic acid can exhibit association leading to dimerization in the vapor phase. Hayden and O'Connell[1] used the virial equation of state truncated after the second term. They developed a correlation for the second virial coefficient of polar, nonpolar, and associating species based on the critical temperature and pressure, the dipole moment, and the mean radius of gyration. Association of like and unlike molecules is described with the adjustable association parameter. Pure component and binary values for the association parameter are available in the Aspen Physical Property System. For detail calculations in the vapor phase, please refer to the above Hayden and O'Connell[1] paper and also to Aspen Physical Property System.[2]

As for the liquid phase, in an ideal liquid solution the liquid fugacity of each component in the mixture is directly proportional to the mole fraction of the component as:

$$f_i^{\mathrm{l}} = x_i f_i^{*,\mathrm{l}} \tag{2.4}$$

where

$x_i$ = mole fraction of component $i$ in the liquid phase
$f_i^{*,\mathrm{l}}$ = liquid phase reference fugacity of component $i$

However, in the systems studied in the following chapters, nonideality in mixtures of unlike molecules is described using a liquid activity coefficient $\gamma_i$ representing the deviation of the mixture from ideality. The above Eq. (2.4) is modified as:

$$f_i^{\mathrm{l}} = x_i \gamma_i f_i^{*,\mathrm{l}} \tag{2.5}$$

The liquid phase reference fugacity $f_i^{*,\mathrm{l}}$ from the above Eq. (2.5) is computed as:

$$f_i^{*,\mathrm{l}} = \varphi_i^{*,\mathrm{V}}(T, P_i^*) P_i^* \theta_i^* \tag{2.6}$$

where

$\varphi_i^{*,\mathrm{V}}$ = fugacity coefficient of pure component $i$ at the system temperature ($T$) and vapor pressures, as calculated from the vapor phase equation of state

$P_i^*$ = liquid vapor pressure of component $i$ at the system temperature

$\theta_i^*$ = Poynting correction for pressure $= \exp\left(\frac{1}{RT}\int_{P_i^*}^{P} V_i^{*,\mathrm{l}}\,\mathrm{d}P\right)$

At low system pressures, the Poynting correction is near unity and can be ignored. Thus the overall vapor–liquid equilibrium (VLE) relationship for most of the mixture systems in the following chapters (excluding acetic acid system) can be described as the following equation:

$$y_i P = x_i \gamma_i P_i^* \tag{2.7}$$

The computation of the liquid activity coefficient $\gamma_i$ of several model forms used in the following chapters will be given in the next section.

### 2.1.2 Liquid–Liquid Equilibrium

For the equilibrium calculation in the decanter, the following relationship is used:

$$x_i^{\mathrm{L1}} \gamma_i^{\mathrm{L1}} = x_i^{\mathrm{L2}} \gamma_i^{\mathrm{L2}} \tag{2.8}$$

where

$x_i^{\mathrm{L1}}$ = mole fraction of component $i$ in the first liquid phase

$\gamma_i^{\mathrm{L1}}$ = activity coefficient of component $i$ in the first liquid phase

$x_i^{\mathrm{L2}}$ = mole fraction of component $i$ in the second liquid phase

$\gamma_i^{\mathrm{L2}}$ = activity coefficient of component $i$ in the second liquid phase

The suitable computations of the liquid activity coefficients of component $i$ in the first and the second liquid phase will be given in Section 2.2.

For certain applications in the heterogeneous azeotropic distillation column, vapor–liquid–liquid equilibrium (VLLE) will occur in the top part of the column. The basic equilibrium relationship to use is to equate the fugacities of all components in all vapor and liquid phases as in the following equation:

$$\varphi_i^{\mathrm{V}} y_i P = x_i^{\mathrm{L1}} \gamma_i^{\mathrm{L1}} P_i^* = x_i^{\mathrm{L2}} \gamma_i^{\mathrm{L2}} P_i^* \tag{2.9}$$

However, phase equilibrium calculations involving vapor and two liquid phases can sometimes lead to an unstable solution. Instead, the technique of Gibbs energy minimization can be used in this situation and always yields a stable solution (tangent plane analysis by Michelsen).[3,4] Aspen Plus can automatically incorporate this calculation technique in the distillation model for the detection of a liquid phase splitting inside of the column. The use of this technique in Aspen Plus will be explained in Chapter 3.

## 2.2 EQUATIONS DESCRIBING PHASE EQUILIBRIUM RELATIONSHIPS

### 2.2.1 Liquid Vapor Pressure

Calculation of the liquid vapor pressure for a pure component is needed in Eq. (2.7) for the VLE relationship. The extended Antoine equation is used to compute liquid vapor pressure as a function of system temperature $T$:

$$\ln P_i^* = C_{1i} + \frac{C_{2i}}{T + C_{3i}} + C_{4i}T + C_{5i}\ln T + C_{6i}T^{C_{7i}} \quad \text{for } C_{8i} \le T \le C_{9i} \tag{2.10}$$

where $C_{1i}$ to $C_{7i}$ are the model parameters, and $C_{8i}$ and $C_{9i}$ represent the valid temperature ranges. Model parameters for many components are available from the pure component databank of the Aspen Physical Property System.

### 2.2.2 Liquid Activity Coefficients

Four model forms will be discussed in this section for the calculation of the liquid activity coefficient as functions of mole fractions and system temperature. They are van Laar, Wilson, nonrandom two-liquid (NRTL), and UNIQUAC. In these four model forms, van Laar and Wilson cannot describe LLE. On the other hand, NRTL and UNIQUAC can be used for the equilibrium of all phases.

***van Laar Model.*** The van Laar model can be used for highly nonideal systems. The equation for the calculation of liquid activity coefficient is:

$$\ln \gamma_i = A_i(1 - z_i)^2\left[1 + C_i z_i\left(z_i - \frac{2}{3}\right) + 2z_i\left(\frac{A_i B_i}{|A_i B_i|} - 1\right)\right] \tag{2.11}$$

where

$$z_i = \frac{|A_i|x_i}{|A_i|x_i + |B_i|(1 - x_i)} \tag{2.12}$$

$$A_i = \sum_j \left(\frac{x_j A_{ij}}{1 - x_i}\right) \tag{2.13}$$

$$B_i = \sum_j \left(\frac{x_j A_{ji}}{1 - x_i}\right) \tag{2.14}$$

$$C_i = \sum_i \left(\frac{x_j C_{ij}}{1 - x_i}\right) \tag{2.15}$$

$$A_{ij} = a_{ij} + \frac{b_{ij}}{T} \tag{2.16}$$

$$C_{ij} = c_{ij} + \frac{d_{ij}}{T} \tag{2.17}$$

$$C_{ij} = C_{ji} \tag{2.18}$$

$$A_{ii} = C_{ii} = 0 \tag{2.19}$$

For a two-component system, there are six parameters that need to be determined from phase equilibrium data. They are $a_{ij}$, $a_{ji}$, $b_{ij}$, $b_{ji}$, $c_{ij} = c_{ji}$, and $d_{ij} = d_{ji}$. These binary parameters are available in the Aspen Physical Property System for a large number of component pairs.

***Wilson Model.*** This model can also be used for highly nonideal systems, especially alcohol–water systems. The equation for the calculation of liquid activity coefficient is:

$$\ln \gamma_i = 1 - \ln\left(\sum_j A_{ij}x_j\right) - \sum_j \frac{A_{ji}x_j}{\sum_k A_{jk}x_k} \tag{2.20}$$

where

$$\ln A_{ij} = a_{ij} + \frac{b_{ij}}{T} + c_{ij}\ln T + d_{ij}T + \frac{e_{ij}}{T^2} \tag{2.21}$$

The above model parameters in Eq. (2.21) are unsymmetrical. The Aspen Physical Property System has a large number of built-in binary parameters for the Wilson model. The binary parameters have been regressed using the vapor–liquid equilibrium data from the Dortmund Databank.[5]

***Nonrandom Two-Liquid (NRTL) Model.*** This model is recommended for highly nonideal chemical systems and can be used for both VLE and LLE applications. The equation for the NRTL model is:

$$\ln \gamma_i = \frac{\sum_j x_j \tau_{ji} G_{ji}}{\sum_k x_k G_{ki}} + \sum_j \frac{x_j G_{ij}}{\sum_k x_k G_{kj}}\left(\tau_{ij} - \frac{\sum_m x_m \tau_{mj} G_{mj}}{\sum_k x_k G_{kj}}\right) \tag{2.22}$$

where

$$G_{ij} = \exp(-\alpha_{ij}\tau_{ij}) \tag{2.23}$$

$$\tau_{ij} = a_{ij} + \frac{b_{ij}}{T} + e_{ij}\ln T + f_{ij}T \tag{2.24}$$

$$\alpha_{ij} = c_{ij} + d_{ij}(T - 273.15\,\mathrm{K}) \tag{2.25}$$

$$\tau_{ii} = 0 \tag{2.26}$$

$$G_{ii} = 1 \tag{2.27}$$

The binary parameters, $a_{ij}$, $b_{ij}$, $e_{ij}$, and $f_{ij}$ are unsymmetrical. The Aspen Physical Property System has a large number of built-in binary parameters for the NRTL model. The binary parameters have been regressed using VLE or LLE data from the Dortmund Databank.[5]

***UNIQUAC Model.*** This model is also recommended for highly nonideal chemical systems and can be used for both VLE and LLE applications. The equation for the UNIQUAC model is:

$$\ln \gamma_i = \ln \frac{\Phi_i}{x_i} + \frac{z}{2} q_i \ln \frac{\theta_i}{\Phi_i} - q'_i \ln t'_i - q'_i \sum_j \frac{\theta'_j \tau_{ij}}{t'_j} + l_i + q'_i - \frac{\Phi_i}{x_i} \sum_j x_j l_j \qquad (2.28)$$

where

$$\theta_i = \frac{q_i x_i}{q_T}; \quad q_T = \sum_k q_k x_k \qquad (2.29)$$

$$\theta'_i = \frac{q'_i x_i}{q'_T}; \quad q'_T = \sum_k q'_k x_k \qquad (2.30)$$

$$\Phi_i = \frac{r_i x_i}{r_T}; \quad r_T = \sum_k r_k x_k \qquad (2.31)$$

$$l_i = \frac{z}{2}(r_i - q_i) + 1 - r_i \qquad (2.32)$$

$$t'_i = \sum_k \theta'_k \tau_{ki} \qquad (2.33)$$

$$\tau_{ij} = \exp\left( a_{ij} + \frac{b_{ij}}{T} + c_{ij} \ln T + d_{ij} T + \frac{e_{ij}}{T^2} \right) \qquad (2.34)$$

$$z = 10 \qquad (2.35)$$

The binary parameters $a_{ij}$, $b_{ij}$, $c_{ij}$, $d_{ij}$, and $e_{ij}$ are all unsymmetrical. The Aspen Physical Property System has a large number of built-in binary parameters for the UNIQUAC model. The binary parameters have been regressed using VLE or LLE data from the Dortmund Databank.[5]

***Binary Parameters: Aspen versus DECHEMA.*** The *DECHEMA Chemistry Data Series* (Gmehling and Onken;[6] Sørensen and Arlt[7]) contains a large number of binary parameters for the above four model forms, which have been obtained by regression using VLE or LLE experimental data. In many cases, the binary parameters of the simplest form of the above models are given. The simplest form for the van Laar model is to have only $b_{ij}$ and $b_{ji}$ parameters, with all other parameters set equal to zero. The situation is the same for the Wilson and UNIQUAC models. As for the NRTL model, besides $b_{ij}$ and $b_{ji}$ parameters, an additional symmetrical $c_{ij}$ parameter needs to be given. Notice that in the *DECHEMA Chemistry Data Series* the binary parameters are called A12, A21, and ALPHA12 (e.g., for NRTL model), where A12 and A21 can be related to $b_{ij}$ and $b_{ji}$ parameters in the above equations by dividing by the gas constant (= 1.987), and the ALPHA12 is equivalent to the $c_{ij}$ parameter in the NRTL model.

## 2.3 DIAGRAMS FOR BINARY SYSTEMS

To analyze a given two-component system, a physical property model must be selected. Aspen Physical Property System has a large number of built-in binary parameters for a number of models. The *Txy*, *yx*, and other binary plots can easily be generated in Aspen with the calculation of the phase equilibrium relationship all hidden.

After selecting valid phases, inputting components, and selecting the property method in Aspen Plus, you need to choose the following route: *Tools* → *Analysis* → *Property* → *Binary* from the Aspen Plus menu bar to go to the dialog box shown in Figure 2.1. In this dialog box, the example system of isopropanol and water is used with the *NRTL* property method selected.

Selecting the analysis type as *Txy*, specifying the system pressure, specifying *Valid Phases* as *Vapor–Liquid* and clicking the *Go* button at the bottom of the dialog box produce the *Txy* plot shown in Figure 2.2. Tabulated binary analysis results like activity coefficients and *K* values are also shown. From the example *Txy* plot of isopropanol–water system, we can see that this system contains an azeotrope at isopropanol mole fraction between 0.6 and 0.8 and azeotropic temperature of about 80°C. To find the exact prediction of the azeotropic composition and temperature, the azeotrope search tool can be used. The use of this tool will be explained later in Section 2.5.2.

At the bottom of the window showing binary analysis results, you can click *Plot Wizard* to generate a *yx* plot or other plots such as activity coefficient or K value versus mole fraction. The predicted *Txy* and *yx* plots by the model can be compared to the experimental data in the literature so that you can have confidence using the Aspen built-in parameters in

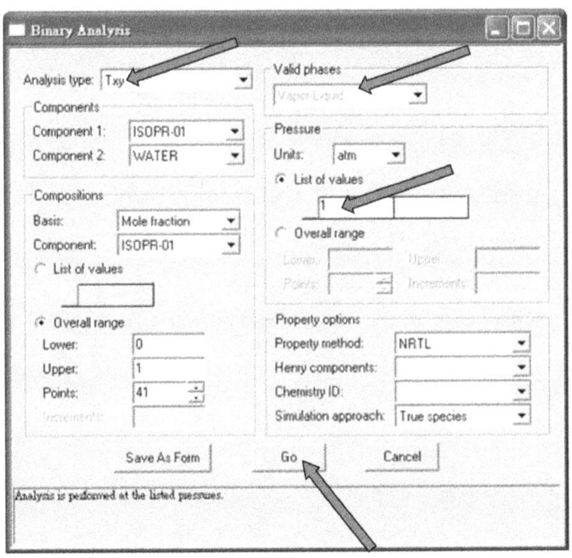


**Figure 2.1** Binary analysis dialog box of the isopropanol–water system.

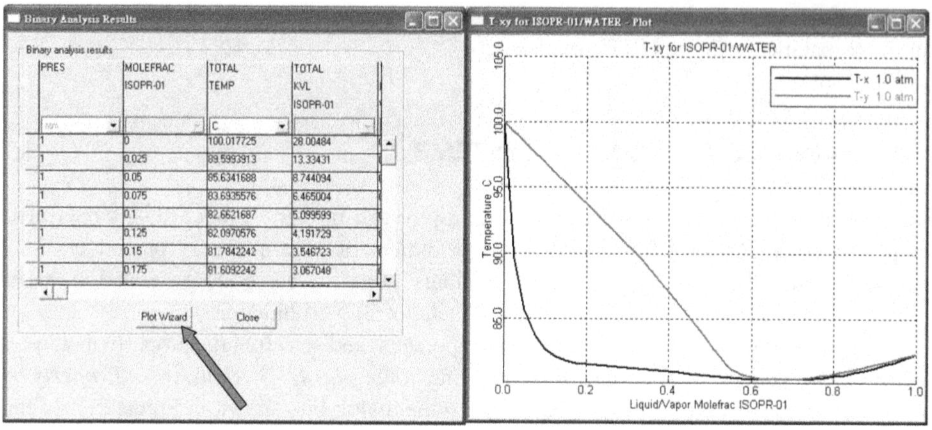

**Figure 2.2** *Txy* plot and the tabulated results of the isopropanol–water system.

the design study. In our experience, these default parameters are quite reliable for predicting the VLE behavior of the systems discussed in this book.

As a second example, binary analysis of the *n*-butanol–water system is shown in Figure 2.3. In this dialog box, the physical property method of *UNIQUAC* is selected, and the system pressure is set at 1 atm. Notice, particularly, that the valid phases should be selected as *Vapor–Liquid–Liquid* to generate a correct plot of this binary system. From Figure 2.4, a heterogeneous binary azeotrope is observed with the azeotropic composition of around 0.24 mole fraction of *n*-butanol, and the azeotropic temperature of about 93°C. This unique feature of this binary system can be used to design a two-column system to separate the mixture into two pure product streams without adding an entrainer. The design and control of this system will be discussed in Chapter 7.

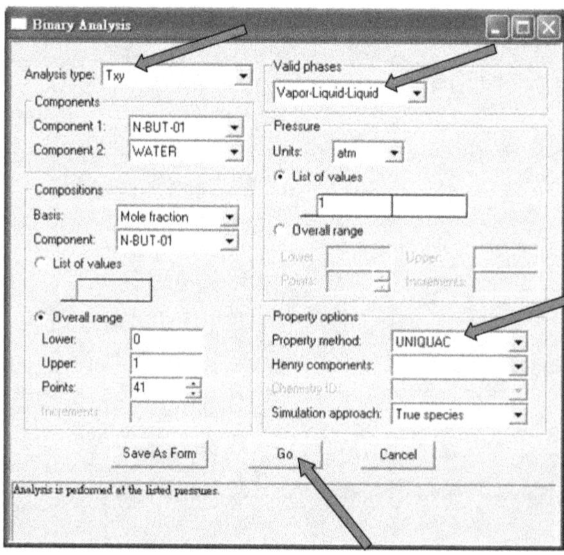


**Figure 2.3** Binary analysis dialog box of the *n*-butanol–water system.

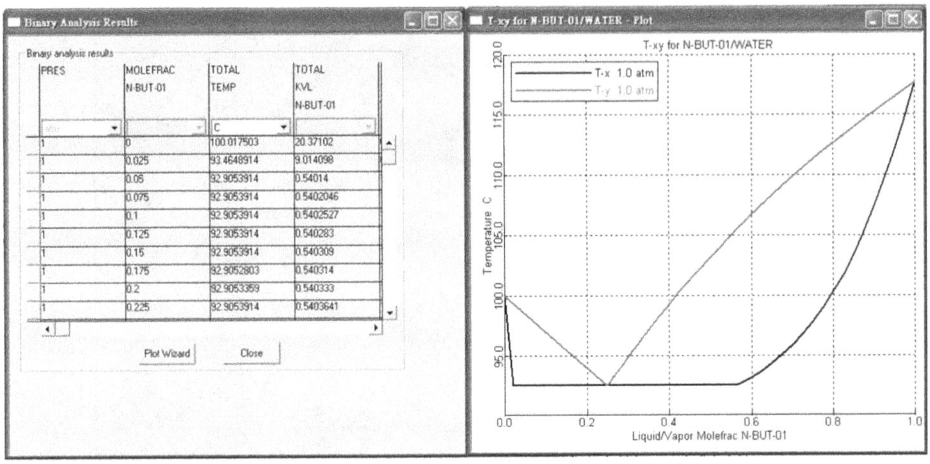


**Figure 2.4** *Txy* plot and the tabulated results of the *n*-butanol–water system.

## 2.4 TERNARY DIAGRAMS

A typically way to separate a binary mixture containing an azeotrope is to add a third component to the system. Aspen can easily generate the following useful ternary diagrams for the purpose of analysis and design.

### 2.4.1 Residue Curves and Distillation Boundaries

Residue curve maps (RCM) are most useful for understanding how to create a distillation sequence to separate a mixture. Doherty and Malone[8] gave an excellent review of RCM and the usage of this tool in the design of distillation systems. In the following we will show the way to easily generate the RCM and distillation boundary of an example system [acetone–chloroform–dimethyl sulfoxide (DMSO)] using the analysis tool in Aspen Plus. After inputting the three components and selecting *UNIQUAC* as the physical property method, the following route from the Aspen Plus menu bar can be selected to go to the dialog box shown in Figure 2.5. The route is *Tools* → *Conceptual Design* → *Ternary Maps*.

In this dialog box, the *VAP–LIQ* phase is selected and the system pressure of 1 atm is chosen. After clicking *Ternary Plot* on the left-hand side of this dialog box, the RCM, with distillation boundary, shown in Figure 2.6, is generated.

Notice that there is a maximum-boiling azeotrope for the binary mixture of acetone and chloroform. The ternary mixture exhibits a distillation boundary from the maximum-boiling azeotrope pointing to the heavy-boiler (DMSO) of the system. This distillation boundary divides the ternary diagram into two distillation regions. The rule for devising feasible products is that the distillate and bottoms products of a distillation column must always be in the same region. If the distillation boundary is linear, the products from the entire sequence of columns must also lie in the same distillation region as the process feed. However, it can be observed in Figure 2.6 that the distillation boundary is curved. The unique feature of this system allows a column sequence to be designed to obtain pure acetone

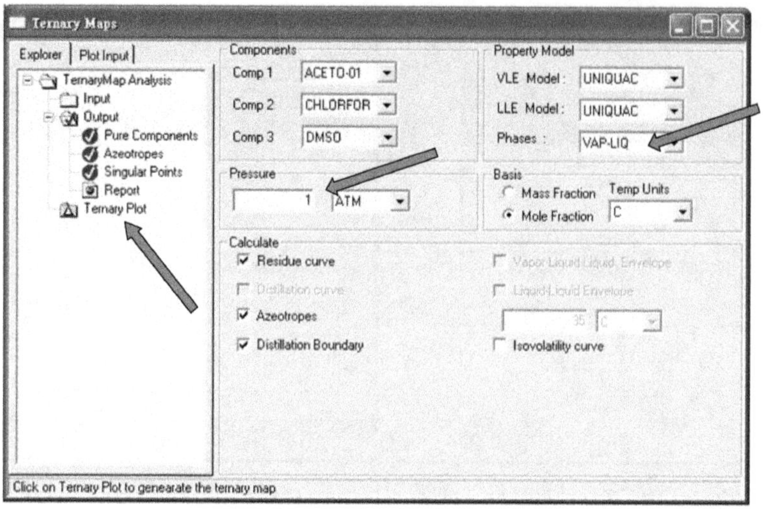


**Figure 2.5** Ternary maps dialog box of the acetone–chloroform–DMSO system.

and chloroform products (although in different distillation regions). The design concept will be explained in detail in Chapter 12.

Any plot generated in Aspen Plus can be exported to other application programs, such as Microsoft Word or PowerPoint, by simply right-clicking on the plot and selecting *Copy* to paste into other programs.

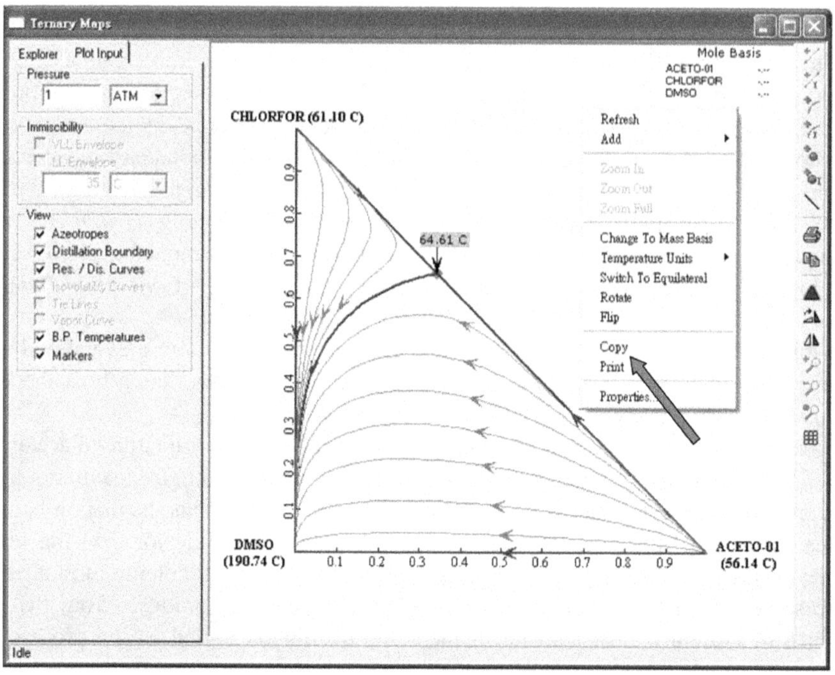


**Figure 2.6** RCM and distillation boundary of the acetone–chloroform–DMSO system.

### 2.4.2 Liquid–Liquid Equilibrium Envelope

In Part 3 of this book an extrainer is added to the system so that liquid–liquid splitting can appear in the top decanter and also maybe in the top few stages of the azeotropic column. The LLE behavior in the decanter, or the VLLE behavior in the top stages of the azeotropic column, can be predicted by Aspen Plus. The system of separating an isopropanol–water mixture using cyclohexane as the entrainer will be used as an example to demonstrate the way to generate a LLE envelope in Aspen Plus.

After inputting the components and selecting the physical property method, the same route presented in Section 2.4.1 can be used to generate ternary diagrams. However, in this case *VAP–LIQ–LIQ* phases should be selected in order to calculate the LLE envelope. Figure 2.7 displays the selection of this dialog box in order to generate the LLE envelope together with the RCM and the distillation boundary. Notice that *NRTL* is selected as property method and *40°C* is chosen in this dialog box to agree with the common operating condition of a decanter. Some options for the advanced input can also be chosen by right-clicking this dialog box. For example, a smoother LLE envelope can be obtained by changing the default step size of the envelope from default of *0.1* to *0.05*.

Clicking *Ternary Plot* on the left-hand side of this dialog box, opens the plot shown in Figure 2.8. The vertical tool bar on the right-hand side of this plot can be used to add more tie lines or more residue curves. Note that this ternary system exhibits four azeotropes. The lightest component in the system is the ternary azeotrope, which is inside of the LLE envelope. This unique feature of this system will be used to design a column sequence to

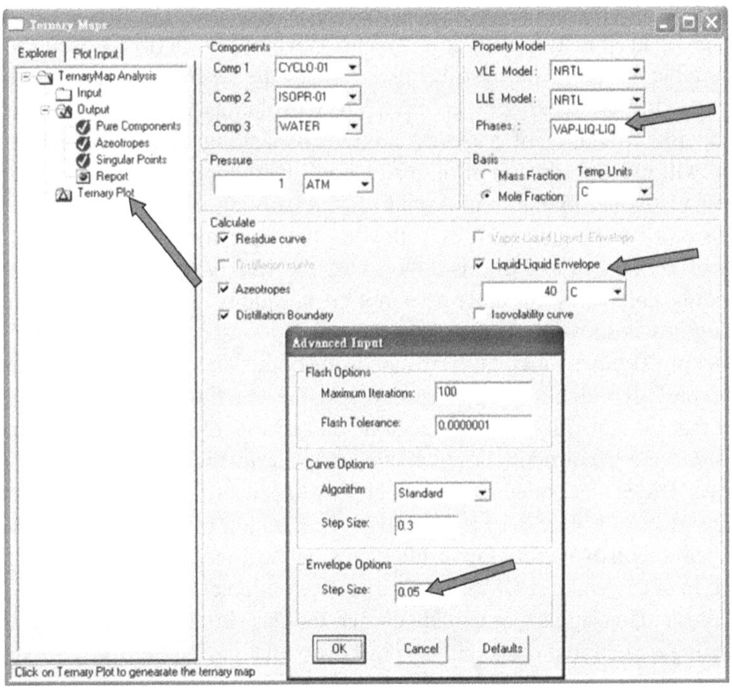


**Figure 2.7** Ternary maps dialog box of the isopropanol–water–cyclohexane system.

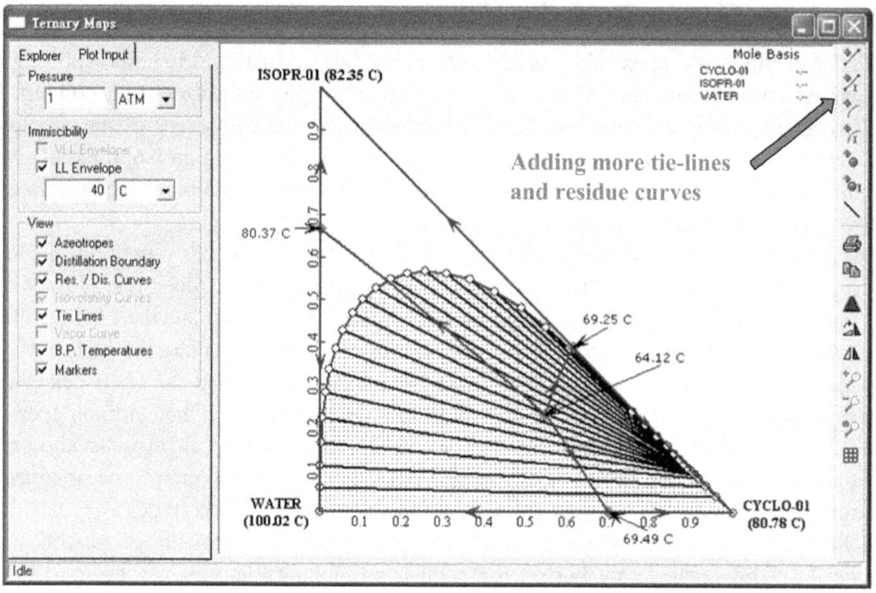


**Figure 2.8** Ternary maps and LLE of the isopropanol–water–cyclohexane system.

separate a mixture containing isopropanol and water. The details of the design concept can be found in Chapter 8.

### 2.4.3 Isovolatility Curve

For the design of an extractive distillation system, isovolatility curves are useful for the selection of a suitable entrainer and also for determining the desirable product in the distillate stream of the extractive distillation column. The isovolatility curve is the curve that traces the unity relative volatility of a binary mixture by adding another component into the system. We will use the separation of acetone and methanol as an example to show how to generate an isovolatility curve in Aspen Plus. Assuming the heavy entrainer selected is DMSO, this plot can be generated by following the same route as in Section 2.4.1 to the dialog box of *Ternary Maps*. In this dialog box, the *VAP–LIQ* phases should be selected because adding this heavy entrainer does not create liquid–liquid splitting. The selection in this dialog box is shown in Figure 2.9.

After clicking *Ternary Plot* on the left-hand side of this dialog box, the isovolatility curve shown in Figure 2.10 will be displayed (the dashed line near the acetone apex of the triangle). Remember that the volatility between acetone and methanol is unity in the binary system when we are at the azeotrope (shown on the ordinate in Fig. 2.10 with a temperature of 55.24°C). As DMSO is added, the isovolatility line moves towards the hypotenuse of the triangle shown in Figure 2.10 (where the concentration of methanol goes to zero). The location where this isovolatility curve intercepts the acetone–DMSO edge of the triangle is important in determining if DMSO is an effective entrainer. Usually, the closer the intersection point is to the original component corner, the less entrainer is required, which means lower operating and capital costs. However, the isovolatility curve is not the only plot we should check in determining which one is a more effective entrainer. We will discuss more of this issue in Chapter 10.

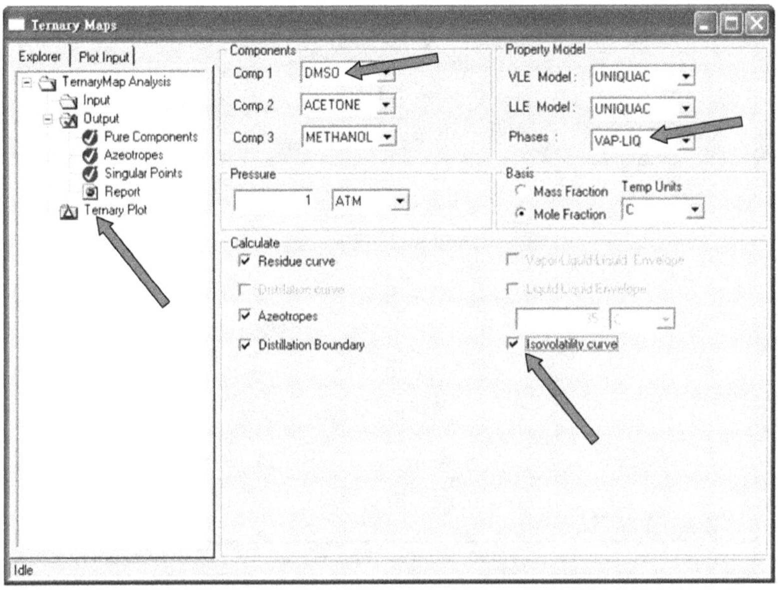


**Figure 2.9** Ternary maps dialog box of the acetone–methanol–DMSO system.

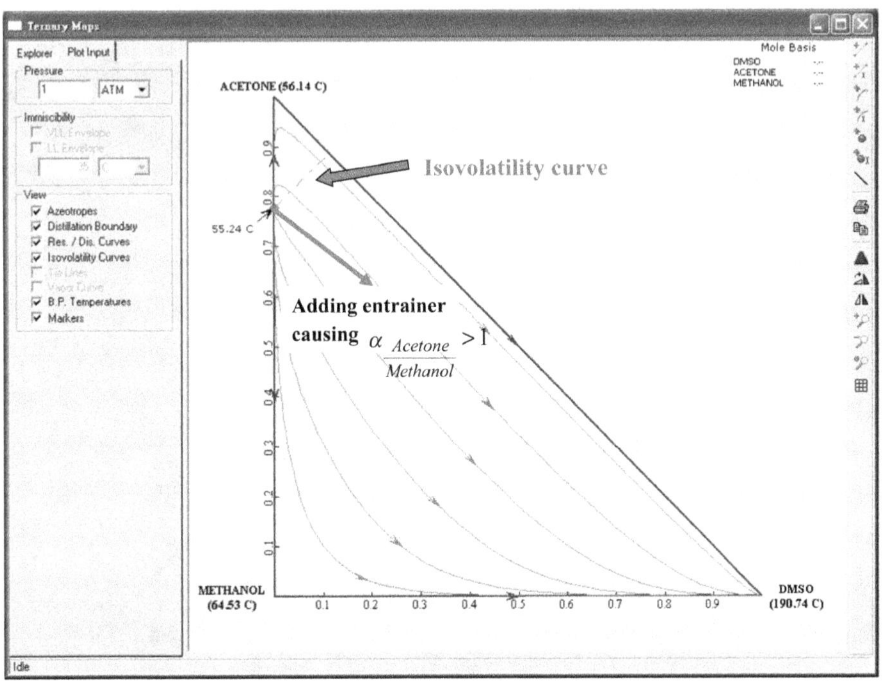


**Figure 2.10** Isovolatility curve of the acetone–methanol–DMSO system.

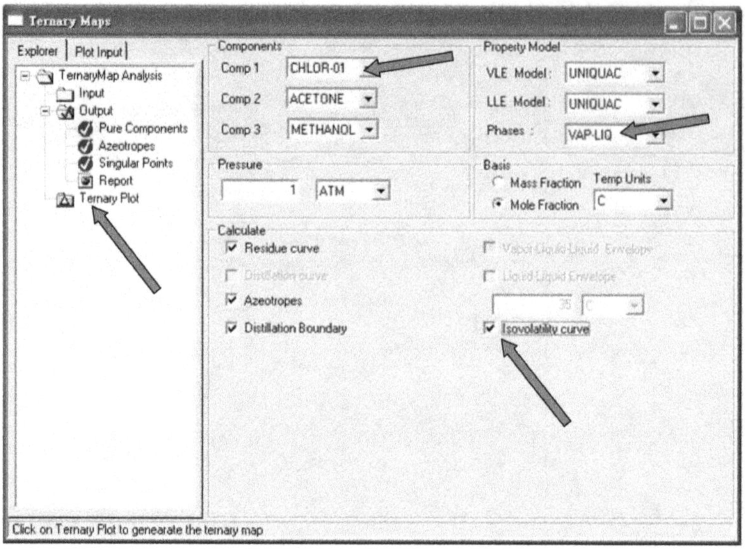


**Figure 2.11** Ternary maps dialog box of the acetone–methanol–chlorobenzene system.

Also, from the isovolatility curve, other useful information can be gained, such as what is the distillate product of this extractive distillation column that can be produced. Since the isovolatility curve in Figure 2.10 intercepts the acetone–DMSO edge of the triangle, it is concluded that acetone is the distillate product. The reasoning is that, by adding DMSO into the system, the relative volatility will become greater than one, causing acetone to go

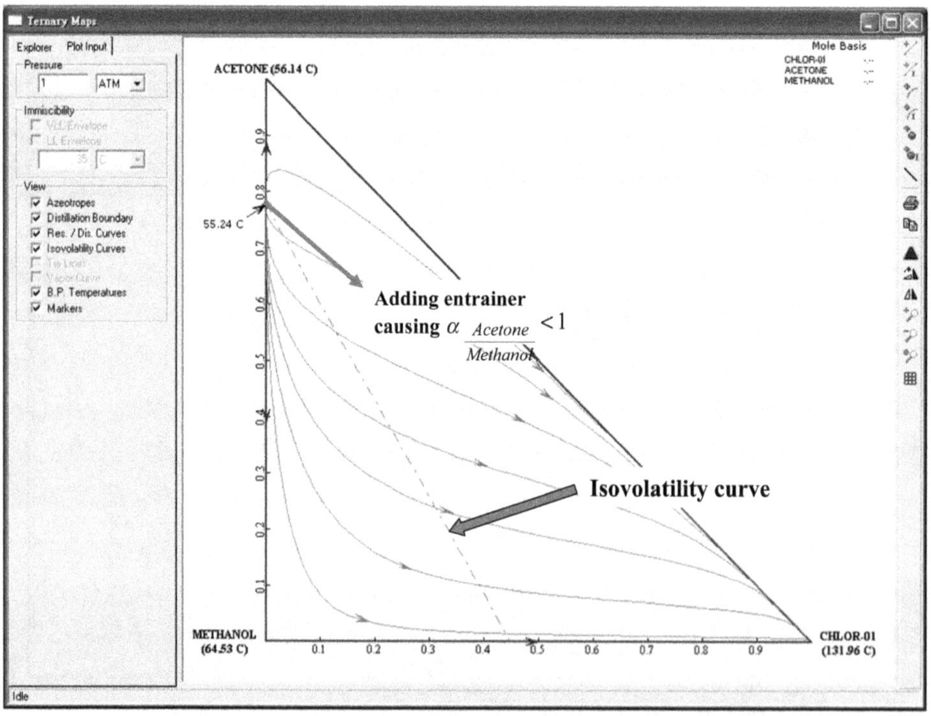


**Figure 2.12** Isovolatility curve of the acetone–methanol–chlorobenzene system.

up the column. Note that the relative volatility between acetone and methanol is greater than one in the region below the isovolatility curve.

Now let us consider another example to show that an isovolatility curve can sometimes intercept another edge of the triangle. Instead of using DMSO, let us select chlorobenzene as the entrainer for the same system of separating acetone and methanol. The selection in the dialog box of *Ternary Maps* is shown in Figure 2.11. After clicking *Ternary Plot* on the left-hand side of this dialog box, the isovolatility curve shown in Figure 2.12 is displayed. In this figure, the isovolatility curve intercepts the methanol–chlorobenzene edge of the triangle. The relative volatility between acetone and methanol is greater than one in the region *below* the isovolatility curve and less than one in the region *above* the isovolatility curve (see Fig. 2.12). This means that by adding chlorobenzene into the system, the relative volatility will be altered to be less than one, and this means that methanol becomes the light component and goes to the top of this column. The use of this tool for comparing alternative entrainers will be demonstrated in Chapter 10.

## 2.5 DATA REGRESSION

For the design study of a particular separation system, we typically start by using the Aspen built-in parameters of a suitable physical property model. The phase equilibrium behavior predicted by the Aspen built-in parameters should be compared with experimental data for validation purpose. It is obvious that an inaccurate description of the phase equilibrium behavior of a separation system will give flowsheet results that do not match the results of the true system. The worst case may be a failure of the separation task in the proposed design flowsheet. Thus, the validation stage is important before doing any design study. The experimental data that can typically be found in literature include the *Txy* and *yx* data, binary and ternary LLE data, VLLE data, and azeotropic information.

### 2.5.1 Sources of Experimental Data

The *Txy* and *yx* data of many binary components can be found in Volume I of the *DECHEMA Chemistry Data Series* (Gmehling and Onken[6]) or in other open literature. A large number of LLE data sources can be found in Volume V of the *DECHEMA Chemistry Data Series* (Sørensen and Arlt[7]) and also in the open literature. Some other VLLE data can also be found in the open literature. These experimental data should be compared to the prediction by using Aspen built-in parameters. The way to generate the prediction of the *Txy*, *yx*, LLE, and VLLE has already been explained in Sections 2.3 and 2.4. Another important source of experimental data is the azeotropic composition and azeotropic temperature, and the status of the phase (homogeneous or heterogeneous). Much binary and ternary azeotrope information can be found in Gmehling.[9] These experimental data should also be compared to the prediction by using the Aspen built-in parameters. We will describe how to obtain this information in the following subsection.

### 2.5.2 Azeotrope Search

The ternary system containing isopropanol, water, and cyclohexane is used as an example to show how to obtain the prediction of azeotropes in Aspen Plus. After inputting components and the physical property method, the following route is selected: *Tools → Conceptual*

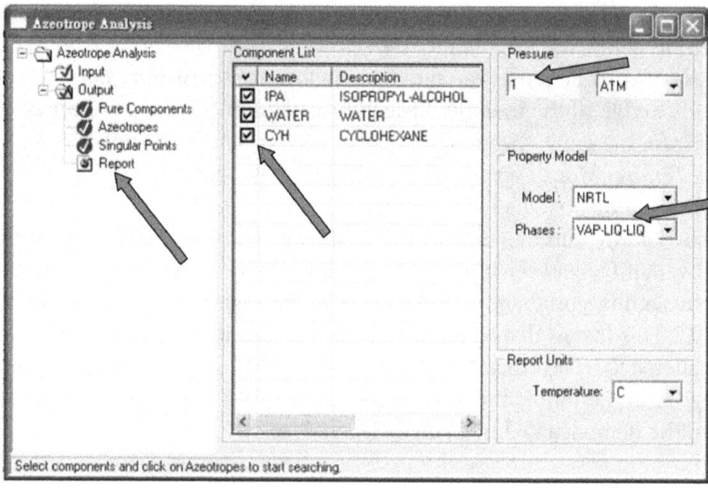


Figure 2.13 Azeotrope analysis dialog box of the isopropanol–water–cyclohexane system.

*Design* → *Azeotrope Search* in the Aspen Plus menu bar. In this dialog box, all components are selected for the search, the operating pressure is selected as 1 atm, and the property method of *NRTL* is also selected. Because liquid–liquid splitting is observed in this system, valid phases of *VAP–LIQ–LIQ* should be selected. The selection is shown in Figure 2.13.

After clicking the *Report* button on the left-hand side of this dialog box, the information of all the azeotropes is generated (see Fig. 2.14). The system contains four azeotropes

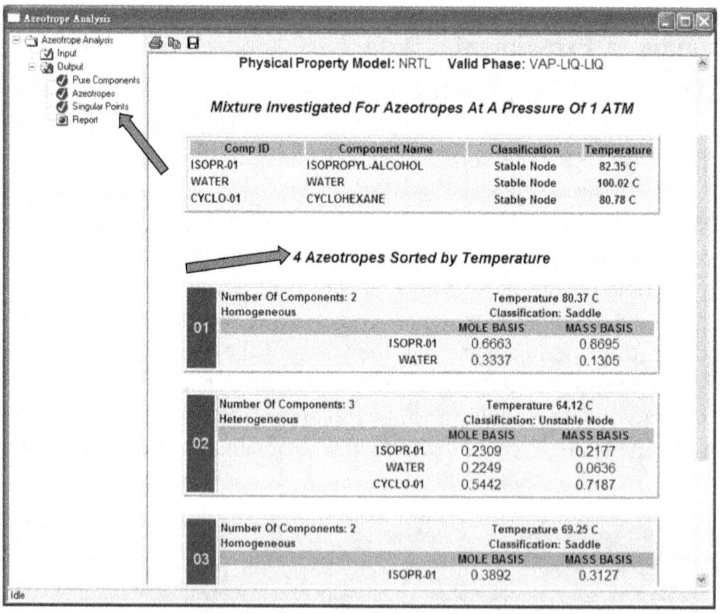


Figure 2.14 Results of the azeotrope search for the isopropanol–water–cyclohexane system.

with three binary azeotropes and one ternary azeotrope. The binary azeotrope of water–cyclohexane and the ternary azeotrope are heterogenous. The other two binary azeotropes are homogeneous. More concise information of all the singular points can be found by clicking *Singular Points* as shown in Figure 2.14.

These prediction results should be compared with the experimental data for validation of the reliability of the thermodynamic model.

### 2.5.3 Steps for Regression

For the cases when Aspen Plus built-in parameters do not exist, or when the prediction of thermodynamic behavior from built-in parameters is not satisfactory, regression to obtain suitable model parameters from experimental data needs to be done. In the following, we will use an example to show the step-by-step procedure to do the data regression.

The particular example is a design study to separate a mixture containing dimethyl carbonate (DMC) and methanol where the built-in parameters cannot be found in Aspen Plus. However, from the paper by Rodriguez et al.,[10] experimental data of *Txy* at 101.3 kPa is available. Therefore, we want to do data regression to obtain suitable binary model parameters to be used in the design study.

The first step for the data regression is to select the route: *Data* → *Setup* at the Aspen Plus menu bar. The *Data Browser* as in Figure 2.15 is shown. In this dialog box, run type of *Data Regression* should be selected. Clicking the *Next* button at the top of *Data Browser* window, the next window for us to input the components is displayed. Clicking *Find* button and inputting the component name as *METHANOL* and then clicking the *Add* button will add this component. Repeat the steps to input *DIMETHYL CARBONATE* into the component selection.

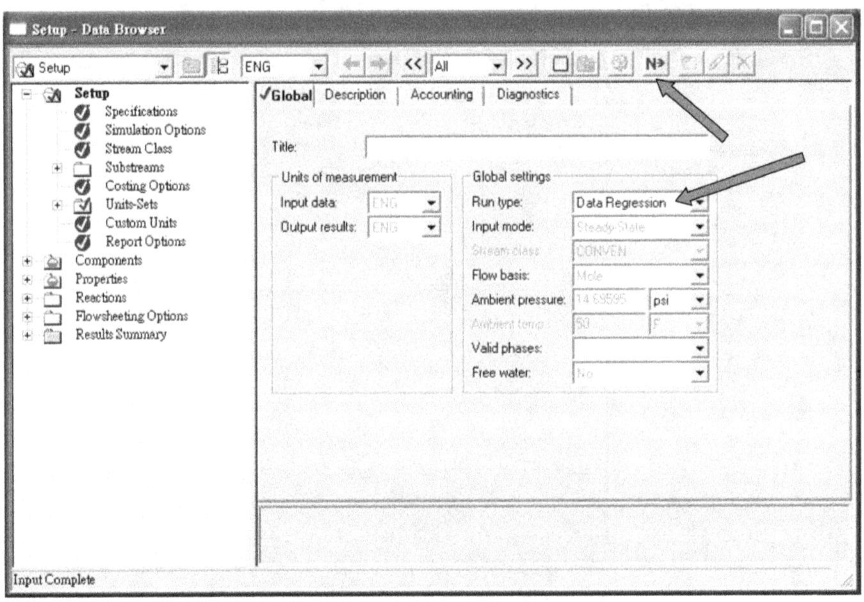

**Figure 2.15** Start data regression steps.

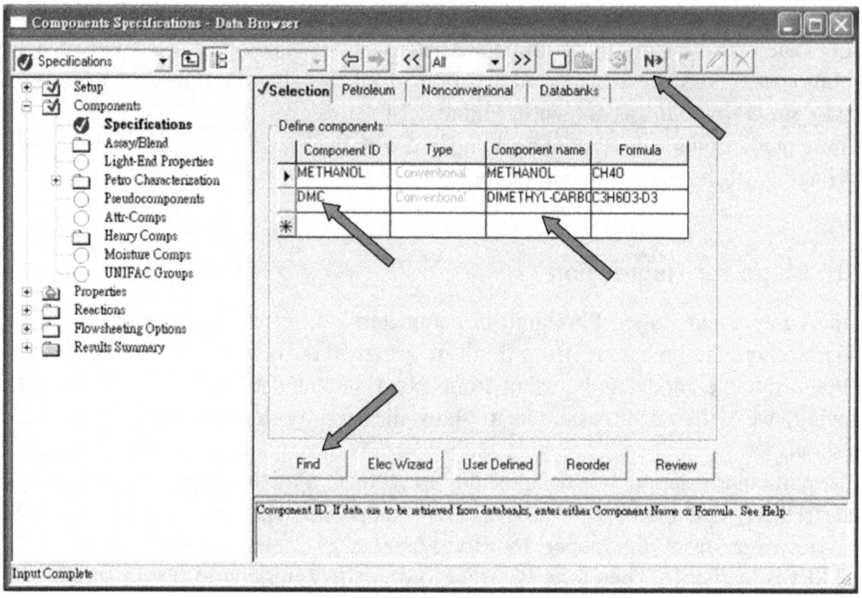


**Figure 2.16** Input components.

Figure 2.16 shows the window with these two components inputted and the Component ID of *DIMETHYL CARBONATE* renamed to an abbreviation of *DMC*. Clicking the *Next* button will automatically go to the next input step. In this *Properties Specifications* window, the *UNIQUAC* base method is selected for the data regression (see Fig. 2.17).

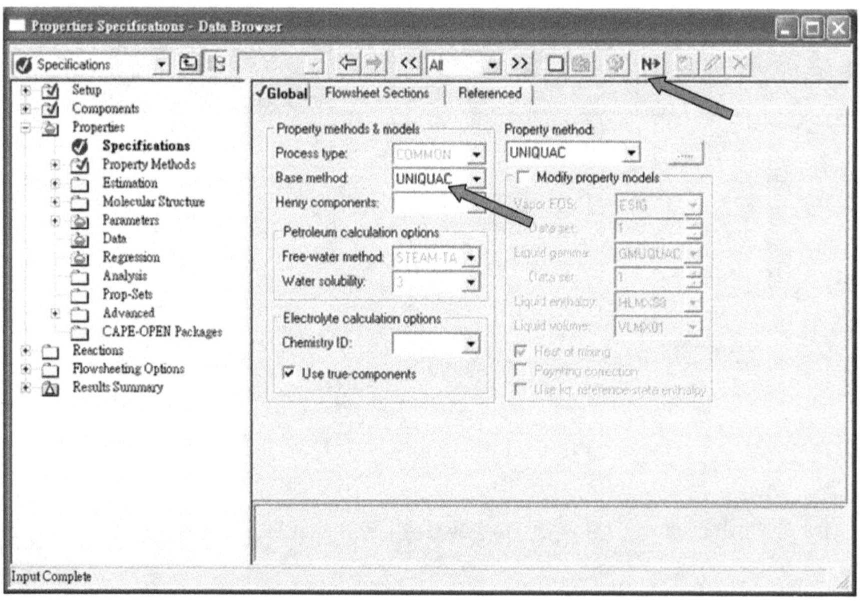


**Figure 2.17** Select base method for the data regression.

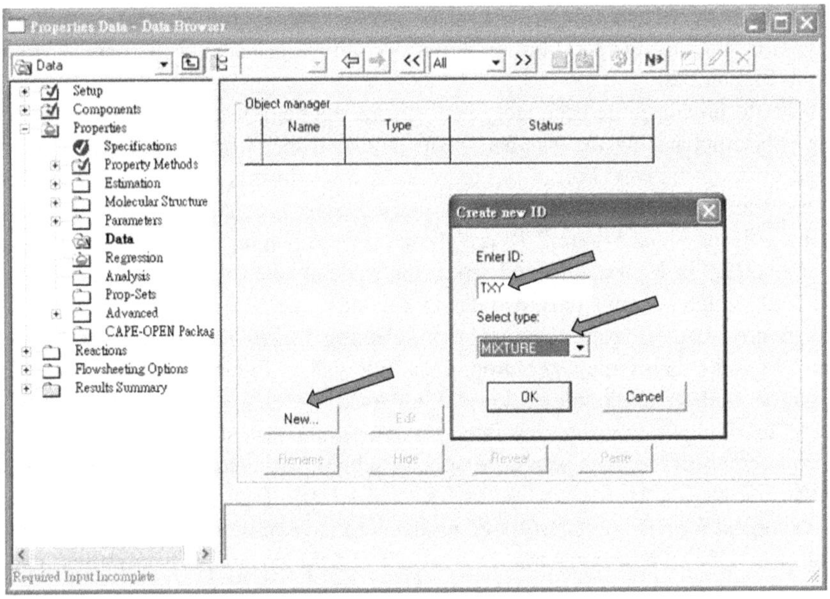


**Figure 2.18** Name of the first dataset.

After Clicking the *Next* button again (taking us to Fig. 2.18), the new experimental dataset is created by clicking *New* and then given the name as *TXY* with type as *MIXTURE*.

Clicking *OK* button in Figure 2.18, the next window (Fig. 2.19) is shown. In this window, the two components are selected by clicking >> to move all available components into selected components cell. In this window, isobaric condition of the experimental data is

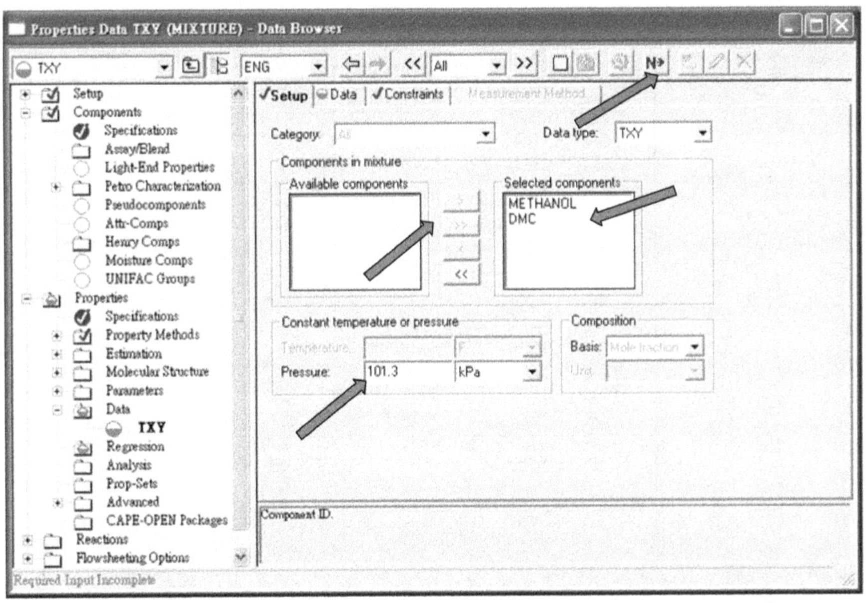


**Figure 2.19** Select components for the dataset and specify isobaric condition.

also selected with operating pressure as *101.3 kPa* to agree with the condition in Rodriguez et al.[10] After clicking the *Next* button again, we are ready to input the experimental data shown in Figure 2.20.

In order to predict more accurately near the azeotrope, a useful tip is to create another dataset only emphasizing the *Txy* data near the azeotrope. After inputting the first set of data and clicking the *Next* button in Figure 2.20, a window as in Figure 2.21 is shown to ask if we want to input more data. Selecting *Enter more data* and clicking the *OK* button, the steps as in previous Figure 2.18 are shown for the naming of the second dataset. We call it *AZEOTXY* in Figure 2.22 and repeat the steps to input the second set of data as in Figure 2.23, which contains only two data points near azeotrope.

We are now ready to go to the next step. Selecting *Specify the data regression cases* in Figure 2.24 and clicking the *OK* button, the window as in Figure 2.25 is shown. Click the *New* button and give the regression ID as *R-1*; the *Setup* window is shown as in Figure 2.26. We put two datasets into this regression task and give heavy weight on the second dataset. The next step is to specify which parameters in UNIQUAC we want to obtain.

The simplest form of the UNIQUAC model is to have only the $b_{ij}$ and $b_{ji}$ parameters in previous Eq. (2.34). These two parameters are defined as element *2* in Aspen Plus. The selection is shown in Figure 2.27 to *Regress* these two binary parameters between components *METHANOL* and *DMC*. Figure 2.28 shows the *Algorithm* dialog box where the algorithm

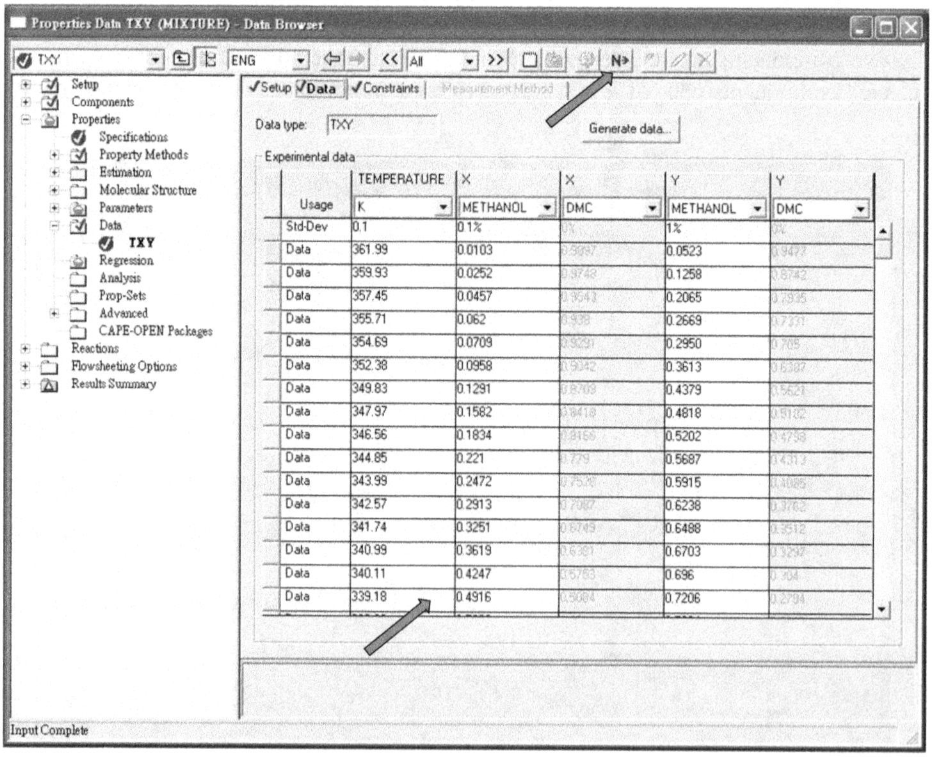

**Figure 2.20** Input *Txy* dataset.

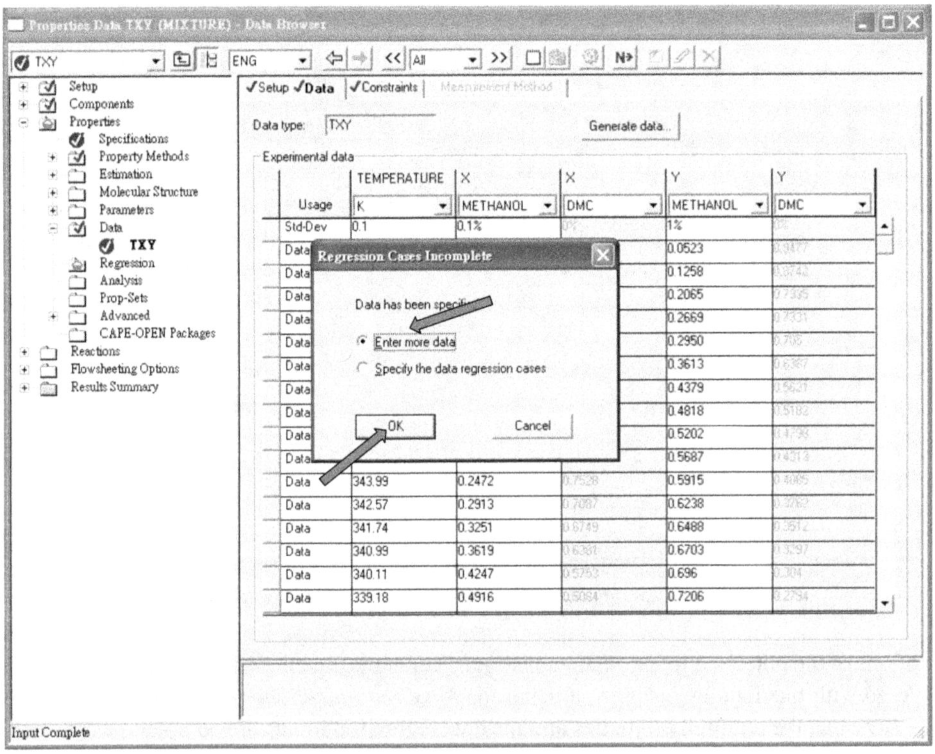


Figure 2.21 Enter more data.

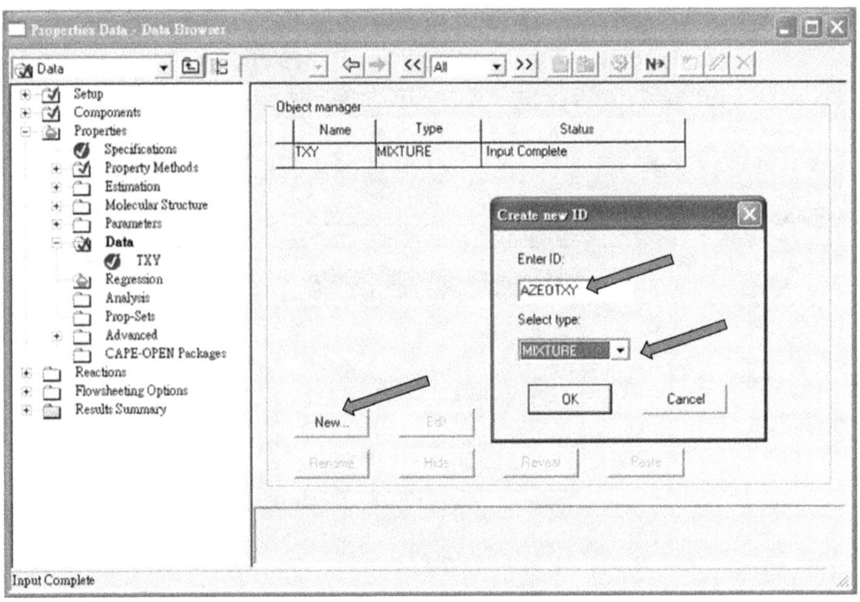


Figure 2.22 Name of the second dataset.

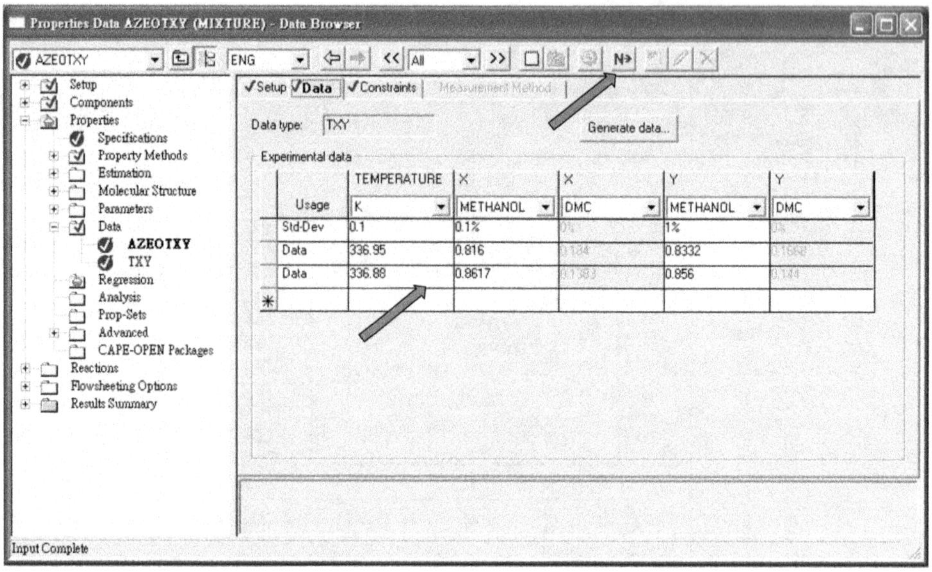

Figure 2.23 Input second *Txy* dataset near azeotropic point.

options of the objective function, the maximum iterations, and the step size can be specified. We go with the default selection of using the *Maximum-Likelihood* method.

Now that the required properties and the data regression inputs are complete, we are now ready to run the regression by clicking the *OK* buttons in Figure 2.29. We select to run regression with the name of *R-1* and then click the *OK* button as in Figure 2.30 to start the regression run. The control panel in Figure 2.31 shows that the data regression run is complete with six iterations.

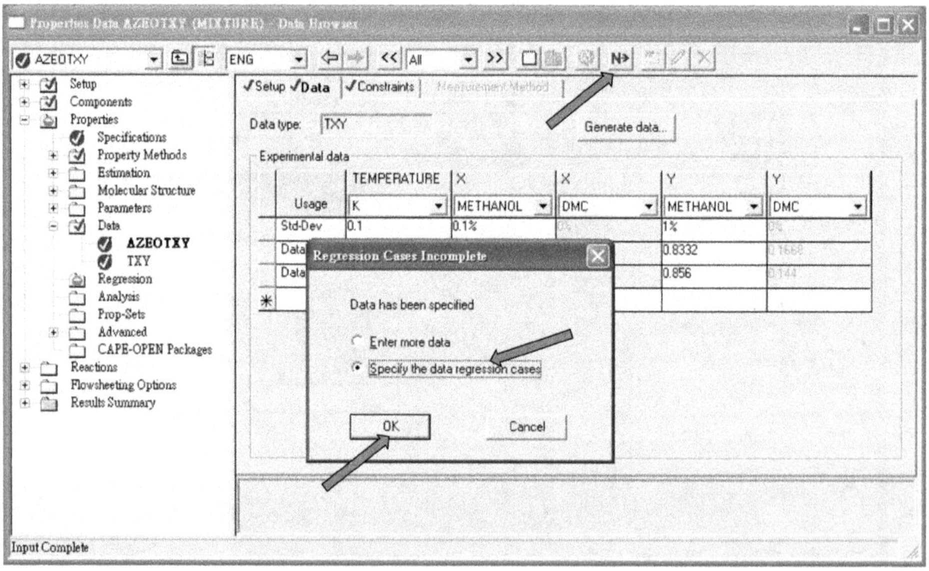

Figure 2.24 Specify the regression case.

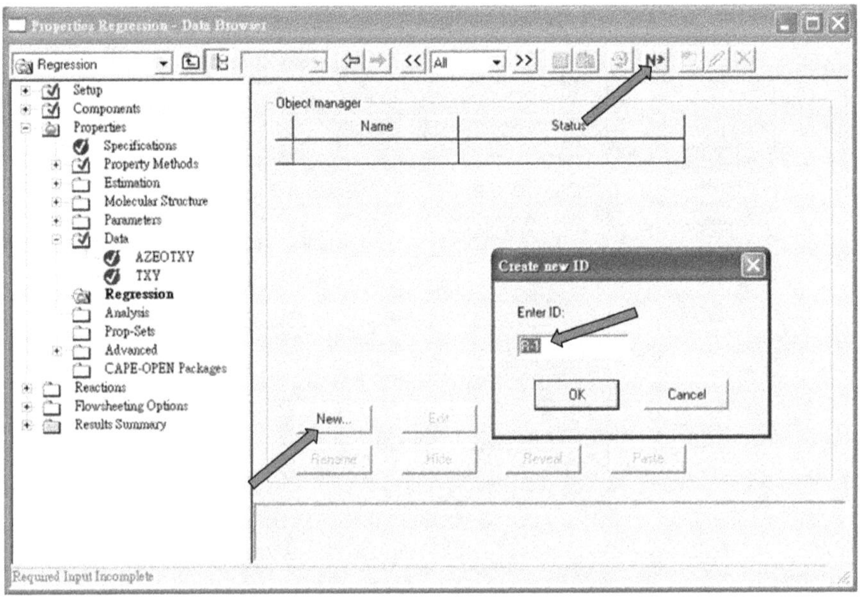

Figure 2.25 Name the regression case.

To know the resulting binary parameters, the folder selection as: *Properties* → *Parameters* → *Binary Interaction* → *UNIQ-1* on the left-hand side of *Data Browser* is selected as shown in Figure 2.32. The value of the $b_{ij}$ parameter is −24.3225 and the value of the $b_{ji}$ parameter is −268.981.

To check on how good the data regression is, *Plot Wizard* in the *Plot* dropdown menu at the top of Aspen Plus main menu can be used. At the first step of *Plot Wizard*

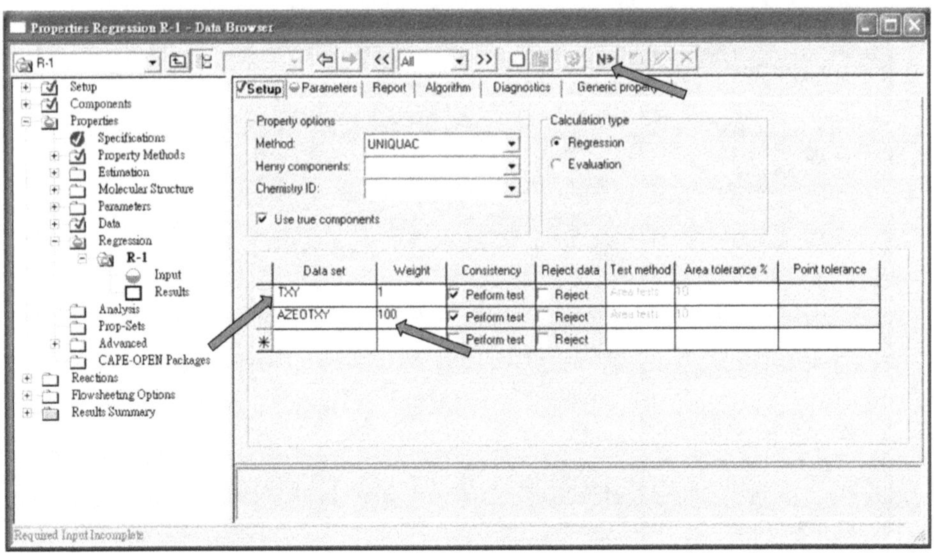

Figure 2.26 Include two datasets with bigger weight on the second dataset.

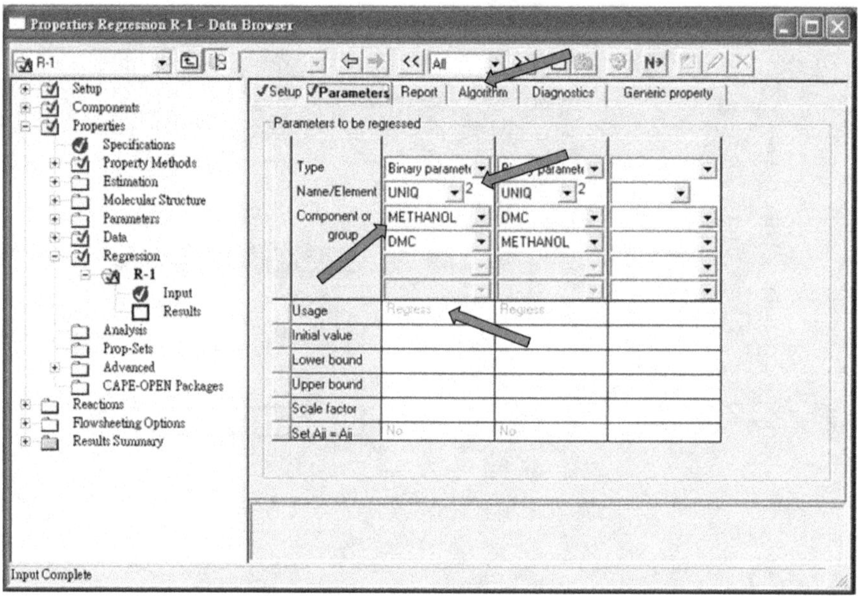

**Figure 2.27** Select $b_{ij}$ *and* $b_{ji}$ parameters to regress.

(see Fig. 2.33), click the *Next* button to go to Figure 2.34. At this figure, selecting *Txy* and then *Next* will go to step 3 of *Plot Wizard* where two datasets are selected. Other selections can be seen in Figure 2.35. The next step is to provide the titles of the plot as well as the axes as in Figure 2.36. After clicking the *Finish* button, the plot as in Figure 2.37 is shown. In this figure, the solid lines are the fitted *Txy* curves and the discrete

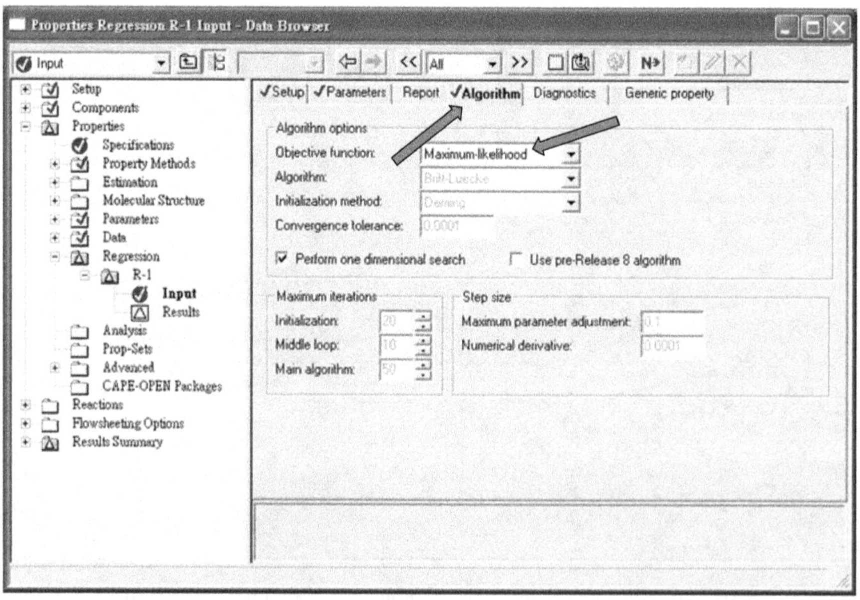

**Figure 2.28** Specify algorithm for the regression.

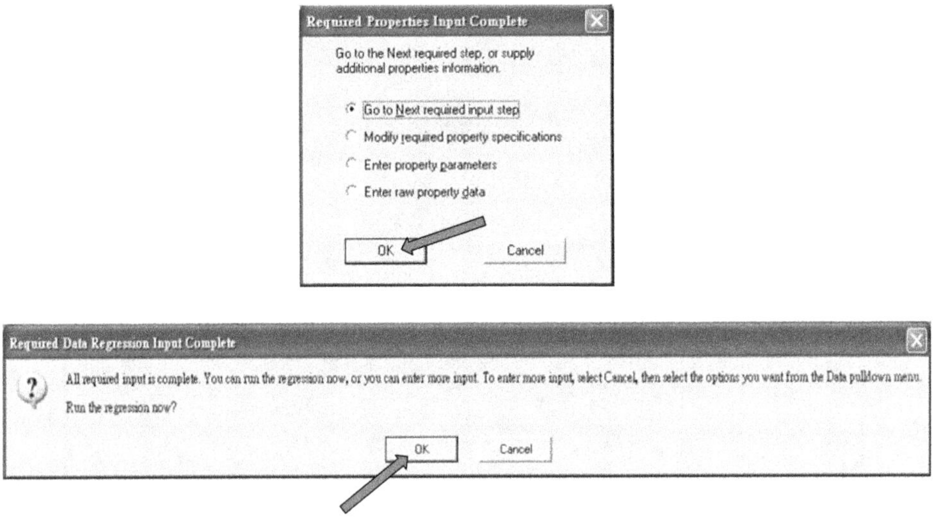


**Figure 2.29** Start the regression run.

data points are the experimental data. It is verified that the model fits the experimental data very well.

The above demonstration is to show the user how to use the data regression capability in Aspen Plus. A more complex situation may be encountered for ternary systems (with three binary pairs) where the binary model parameters not only need to fit the corresponding VLE data, but also need to fit, for example, a ternary LLE envelope. In this more complex case, some sacrifice always needs to be made. This can be accomplished by adjusting the weight of individual dataset as in Figure 2.26.

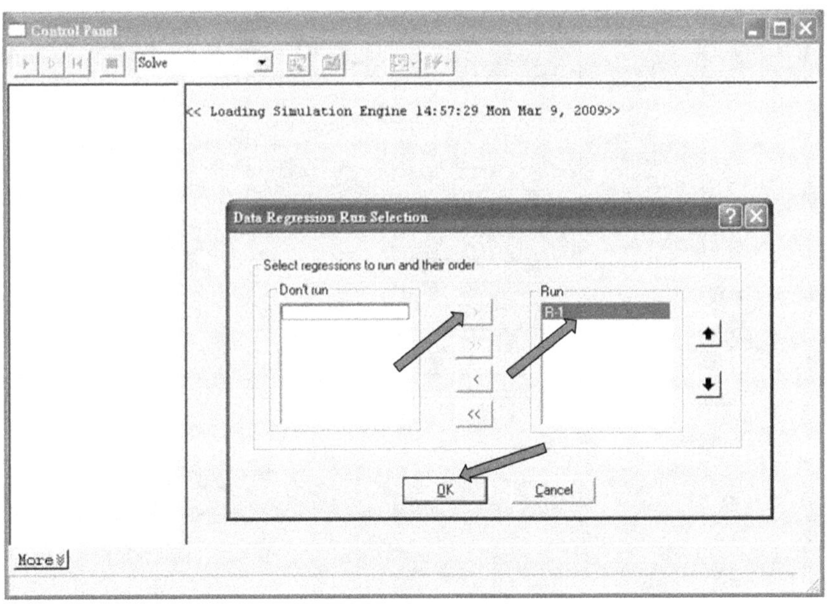


**Figure 2.30** Regression run selection.

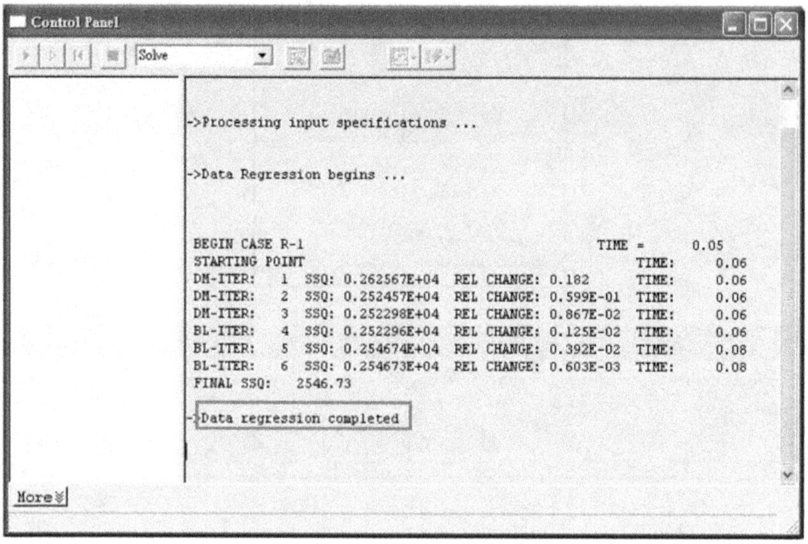


**Figure 2.31** Regression completed.

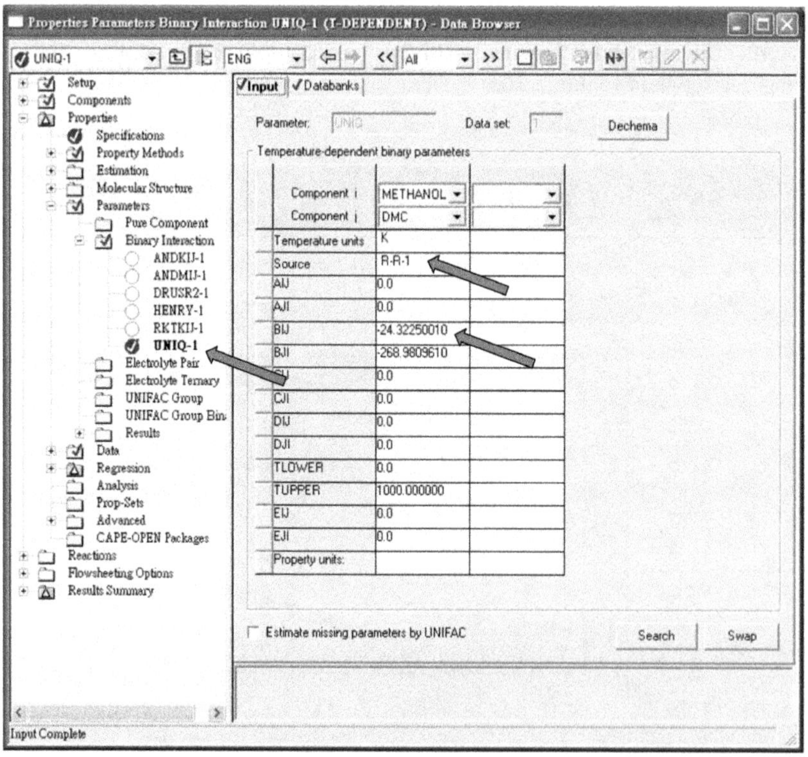


**Figure 2.32** The resulting model parameters.

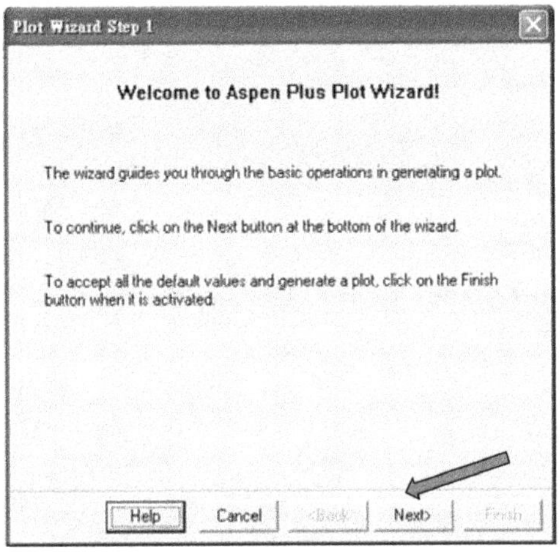


Figure 2.33 Plot the regression result.

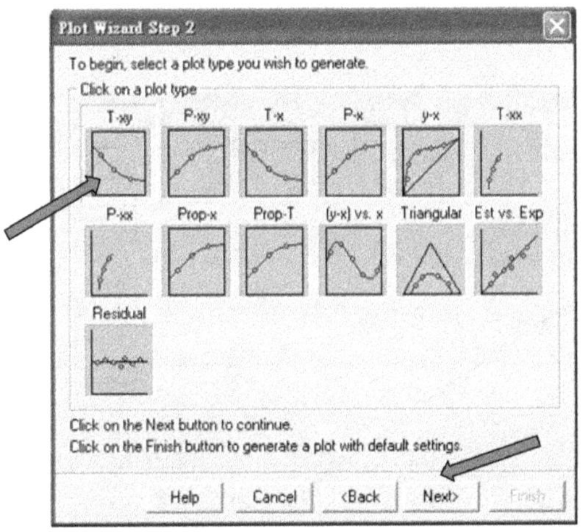


Figure 2.34 Select *Txy* plot.

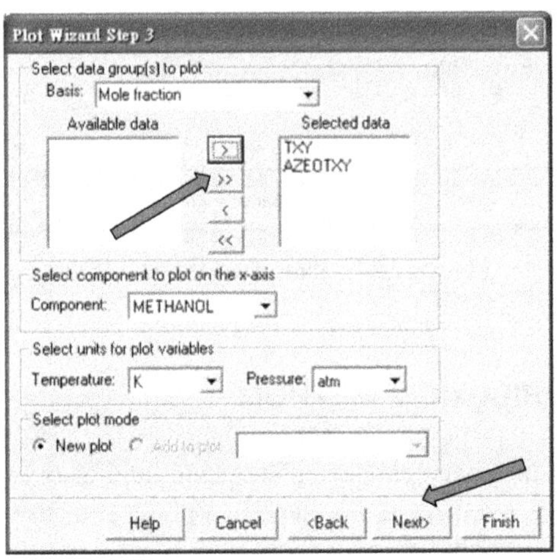


Figure 2.35 Select two datasets.

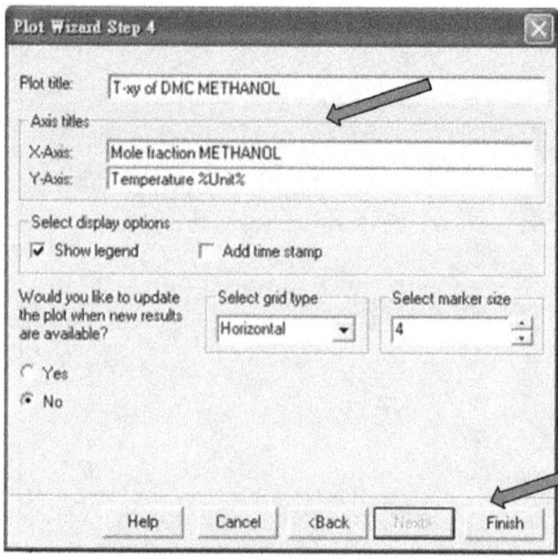


**Figure 2.36** Fill in plot and axis titles.

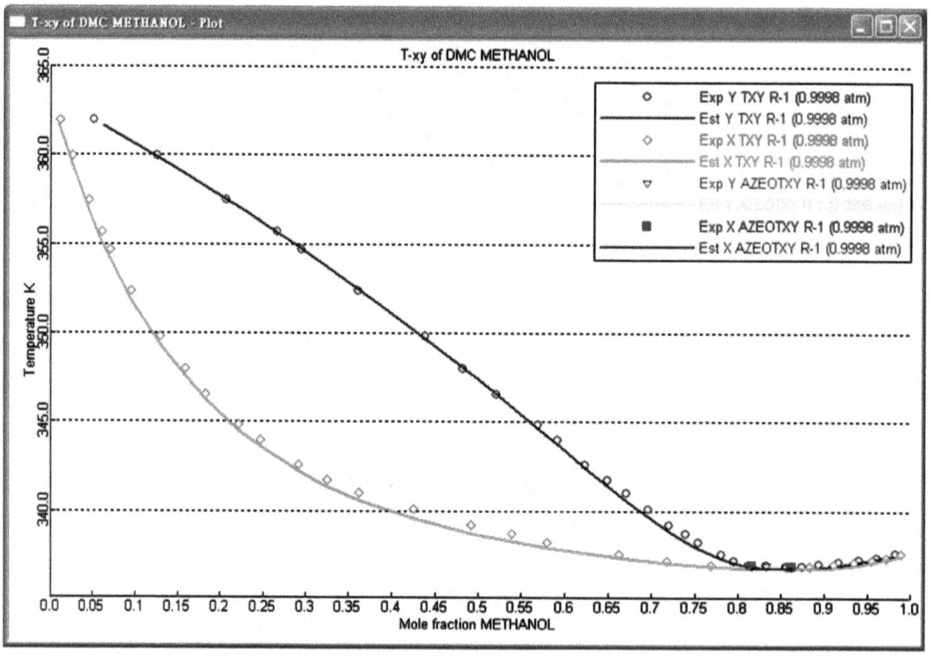


**Figure 2.37** Plot of regression result.

## 2.6 GROUP CONTRIBUTION METHOD

When there is no built-in binary parameter in Aspen, and also when we cannot find any experimental data in the open literature, another option is to use the group contribution model in Aspen to predict liquid activity coefficient for this binary pair or the entire component system. UNIFAC is an activity coefficient model, like NRTL or UNIQUAC, but it is based on group contributions, rather than molecular contributions. UNIFAC uses group parameters and group–group interaction parameters determined from a limited, well chosen set of experimental data to predict activity coefficients between almost any pair of components. All published group parameters and group-binary parameters are stored in the Aspen Physical Property System.[2] The parameter sets for all UNIFAC models are regularly revised and extended.

There are two ways to use the UNIFAC model. One is to use the UNIFAC model to describe all binary pairs in the whole component system. The other is to only use the UNIFAC model to estimate the binary pair parameters for the particular pair where the built-in parameters are not available in Aspen. We will discuss these two situations in the following.

In the first case, after inputting all the components in the studied system, at the *Data Browser* of the *Properties Specifications*, we just need to select UNIFAC or two other modifications (UNIF-DMD or UNIF-LBY) in the base property method. After this selection, we are ready to do the simulation once the flowsheet and unit operation conditions are specified. A demonstration of this choice is shown in Figure 2.38.

For illustrating of the second case, the component system of a reactive distillation system for the production of butyl acrylate (see Zeng et al.[11]) will be used here as an example. Butyl acrylate is produced by an esterification reaction of acrylic acid and butanol with a byproduct of water. The total system includes four components: acrylic acid (AA), butanol (BuOH), butyl acrylate (BA), and water. The base-property method was chosen as *NRTL-HOC* because AA is known to have vapor association. There are a total of six binary pairs in the whole component system. In these six binary pairs, built-in model parameters can be found for the four pairs of AA–BuOH, AA–$H_2O$, BuOH–BA, and BuOH–$H_2O$. The model parameters of the remaining two pairs (AA–BA and BA–$H_2O$) need to be estimated by using the group-contribution method of UNIFAC before any simulation run can be done.

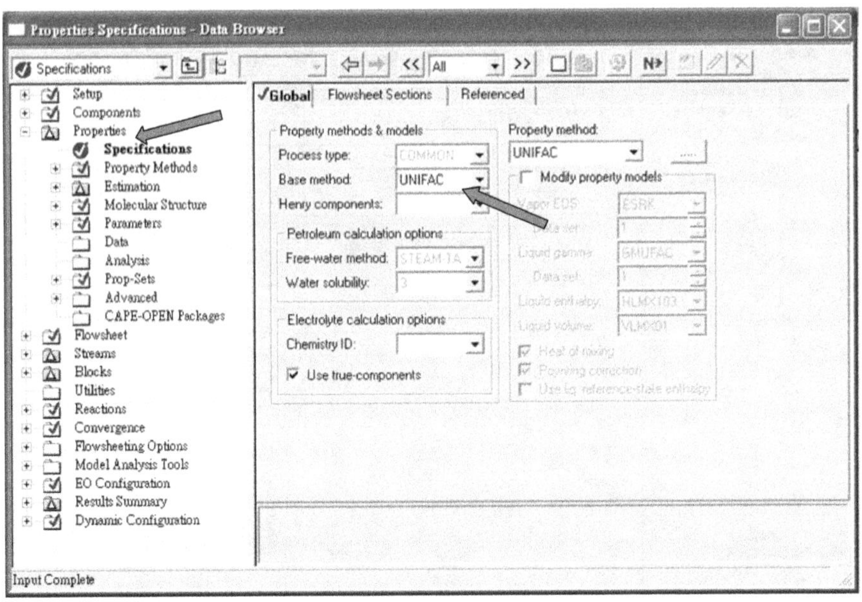


**Figure 2.38** Example of selecting UNIFAC as base method.

To show the Aspen built-in binary parameters, the folder selection as: *Properties → Parameters → Binary Interaction → NRTL-1* on the left-hand side of *Data Browser* is selected as shown in Figure 2.39. In this figure, model parameters of four binary pairs are shown. Notice that the source of the parameters is from *VLE-HOC*, which means they are Aspen built-in parameters. Since model parameters of two binary pairs are missing, we select *Estimate missing parameters by UNIFAC* in this window.

The next step is to go to *Setup → Specifications* and change the run type into *Property Estimation* (see Fig. 2.40). After clicking the *Next* button, the property estimation is ready to run as in Figure 2.41. Figure 2.42 shows that the estimation run is complete and asks if we want to replace the existing model parameters of the four binary pairs with the new estimated parameters. Since the built-in parameters are obtained from the fitting of known experimental data and should be more reliable, we select *No to all.*

In order to exhibit the entire binary parameters for this system, we go back to *Setup → Specifications* and change the run type back to *Flowsheet* (see Fig. 2.43). At the folder route: *Properties → Parameters → Binary Interaction → NRTL-1*, all six binary pairs are shown as in Figure 2.44. In that, the last two pairs are estimated by UNIFAC. This is identified because the source is *R-PCES*, which stands for Property Constant Estimation System. After finishing the above step, the whole thermodynamic model is given and we are ready to do the rest of the simulation.

One final note about the UNIFAC is that, although Aspen is regularly revised and extended with the group parameters and group–group interaction parameters from published

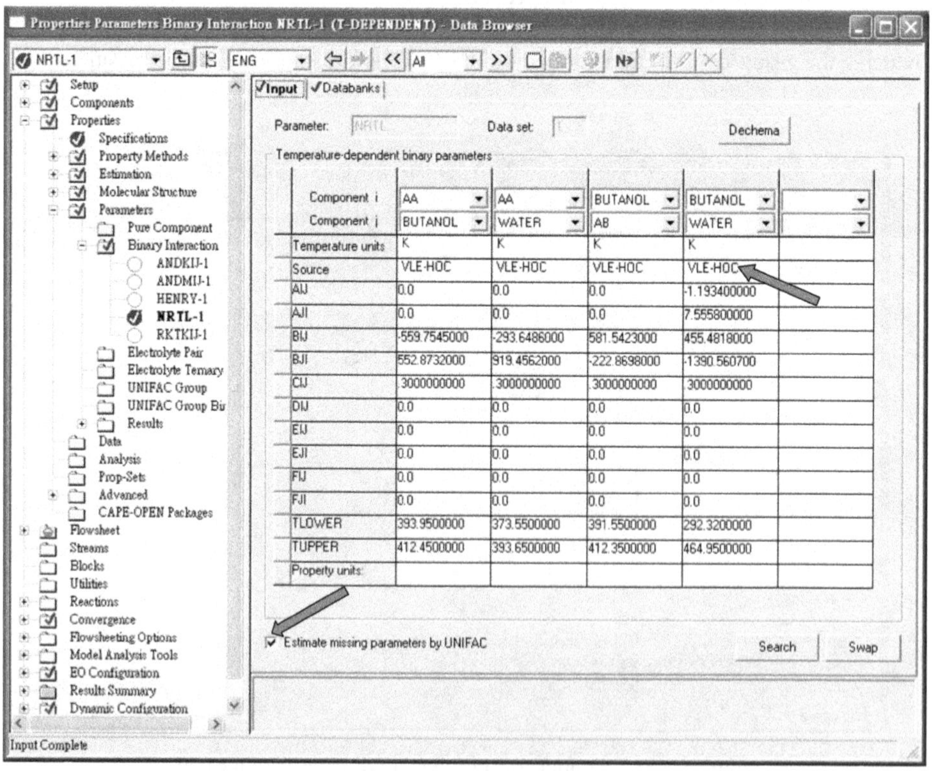

**Figure 2.39** Estimate missing parameters with UNIFAC.

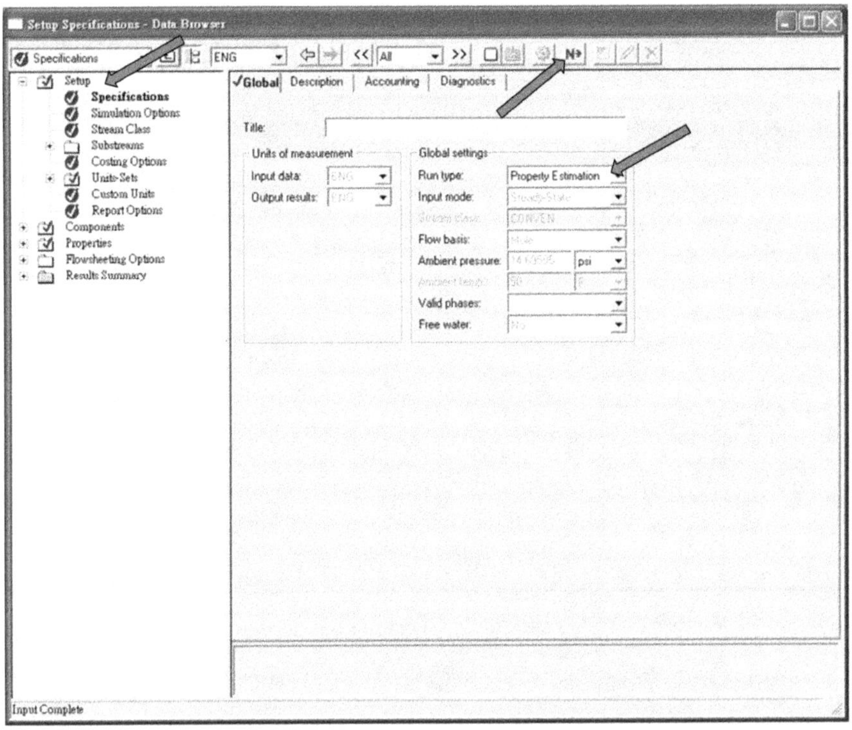


**Figure 2.40** Select run type as property estimation.

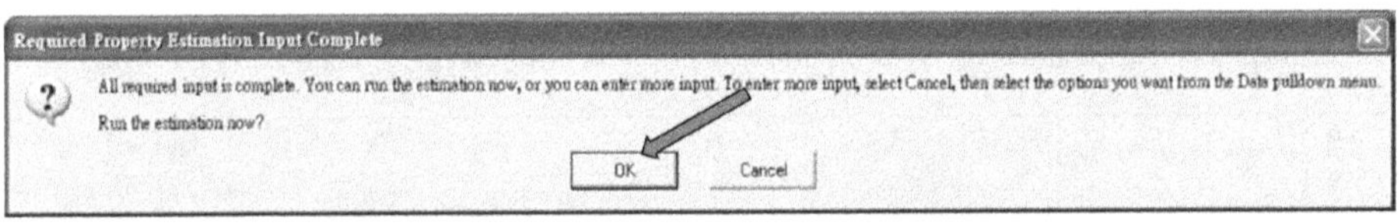


**Figure 2.41** Start the property estimation.

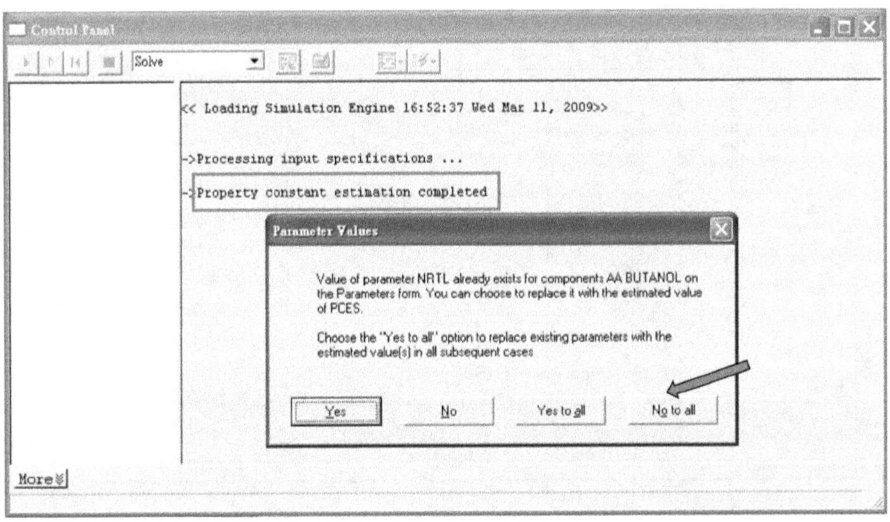


**Figure 2.42** Property estimate completed and select to retain built-in parameters.

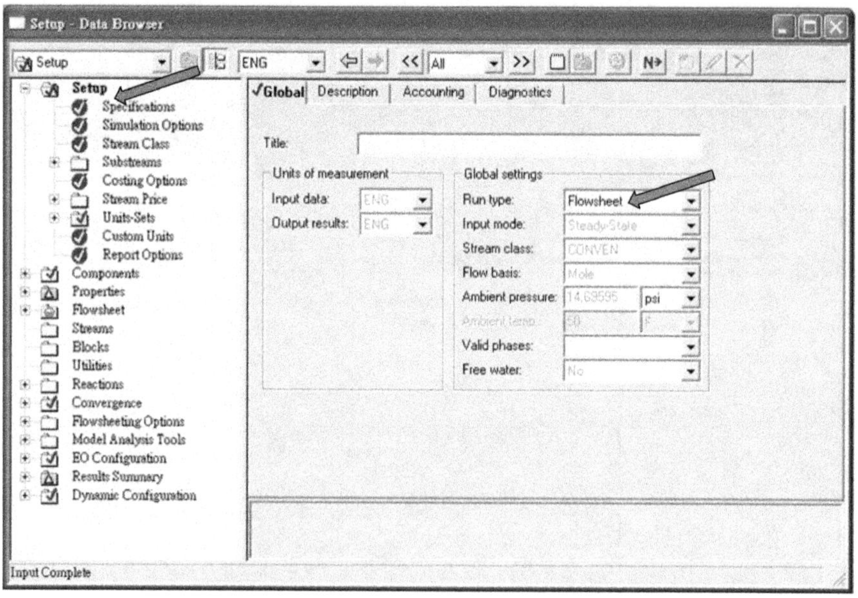

**Figure 2.43** Go back to flowsheet run type to start simulation.

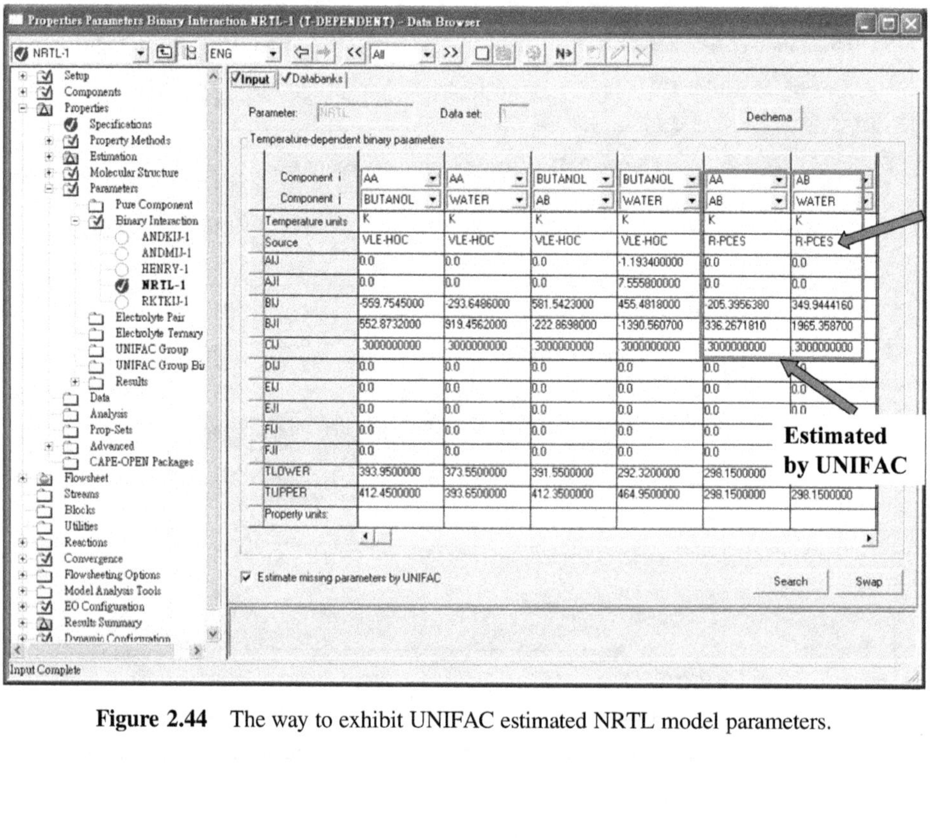

**Figure 2.44** The way to exhibit UNIFAC estimated NRTL model parameters.

literature, there are still some situations where the parameter sets are lacking for particular function groups. In this rare situation, another predictive thermodynamic method based on conductor-like screening models (COSMO) can be used to complement the UNIFAC group contribution method. Aspen also provides the use of this model called COSMO–SAC in the Aspen Physical Property System.[2] For more detailed information of this model, please refer to Mullins et al.[12]

## 2.7 CONCLUSION

In this chapter, the background of phase equilibrium was reviewed. The useful binary and ternary tools in Aspen were also illustrated with examples. Data regression capability in Aspen for obtaining thermodynamic model parameters from experimental data was also demonstrated. For the cases where there is no experimental data, how to use UNIFAC to estimate physical properties was also given in this chapter.

## REFERENCES

1. Hayden J. G. and J. P. O'Connell, A generalized method for predicting second virial coefficients, *Ind. Eng. Chem. Process Des. Dev.*, **14**, 209–216 (1975).
2. Aspen Physical Property System, Version 2006.5, Aspen Technology, Inc., Cambridge, MA, 2007.
3. Michelsen M. L., The isothermal flash problem. Part I. Stability, *Fluid Phase Equilibria*, **9**, 1–9 (1982).
4. Michelsen M. L., The isothermal flash problem. Part II. Phase-split calculation, *Fluid Phase Equilibria*, **9**, 21–40 (1982).
5. Dortmund Databank (DDB). http://www.ddbst.de/.
6. Gmehling J. and U. Onken, Vapor–liquid equilibrium data collection. In *DECHEMA Chemistry Data Series*, Eds. D. Behrens and R. Eckermann, DECHEMA Publisher, Frankfurt, Germany, 1977.
7. Sørensen J. M. and W. Arlt, Liquid–liquid equilibrium data collection. In *DECHEMA Chemistry Data Series*, Eds. D. Behrens and R. Eckermann, DECHEMA Publisher, Frankfurt, Germany, 1979.
8. Doherty M. F. and M. F. Malone, *Conceptual Design of Distillation Systems*, McGraw-Hill, New York, 2001.
9. Gmehling J., *Azeotropic Data*, Wiley-VCH, Weinheim, 2004.
10. Rodriguez A., J. Canosa, and A. Dominguez, Vapor–liquid equilibrium of dimethyl carbonate with linear alcohols and estimation of interaction parameters for the UNIFAC and ASOG method, *Fluid Phase Equilibria*, **201**, 187–201 (2002).
11. Zeng K. L., C. L. Kuo, and I. L. Chien, Design and control of butyl acrylate reactive distillation column system. *Chem. Eng. Sci.*, **61**, 4417–4431 (2006).
12. Mullins E., R. Oldland, Y. A. Liu, S. Wang, S. I. Sandler, C. C. Chen, M. Zwolak, and K. C. Seavey, Sigma-profile database for using COSMO-based thermodynamic methods, *Ind. Engng Chem. Res.*, **45**, 4389–4415 (2006).

# CHAPTER 3

# STEADY-STATE DESIGN IN ASPEN PLUS

The main purpose of this chapter is to demonstrate how to build a steady-state simulation in Aspen Plus for design and optimization studies. This steady-state model can also be exported to Aspen Dynamic for control studies. The step-by-step procedure for setting up a steady-state simulation will be shown in this chapter, while the procedure for setting up a dynamic simulation will be shown in Chapter 4. We will begin in Section 3.1 by taking a simple distillation example and go through all the details of setting up a simulation of this system in Aspen Plus using the rigorous distillation column simulation *RadFrac*. In Section 3.2, we will describe all unit operation blocks in Aspen Plus, which will be used in various example systems in this book. These include: rigorous distillation column (*RadFrac*), liquid–liquid separation (*Decanter*), and heat exchangers (*Heater*, *HeatX*). For situations in which one or more components cannot be found in the Aspen databank, there is a way to study the steady-state design by adding a nondatabank component into Aspen Plus. The detailed steps, with an example, will be given in Section 3.3.

## 3.1 BUILDING A STEADY-STATE MODEL

In this section we use an extractive distillation column as an example to demonstrate how to build a steady-state simulation. This is one column of an overall two-column system for separating isopropanol and water. The detailed design and control of the overall distillation system will be given in Chapter 10.

The column configuration is shown in Figure 3.1. There are two feed streams entering this column: a fresh feed (mixture of IPA and water) is fed into Stage 35, and a solvent entrainer stream (dimethyl sulfoxide) is fed into Stage 7. The column has total of 41 stages (including total condenser and reboiler). Other operating condition can be found in Figure 3.1.

*Design and Control of Distillation Systems for Separating Azeotropes.* By William L. Luyben and I-Lung Chien

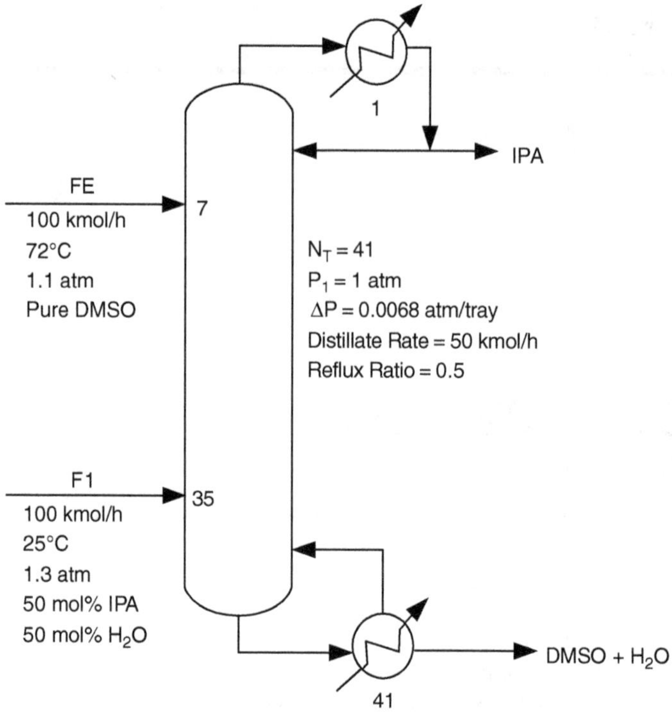


**Figure 3.1** Extractive distillation example.

The purpose of the solvent entrainer in this extractive distillation column is to alter the relative volatility between IPA and water, making IPA go to the top of the column and water go to the bottom of the column. The upper section of the column (above the entrainer feed location) is called the rectifying section, and its purpose is to separate the IPA and the entrainer. The middle section of the column (between the entrainer feed stage and the fresh feed stage) is called the extractive section. The purpose of this section is to suppress water from going up the column. The bottom section of the column (below the fresh feed location) is called the stripping section, and its purpose is to keep IPA from going down the column. The bottoms product of the column is the mixture of water and the entrainer, and it is fed to another downstream entrainer recovery column to separate these two components, so the entrainer can be recycled back to the extractive distillation column.

### 3.1.1 Defining the Flowsheet

Open up a blank flowsheet by going to *Start* and *Programs* and then clicking sequentially on *Aspen Tech, Aspen Engineering Suite, Aspen Plus 2006.5, Aspen Plus User Interface*. The window shown in Figure 3.2 opens up. Selecting the *Blank Simulation* button and clicking *OK* opens up the blank flowsheet as in Figure 3.3. The page tabs along the bottom let us choose which unit operations to place on the flowsheet. Clicking the *Columns* page tab, the list of available distillation columns appears as a row of icons with descriptions appearing at the bottom of the window. Reading the prompt for the *RadFrac* block and clicking the arrow just to the right of this block opens the window shown in Figure 3.3. This contains

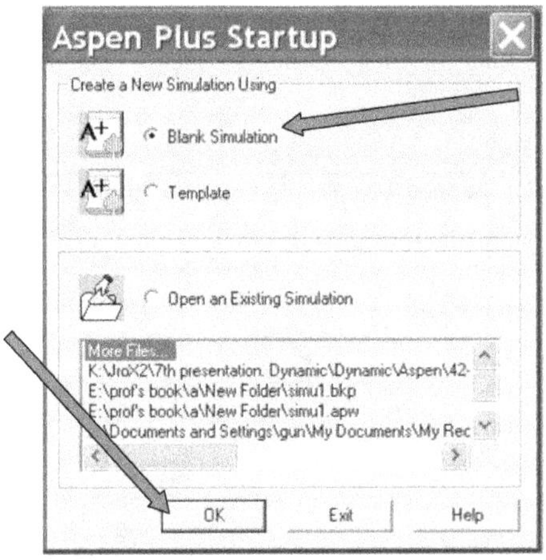


**Figure 3.2** Aspen Plus startup.

several types of columns, including full columns, strippers (with a reboiler but no condenser), rectifiers (with a condenser but no reboiler), absorbers (with neither), and more complex columns with side pump-around. Click the full-column button on the top row, second from the left, and move the cursor to the blank flowsheet. The cursor becomes a cross. If we click on the flowsheet, a column icon appears, as shown in Figure 3.4.

The next job is to add streams to connect all the pieces in the flowsheet. This is achieved by moving the cursor all the way to the left at the bottom of the window and clicking the *Material STREAMS* button. Then move the cursor to the flowsheet. A number of arrows appear as shown in Figure 3.5 that show all the possible places where a material stream

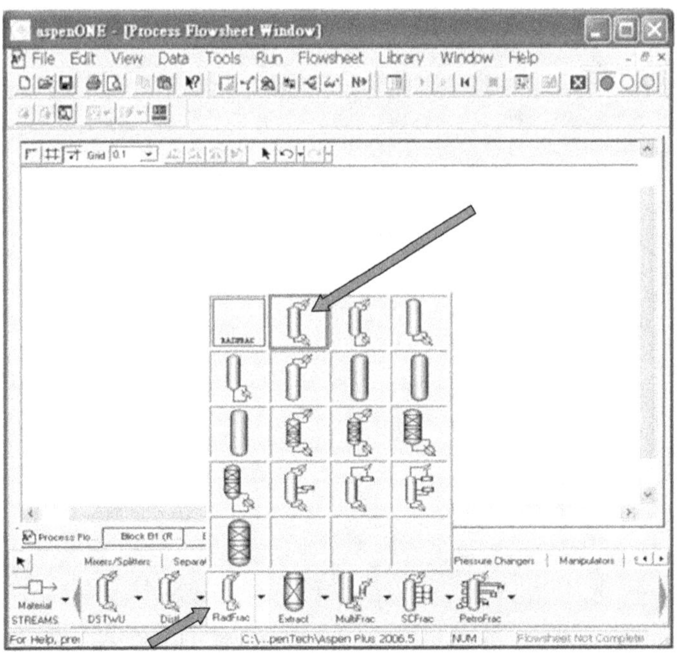


**Figure 3.3** Selecting type of column.

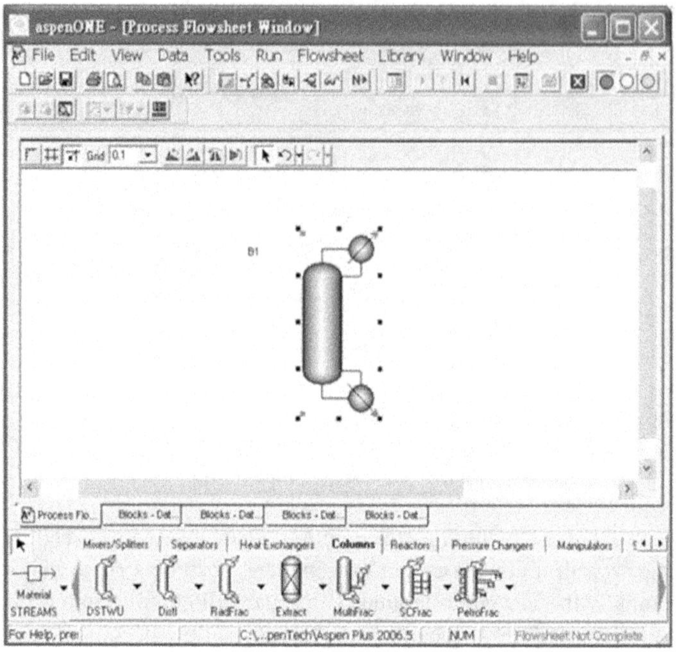


**Figure 3.4** Paste the column icon on the flowsheet.

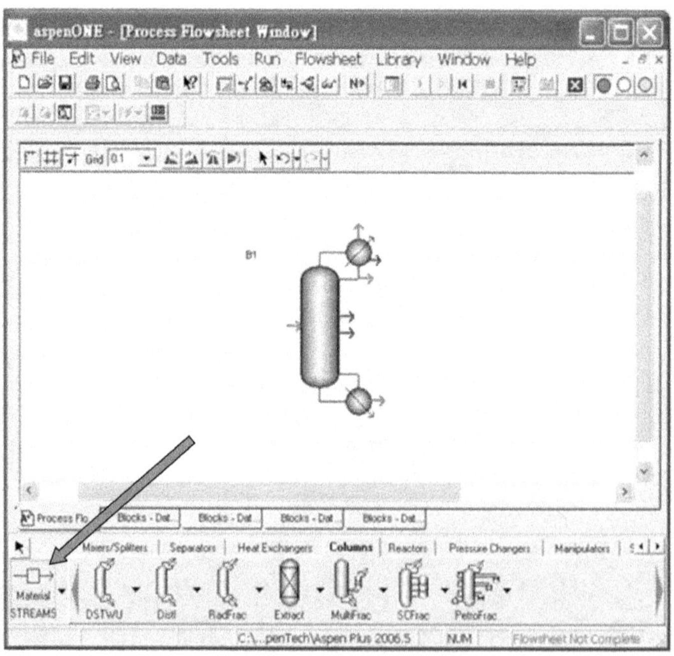


**Figure 3.5** Possible connections displayed.

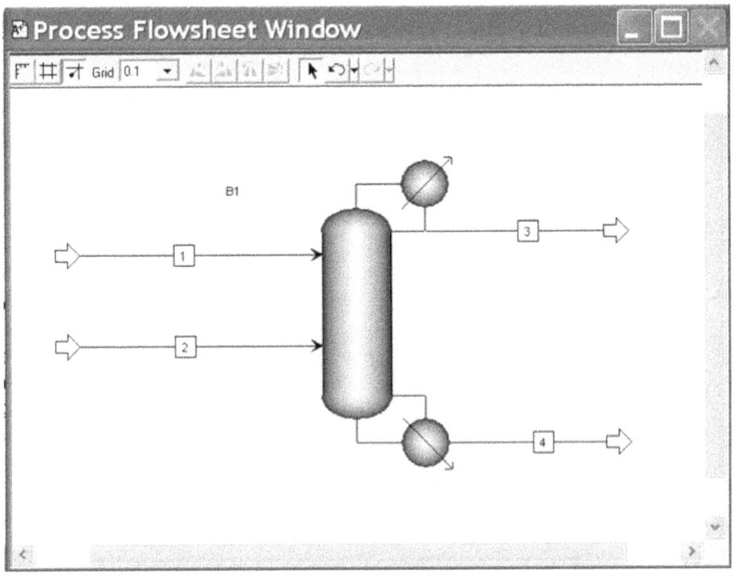


**Figure 3.6** Flowsheet with all streams attached.

can be located as an input stream or an output stream from the distillation column. Place the cursor on one arrow and left click. Then place the cursor where you want to connect the stream, and click. The four streams associated with the column (two feed streams, the distillate, and the bottoms) are connected in a similar manner with the result shown in Figure 3.6. Note that the distillate stream has been connected at the arrow that is below the condenser symbol. This gives a liquid distillate product. If the stream had been connected to the arrow coming out of the top of the condenser, the distillate would be a vapor and the condenser would be a partial condenser.

At this point, the flowsheet has been configured. It is a good idea to rename the various pieces of the equipment in the flowsheet and the streams to help keep track of the identity of each of these components. To rename a block, left click its icon and right click to get a drop-down menu. Select *Rename* and type in the desired new name. The same procedure is used to rename streams. Figure 3.7 gives the flowsheet with the block and streams rename to correspond to conventional distillation terminology. The fresh feed stream is *F*1, the entrainer feed stream is *FE*1, the distillate stream is *D*1, and the bottoms stream is *B*1. Some logical scheme for renaming units and streams is essential in a large plantwide simulation with many units and many streams.

### 3.1.2 Entering Components

The structure of the flowsheet is now completely specified. Next we must define the chemical components involved in the separation. We can use the *Data Browser* input sheets to enter the remaining required information for this run. The Aspen Plus *Next* function displays the required input sheets automatically. After clicking the *Next* button, Aspen Plus displays the *Flowsheet Complete* dialog box indicating that the flowsheet is complete and that we need to provide the remaining problem specifications. Click *OK* to display the *Data Browser* window as shown in Figure 3.8. This is the window that is used to look at all aspects of

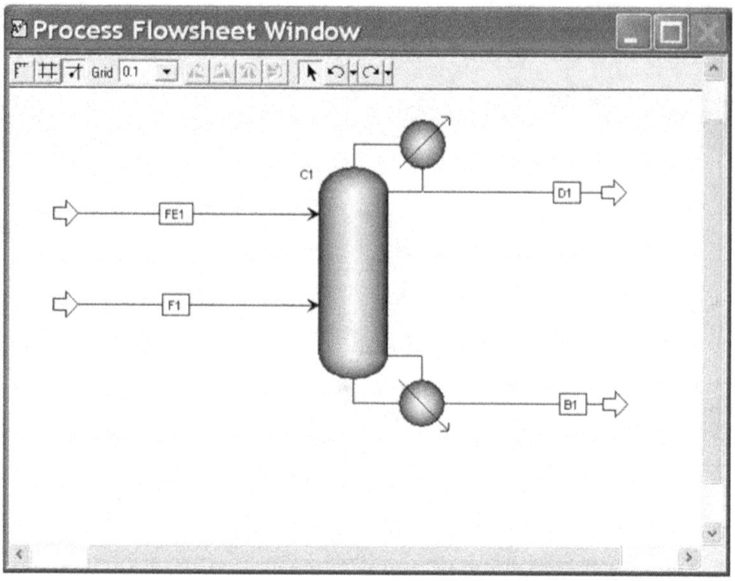


**Figure 3.7** Renamed flowsheet.

the simulation. It is used to define components, set physical properties, specify the parameters of the equipment (e.g., the number of stages in the column, the feed location, and the column pressure, etc.), and specify properties of various streams (e.g., flowrate, composition, temperature, and pressure).

A couple of preliminary items should be done first. In the middle of the window there are two boxes in which we can specify the units to be used in the simulation. They are: ENG (English engineering), MET (metric), and SI (Système International). We will use SI

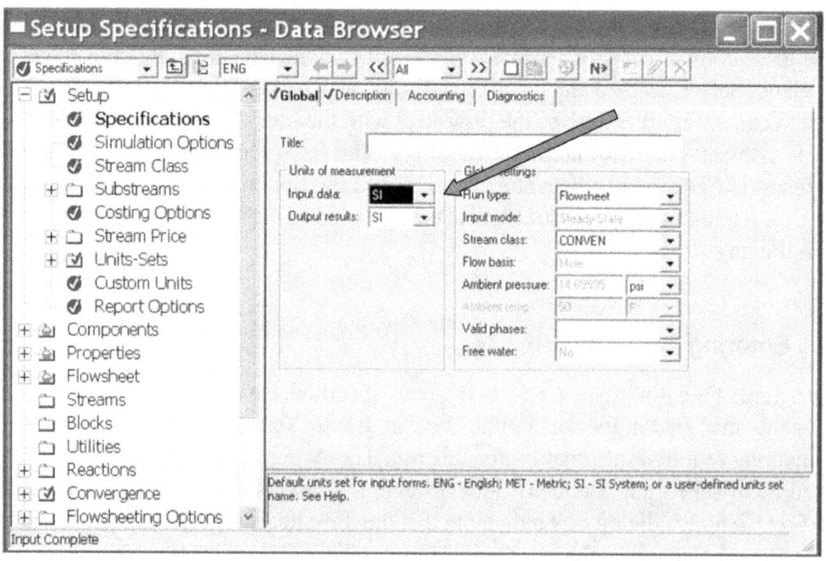


**Figure 3.8** Data Browser window.

units in most of the examples in this book. However, we will make one departure from regular SI units. In the SI system, pressures are expressed in pascals ($N/m^2$), which are quite inconvenient for most chemical processes because typical pressures are very large numbers in pascals (1 atm = 101,325 Pa). Therefore we will use pressures in atmospheres in most of the examples. However, make sure that you select the correct units when you enter data.

The second preliminary item is to indicate what properties we want to see for all the streams. The defaults do not include compositions in mole fractions, which are very useful in distillation calculations. To include mole fractions in the list of the stream properties, click on the last item on the *Setup* list at the left of the window, which is labeled *Report Options*. Note that if the *Report Options* item is not visible, click the + symbol next to the *Setup* folder to expand it. After you click the *Report Options* item, click the *Stream* tab on the top of the window. This opens the window shown in Figure 3.9. Select the *Mole* checkbox under the *Fraction basis* column in the middle of the window.

With the bookkeeping issues out of the way, we can select the chemical components by clicking *Components* on the left of the *Data Browser* window. This opens the window shown in Figure 3.10. Clicking the *Find* buttons near the bottom of the window opens another window. Type in *isopropanol* and click *Find now* as shown in Figure 3.11. A list of components is opened at the bottom of the window. Click *ISOPROPYL ALCOHOL* and click the *Add* button at the bottom of the window. Repeat for *Water* and *dimethyl sulfoxide* and click *Close*. Note that if we need to search for components based on molecular weight range, boiling point range, or CAS numbers, we can select the *Advanced* button in the *Find* dialog box in Figure 3.11.

The three components have now been selected as shown in Figure 3.12. It is often desirable to change the name of a component. For example, suppose that we want to use "IPA" for isopropyl alcohol and "DMSO" for dimethyl sulfoxide. This can be accomplished by

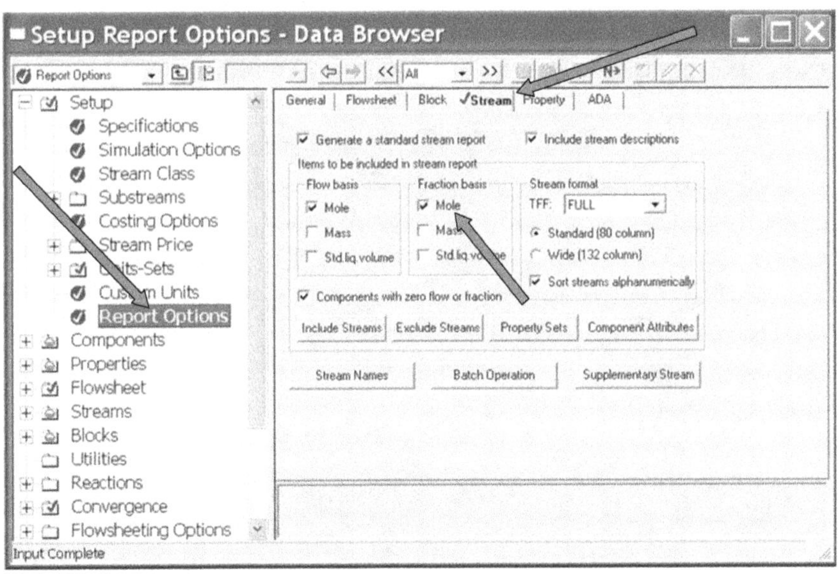


**Figure 3.9** Specifying mole fractions.

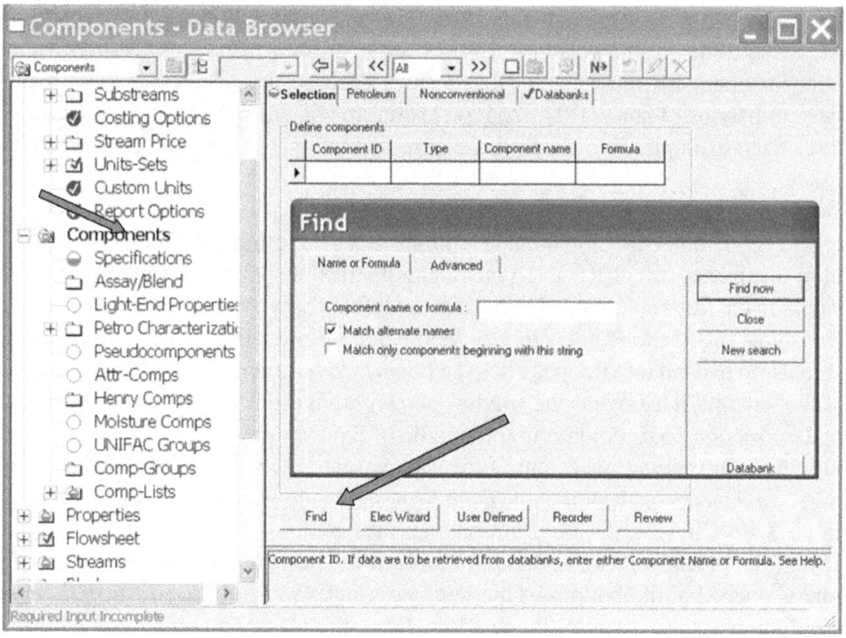


**Figure 3.10** Finding chemical components.

highlighting the name listed under *component ID* and typing the desired name. Click any place on the window and select *rename*; the names will be changed.

If any component in the simulation cannot be found in Aspen Plus, a method to include a nondatabank component will be demonstrated in Section 3.3 with an example.

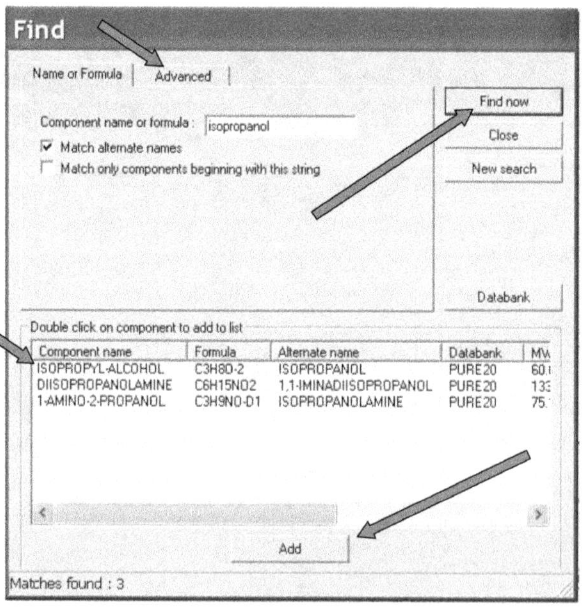


**Figure 3.11** Type in components.

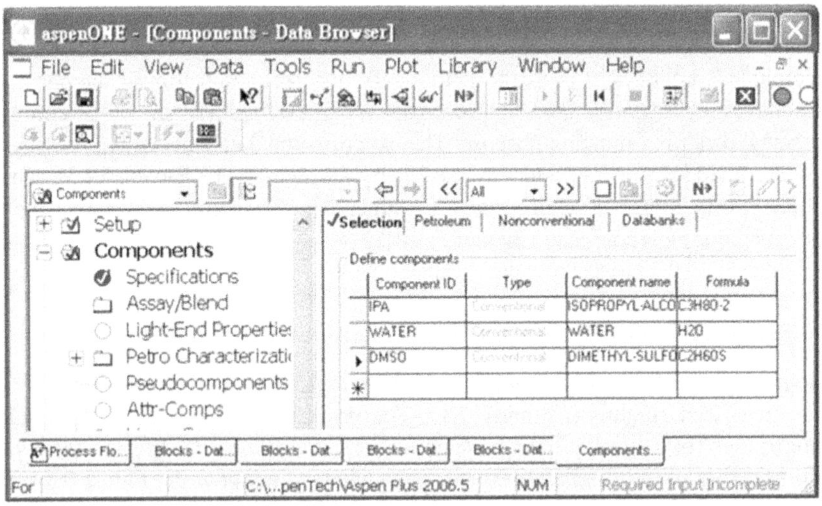


Figure 3.12 Three components selected.

### 3.1.3 Selecting Physical Property Method

Now we are ready to select a physical property package. Click *Properties* and then *Specifications* on the left-hand side of the *Data Browser* window. Figure 3.13 shows the window that opens. Under *Property methods & models* click the arrow on the right of *Base method*; a long list of alternatives appears as shown. Scroll down and select *NRTL* for this system.

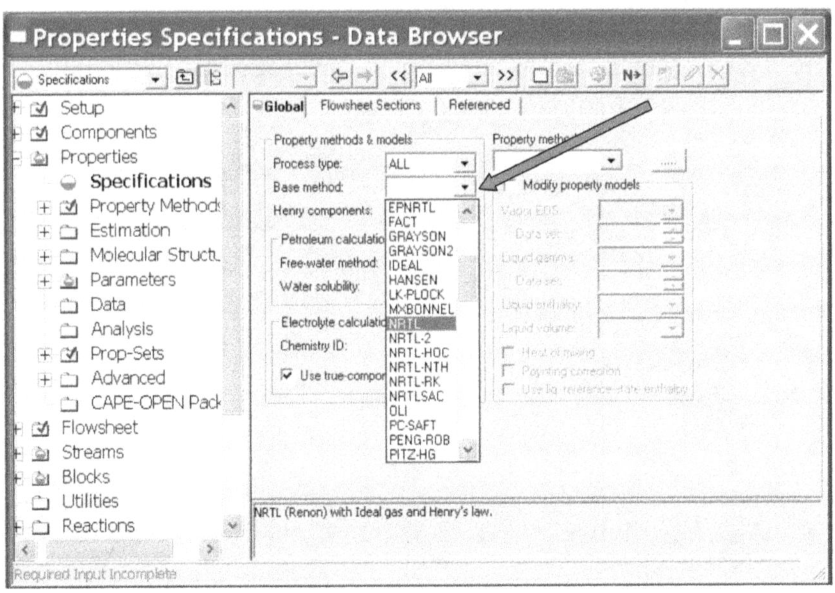


Figure 3.13 Specifying property base method.

Note that selecting a proper physical property method is extremely important for obtaining reliable simulation results of a process flowsheet. Aspen Physical Property System[1] gives the recommended classes of property methods for different applications. The phase equilibrium behavior predicted by the selected physical property method should be compared to experimental data for validation purpose. The experimental data that can typically be found in literature includes: the *Txy* and *yx* data; binary and ternary LLE data; VLLE data; and azeotropic information. For details of how to validate the prediction of the selected physical property method, please refer to Chapter 2.

### 3.1.4 Entering Feed Stream Data

The input streams to the process must be specified. In this example, as shown in Figure 3.1, there are two feed streams (*F*1 and *FE*1). The flowrate, composition, temperature, and pressure of these two streams must be specified. Clicking *Streams*, *F*1, and then *Input* on the left-hand side of the *Data Brower* opens the window, as shown in Figure 3.14.

In distillation calculations, molar flowrates and compositions are usually employed. Let us assume that the fresh feed flowrate is 100 kmol/h, the feed temperature is 25°C, and the feed pressure is 1.3 atm. These are entered in the middle of the window. The feed composition is 50 mol% IPA and 50 mol% water to represent a typical waste stream from semiconductor industry. The composition can be entered in terms of mole or mass fractions, or it can be entered in terms of component molar or mass flowrates. In our example, we use the drop-down arrow to change to *Mole-Frac* and enter the appropriate values. Repeat the same procedure to enter the feed-stream data of the entrainer feed (*FE*1) as can be seen in the flowsheet given in Figure 3.1.

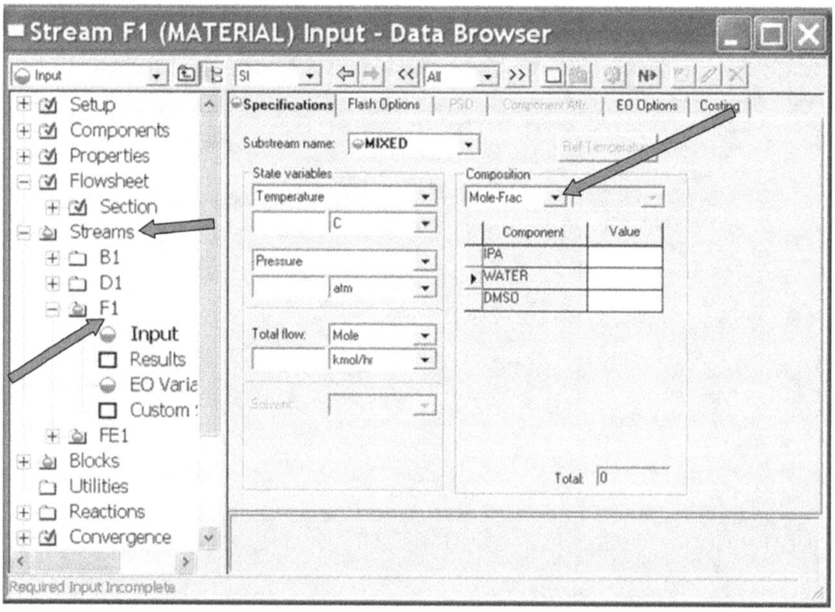


**Figure 3.14** Input feed stream data.

### 3.1.5 Entering Unit Operation Block Data

The parameters for the distillation column must be specified next. Clicking on *Blocks*, *C1*, and *Setup* on the left-hand side of the *Data Browser* opens the window shown in Figure 3.15. There are several page tabs. The first is *Configuration*, on which the total number of stages, the type of condenser, the type of reboiler, the numerical convergence method, and two other variables need to be specified. We consider each of these below.

1. *Number of Stages*. The rigorous way to select the number of stages is to perform an economic optimization of this example. We will discuss this in detail in the design part of this example in Chapter 10. For the moment let us select a column with 41 stages. Aspen uses the tray numbering convention of defining the reflux drum as Stage 1. The top tray is Stage 2 and so forth down the column. The base of the column in this example is Stage 40 with the reboiler as Stage 41. Therefore, this column has 39 trays.
2. *Condenser*. Use the dropdown menu to select *Total*. If the distillate were removed as a vapor, *Partial-Vapor* should be selected.
3. *Reboiler*. Both the kettle and the thermosyphon reboilers are partial reboilers (the vapor from the reboiler is in equilibrium with the liquid bottoms product withdrawn), so it does not matter which you select.
4. *Convergence*. The *standard* method works well in hydrocarbon systems. An alternative method must be used in highly nonideal systems. We select *Strongly non-ideal liquid* for this example.
5. *Operating Specifications*. A standard distillation column with two product streams has two degrees of freedom once the feed, pressure, number of stages, and feed stage location have been fixed. There are several alternative ways to select these

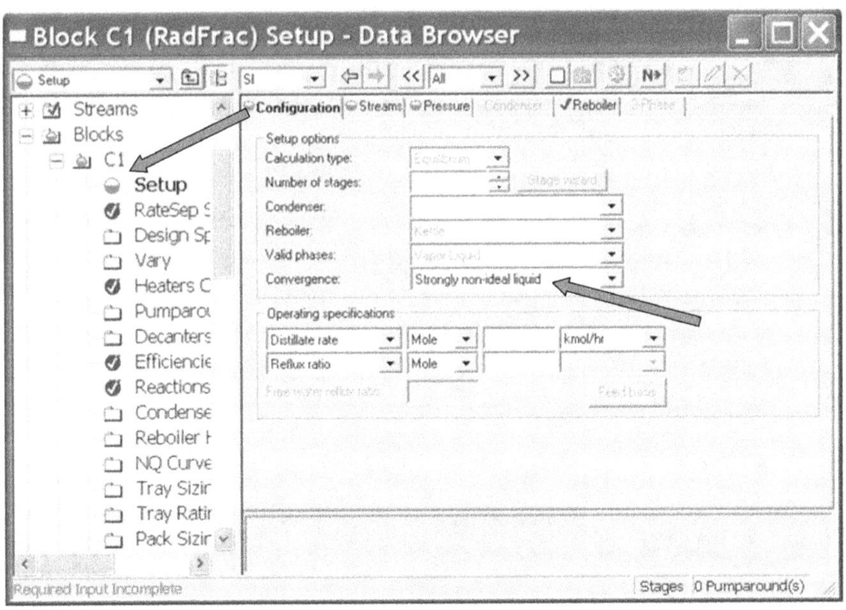


**Figure 3.15** Column set-up configuration page.

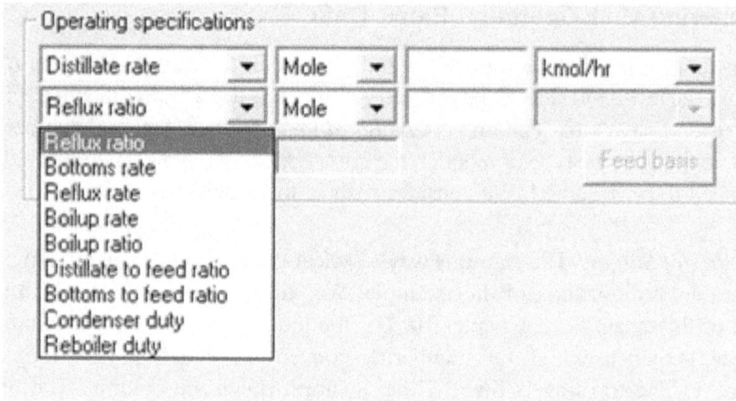


**Figure 3.16** Alternative choices of operating specifications.

two degrees of freedom, as shown in Figure 3.16. At this stage in our simulation, the usual approach is to fix the distillate flowrate and the reflux ratio.

Later, once we obtain a converged solution, we will change the specified variable so that the product specifications are met. We will discuss later in this chapter the use of the *Design Spec/Vary* feature in Aspen Plus to meet product specifications exactly. For now let us fix the distillate flowrate at 50 kmol/h since we know that this is the molar flowrate of IPA in the feed. In addition, let us select a reflux ratio of 0.5. Figure 3.17 shows the *Configuration* page with all these data inserted. Note that the red dot on the *Configuration* page becomes a blue checkmark when all the required input data have been provided.

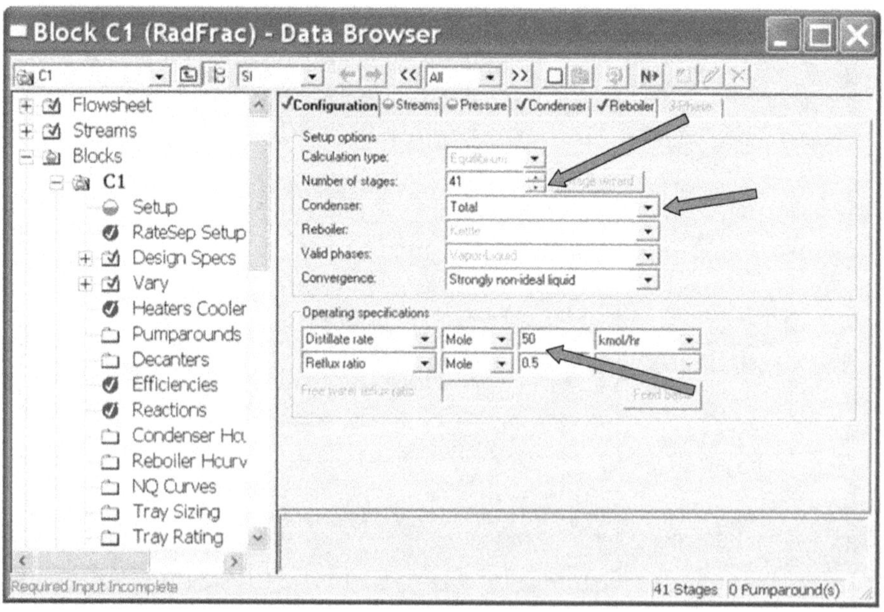


**Figure 3.17** Configuration page with all data.

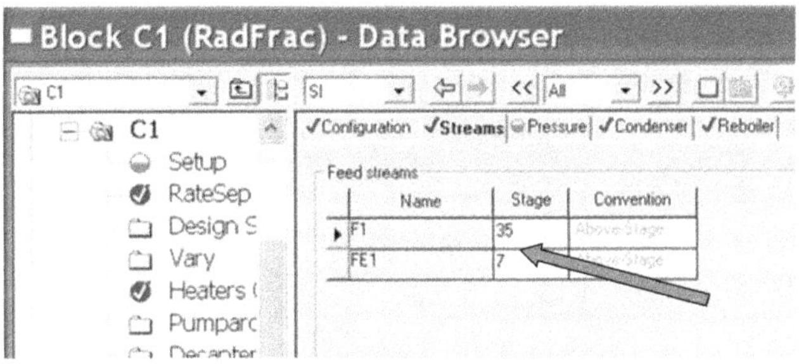


**Figure 3.18** Specifying two feed stages.

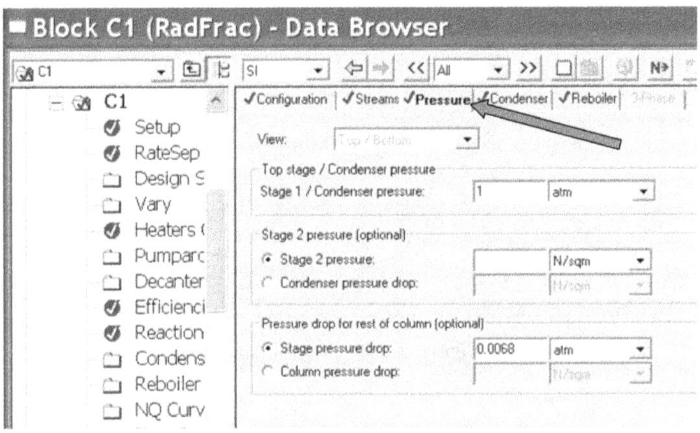


**Figure 3.19** Specifying pressure and tray pressure drop.

Now click the *Streams* page tab. A window opens on which the location of the feed stage must be given. For now we select the fresh feed on Stage 35 and entrainer feed on Stage 7 (see Fig. 3.18). Later in Chapter 10 we will return to this example and determine the "optimum" feed stage location by minimizing the total annual cost.

The last page tab that needs to be determined is *Pressure*. Clicking it opens the window shown in Figure 3.19, in which we specify the pressure in the reflux drum (condenser) and the pressure drop through each of the stages in the column. For the moment, we set the reflux drum pressure at 1 atm (be careful to change from $N/m^2$ to atm). A reasonable tray pressure drop is at 0.0068 atm per tray (0.1 psi per tray).

All the items in the *C*1 block are now blue, so the column is completely specified, and we are ready to run the simulation.

### 3.1.6 Running the Simulation

The blue *N* button ("next") at the top right-hand side of the *Data Browser* window is clicked to run the simulation. If any information is needed, the program will go to that location on the window and display a red symbol. If everything is ready to calculate, the *Required Input Complete* dialog box will appear, and you should click *OK*. The *Control Panel* window shown in Figure 3.20 opens and indicates that the column has been successfully converged. It took 11 iterations to converge the column.

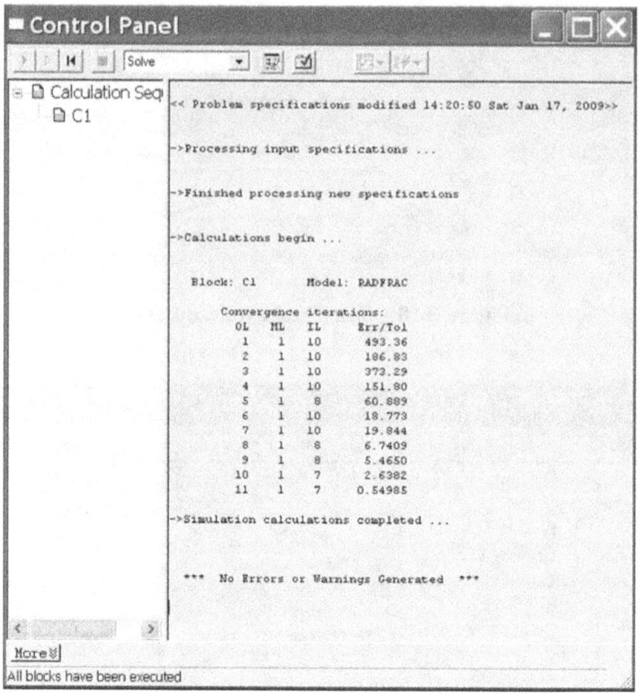


**Figure 3.20** Control panel.

Now we want to look at the composition of the product streams leaving the column to see if they satisfy their desired purities. We assume that the specification of the purity in the distillate is very high at 99.9999 mol% IPA to satisfy the stringent purity requirement in the semiconductor industry. The specification of the bottoms product is to let the mole fraction ratio of IPA to the sum of IPA and water be 0.001 so that high-purity water product can be obtained in the downstream entrainer recovery column. To look at the properties of these streams, we open the *C*1 block in the *Data Browser* window and click the item *Custom Stream Results*, which is at the very bottom of the list.

There are four streams in the table shown in Figure 3.21. Stream *F*1 is the fresh feed to the column, and stream *FE*1 is the entrainer feed to the column. Stream *D*1 is the liquid distillate leaving the reflux drum. Stream *B*1 is the liquid bottoms leaving the base of the column. We can see that there is 99.9769 mol% IPA in the distillate and the mole fraction ratio of IPA to the sum of IPA and water in the bottoms stream is purer than 0.001. Distillate purity is too low so we need to change the operating specifications or add more stages to get a better separation.

If we go back to *Setup* in the *C*1 block, change the reflux ratio to 0.7, and click the *N* button, the simulation converges with the results shown in Figure 3.22. The distillate purity has increased to 99.9974 mol% IPA and bottoms now has the mole fraction ratio of IPA to the sum of IPA and water to be even purer than previous run. We could continue to manually change the reflux ratio and the distillate flowrate to attempt to achieve the two desired product purities by trial and error. However, there is a much easier way, as discussed in the next section.

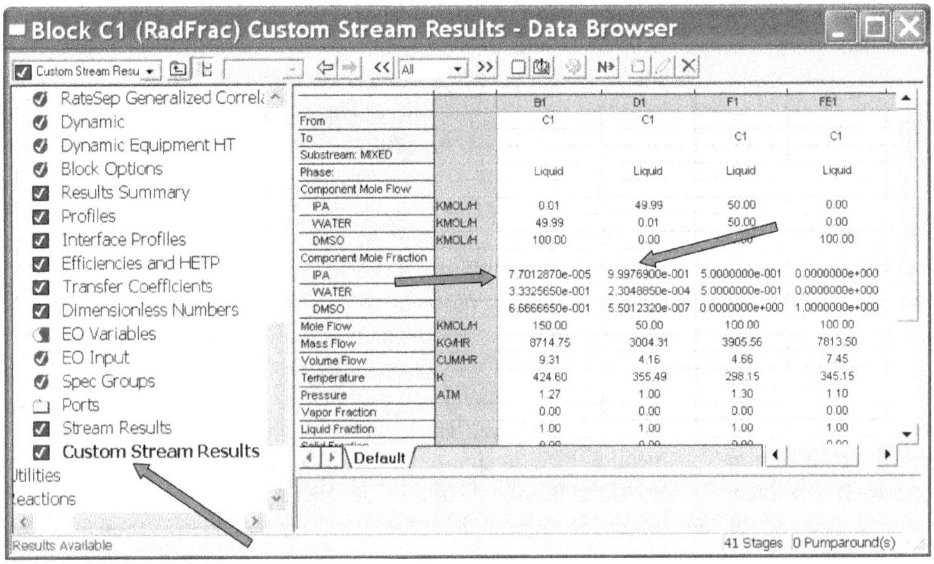


**Figure 3.21** Custom stream results with reflux ratio = 0.5.

### 3.1.7 Using *Design Spec/Vary* Function in *RadFrac*

The two product specifications are 99.9999 mol% IPA in the distillate and the mole fraction ratio of IPA to the sum of IPA and water in the bottoms stream to be 0.001. To achieve these two precise specifications, Aspen Plus uses the *Design Spec/Vary* function. A desired value of some "controlled" variable is specified, and the variable to be manipulated is also

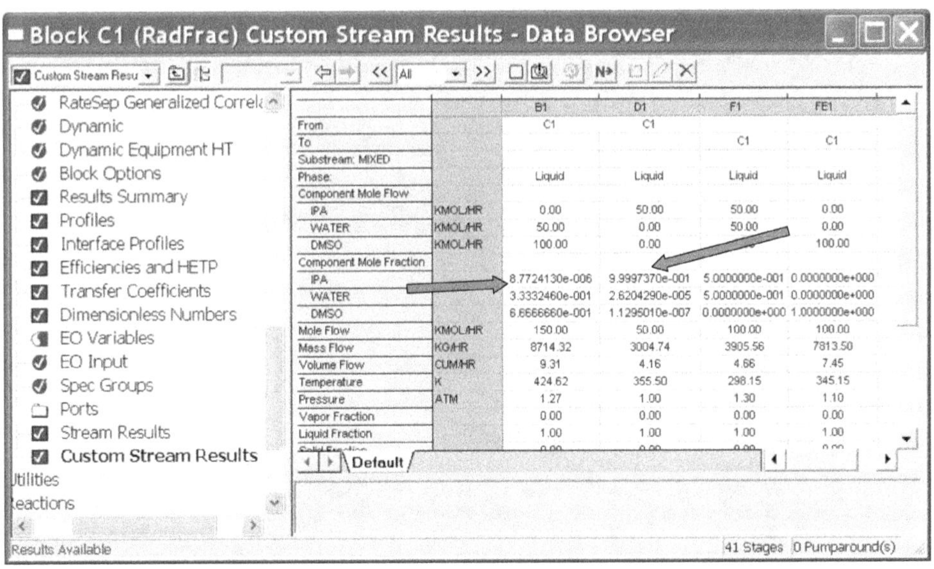


**Figure 3.22** Custom stream results with reflux ratio = 0.7.

specified. The simulation attempts to adjust the manipulated variable in such a way that the specified value of the controlled variable is achieved.

A word of caution might be useful at this point. The solution of a large set of simultaneous, nonlinear, algebraic equations is very difficult. There is no guarantee that a solution will be found because of numerical problems. In addition, if good engineering judgment is not used in selecting the target values, there may be no physically realizable solution. For example, if the specified number of stages is less than the minimum required for the specified separation, there is no value of the adjusted variable that can produce the desired result.

Another possible complication is multiplicity. Since the equations are nonlinear, there may be multiple solutions. Sometimes the program will converge to one solution and at other times it will converge to another solution, depending on the initial conditions.

It is usually a good idea to start by converging only one variable at a time instead of trying to handle several simultaneously. In our example, we will converge the distillate specification first by adjusting the distillate flowrate. Then, with this specification active, we will converge the bottom specification by adjusting the reflux ratio. The order of this sequential approach is deliberately selected to use the distillate flowrate first because the effect of this manipulated variable on compositions throughout the column is much larger than the effect of the reflux ratio.

To set up the *Design Spec/Vary* function, click on *Design Spec* under the *C*1 block in the *Data Browser* window. The window shown in Figure 3.23 opens up. Clicking the *New* button opens the window shown also in Figure 3.23. Select this Design Spec number as *1* and click *OK*. Another window opens that has several tabs (Fig. 3.24). On the first one, *Specifications*, you can specify the type of variable and what its desired value is. Clicking the dropdown menu under *Design Specification* and *Type* opens a long list of possible types of specifications. Select *mole purity*. Go down to *Target* and type in "0.999 999". This is the desired mole fraction of IPA in the distillate. Then click the second page tab

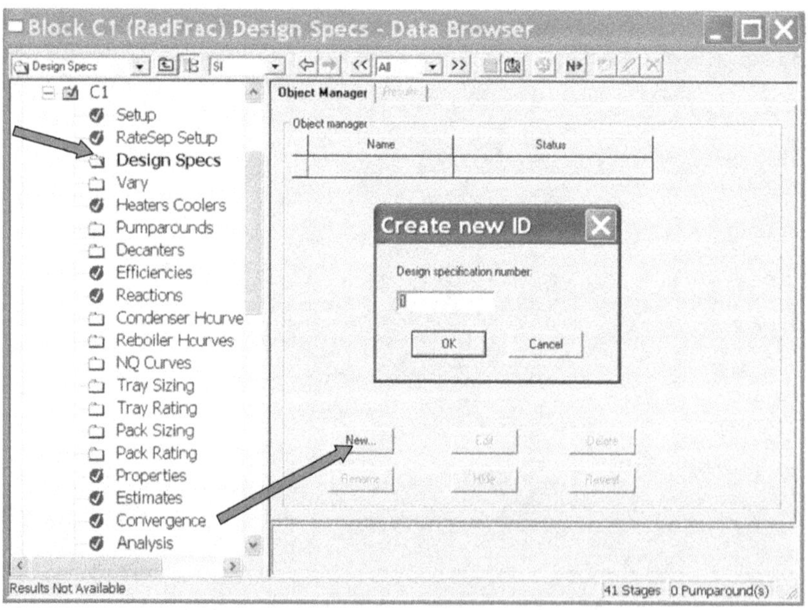


**Figure 3.23** Setting up the first *Design Specs*.

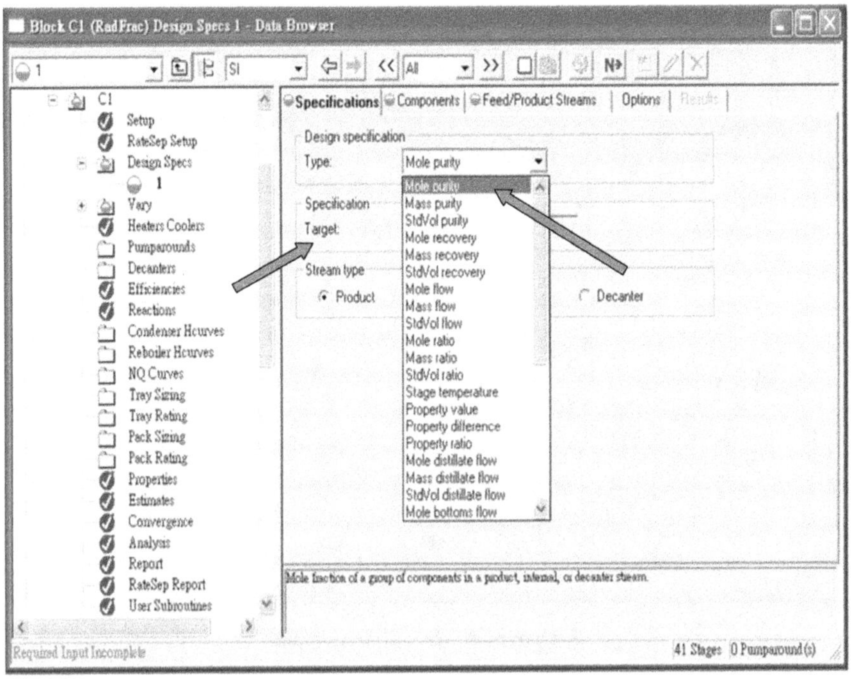


**Figure 3.24** Specifying the controlled variable as mole purity.

*Components* (Fig. 3.25). Click the *IPA* in the left column under *Available components*. Clicking the ">" button moves IPA over to the right *Selected components* column.

Click the third page tab *Feed/Product Streams*, select *D*1 in the left column, and click the ">" button to move it over to the right *Selected stream* column. Figure 3.26 shows the

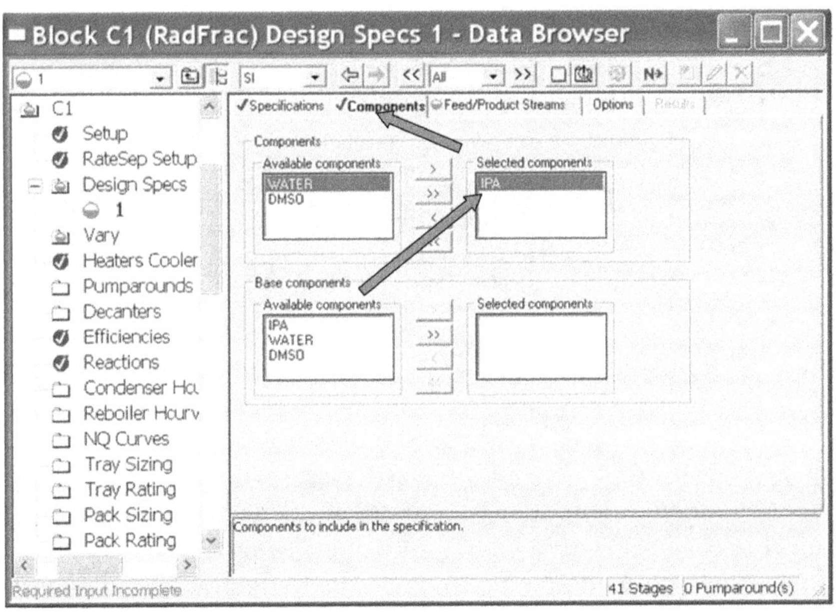


**Figure 3.25** Selecting the component.

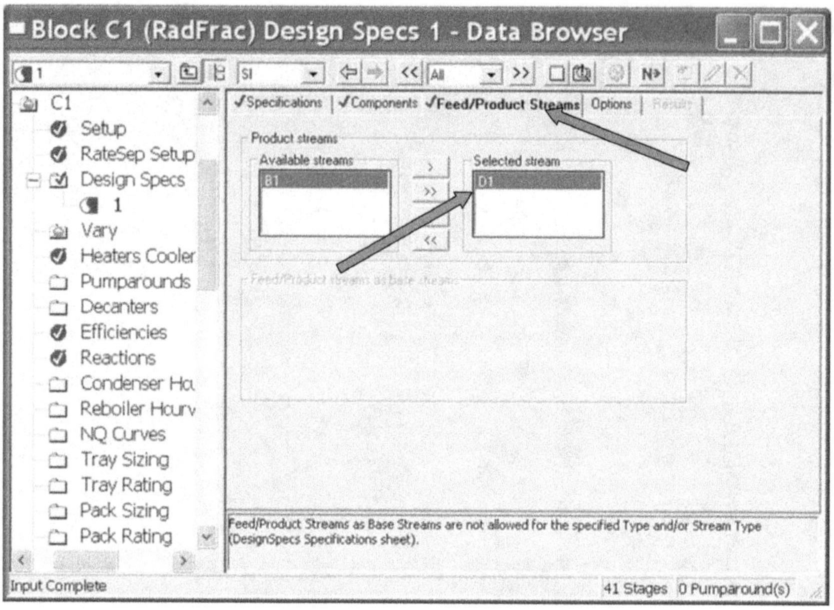

**Figure 3.26** Specifying stream.

results of this selection. The Design Spec is now completed. Note that the number "1" on the left side of the data browser under *Design Spec* in Figure 3.26 is blue. Now we must specify what variable to adjust. Clicking *Vary* under the *C*1 block opens the window shown in Figure 3.27. Clicking the *New* button and specify the number as "1" opens the window shown in Figure 3.28, where the manipulated variable can be defined.

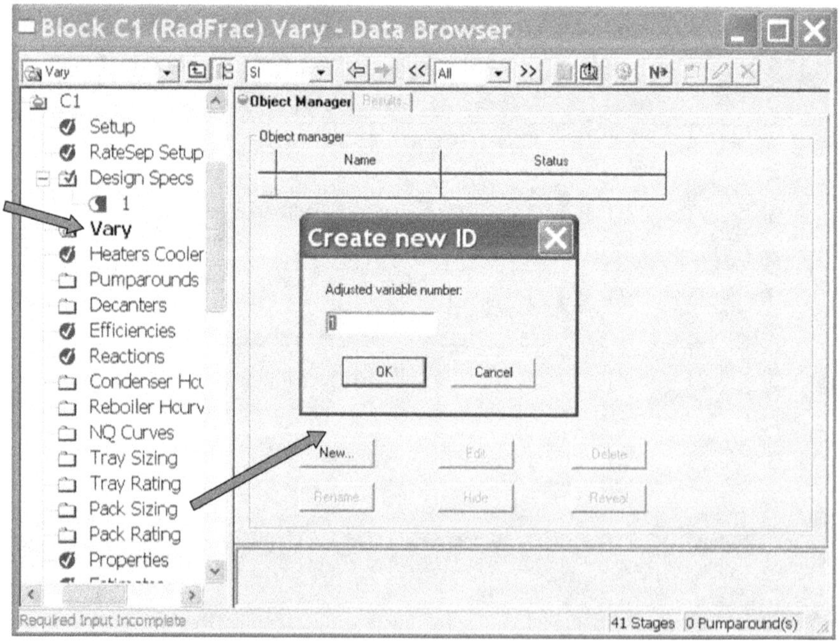

**Figure 3.27** Setting up the first *Vary*.

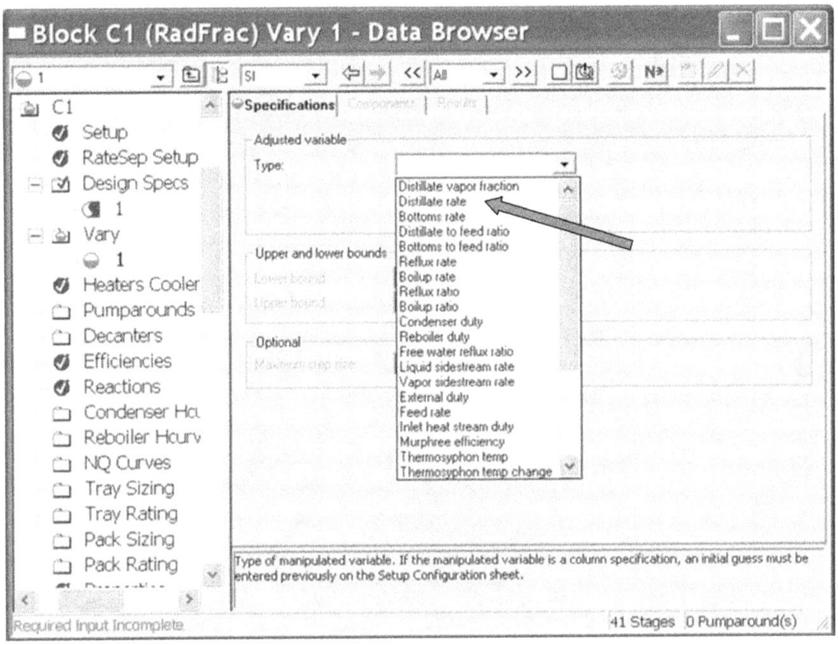

Figure 3.28 Defining the manipulated variable as distillate rate.

Opening the dropdown menu under *Adjusted variable* and *Type* produces a long list of possible variables. We select the *Distillate rate*, which opens several boxes (Fig. 3.29) in which the range of changes in the distillate flowrate can be restricted. We set the lower bound at 0.01 kmol/s (which corresponds to 36 kmol/h) and the upper bound at 0.02 kmol/s (which corresponds to 72 kmol/h).

Note that all the items in the *Data Browser* window are blue, showing that the simulation is ready to run; therefore, we click the blue *N* button and run the program. The *Control Panel* window opens and tells us that it has taken 19 iterations to converge (Fig. 3.30). Going down to *Custom Stream Results* at the bottom of the list (Fig. 3.31) under the *C*1 block lets us look

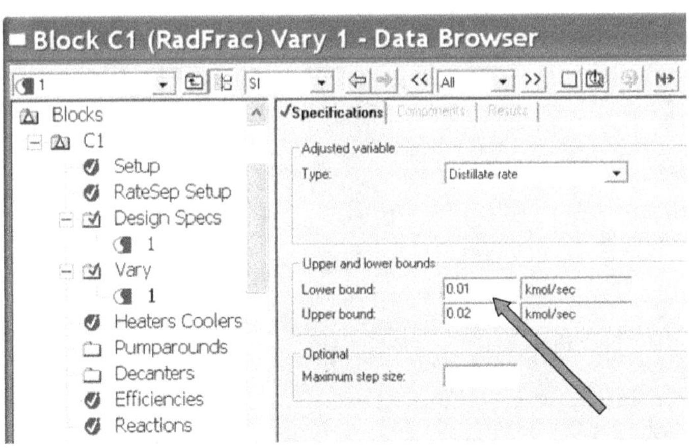

Figure 3.29 Setting limits on distillate flowrate.

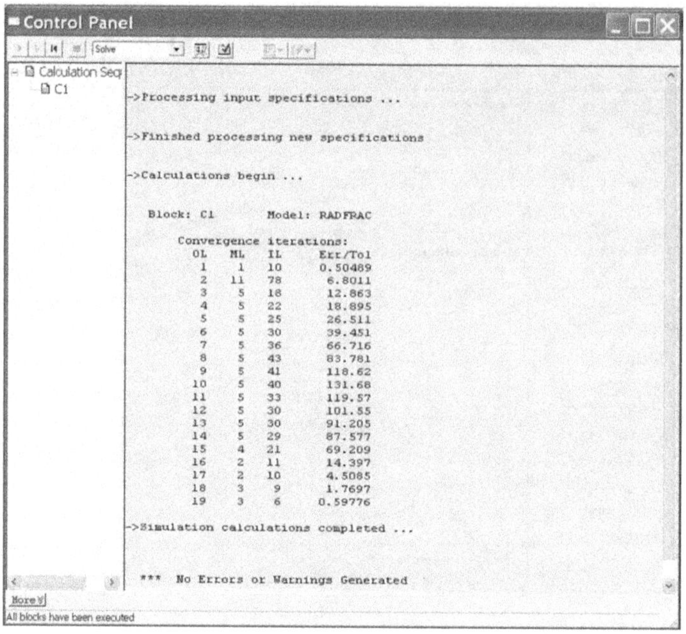


**Figure 3.30** Control panel showing the run is completed.

at the new values of the stream properties. Figure 3.31 shows that mole fraction of IPA in $D1$ is 0.999 999 00, which exactly meets the desired mole fraction specification. Note that the flowrate of $D1$ has changed to 49.98 kmol/h.

The second *Design Spec/Vary* is set up in a similar way. Clicking *Design Spec* opens a window in which you specify a new design spec ("2"). Then the mole fraction ratio of IPA to the sum of IPA and water in the bottoms stream is set to be 0.001. The reason we set the IPA

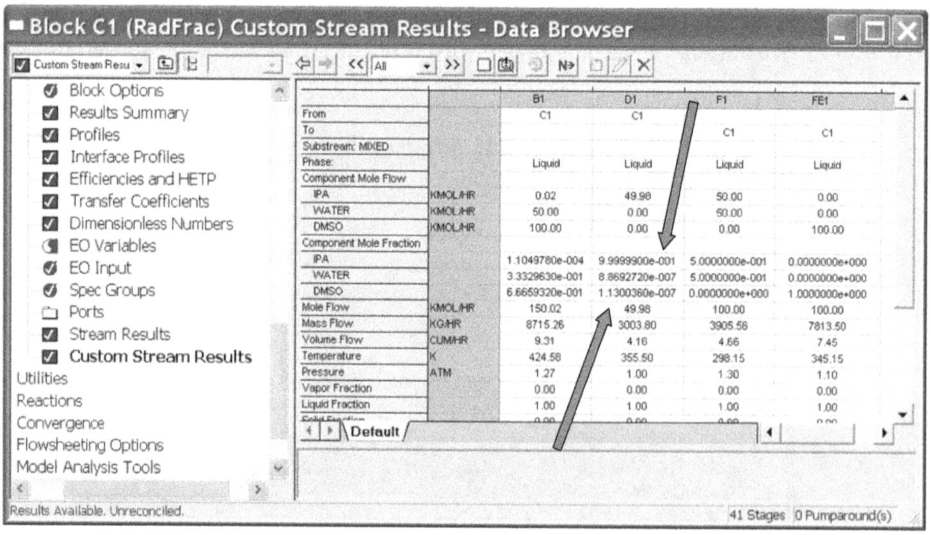


**Figure 3.31** Custom stream results showing the Design Spec is met.

impurity ratio at the bottoms is because the distillate product of the downstream entrainer recovery column will contain only water with impurity of IPA (DMSO is a much heavier component). Setting this ratio indirectly specifies the purity in the distillate product of this downstream column. Notice that this kind of design specification (with ratio) is also possible by selecting *Mole Purity* in the dropdown menu of *Type*, and then choose *IPA* component as *selected component* and also choose the *base components* to be *IPA* and *WATER*. The definition in Aspen is to treat the *selected component* as numerator and the *base components* as denominator in the calculation. See Figure 3.32 for this step of setting a ratio specification, and compare with the one in Figure 3.25 where *base components* were not selected at all.

Next a second *Vary* ("2") is set up similarly with reflux ratio selected. The upper and lower bounds are set at 0.1 and 1.0, respectively. Everything is ready to run again. Clicking the blue *N* button executes the program. The simulation converges in 16 iterations.

Figure 3.33 shows the new stream results. The two product specifications are both met. Of course, the distillate flowrate and the reflux ratio have been changed to produce the desired product purities. The stream results show that the flowrate of *D*1 is 49.95 kmol/h. To ascertain the reflux ratio, click on *Results Summary* under the *C*1 block. The window shown in Figure 3.34 opens, in which the condition at the top of the column are given. The reflux ratio is 0.6835.

The other important pieces of information in the window are the condenser heat removal [−920 kW (kilowatts)] and the reflux drum temperature (355 K) at the 1 atm pressure we specified. Note that, because the cooling water inlet temperature is usually assumed to be at 305 K, it could be used in the condenser. The condenser heat removal can be used to estimate the heat transfer area of the condenser by the following equation:

$$A_c = \frac{Q_c}{U_c \cdot \Delta T_c} \tag{3.1}$$

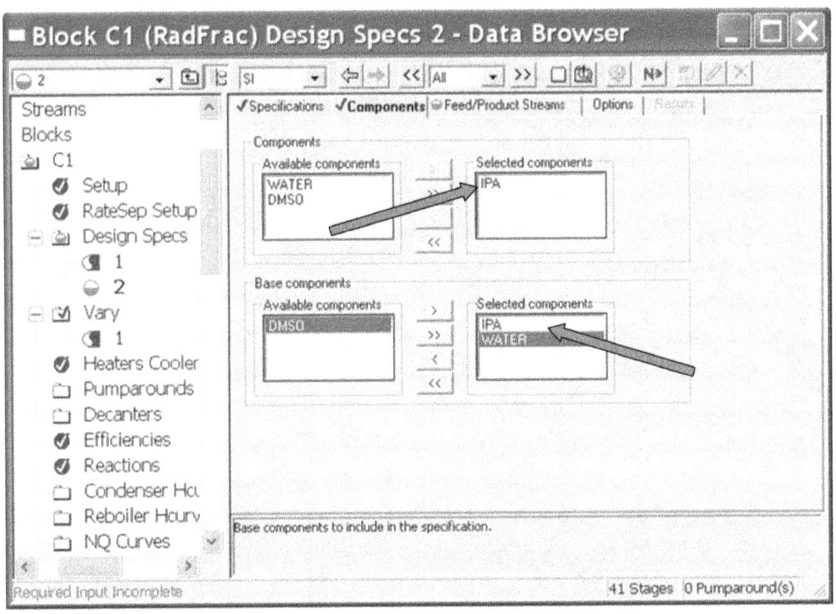


**Figure 3.32** The way to set up the second *Design Specs.*

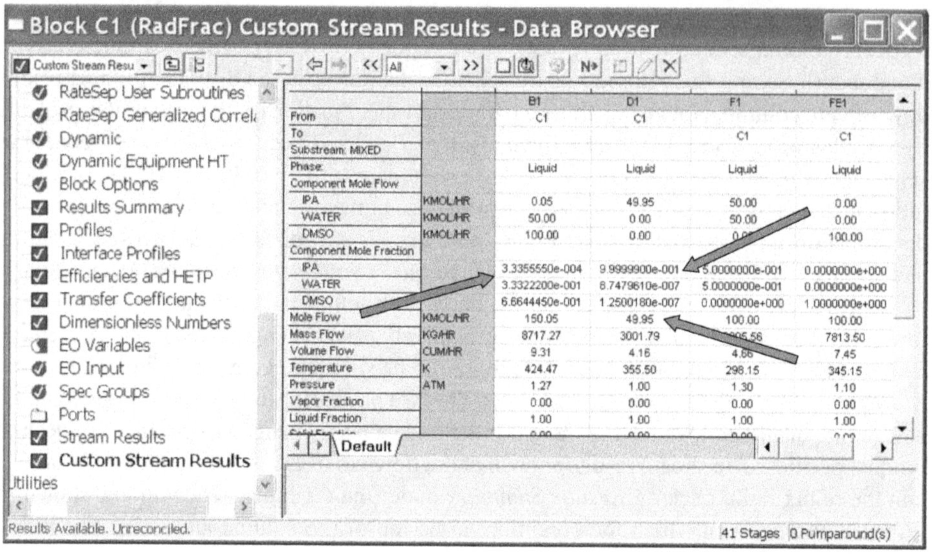


**Figure 3.33** Results where two Design Specs are met.

where the overall heat transfer coefficient is estimated according to Table A.7 of Douglas,[2] to be 150 Btu $ft^{-2} h^{-1} °F^{-1}$ and the temperature difference between the process fluid and the cooling medium is:

$$\Delta T_c = \frac{(120 - 90)}{\ln(T_{top} - 90/T_{top} - 120)} \tag{3.2}$$

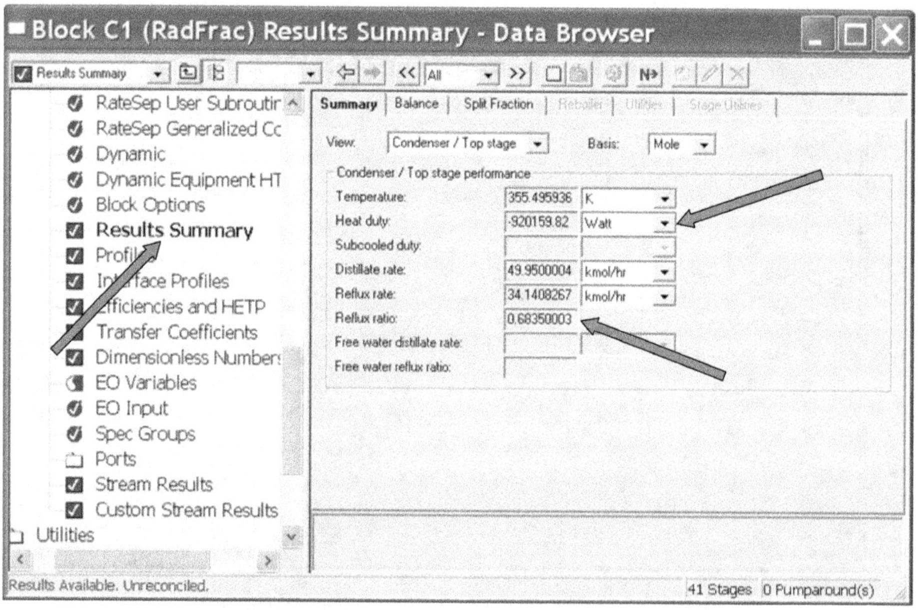


**Figure 3.34** Results for top of column.

In Eq. (3.2), cooling water inlet and outlet temperatures are assumed to be 90°F and 120°F, respectively, and $T_{top}$ is the reflux drum temperature in °F. With a known heat transfer area, the installed cost of the condenser can be calculated by using the following formula taken from Appendix E of Douglas:[2]

$$\text{Installed Cost[\$]} = \left(\frac{\text{M\&S}}{280}\right) 101.3 A_c^{0.65} (2.29 + F_c) \tag{3.3}$$

Note that in this equation, the heat transfer area must be in the English unit of ft$^2$. M&S is the Marshall and Swift index and $F_c$ is a correction factor to do with the condenser type, operating pressure, and the material of construction. Please refer to Douglas[2] to obtain the estimated value of the correction factor. With the known condenser heat removal and the inlet and outlet temperature difference on the cooling water side, the annual cost of the cooling water can also be easily estimated.

To find the conditions at the base of the column, we use the dropdown menu that is next to *View* on the *Results Summary* window and select *Reboiler/Bottom stage*. Figure 3.35 shows the information obtained. The most important pieces of information are the reboiler heat input at 1527 kW and the base temperature of 424 K.

The reboiler heat input can be used to estimate the heat transfer area of the reboiler by the following equation:

$$A_R = \frac{Q_R}{U_R \cdot \Delta T_R} \tag{3.4}$$

where the overall heat transfer coefficient is estimated according to Table A.7 of Douglas[2] to be 250 Btu ft$^{-2}$ h$^{-1}$ °F$^{-1}$ and the temperature difference between the process fluid and the heating medium can reasonably be estimated to be 45°F (25°C). With this information,

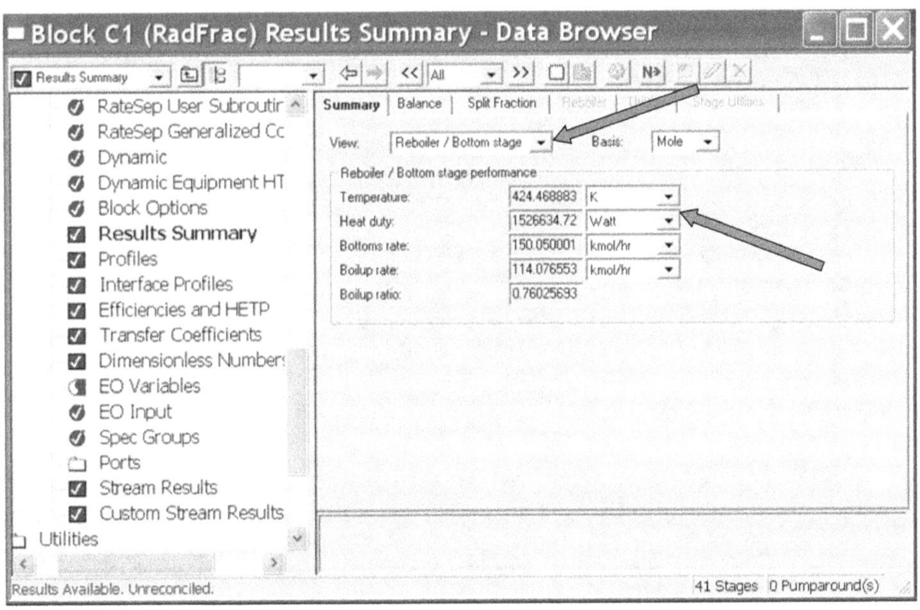


**Figure 3.35** Results for base of column.

the heat transfer area (in ft$^2$) of the reboiler can be calculated once the reboiler heat input (in Btu h$^{-1}$) is known. Further, the installed cost of the reboiler can be estimated by the same formula in Eq. (3.3).

The base temperature can be used to estimate the required steam temperature (with temperature difference estimated to be 45°F) and further estimate the steam latent heat of vaporization ($\lambda_v$) by referring to the steam tables. The annual cost of the steam can thus be estimated with this information, together with the reboiler heat input.

The column temperature and composition profiles can be obtained by selecting *Profiles* in the *C*1 block. The window that opens is shown in Figure 3.36. There are several page tabs. The first *TPFQ* (temperature, pressure, flow, heat) gives the temperature and pressure on each stage. Selecting the second page tab *Compositions* opens the window shown in Figure 3.37, in which *Liquid* has been selected from the dropdown menu in the *View* box. Using the Plot Wizard program makes generating plots of these profiles quite easy. Click on *Plot* at the toolbar of the Aspen Plus simulation window, and then click *Plot Wizard*. This opens the window shown in Figure 3.38. Clicking *Next* opens the window shown in Figure 3.39. Clicking on the upper left picture labeled *Temp* produces the temperature profile plot given in Figure 3.40. Clicking on the picture labeled *Comp* and then clicking *Next* opens the window on which you can select what components to plot and what phase (liquid or vapor compositions). Figure 3.41 gives the liquid composition profile.

### 3.1.8 Creating Reports, Saving Files, and Opening an Existing Simulation

Aspen Plus allows you to generate a report file containing the simulation inputs and calculated results. From the Aspen Plus menu bar, select *File* and then select *Export*. The export

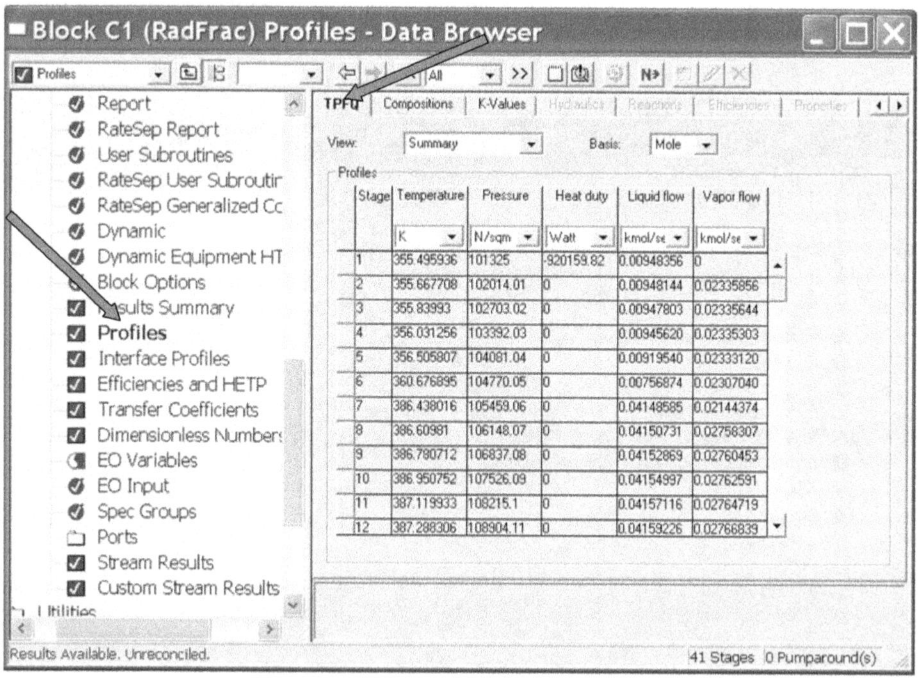

Figure 3.36 Temperature, pressure, and flow profiles.

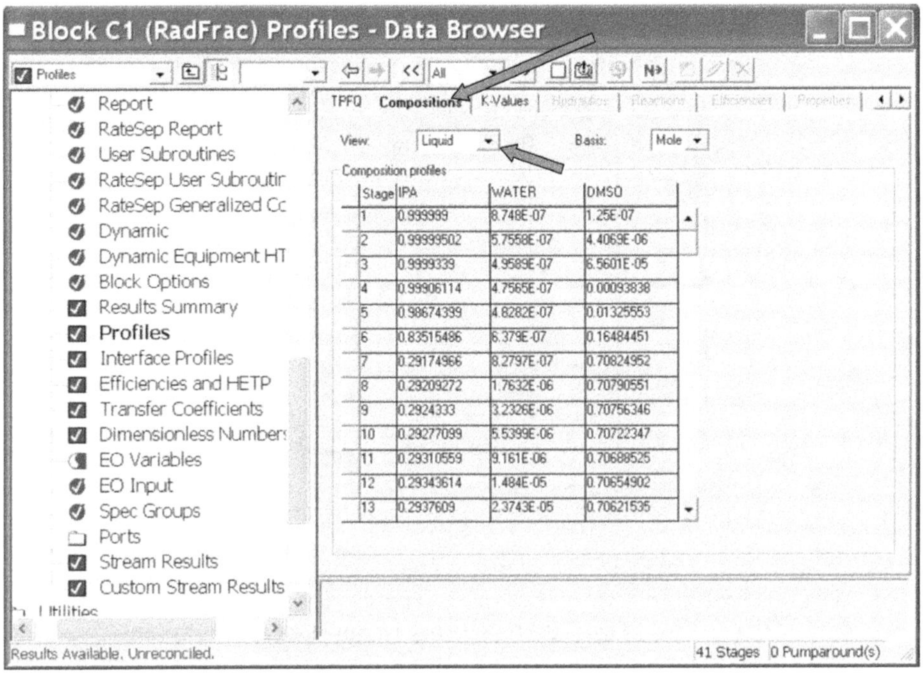


**Figure 3.37** Liquid composition profiles.

dialog box appears (Fig. 3.42). Clicking the dropdown menu *Save as type* opens a long list of possible types to save as. Select *Report Files* (*.*rep*). In the *File name* list, type in the name of the report file you want to save (for example, *IPA*). Click *Save* to generate the report file with file extension as ".rep". By default, this report file is saved in your working directory. You

**Figure 3.38** Plot wizard.

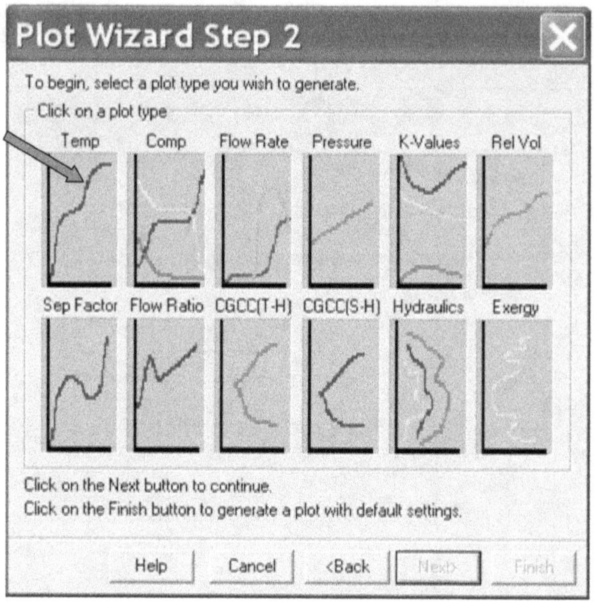


**Figure 3.39** Selecting type of plot.

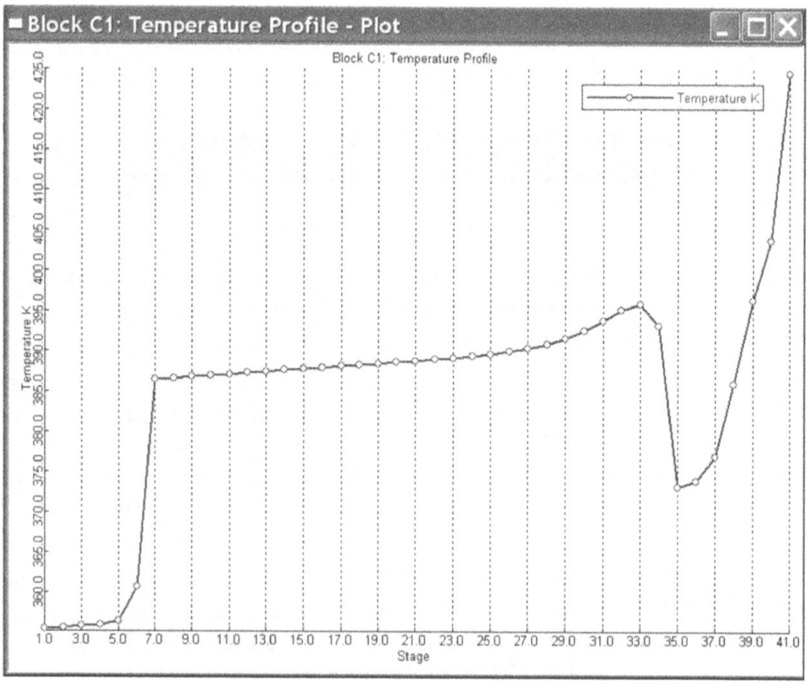


**Figure 3.40** Temperature profile.

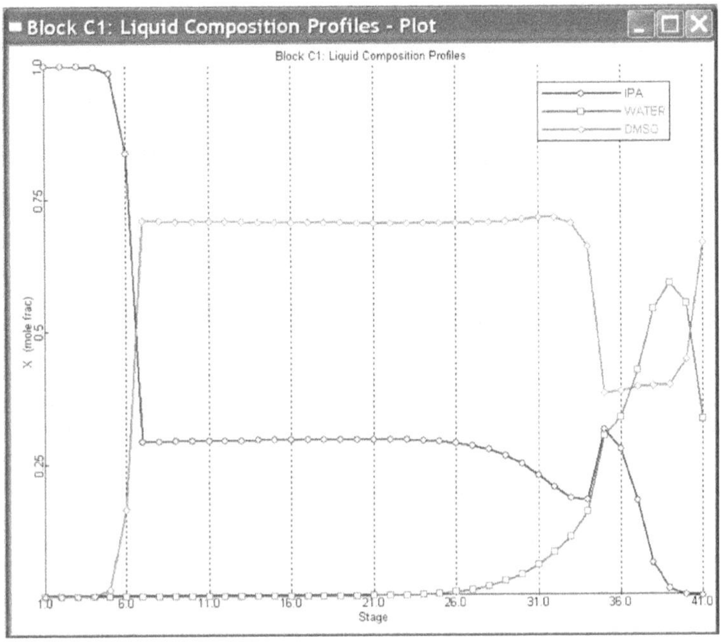


**Figure 3.41** Liquid composition profile.

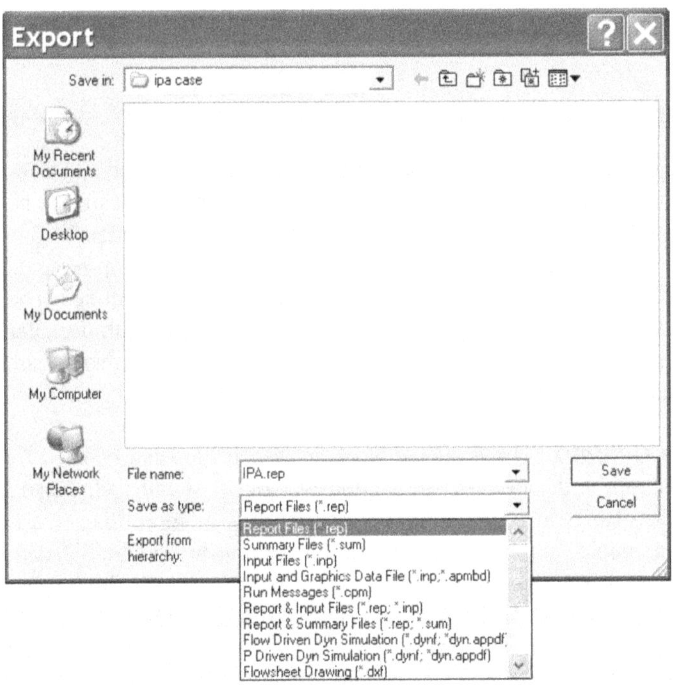


**Figure 3.42** Export the report files.

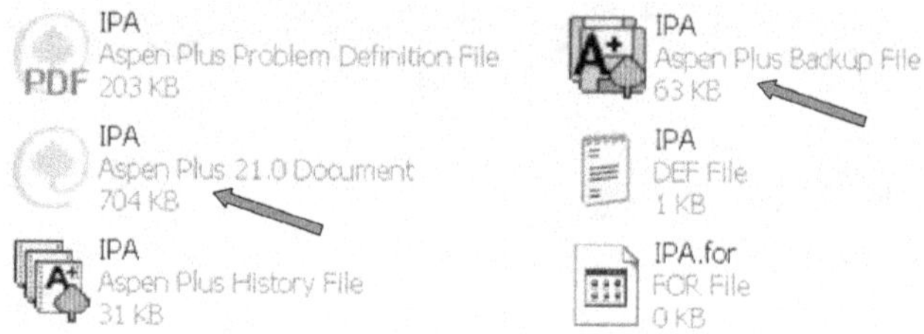


**Figure 3.43** Six files are saved.

can select another directory by navigating to it. This report file is ready to be opened with a text editor.

To save the simulation program for later usage, we select *File* and then *Save as* from the Aspen Plus menu bar. In the *File name* field, type the name we want to save (for example, *IPA*). Make sure that the *Save as type* field reads *Aspen Plus Documents* (**.apw*) and then click *Save*. This will save together six simulation programs in your working folder as in Figure 3.43.

At a later time, we can then open this saved simulation file by navigating to the folder containing the files and then double clicking *IPA.apw* or *IPA.bkp*. The *Process Flowsheet* window for this IPA dehydration example will appear, and we can make any changes in the simulation.

## 3.2 UNIT OPERATION BLOCKS USED IN THIS BOOK

In the following, the unit operation blocks in Aspen Plus that are used in this book will be explained.

### 3.2.1 *RadFrac*

The *RadFrac* block can be used for the ordinary distillation column and also for the extractive distillation column as shown in the example in Section 3.1. It can also be used as strippers (with a reboiler but no condenser), rectifiers (with a condenser but no reboiler), absorbers (with neither), and more complex columns with side pump-around. In the following, several columns other than the ordinary distillation will be outlined. The *RadFrac* can also be used as a heterogeneous azeotropic distillation column with decanter replacing the reflux drum at top of this column. The vapor–liquid–liquid calculation can be performed inside the column if needed.

***Sidestream Column.*** The *RadFrac* block in Aspen Plus can be used for the ordinary distillation column with one feed stream and two outlet streams (distillate and bottoms). This block can also be used for multiple feed streams or more than two ordinary outlet streams. For example, the extractive distillation column in Section 3.1 demonstrated that two feed streams can be connected to the *RadFrac* block. In a similar way, more outlet streams can be drawn out of the *RadFrac* block as a sidestream column. The sidestream draw-off can be either a liquid stream or a vapor stream. All these varieties of different columns can easily be configured by adding material streams to the flowsheet with the *RadFrac* block (see previous Fig. 3.5).

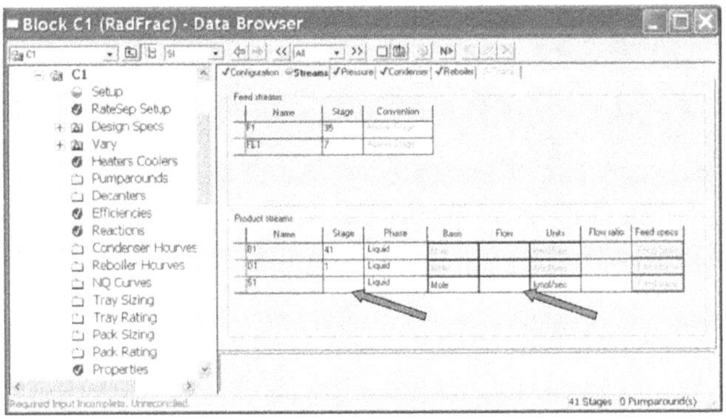

**Figure 3.44** Setting up the sidestream.

Notice that the number of the operating specifications increases if your column configuration has a sidestream. For an ordinary distillation column or the extractive distillation column, there are two degrees of freedom. The degrees of freedom increase to three if you have one additional sidestream. An example of where to set up the third degree of freedom for a sidestream column can be seen in Figure 3.44 where, besides specifying the location of the sidestream, the sidestream flowrate also needs to be specified. Of course, alternative ways to select the other two degrees of freedom are available at *operating specifications* of the *configuration* page tab.

***Stripper.*** In Chapter 7 of this book, two strippers are used together with a decanter to separate a mixture containing *n*-butanol and water. The configuration of a stripper is to have feed going into this column at Stage 1 and the top vapor goes out of the column with no condenser. The bottom part of the stripper is the same as an ordinary distillation column with a reboiler.

A stripper can easily be configured in Aspen Plus by also using the *RadFrac* block. As a demonstration, open up a blank flowsheet in the usual way as described in Section 3.1. Clicking the *Columns* page tab and clicking the arrow just to the right of the *RadFrac* opens a window. Instead of picking the full column button as in Figure 3.3, the stripper button is selected as shown in Figure 3.45. Next, follow the same step-by-step procedure as in Section 3.1 to connect material streams, to specify chemical components, physical properties, and feed stream properties. After that, clicking on the stripper block opens a window with a long listing of items. The top subitem is labeled *Setup*. Clicking it opens the window shown in Figure 3.46. There are several page tabs. The first is *Configuration* on which set-up options and the operating specifications need to be set. Notice that in this case the condenser field is set to be *None*. After setting the first operating specification, no additional specifications are allowed. In the second *Streams* page tab, the feed stream must be set to be in the first stage as shown in Figure 3.47, and the top product stream should be vapor phase. All other set-up steps are the same as in the extractive distillation example in Section 3.1 except that only one *Design Spec/Vary* function can be set in a stripper.

***Heterogeneous Azeotropic Distillation.*** One way to separate azeotropes is to add a light entrainer that forms two liquid phases and use the liquid-splitting nature of the

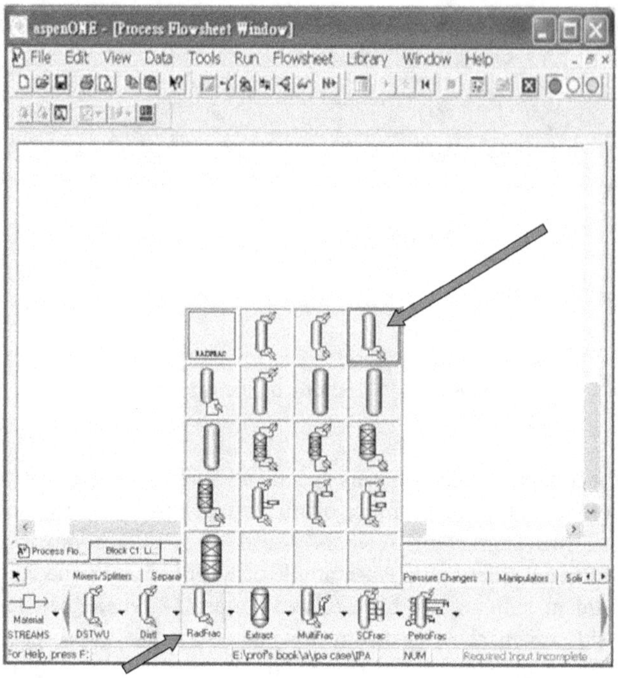


**Figure 3.45** Selecting stripper.

system to aid in the separation. In the following, the separation of IPA and water will be used as an example to show how to configure a heterogeneous azeotropic distillation column in Aspen Plus.

Figure 3.48 shows the configuration of this column with a decanter for a reflux drum. Assume that the feed composition is close to the azeotropic composition of IPA + $H_2O$ mixture. This feed stream may be coming from the distillate of a preconcentrator column with its

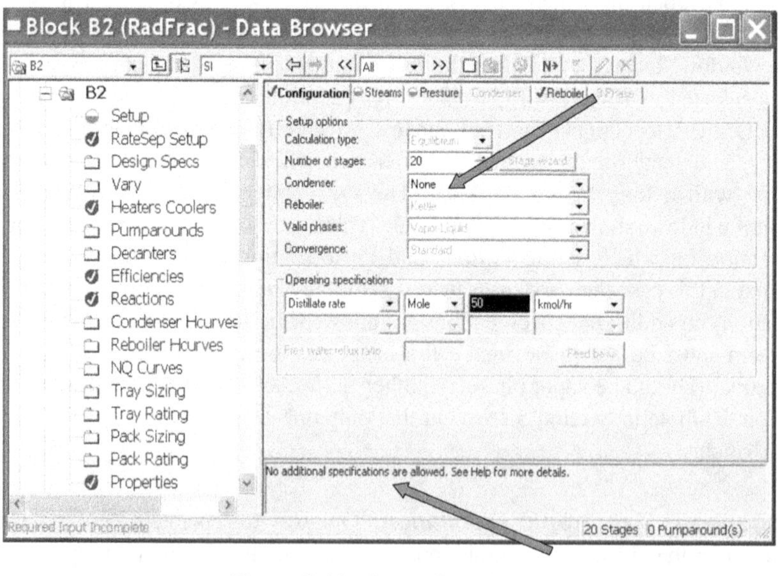


**Figure 3.46** Set-up for stripper.

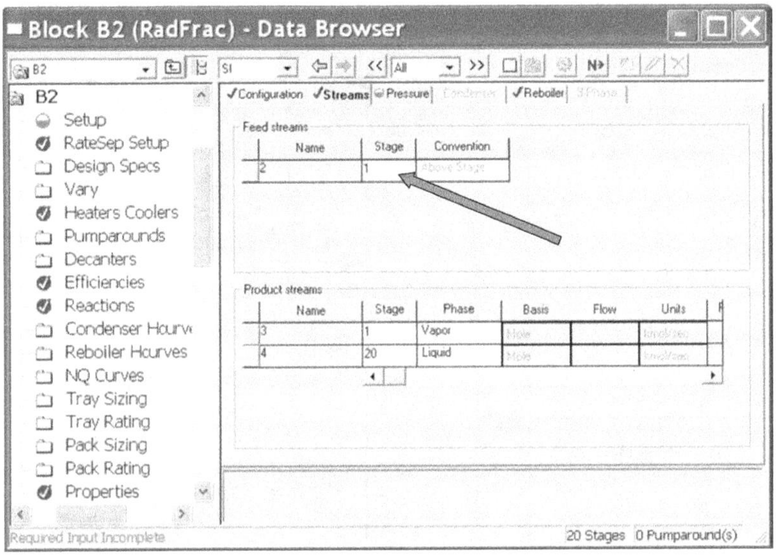


**Figure 3.47** One of the feed streams has to be at first stage.

feed composition containing more water. With an extra component, cyclohexane (CyH), going into the heterogeneous azeotropic distillation column through the organic phase of a decanter at the top of this column, the top vapor composition should reach the minimum temperature (ternary azeotrope) of the system, and the bottoms composition should reach the pure IPA corner of the ternary diagram.

The conceptual design of this column can be explained by Figure 3.49. With the inner material balance envelop as shown in Figure 3.48, the material balance lines of adding $F1$ and OR to produce the split of V1 at column top and pure IPA at column bottoms are illustrated in Figure 3.49. In the same figure, the outer material balance lines of adding $F1$ and pure CyH to produce the split AO (aqueous outlet) and pure IPA can also be illustrated.

The aqueous phase still contains much of the IPA and should go to a recovery column to obtain pure water product. The overall flowsheet of the complete design will be explained in Chapter 8. For now, we will demonstrate how to configure the heterogeneous azeotropic distillation column system shown in Figure 3.48. The procedure is similar to that outlined in Section 3.1 up to the point where you need to enter the unit operation block data in Section 3.1.5. In order to configure the whole system shown in Figure 3.48, containing a heterogeneous azeotropic distillation column and a decanter, *RadFrac* can still be used.

Clicking on *Blocks* and *B*1 and *Setup* on the left-hand side of the *Data Browser* opens the window shown in Figure 3.50. This is similar to Figure 3.17; however, in this case the *valid phases* under *Setup Options* should be selected as *Vapor–Liquid–Liquid*. Also, the condenser should be selected as *None* because the top vapor of this column will be subcooled to a liquid phase in an external decanter, which will be specified a little later. Since the condenser is selected as *None*, there is only one operating specification.

In the *Streams* page tab, the feed stream is specified the same way as in the extractive distillation example except, in this case, the second feed stream is the makeup stream that should come into the column at the first stage. For the naming convention of this heterogeneous azeotropic distillation column, the first stage is the decanter and the last stage is the reboiler. The flowsheet of this *RadFrac* simulation with a built-in decanter as the first stage is shown in Figure 3.51.

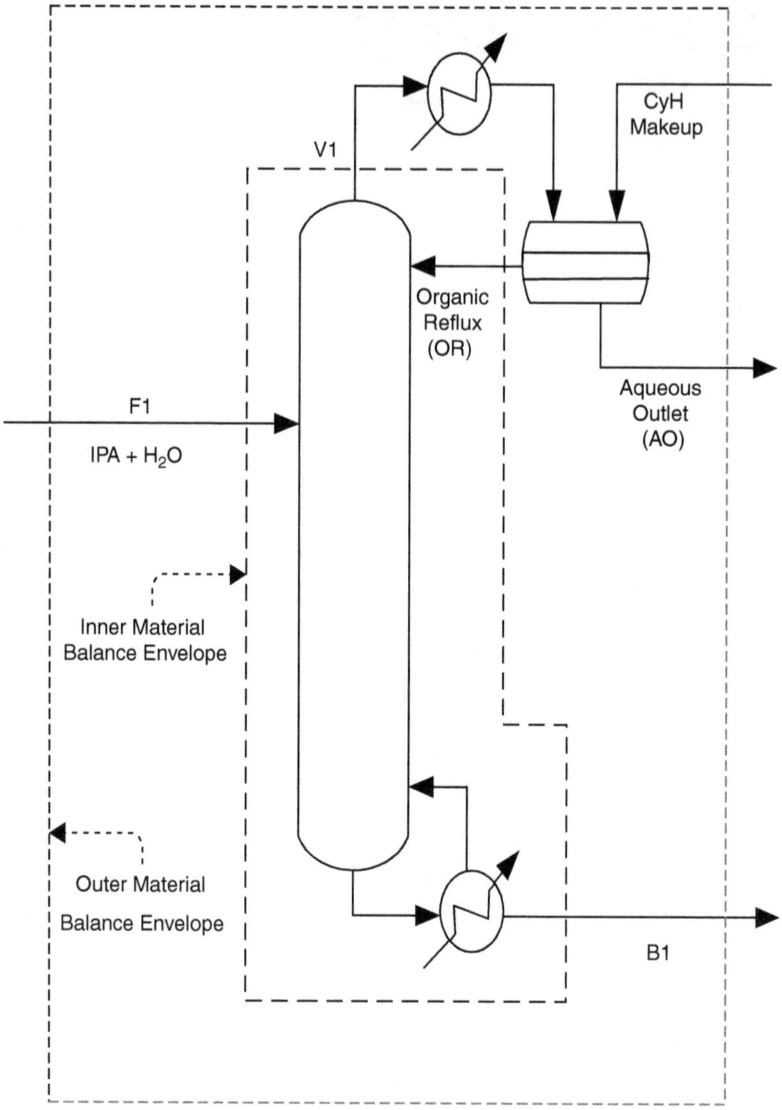


**Figure 3.48** Flowsheet for a heterogeneous azeotropic column.

After setting up the operating pressure and pressure drop per tray of this column in the *Pressure* page tab, a new page tab called *3-Phase* should be selected. Clicking on this page tab, a window as in Figure 3.52 showing Stages 1 to 5 are chosen to test for two liquid phases. Notice that after completing the set-up of this simulation and the result shows that the fifth stage actually has two liquid phases, the ending stage in this window should be extended to make sure all stages with two liquid phases are allowed to be tested. In this window, *WATER* is also selected as the key component to identify a second liquid phase.

Next, *Decanters* in *B*1 block on the left-hand side of the *Data Browser* is clicked to open the window as shown in Figure 3.53. In this window, the fractions of the decanter liquid

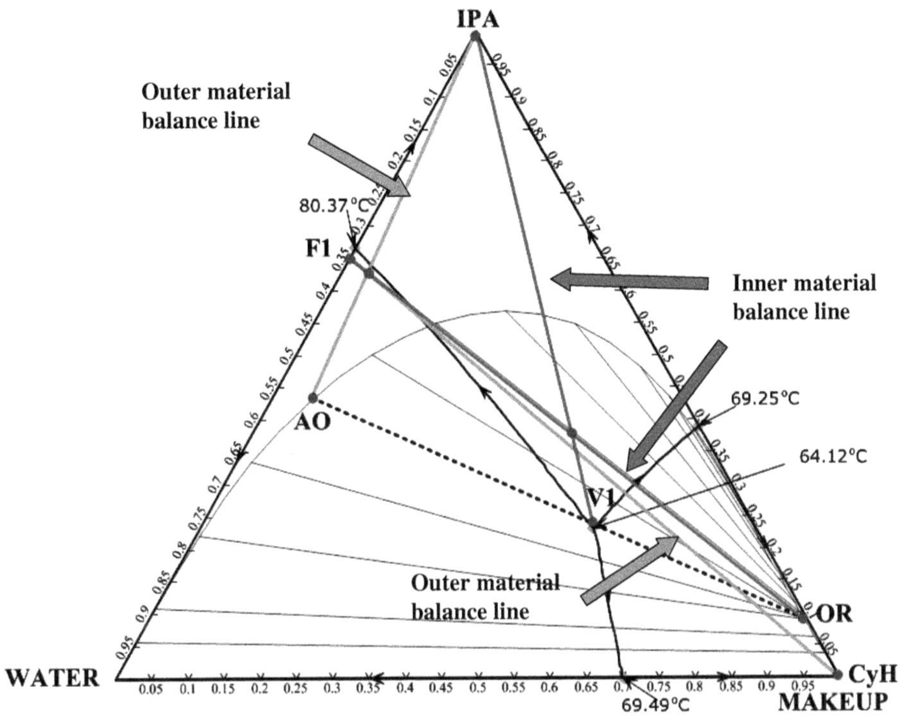


**Figure 3.49** Conceptual design of IPA dehydration column.

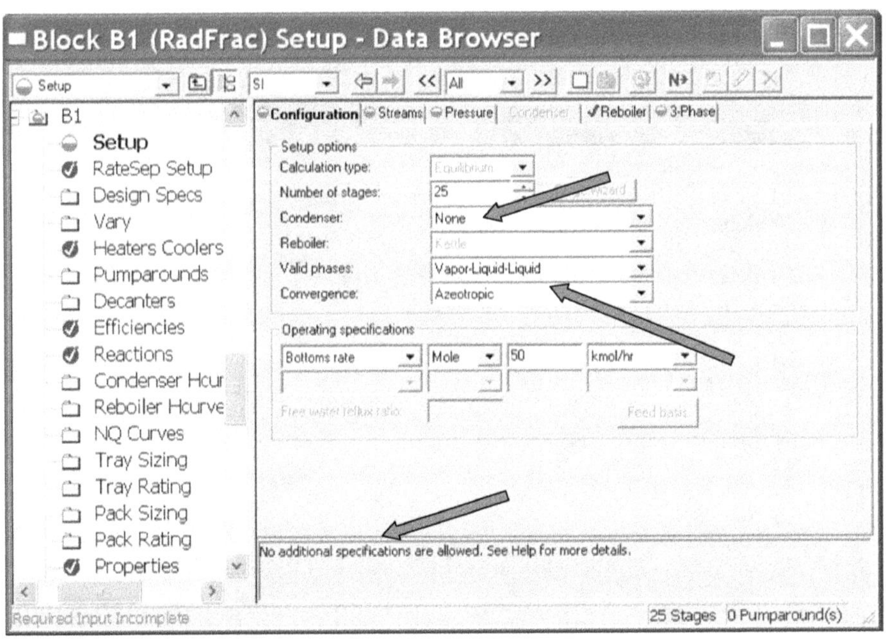


**Figure 3.50** Set-up of the heterogeneous azeotropic column.

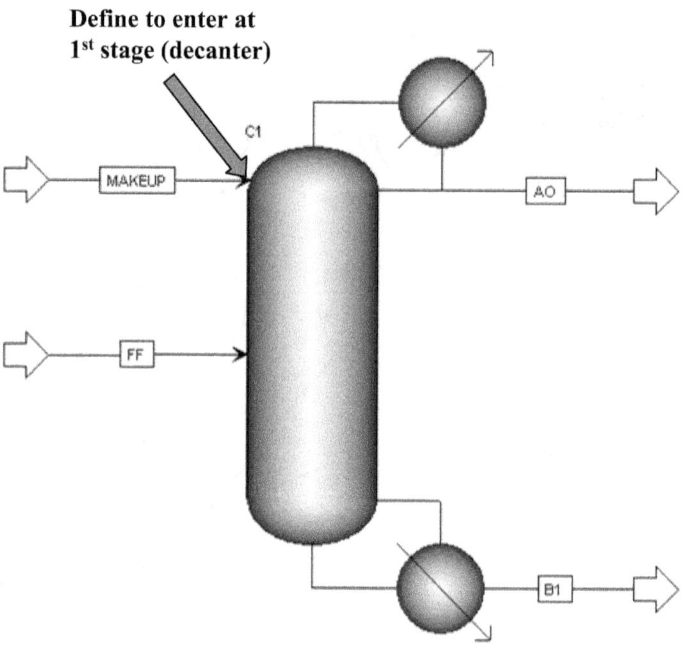


**Figure 3.51** Flowsheet of only one *RadFrac* with built-in decanter.

flows returned to column are specified. From Figure 3.48, all the organic phase is returned to the heterogeneous azeotropic distillation column, while none of the aqueous phase is returned. Thus in Figure 3.53, "1" is chosen for the first liquid and "0" is chosen for the second liquid. In this window, the subcooled temperature in the decanter can also be selected. By using cooling water as the cooling medium, a reasonable decanter temperature is 40°C.

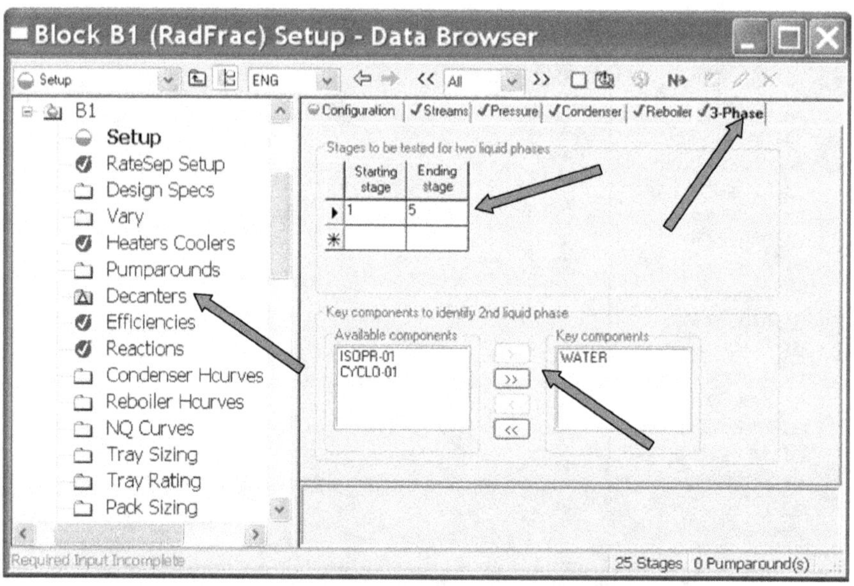


**Figure 3.52** Specifying three-phase calculation inside the column.

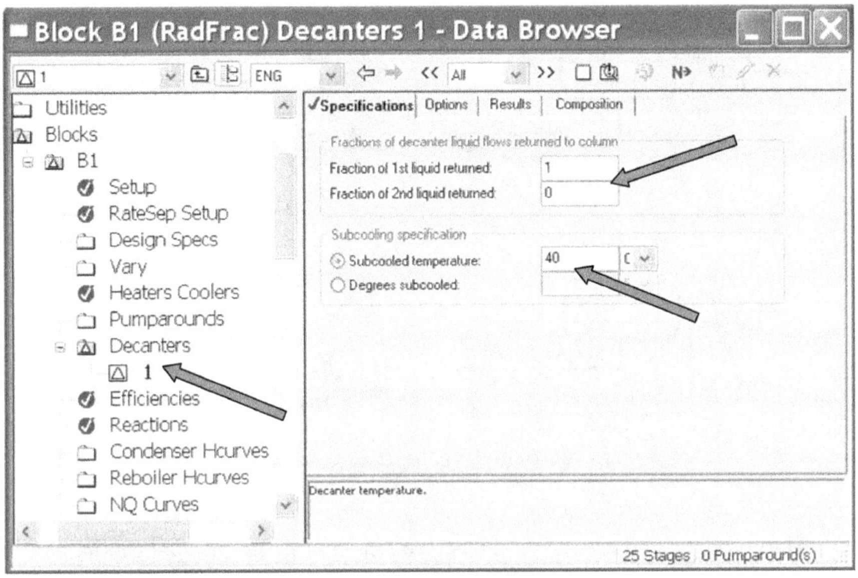


**Figure 3.53** Specifying fractions of decanter liquid flows returned to column.

This heterogeneous azeotropic distillation column system is ready to run. There are two degrees of freedom needed to be determined in this system. One is at the operating specification in the configuration page tab of the column set-up. Another external degree of freedom is the flowrate of the makeup stream, which is set in the *Streams* page tab of the column set-up.

The recommended operating specification to use for the first time is the bottom flowrate. The reasonable bottom flowrate, and also the makeup flowrate, can roughly be estimated by utilizing the total material and IPA component balances of the outer material balance envelop. Additional ways to estimate the stream information will be discussed in Chapter 8.

Once a converged simulation result is achieved, the *Design Spec/Vary* function can be used to achieve the exact bottom composition. Notice that only one *Design Spec/Vary* can be set up with a particular flowrate of the makeup stream. The optimum makeup flowrate is usually set by minimizing the reboiler duty while achieving a specific bottom composition specification.

It is a good time right now to talk about the different configurations of the decanter in a heterogeneous azeotropic column. Figure 3.54 shows four different configurations that have been used in various applications. The first one (Fig. 3.54*a*) with all organic phase refluxing and all aqueous phase going out of the system has been demonstrated in the IPA dehydration process shown in Figure 3.48. The second one (Fig. 3.54*b*) with all organic phase refluxing and a partial aqueous phase also refluxing has been used in acetic acid dehydration, which will be explained in Chapter 9. The purpose of the extra aqueous reflux is to limit the acetic acid loss through the AO stream. The third configuration (Fig. 3.54*c*) with partial organic phase refluxing and all aqueous phase going out of the system has been reported in the vinyl acetate process (see Luyben and Tyreus).[3] The additional organic outlet stream is to withdraw the vinyl acetate product. The fourth application, as in Figure 3.54*d*, is an unusual one with a partial aqueous phase refluxing and an organic phase going out of the system. This is an application of utilizing a feed stream with two alcohols to react with

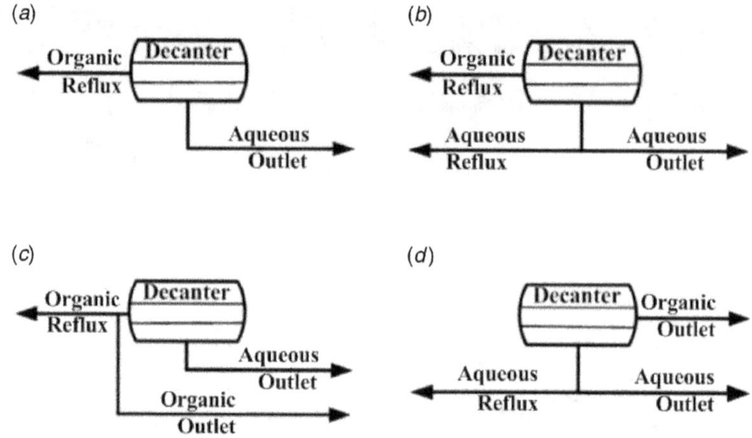


**Figure 3.54** Possible decanter configurations.

acetic acid to produce two acetate products (see Lee et al.).[4] In this application, part of the aqueous phase is used to go back to a reactive distillation column to carry one of the acetate products going to the overhead and out through the organic outlet stream. Further processing of this organic outlet stream is needed to obtain purified acetate product.

All of the above various applications can easily be simulated with Aspen Plus using the *RadFrac* operation block. If there are two outlet streams from the decanter (e.g., both organic and aqueous), two top outlet steams should be added to the flowsheet. One is connected at the arrow that is below the condenser symbol as the outlet stream for the first liquid phase. Another stream should be connected to the arrow coming out right at the condenser as the outlet stream for the second liquid phase. Please refer to previous Figure 3.5 for these two locations. The fractions of the organic phase and aqueous phase liquid streams returned back to the column are given on the *Specifications* page tab under *Decanters* in the column operation block (see Fig. 3.53). Remember that, with an additional fraction specified, an addition degree of freedom needs to be determined in the column design.

### 3.2.2 Decanter

Another way to configure a heterogeneous azeotropic column is to use two unit operation blocks in Aspen – a stripper and a decanter – and then connect them together in a flowsheet simulation. Building a heterogeneous azeotropic column in this way is particularly useful when we want to conduct control studies. The main reason is that the selection of the manipulated variable is more flexible. For example, the column top vapor flow cannot be selected as a manipulated variable using only *RadFrac* to simulate the heterogeneous azeotropic column as in previous Section 3.2.1. However, using only *RadFrac* to simulate the heterogeneous azeotropic column has its advantage in faster convergence of various design cases because of there being no external recycle stream in the flowsheet simulation. Thus, we would use the method in Section 3.2.1 to simulate the heterogeneous azeotropic column for doing the column design study. Once the optimal design flowsheet is found, we would then convert the optimal design flowsheet into another flowsheet simulation by using a stripper and a decanter for conducting further control studies.

We will discuss the use of a decanter in this section. The steps given in Section 3.1.1 are followed to the point where we need to choose a unit operation. Clicking the *Separators* page tab, a list of available separators appears as a row of icons with the description to appear at the bottom of the window. Selecting the prompt for the *Decanter* block and clicking the arrow just to the right of this block opens the window shown in Figure 3.55. This contains several types of decanters. Click the horizontal decanter button on the top right-hand corner of the selection, and move the cursor to the blank flowsheet. The cursor becomes a cross. If we click on the flowsheet, a decanter icon appears in the process flowsheet window.

The next step is to follow the steps in Section 3.1.1 to connect the material streams in this decanter simulation. The decanter operation has one feed stream and two outlet streams (organic and aqueous). For simulating different configurations from the decanter to the heterogeneous azeotropic column as in Figure 3.54, another unit operation block is needed to connect to the two outlet streams of the decanter. For the configuration with any one of the two outlet streams having flow-split (one goes to the heterogeneous azeotropic column and another one out of the system), *Fsplit* in *Mixer/Splitter* page tab along the bottom of the process flowsheet window needs to be selected and added to the process flowsheet with specified split fraction.

The next steps are entering components and selecting physical property methods as in previous Sections 3.1.2 and 3.1.3. Notice that only *NRTL*, *UNIQUAC*, or *UNIFAC* should be selected because we need to describe LLE in this decanter operation.

After entering feed stream data, the next step is to enter data needed in the decanter block. Clicking on *Blocks* and *DECANTER* and *Input* on the left-hand side of the *Data Browser* opens the window shown in Figure 3.56. The first page tab is *Specifications*, on which operating pressure and another one decanter specification need to be specified. We select the

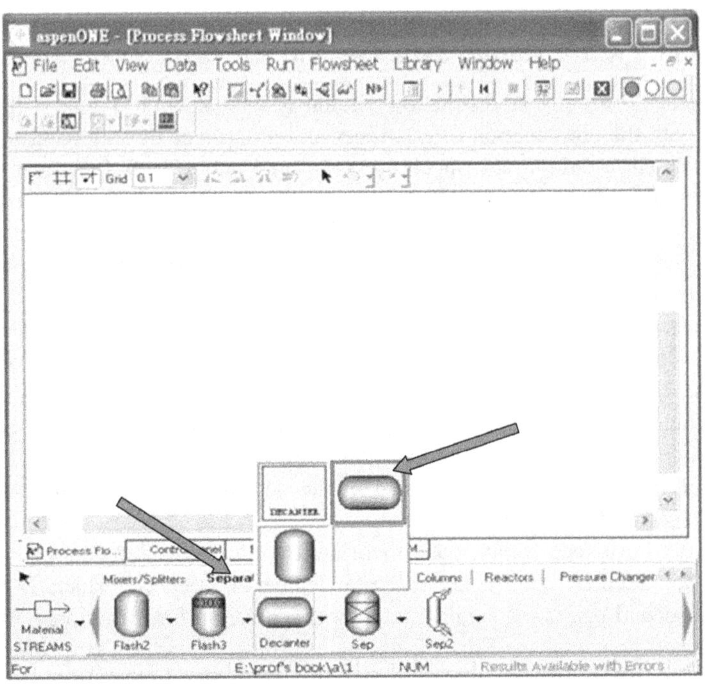


**Figure 3.55** Selecting decanter.

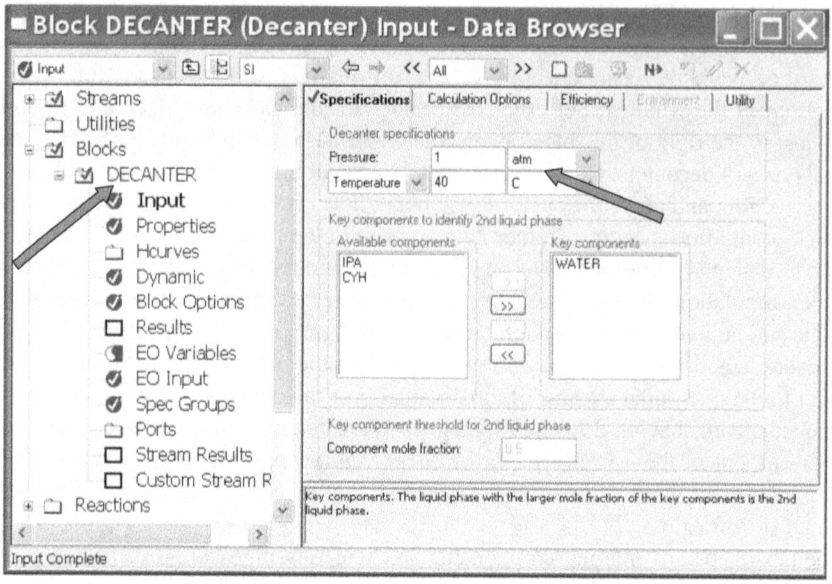


**Figure 3.56** Decanter input specifications.

decanter operating pressure at 1 atm and decanter temperature at 40°C so that cooling water can be used as the cooling medium. In this *Specifications* page tab, the key component to identify the second liquid phase should also be selected. All the information needed is now complete, and we are ready to run the simulation.

### 3.2.3 Column with External Decanter and Recycle

The heterogeneous azeotropic distillation column system shown in Figure 3.48 can also be configured with the combination of a stripper and a decanter by following the steps previously explained for adding a stripper and a decanter into a blank flowsheet. Note again, the stripper is added by clicking the *Columns* page tab and then clicking *RadFrac*, followed by selecting the stripper button as in Figure 3.45. The decanter is added by clicking the *Separators* page tab and then clicking *Decanter* followed by selecting the horizontal decanter button as in Figure 3.55.

All the material streams are connected as shown in Figure 3.57, with two feed streams for the stripper. One feed stream to the stripper must enter on Stage 1 (this is the organic outlet stream of the decanter). Another feed stream (fresh feed) can be entered at any optimal feed location as determined by the design study. The top vapor-outlet stream of the stripper is connected as one feed stream of the decanter. Another feed stream of the decanter is the entrainer makeup stream. The AO stream of the decanter (defined as the second liquid phase) is designed to go out of the system.

Besides the fresh feed stream data, there are two degrees of freedom for this system needed to be specified. One is an external degree of freedom, which is the makeup stream. The second one is the single operating specification for the stripper. For example, as demonstrated previously in Figure 3.46, the distillate (top vapor) rate can be specified. Alternatively, the bottoms rate can be specified instead of the distillate rate. The overall flowsheet includes a recycle loop between two unit operations (stripper and decanter).

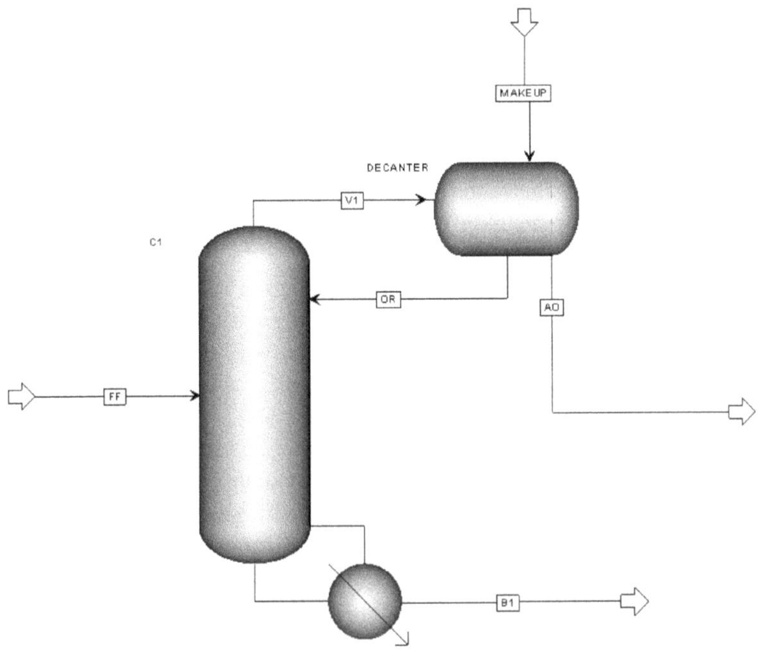


**Figure 3.57** Flowsheet of heterogeneous azeotropic column using a stripper and a decanter.

Therefore, a good guess of the above two specifications will help the convergence of the flowsheet simulation. For an educated guess of these two specifications, the conceptual design, shown in Figure 3.49, can be used.

Assuming that the composition of the top vapor stream reaches the ternary azeotrope (with the minimum temperature of the whole system), the compositions of both the AO and the OR can be estimated. With the known fresh-feed flowrate and the lever rule, the flowrate of the OR and the makeup flowrate can be estimated by the inner and outer material balance envelopes, respectively. Second, from the tie-line information and the lever rule, the flowrate of the AO stream can also be estimated. Thirdly, by using the outer material balance envelop and the lever rule, the bottom flowrate can further be estimated. Finally by using the inner material balance envelop and the lever rule, the top vapor flowrate can be estimated.

For the flowsheet simulation of Figure 3.57, besides giving the two specifications, the other estimated compositions and flowrates in the recycle loops are preferably given in the flowsheet as an initial guess to help the convergence of the calculations. The way to give the initial guess is to double click the material stream and type in the known estimated information.

With a converged flowsheet simulation run, the *design spec/vary* function in *RadFrac* can be used to set the purity of the bottoms product stream. Note that, because there is only one operating specification for the stripper, only one *design spec/vary* can be used.

### 3.2.4 Heat Exchangers

With the use of the *RadFrac* unit operation block in Aspen Plus, the heating (reboiler) and the cooling (condenser) equipment of a distillation column are built in with the column simulation. In other separation system applications, additional heaters, coolers, or process-to-process heat exchangers may need to be simulate. We will now discuss how to build a steady-state simulation of various heat exchangers.

***Heater or Cooler.*** This is the unit operation where a process fluid exchanges heat with a utility stream. For example, the heater is the one where the process fluid receives the heat from steam or other heating medium, and the cooler is the one where process fluid gives out the heat to cooling water or other cooling medium.

The way to do the simulation is to first follow the steps in Section 3.1.1 to the point where a unit operation needs to be picked. By clicking the *Heat Exchangers* page tab, the list of available heat exchangers appears as a row of icons, with the descriptions appearing at the bottom of the window. Selecting the *Heater* block and clicking the arrow just to the right of this block opens the window shown in Figure 3.58.

This contains several types of heater. Click the heater button on the top row, second from the left, and move the cursor to the blank flowsheet; the cursor becomes a cross. If we click on the flowsheet, a heater icon appears in the process flowsheet window.

Next, follow the same step-by-step procedure as in Section 3.1 to connect inlet and outlet material streams, to specify chemical components and physical properties, and to specify feed stream properties. After that, clicking on the *Heater* block opens a window with a long list of items. The top subitem is labeled *Input*. Clicking it opens the window shown in Figure 3.59. The first page tab is *Specifications*, on which the operating specifications must be entered. We usually select the operating pressure and also the outlet temperature as shown in Figure 3.59. The heat duty required for this operation, and other properties

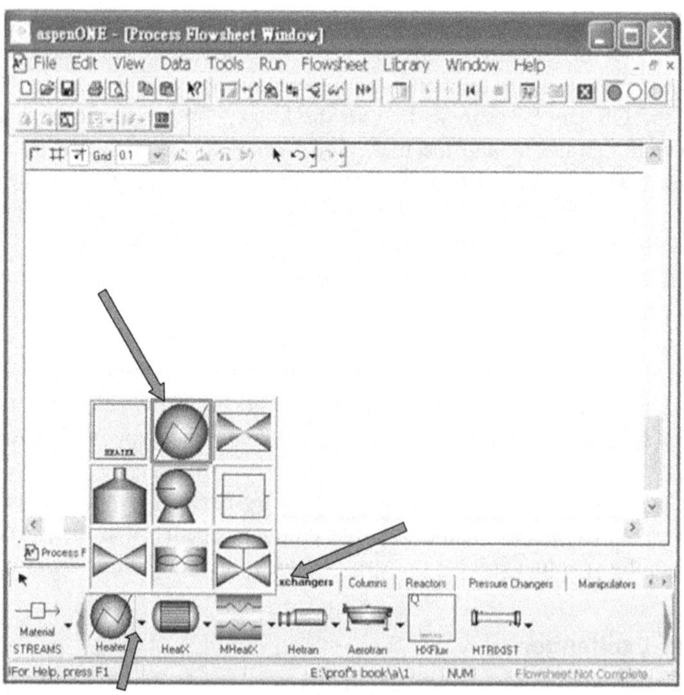


**Figure 3.58** Selecting heater.

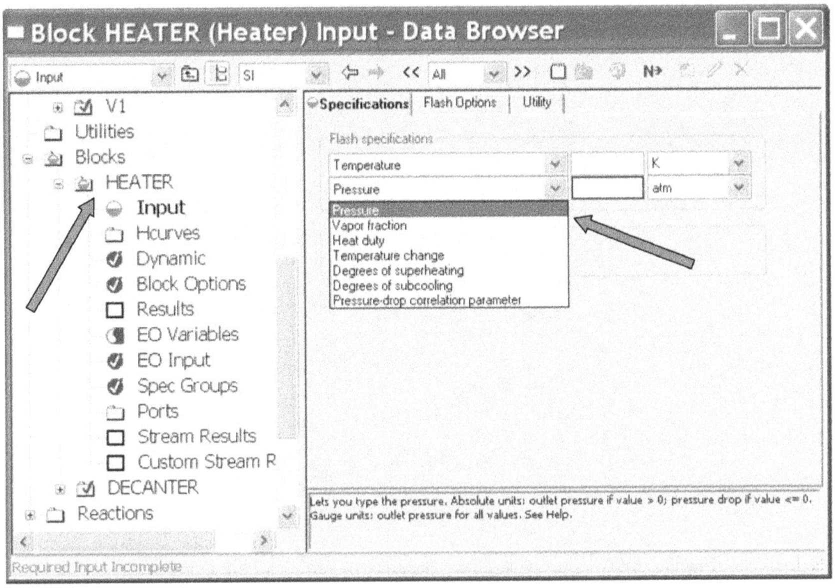


**Figure 3.59** Heater input specifications.

for the outlet stream, can then be calculated. Other alternative specifications can also be made. It is often useful to specify the vapor fraction leaving the heater in order to set the phase of the exit stream. For complete vaporization, the vapor fraction is set to 1. For complete condensation, the vapor fraction is set to 0.

Note that this block can be used for simulation of a heater and also for a cooler. The sign of the heat transfer will be positive for a heater when heat is added and negative for a cooler when heat is removed.

***Process-to-Process Heat Exchanger.*** Another way to describe a unit operation with heat transfer, which is widely used for heat integration studies, is a process-to-process heat exchanger. This is used in situations when a hot stream in a process flowsheet needs to be cooled and another cold stream needs to be heated. By coupling these two process streams together, saving of energy for the overall process can be achieved by reducing of the heat and cooling media usage. Of course the two feed inlet temperatures need to be suitable for transferring the heat from the hot stream to the cold stream. Also, the heating and cooling demands of the two process streams usually will not be exactly the same, so the unit operation of a heater and a cooler cannot be perfectly replaced by this heat exchanger. Another consideration is that the dynamic controllability of the heat integration system may require the existence of the extra heater or cooler for a more robust operation in the face of various disturbances. An example of the design of heat integration for pressure-swing azeotropic distillation systems will be given in Chapter 6.

The way to do the simulation is the same as for the heater or cooler, except, after clicking the *Heat Exchangers* page tab, the *HeatX* block instead of the *Heater* block is selected. Click the process-to-process heat exchanger button within the *HeatX* block and move a heat exchanger icon on the flowsheet, as shown in Figure 3.60. Then connect two inlet streams (cold in and hot in) and two outlet streams (cold out and hot out), making sure that the

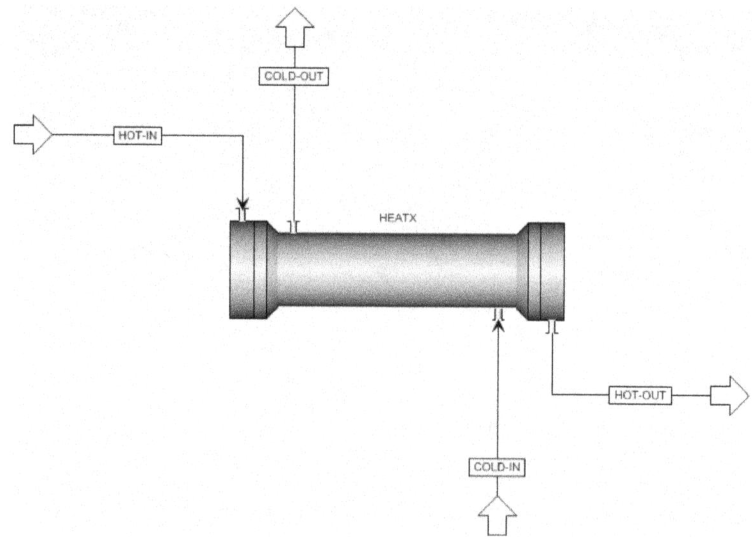


**Figure 3.60** Flowsheet of a process-to-process heat exchanger.

material streams are connect to the correct inlet and exit connections on the heat exchanger. Next, follow the same step-by-step procedure as in Section 3.1 to specify chemical components, physical properties, and two feed stream properties. The next step is to enter the unit operation block data. Clicking on the *HeatX* block opens a window with a long list of items. The top subitem is labeled *Setup*. Clicking it opens the window shown in Figure 3.61. There are several page tabs. The first is *Specifications*, on which the calculation type, the flow arrangement, and the exchanger specifications are entered. In this figure, we go with the default selection that uses shortcut calculations and a counter-current flow

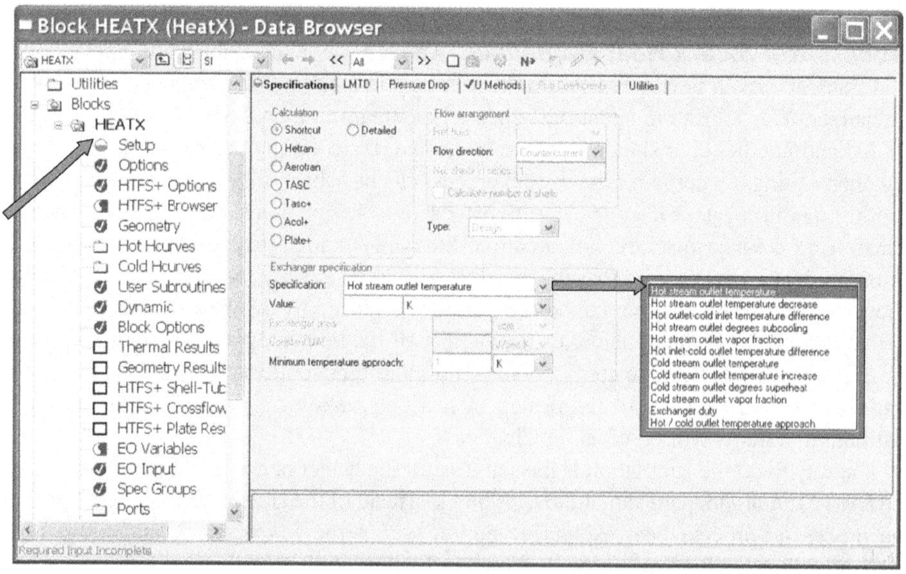


**Figure 3.61** *HeatX* setup specifications.

arrangement. For the exchanger specification, we usually select temperature of one of the outlet streams (hot or cold). After we input the value of the hot-stream outlet temperature, the heat exchanger simulation is ready to run.

## 3.3 ADD A NONDATABANK COMPONENT

In some column simulations, we may encounter a situation when a particular component cannot be found by following the steps to enter components as in Section 3.1.2. Fortunately, in this situation, Aspen Plus provides a way to add a nondatabank component to the flowsheet simulation. We will use an example to demonstrate how to add a nondatabank component in Aspen Plus simulation.

The example is a reactive distillation study for the production of diethyl carbonate (DEC). This chemical can be produced by the transesterification reaction of ethanol with dimethyl carbonate (DMC). The reaction intermediate for the two-step reaction is a component called ethyl-methyl carbonate (EMC), which cannot be found in the Aspen Plus databank. Thus, for simulations involving this component, the following steps must be performed.

Follow the steps given in Section 3.1 to start up Aspen Plus. Instead of building up a new flowsheet, we need to first add EMC into the simulation. From the dropdown menu in *Data*, select *Setup* from the list as shown in Figure 3.62. The *Data Browser* window opens (Fig. 3.63). Previously we selected *Flowsheet* as the run type (Fig. 3.8). Instead, now we select *Property Estimation* as the run type. The next step is to go to the left-hand side of the *Data Browser* window and click *Specifications* under the *Components* list

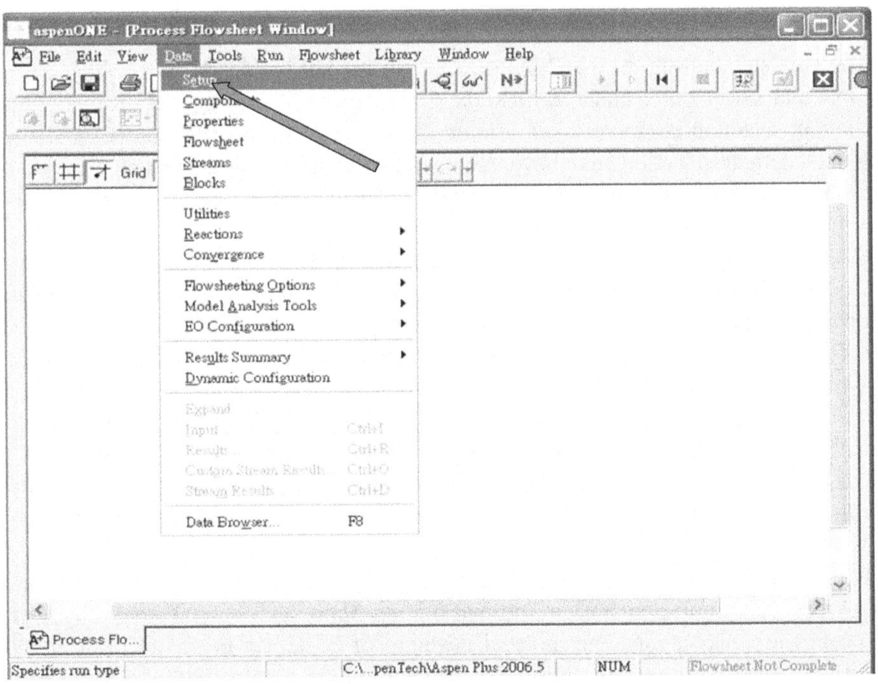


**Figure 3.62** Selecting *Setup* in *Data* dropdown lists.

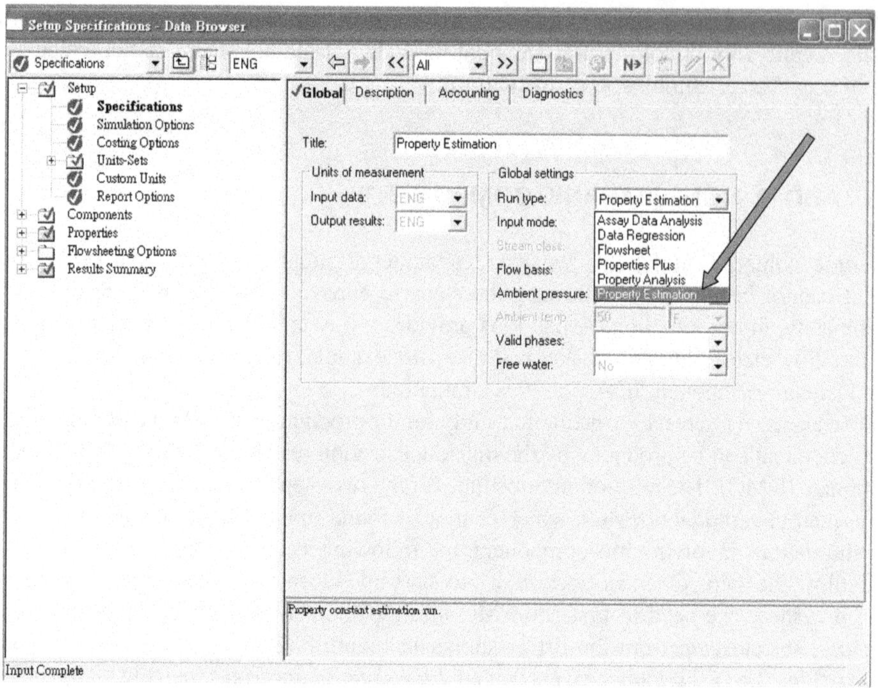

**Figure 3.63** Selecting run type as *Property Estimation.*

to add the nondatabank component. As shown in Figure 3.64, a new component ID *EMC* is added.

The next step is to go to the *Properties* list on the left-hand side of the *Data Browser* window and select *Input* under *Estimation*. Clicking the *Setup* page tab in this window, the default selection is *Estimate all missing parameters* as shown in Figure 3.65. We go with this default selection and have Aspen estimate all missing parameters. The next step is to enter the molecular structure of EMC and also to enter as much physical property information about this new component as is known.

One way to input molecular structure of the new component is to enter individual atoms and bonds. However, a simpler way is to import the molecular structure from other component databanks. A very useful resource is the NIST (National Institute of Standards and Technology) Chemistry WebBook.[5] The information for this new component is shown in Figure 3.66. Click the molecular structure file, *2d Mol file*, and save in a known directory with a given file name. The next step is to go to *Molecular Structure* of *EMC* under *Properties* and Click *Import Structure* to import this file into Aspen Plus (see Fig. 3.67). After this step, ask Aspen Plus to calculate bonds by clicking *Calculate Bonds*. The graphical structure of EMC is successfully imported into Aspen Plus as shown in Figure 3.68.

For demonstration purposes, other physical properties of this component are entered into Aspen Plus: molecular weight, normal boiling point, and Antoine coefficients in the vapor pressure calculation. The molecular weight of EMC can be found from the NIST Chemistry WebBook (see Fig. 3.66). The normal boiling point of 107.9°C is taken from Mueller et al.[6] and the Antoine coefficents are obtained from Luo et al.[7]

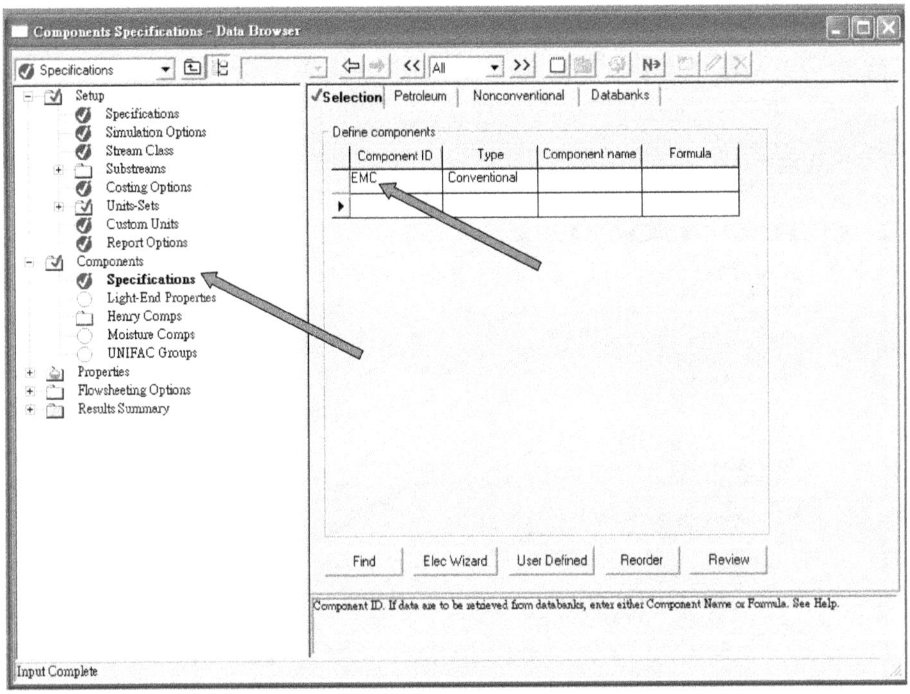


**Figure 3.64** A new component ID of *EMC* is added.

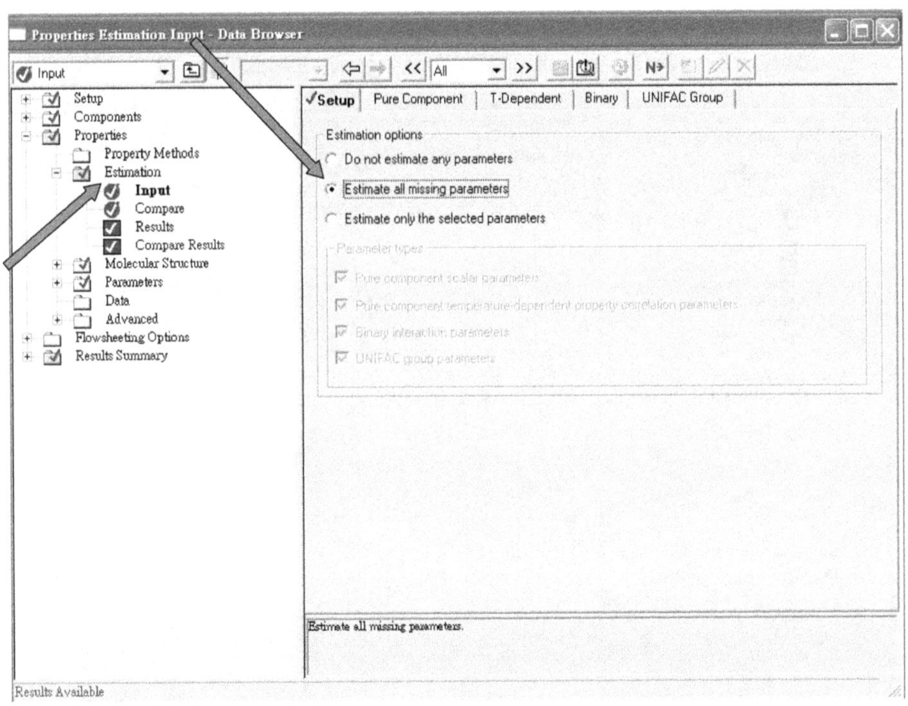


**Figure 3.65** Estimate all missing parameters.

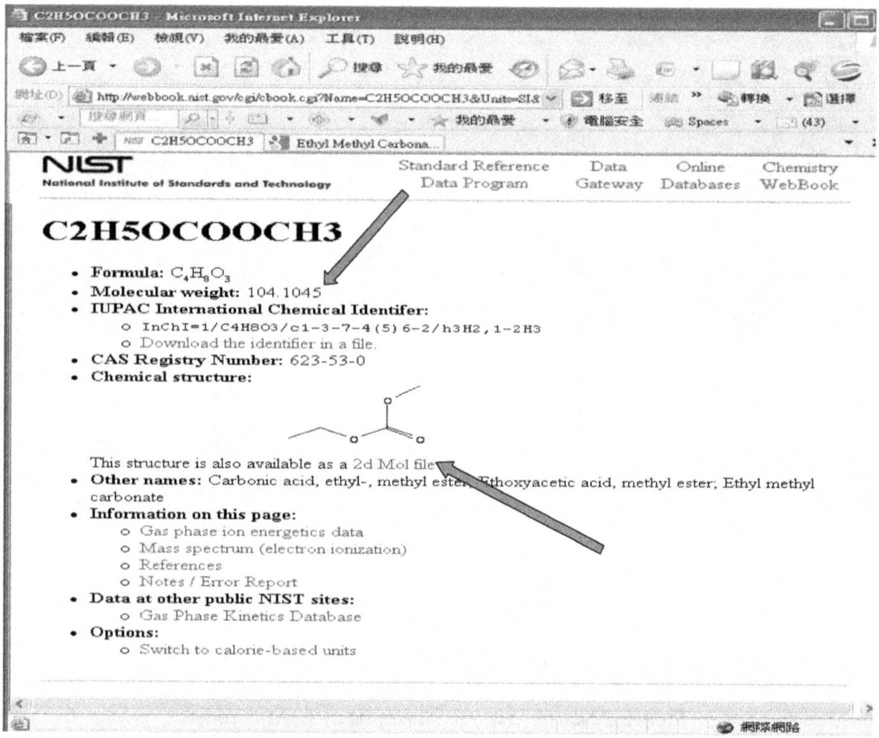

Figure 3.66 EMC in NIST Chemistry WebBook.

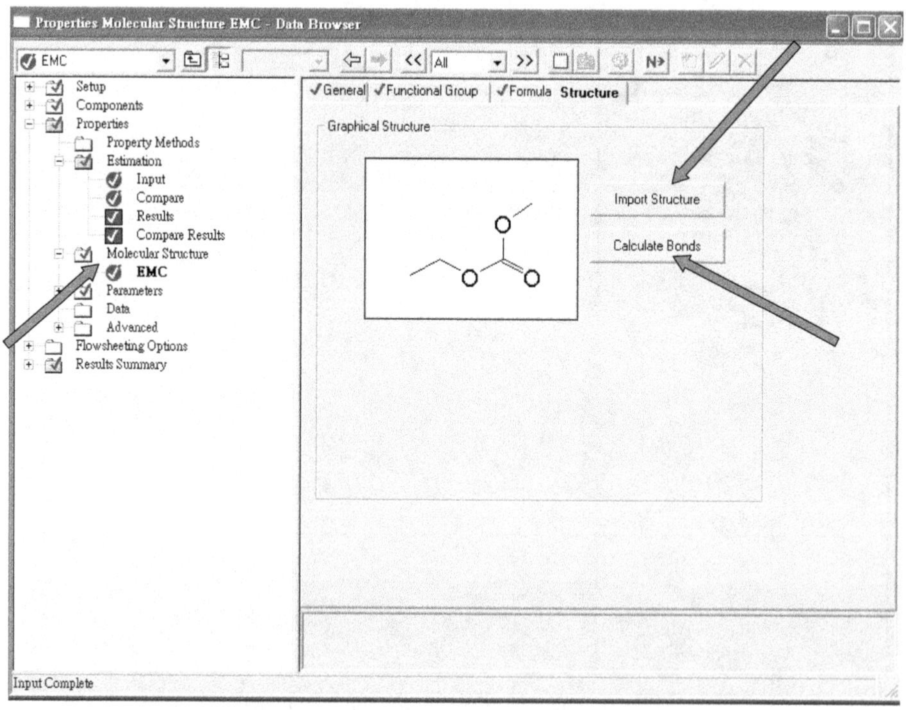

Figure 3.67 Import molecular structure and then calculate bonds.

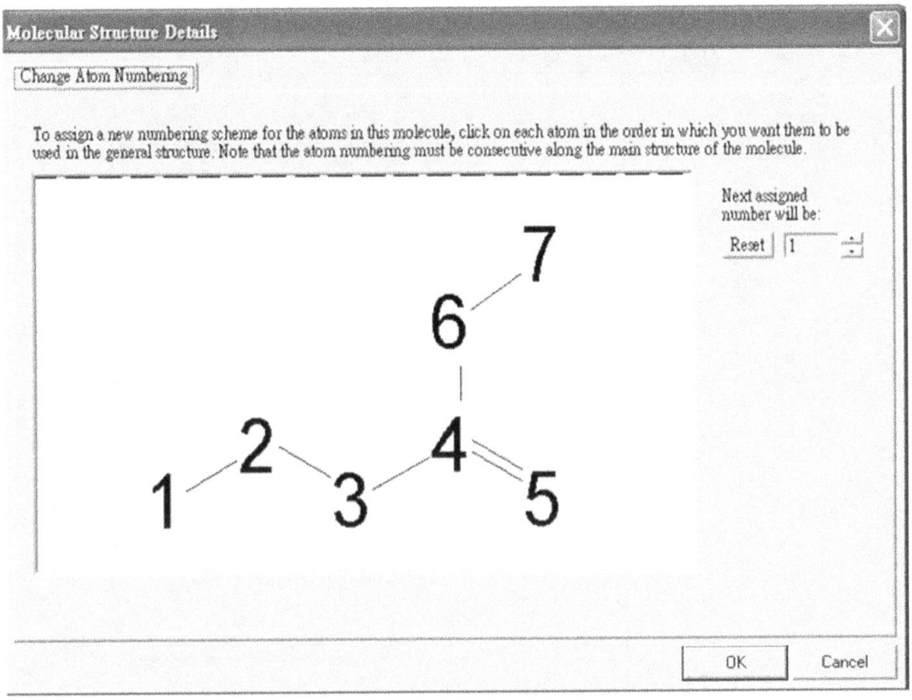


**Figure 3.68** The results after calculate bonds.

To enter normal boiling point and molecular weight, select *Pure Component* under *Parameters* in the *Properties* folder on the left-hand side of the *Data Browser* window and then click *New* to create new parameters. As shown in Figure 3.69, select *Scalar* and enter new name as *TBMW* and then click *OK*. A window as in Figure 3.70 appears for us to enter the normal boiling point and molecular weight of the new component. Clicking the *Parameters* field, a list of choices appears. Select *TB* as normal boiling point and select units as *C* and then enter the value of 107.9. Repeat the similar step to enter MW (molecular weight).

For adding Antoine coefficients of this new component, follow the same step as above to add new parameters. Instead of selecting *Scalar*, this time *T-dependent correlation* is selected. With this selection, a list of choices appears as shown in Figure 3.71. Under *Liquid vapor pressure*, select *PLXANT-1* for the extended Antoine vapor pressure equation. After clicking *OK*, a window as in Figure 3.72 appears for the user to input Antoine coefficients. The extended Antoine equation in Aspen Plus is defined as:

$$\ln p_i^{\mathrm{v}} = c_1 + \frac{c_2}{T + c_3} + c_4 T + c_5 \ln T + c_6 T^{c_7} \quad \text{for } c_8 \leq T \leq c_9 \tag{3.5}$$

After inputting all the Antoine coefficients, we are ready to run the property estimation. Click the *Next* button in Figure 3.72; the message as shown in Figure 3.73 indicates that the required input is complete. Click the *OK* button, and the property estimation is executed. After the run is complete, the status messages in the *Control Panel* as in Figure 3.74 are displayed. There are three warning messages to do with the functional group generation

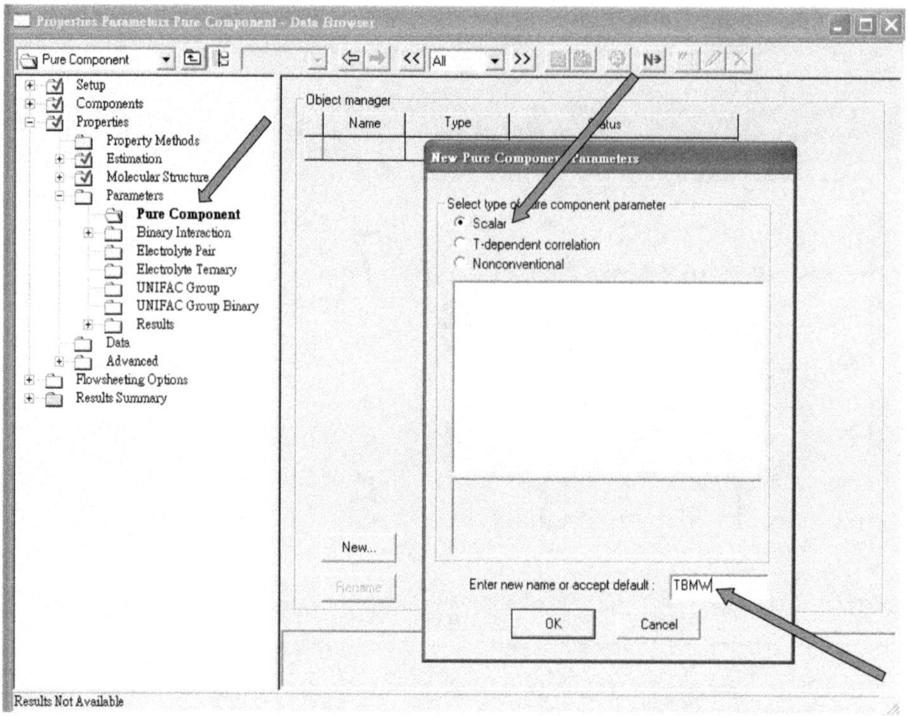

**Figure 3.69** Input scalar pure component parameters.

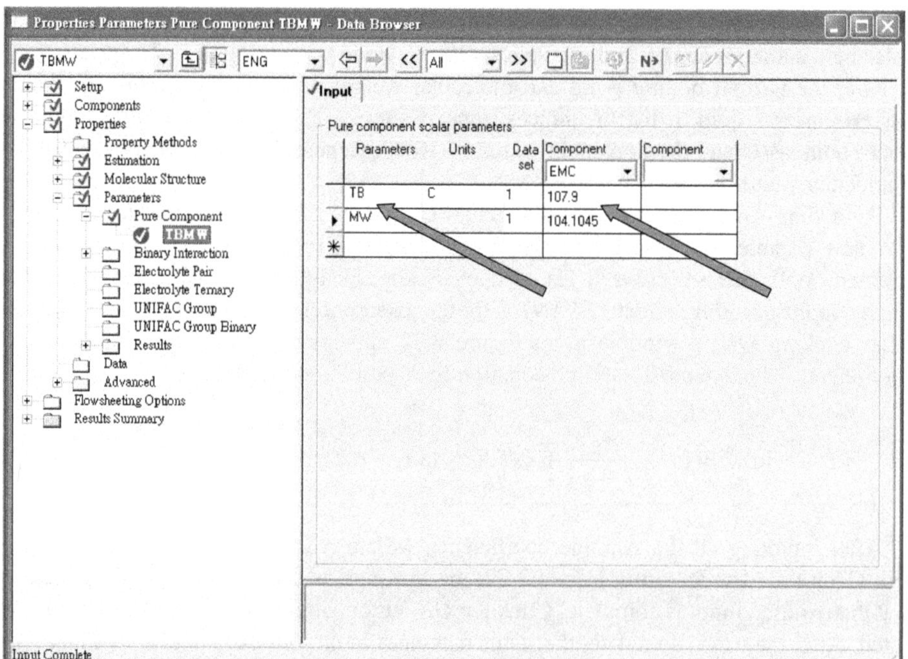

**Figure 3.70** Input normal boiling point and molecular weight.

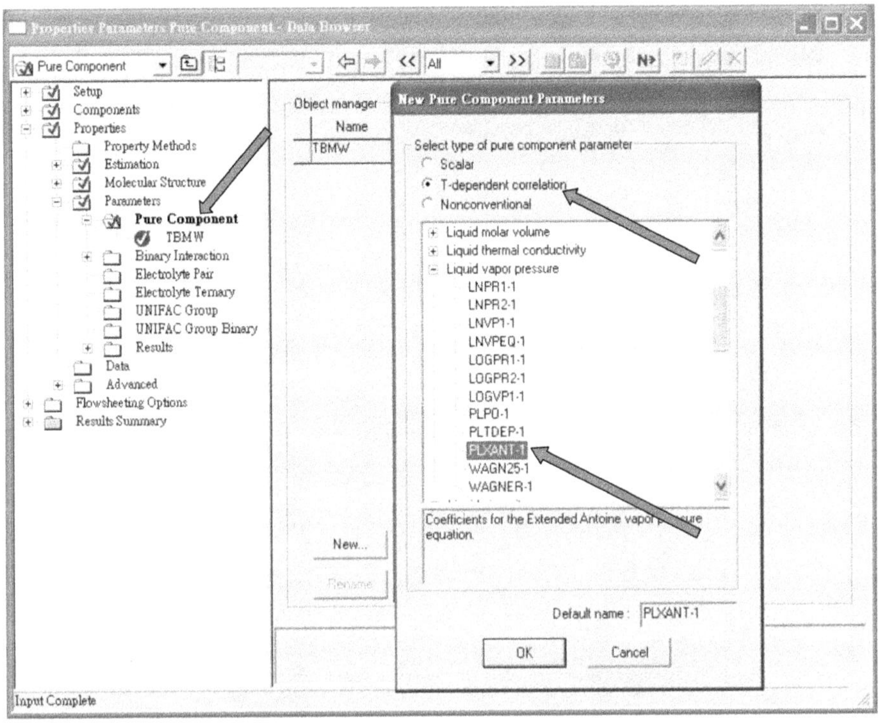


**Figure 3.71** Input temperature-dependent correlation.

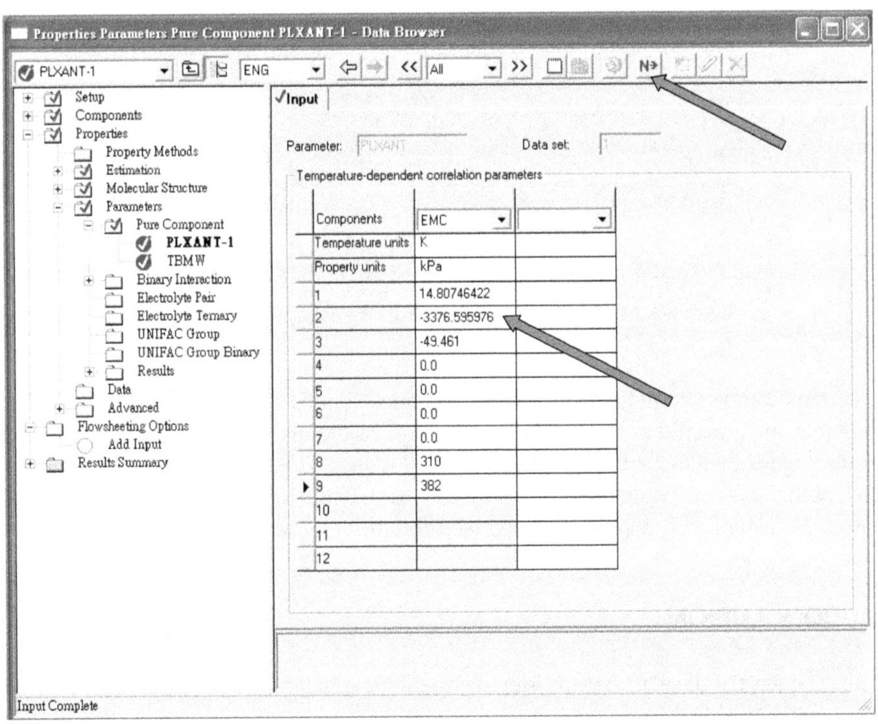


**Figure 3.72** Input Antoine coefficients.

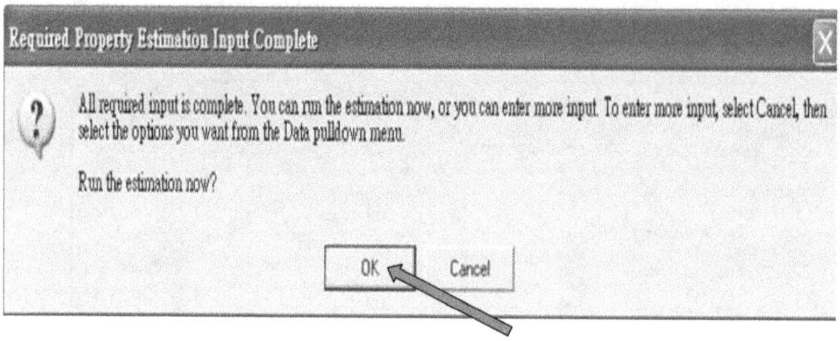


**Figure 3.73** Ready to run the estimation.

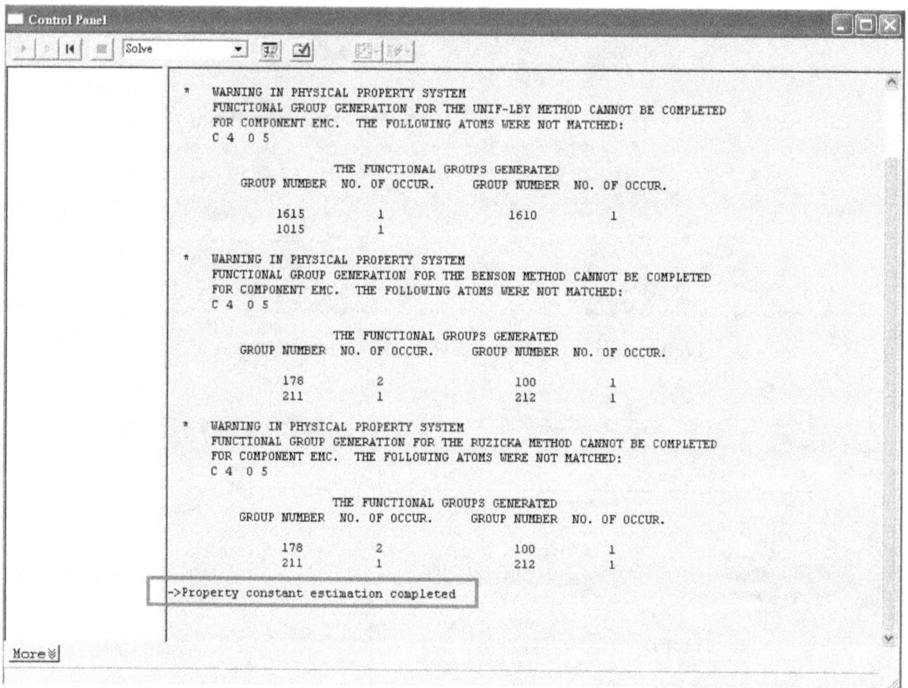


**Figure 3.74** Message showing property estimation completed.

for certain estimation methods. In this example, ignoring these warning messages is acceptable since we did not use functional groups in the molecular structure specification. The message shown at the bottom of Figure 3.74 (*Property constant estimation completed)* indicates that this new component is successfully added into the Aspen simulation. The usual steps, as in Section 3.1, can be followed next to build the steady-state model.

## 3.4 CONCLUSION

In this chapter, the detailed step-by-step instructions to build a steady-state model are given with an extractive distillation column example. The unit operation blocks such as *RadFrac*,

*Decanter*, *Heater*, and *HeatX* used in this book are also explained. For the situation when a particular chemical component cannot be found in Aspen, a way to add a nondatabank component into an Aspen simulation is also demonstrated with an example. Following step-by-step instructions in this chapter, the reader should have a basic knowledge of building a steady-state model.

## REFERENCES


1. Aspen Physical Property System, Version 2006.5, Aspen Technology, Inc., Cambridge, MA, 2007.
2. Douglas J. M., *Conceptual Design of Chemical Processes*, McGraw-Hill, New York, 1988.
3. Luyben M. L. and B. D. Tyreus, An industrial design/control study for the vinyl acetate monomer process, *Comput. Chem. Engng.*, **22**, 867–877 (1998).
4. Lee H. Y., L. T. Yen, I. L. Chien, and H. P. Huang, Reactive distillation for esterification of an alcohol mixture containing *n*-butanol and *n*-amyl alcohol, *Ind. Engng. Chem. Res.*, **48**, 7186–7204 (2009).
5. NIST Chemistry WebBook. http://webbook.nist.gov/chemistry/
6. Mueller I., C. Pech, D. Bhatia, and E. Y. Kenig, Rate-based analysis of reactive distillation sequences with different degrees of integration, *Chem. Eng. Sci.*, **62**, 7327–7335 (2007).
7. Luo H. P., W. D. Xiao, and K. H. Zhu, Isobaric vapor-liquid equilibria of alkyl carbonates with alcohols, *Fluid Phase Equilibria*, **175**, 91–105 (2000).

# CHAPTER 4

# DYNAMICS AND CONTROL IN ASPEN DYNAMICS

The previous chapter discussed the methods and techniques for using Aspen Plus simulation software to develop and optimize steady-state designs for azeotropic distillation systems. Once the steady-state design is complete, the dynamic controllability of the process should be explored. Only looking at the steady state does not tell you whether the process is operable. Dynamic simulations and the development of an effective control structure are vital parts of process development.

To perform a dynamic simulation, additional information about the physical size of equipment is required because the larger the vessel for a given flowrate of material through it, the slower the process will respond and the larger its time constant will be. The sizes of reactors in the process are, of course, known from the steady-state design, since holdup (amount of catalyst) affects reaction rates. The sizes of surge tanks, reflux drums, flash vessels, and other pieces of equipment that do not involve reactions, do not affect the steady-state design, except of course their impact on capital investment. However, the sizing of these units is critical to the dynamic controllability of the process; therefore, these units need to be sized in a rational way.

Realistic dynamic simulations should be "pressure driven" (as opposed to "flow driven") since this represents what really occurs in the operation of the plant. Adequate pump and compressor heads must be incorporated into the design, and control valves must be designed to ensure that flows can be changed sufficiently to handle the inevitable disturbances that will occur during operation of the plant. Control valve sizing is very important; Valves must be designed with sufficient pressure drop at design conditions so that flowrates can be changed over a range of values and valve saturation does not occur. All of these issues are covered in this chapter.

Once these issues have been resolved in Aspen Plus, the procedure is to "export" the Aspen Plus file into Aspen Dynamics. Controllers must then be added to implement the

*Design and Control of Distillation Systems for Separating Azeotropes.* By William L. Luyben and I-Lung Chien

desired control structure (choice of controlled and manipulated variables and their pairing). These controllers must then all be tuned in a convenient, consistent, and rational way. These steps are also covered in detail in this chapter.

## 4.1 SIZING OF PROCESS VESSELS

Reactors must be sized during steady-state design. All other vessels must be sized before going into dynamics.

### 4.1.1 Distillation Columns

The diameter of each column, along with a weir height and tray spacing, must be specified. The diameter can easily be found in Aspen Plus using the *Tray Sizing* feature under the column block, as shown in Figure 4.1. In the example shown, the calculated diameter is 1.14 m. A typical weir height is 1–2 inches, and a typical tray spacing is 2 ft.

We must also calculate the size of the reflux drum and the column base. These provide liquid surge capacity, which helps to filter disturbances in both flow and composition to downstream units. They also permit the column to ride through large disturbances without upsetting the column to the point where liquid or vapor hydraulic limitations are encountered (flooding or weeping), which can result in the loss of separation and the production of off-specification products.

Distillation control wisdom suggests that these surge volumes should be large enough to provide 5 min of holdup when half full, using the total liquid entering or leaving these locations. Figure 4.2 shows how the volumetric flowrates of liquid from the reflux drum and liquid into the base of the column can be easily determined. The first thing to do is to go to *Report* in the column block, select the page tab call *Property Options* and check the *Include hydraulic parameters* box (Fig. 4.2*a*). When the program is run, the desired flowrates can be found by going to *Profiles* in the column block, selecting the *Hydraulics* page tab and looking at the column *Volume flow liquid from* (Fig. 4.2*b*). Stage 1 is the reflux drum and shows a volumetric flowrate of 0.225 $m^3/min$ (Fig. 4.2*b*). To have 10 min of total holdup when the drum is 100% full, the volume of the reflux drum must be:

$$(0.225\,\mathrm{m^3/\,min})(10\,\mathrm{min}) = 2.25\,\mathrm{m^3} \tag{4.1}$$

Assuming the reflux drum has an aspect ratio (length over diameter) $L/D$ of 2, the total volume is

$$\frac{\pi D^2}{4}L = \frac{\pi D^2}{4}(2D) = \frac{\pi D^3}{2} = 2.25\,\mathrm{m^3} \tag{4.2}$$

Thus the reflux drum diameter is $D = 1.13$ m and its length is 2.26 m.

These values are entered by clicking the *Dynamics* button on the top toolbar, as shown in Figure 4.3. The *Dynamic* item under the column block turns red, indicating input data is required. Selecting the *Reflux Drum* page tab opens the window shown in Figure 4.3 where the diameter and the length of the reflux drum are entered.

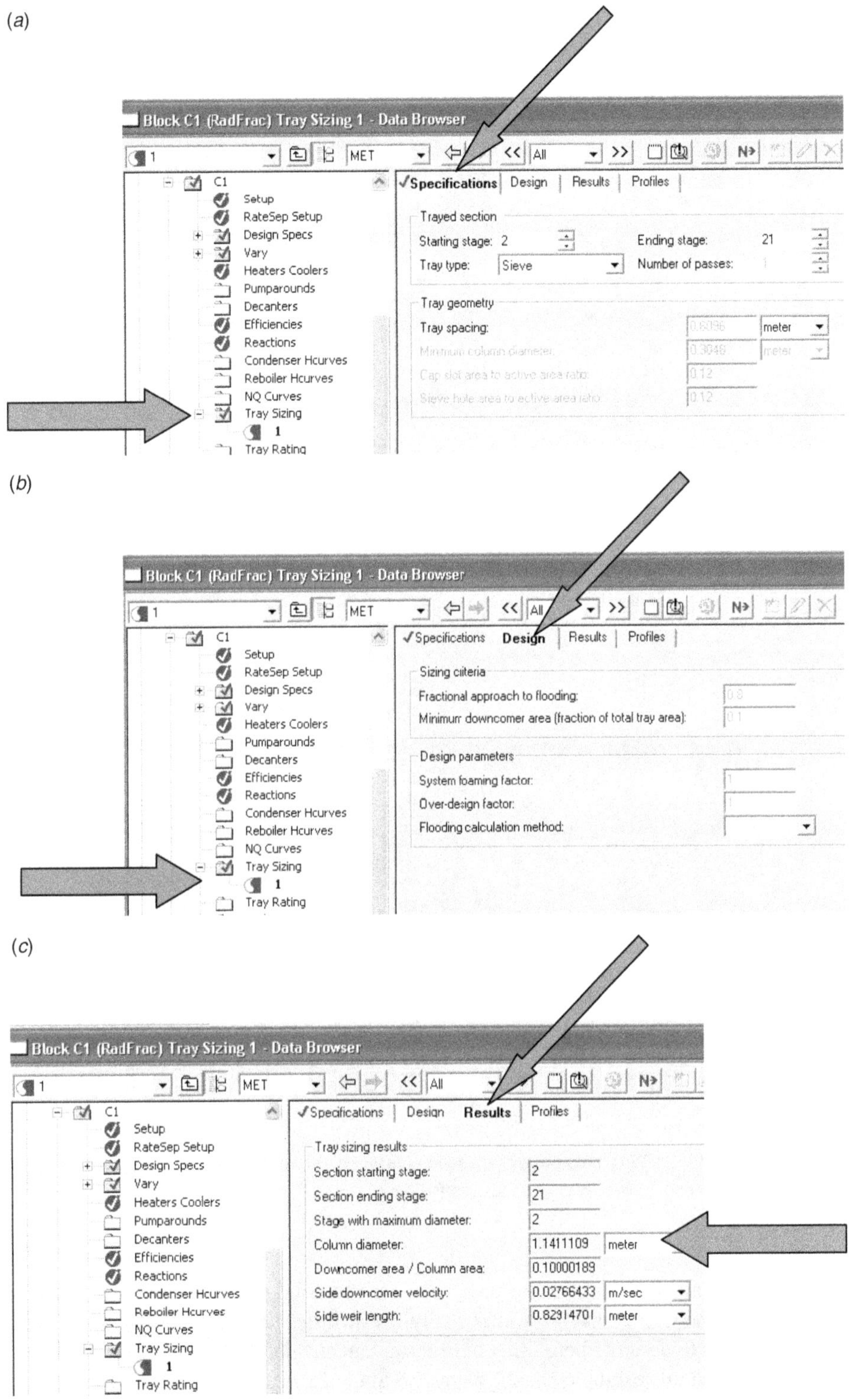


**Figure 4.1** (*a*) Aspen Plus *Tray Sizing*; *Specifications*. (*b*) Aspen Plus *Tray Sizing*; *Design* parameters. (*c*) Aspen Plus *Tray Sizing*; *Results*.

(*a*)

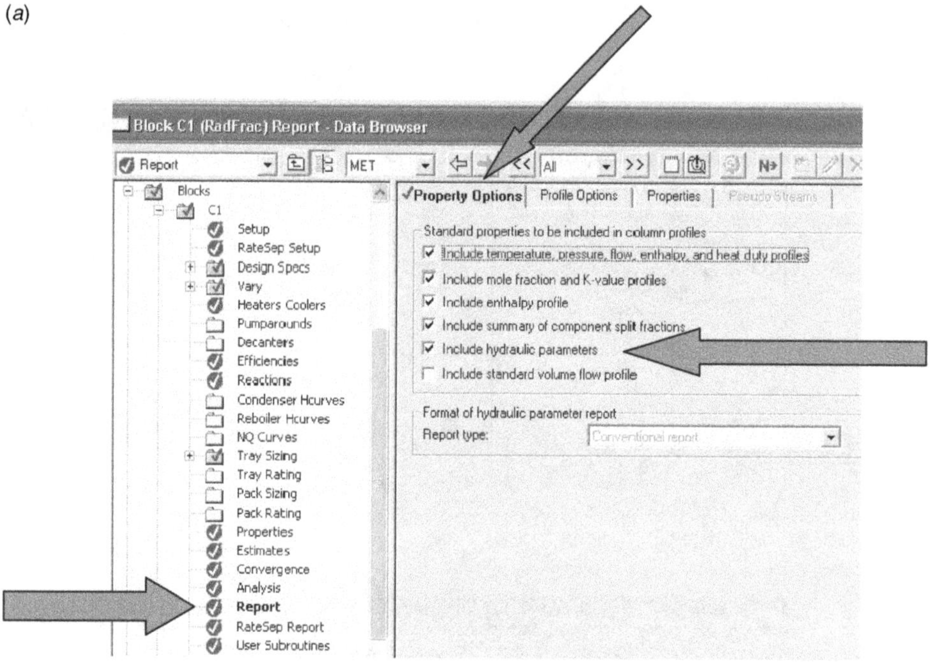

(*b*)

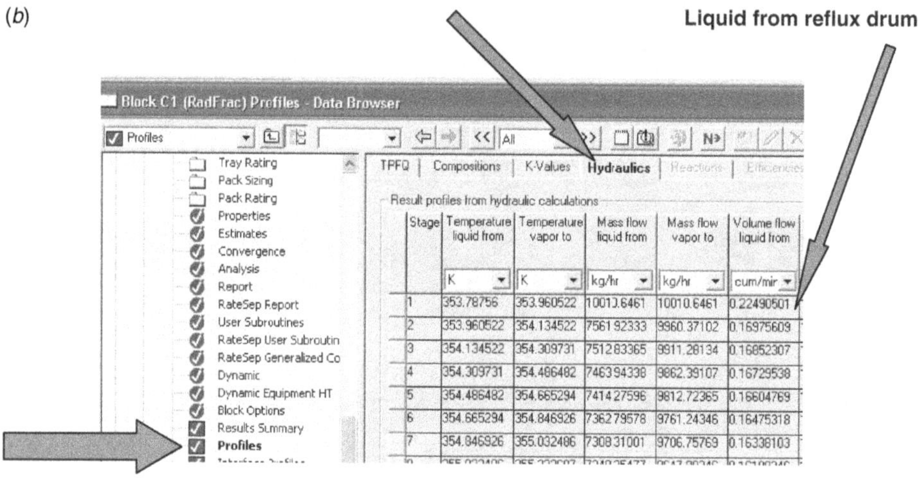


**Figure 4.2** (*a*) Aspen Plus Hydraulics; *Property Options.* (*b*) Aspen Plus *Hydraulics*; results for reflux drum. (*c*) Aspen Plus *Hydraulics*; results for base.

It should be noted that there are page tabs for the condenser and reboiler. Since the dynamics of these heat exchangers are usually much faster than those of the column, the default setting is to assume these units have instantaneous responses (no dynamic lags).

The column in this example has 22 stages, so Stage 22 is the column base (or "sump"). Figure 4.2*c* shows a volumetric flowrate of 0.320 $m^3$/min leaving Stage 21 and entering the

(c)

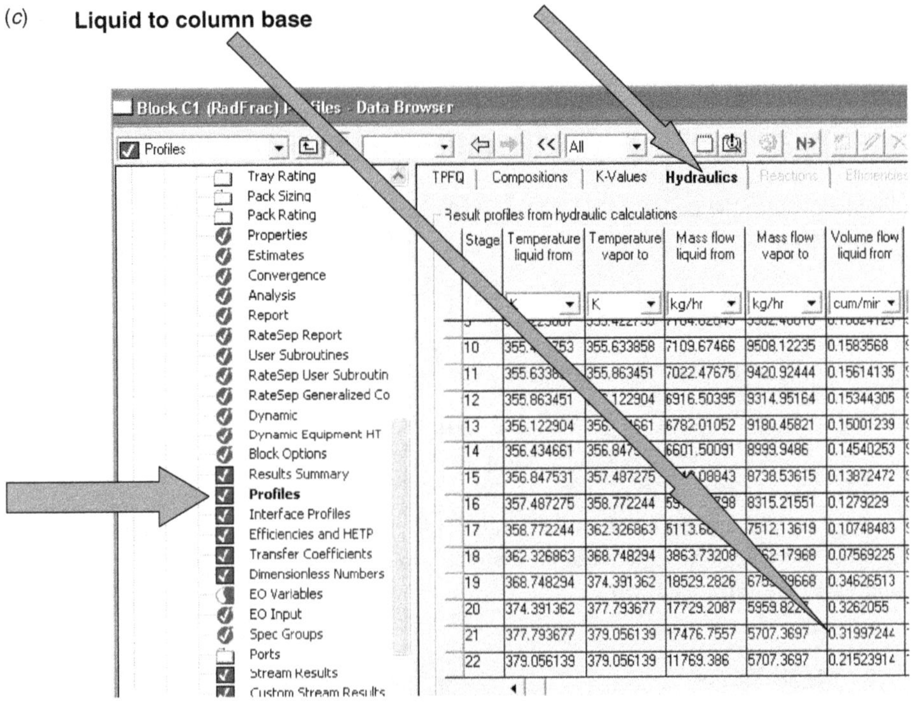


**Figure 4.2** *Continued.*

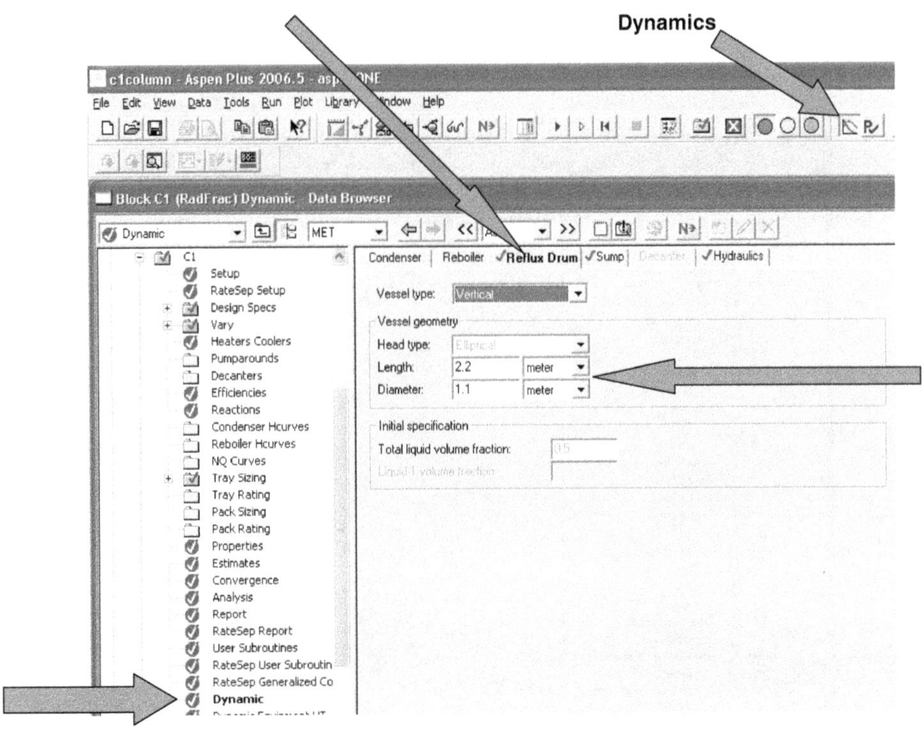


**Figure 4.3** Specifying *Reflux Drum* size.

column base. We do not use the liquid *leaving* the base because this is less than that entering since there is vapor boil-up. To have 10 min of total holdup when the sump is 100% full, the volume of the base must be:

$$(0.320\,\mathrm{m^3/min})(10\,\mathrm{min}) = 3.20\,\mathrm{m^3} \tag{4.3}$$

If the diameter of the column is $D_C$ (1.14 m for the numerical example), the total volume is:

$$\frac{\pi(D_C)^2}{4}L_B = \frac{\pi(1.14)^2}{4}L_B = 3.20\,\mathrm{m^3} \tag{4.4}$$

Thus the total height $L_B$ of the liquid surge volume in the base of the column is

$$L_B = \frac{4(3.20)}{\pi(1.14)^2} = 3.14\,\mathrm{m} \tag{4.5}$$

These values are entered, as shown in Figure 4.4, by selecting the *Sump* page tab.

Figure 4.5 shows how the column diameter, tray spacing, and weir height are entered, using the *Hydraulics* page tab.

### 4.1.2 Flash Tanks

The size of a flash tank can be set by one of two possible criteria. The first is to have a liquid holdup time of at least 5 min with the tank half full, based on the liquid leaving the tank. The

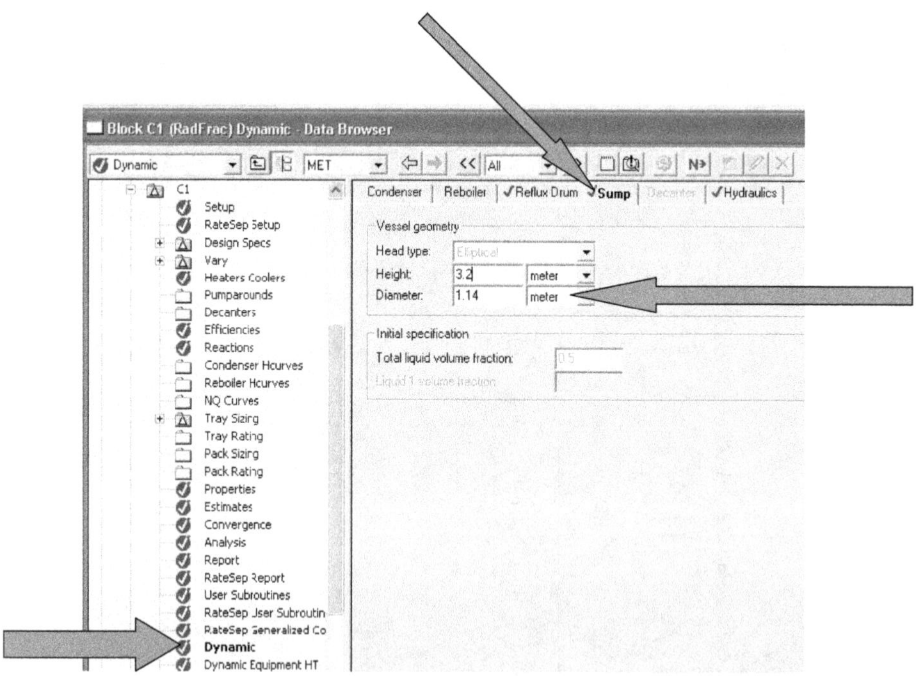


**Figure 4.4** Specifying column base size.

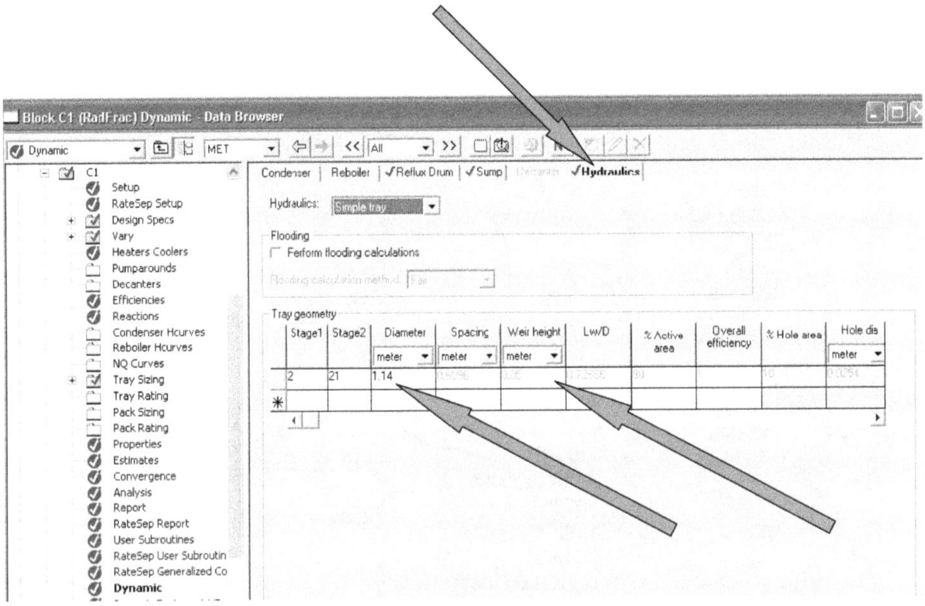


**Figure 4.5** Specifying column diameter and weir height.

second criterion is to have a vapor velocity up through the top of the tank that is below a maximum $V_{max}$ so that liquid entrainment is avoided. The maximum vapor velocity is calculated using an F-Factor of 0.5 (in English engineering units) and the density of the vapor $\rho_V$.

$$\text{F-Factor} \equiv V_{max}(\text{ft/sec})\sqrt{\rho_V(\text{lb/ft}^3)} = 0.5 \tag{4.6}$$

To illustrate the calculations, let us take a numerical example in which a mixture of hydrocarbons at high pressure is adiabatically flashed into a drum. Figure 4.6 gives the flowsheet with stream conditions. The liquid feed at 300 psia and 225°F is fed into a drum at 75 psia. The results of the adiabatic flash are a temperature of 132.6°F, a vapor flowrate of 4979 lb/h (vapor density $\rho_V = 0.7243$ lb/ft$^3$) and a liquid flowrate of 693 lb/h (liquid density $\rho_L = 34.83$ lb/ft$^3$).

First we calculate the size of the tank based on a liquid holdup time of 5 min when half full. An aspect ratio of 2 is used.

$$\begin{aligned}
(693\text{ lb liquid/h})\left(\frac{\text{ft}^3}{34.83\text{ lb}}\right)\left(\frac{\text{h}}{60\text{ min}}\right) &= 0.3316\text{ ft}^3/\text{min} \\
(0.3316\text{ ft}^3/\text{min})(10\text{ min}) &= 3.316\text{ ft}^3 \quad \text{total volume} \\
\left(\frac{\pi D^2}{4}\right)(L) = \left(\frac{\pi D^2}{4}\right)(2D) = \frac{\pi D^3}{2} &= 3.316\text{ ft}^3 \\
D &= 1.28\text{ ft}
\end{aligned} \tag{4.7}$$

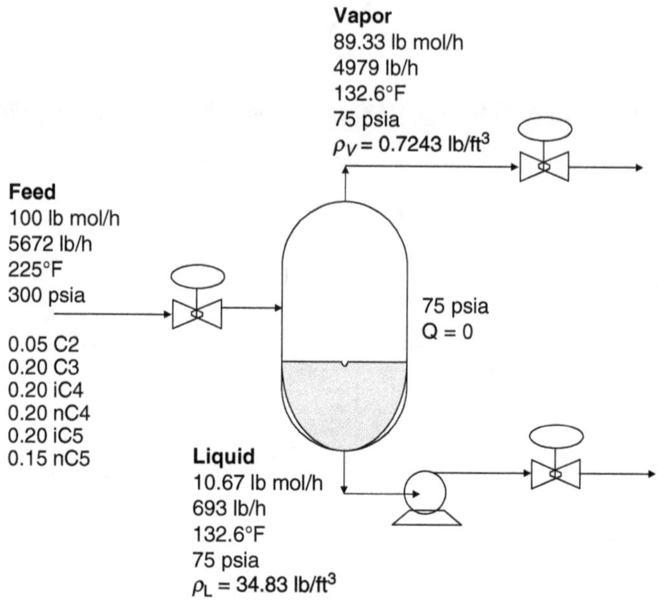


**Figure 4.6** Flash tank sizing.

Then we calculate the size of the tank based on the vapor flowrate and density.

$$\text{F-Factor} \equiv V_{\max}(\text{ft/s})\sqrt{\rho_V(\text{lb/ft}^3)} = V_{\max}\sqrt{0.7243} = 0.5$$

$$V_{\max} = 0.5875\ \text{ft/s}$$

$$(4979\ \text{lb vapor/h})\left(\frac{\text{ft}^3}{0.7243\ \text{lb}}\right)\left(\frac{\text{h}}{3600\ \text{s}}\right) = 1.910\ \text{ft}^3/\text{s}$$

$$\frac{1.910\ \text{ft}^3/\text{s}}{0.5875\ \text{ft/s}} = 3.251\ \text{ft}^2 \tag{4.8}$$

$$\left(\frac{\pi D^2}{4}\right) = 3.251\ \text{ft}^2$$

$$D = 2.03\ \text{ft}$$

Since the diameter calculated from the vapor velocity criterion is larger than that calculated from the liquid holdup criterion, the tank diameter of 2.03 ft is selected. The tank height is 4.06 ft.

### 4.1.3 Decanters

The conventional heuristic used for sizing a decanter is to specify at least 20 min of holdup time. This is larger than for simple surge drums because the small difference in liquid densities between the two liquid phases requires a large settling time. The inventory of both phases must be measured and controlled.

The conventional configuration is to measure the level of the light liquid in the decanter, which floats on top of the heavy liquid. This is a liquid–gas interface, so the density difference is large and a level sensor, such as differential-pressure or displacement transmitters, can be effectively used. The liquid–liquid interface between the light and heavy liquid phases must also be measured. The density difference between these liquid phases is much smaller than between liquid and gas, so this measurement is more difficult.

The conventional set-up, shown in Figure 4.7*a*, also has the problem of interaction between the two level control loops. A change in the flowrate of the light liquid leaving the upper part of the decanter only affects the top level measurement and does not affect the interface level. However, a change in the flowrate of the heavy liquid leaving the bottom of the decanter affects both the top level and the interface level. Figure 4.7*b* shows the conventional control structure with interaction between the two liquid level loops (top liquid level and interface level). Figure 4.7*c* shows a control structure in which the interaction is eliminated by using the difference between the two level signals as the process measurement input to the top level controller. The interface level $h_{hvy}$ is subtracted from the top level signal $h_{light}$, which gives a signal that depends only on the amount of light liquid in the decanter.

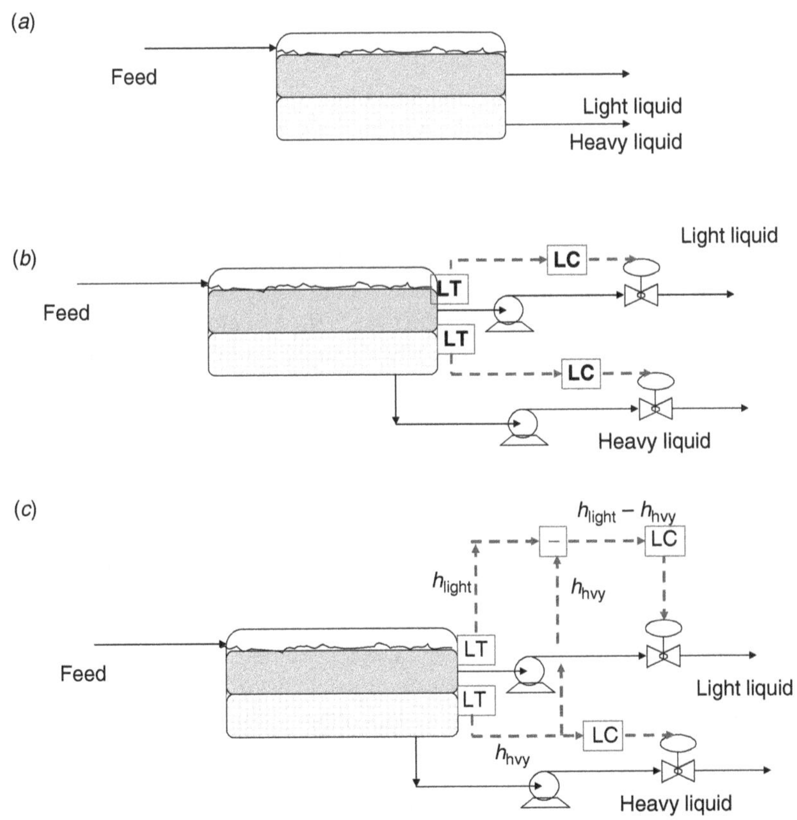


**Figure 4.7** Decanter design. (*a*) Conventional design. (*b*) Control structure with interaction. (*c*) Control structure without interaction.

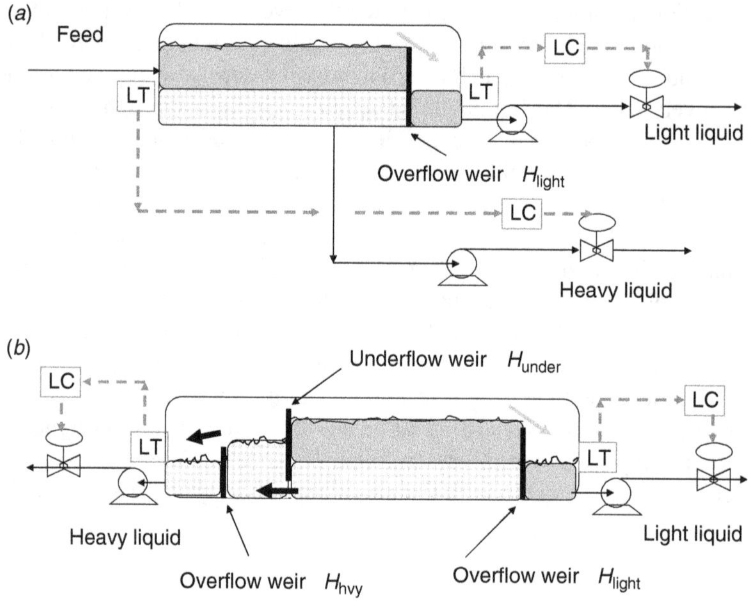


**Figure 4.8** Modified decanter design. (*a*) Overflow weir for light liquid. (*b*) Overflow and underflow weirs.

The interaction can also easily be eliminated by a simple process change, as shown in Figure 4.8*a*. An overflow weir is installed at one end of the decanter. The lighter liquid flows over this weir into a surge volume whose level can easily be measured and controlled. Measurement of the interface level is still required with this system.

An alternative modification to the design of the decanter can eliminate both interaction and interface measurement. As shown in Figure 4.8*b*, two overflow weirs and one underflow weir are installed in the decanter. The lighter liquid flows over weir $H_{light}$ into a surge volume. The heavier liquid first flows under weir $H_{under}$ before flowing over weir $H_{hvy}$ into its own surge volume. The heights of the two overflow weirs are set so that the interface level is approximately half way up the light overflow weir. This requirement is satisfied if:

$$\frac{H_{hvy}}{H_{light}} = \frac{\rho_{hvy} + \rho_{light}}{2\rho_{hvy}} \tag{4.9}$$

where the densities of the two liquid phases are $\rho_{hvy}$ and $\rho_{light}$. Notice that the two liquid phases have independent level measurements, which are based on the differences between liquid and vapor densities. No difficult interface measurement is required.

While we are on the subject of decanters, it is a good time to discuss how to simulate a decanter using Aspen software. There are two models that can be used. These are found under the page tab of *Separators* on the Aspen Plus window. One is labeled *Decanter*. The other is labeled *Flash3*. They have distinct differences that need to be understood.

The *Decanter* model assumes that only two liquid phases exist in the vessel. There is no vapor phase. Two specifications can be made, as shown in Figure 4.9. Typically these are pressure and temperature. The pressure does not change during dynamic simulations in Aspen Dynamics even though compositions and temperature change, which is physically

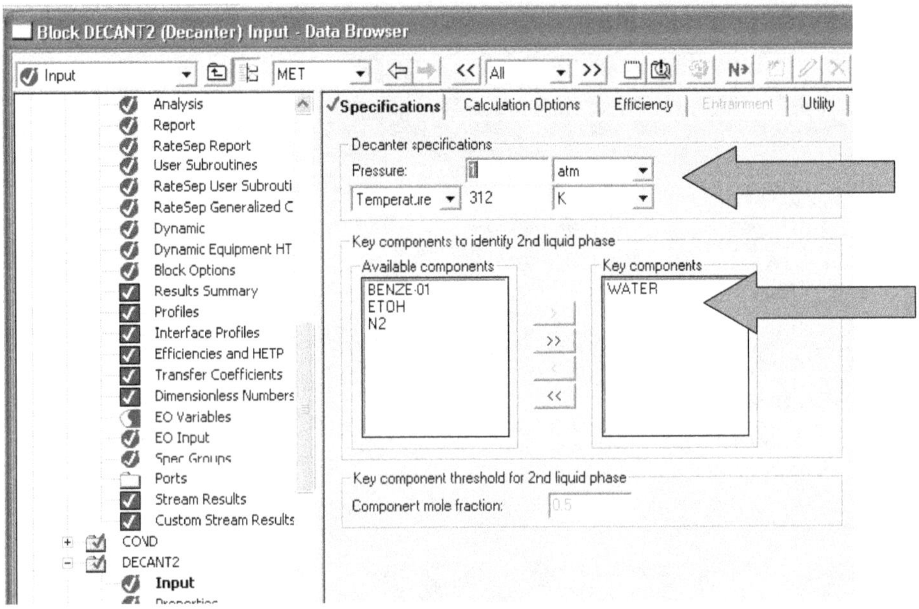


**Figure 4.9** Decanter specifications.

unrealistic. The liquid is not at its bubblepoint pressure. At the steady-state design stage in Aspen Plus, heat is transferred in the *Decanter* model to attain the specified temperature. This is also unrealistic since the decanter is usually just a horizontal drum with no internal cooling coils or external cooling jacket. The key component to identify the second liquid phase must also be selected on the same page tab, and water is usually selected (Fig. 4.9).

The *Flash3* model is more rigorous and simulates a decanter in a more realistic way. A vapor phase is present, and the pressure in the vessel changes during the dynamic simulation. Many decanters have an inert gas introduced into the vapor phase (a nitrogen supply or an atmospheric air vent) that affects the pressure. The *Flash3* model permits modeling this type of set-up. Figure 4.10 shows the windows that open for each of the two page tabs. If the heat transfer is set at zero, the temperature in the decanter will depend on the conditions in the upstream condenser. A condenser exit temperature is usually specified, and the required heat-removal rate in the condenser is calculated. A small stream of nitrogen can be added to the decanter to produce a small flow of vapor leaving the top of the decanter. Figure 4.11 gives an example of a two-column, decanter azeotropic process with an inert-gas stream fed to the decanter. Pressure is controlled by the exit gas flow. A frequently used alternative control structure is to use a vent–bleed system in which the valves on the inert-gas feed and the gas from the decanter are "split-ranged" so the inlet valve opens when pressure is low and the exit valve opens when the pressure is high.

The two different models also use different terminology when attaching the two exit streams (organic and aqueous).The *Decanter* model calls these *Level 1* and *Level 2*, and it is not obvious which corresponds to which liquid phase. The first is the organic phase. The second is the aqueous phase. The *Flash3* models used the notation *Interface* and *Level* for the aqueous and organic levels, which avoids the guesswork of trying to attach a stream to the correct level.

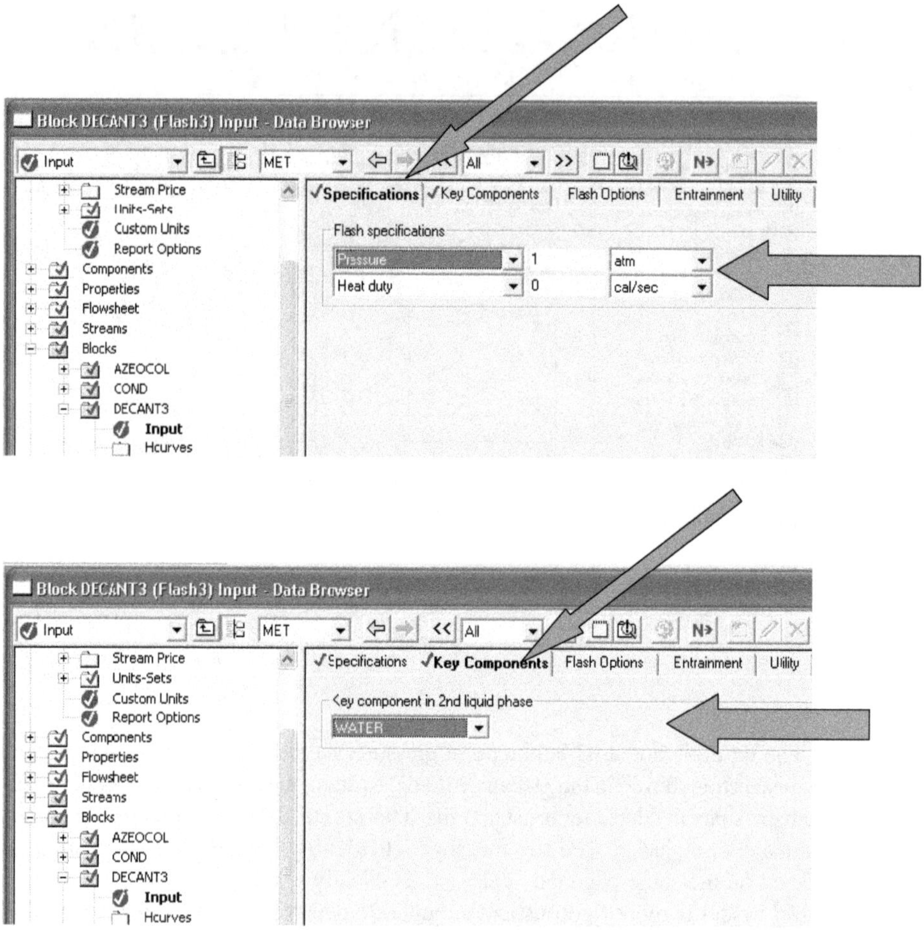


**Figure 4.10** *Flash3* specifications.

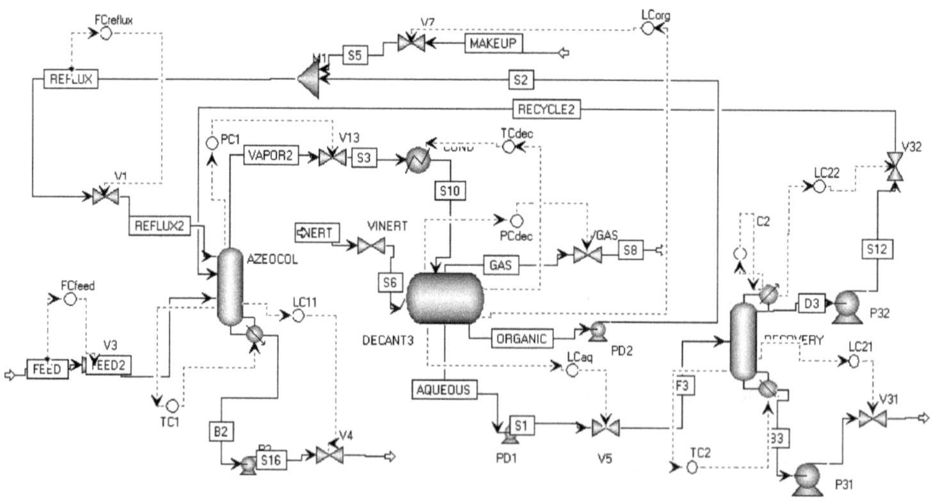


**Figure 4.11** Decanter using *Flash3* with inert gas.

## 4.2 SIZING OF PUMPS, COMPRESSORS, AND CONTROL VALVES

In a pressure-driven simulation, just like in a real-life process, fluids only flow from regions of high pressure into regions of low pressure. *Water does not flow uphill.* The pressure driving force is provided by pumps (if the fluid is liquid) or by compressors (if the fluid is gas). Of course, compression is much more energy intensive by several orders of magnitude than pumping, so the design of the process should avoid compressing if possible. If a vapor stream can be condensed and then pumped up to the required pressure, the resulting energy requirements are usually much lower than using a compressor for the same job.

Increasing the pressure of a stream may be required for a number of reasons.

1. To get the fluid to flow from a low-pressure vessel into a high-pressure vessel.
2. To overcome pressure drops in equipment such as heat exchangers.
3. To provide pressure drops through control valves.

The first item is self explanatory. The pressures are set by the operating pressures in the vessels. Reactors and distillation columns, for example, run at different operating pressures so as to optimize the performance of each unit.

The second item usually involves the engineering trade-off between heat exchanger area (capital investment) and energy consumption in pumps and compressors. The higher the pressure drop in a heat exchanger, the larger the heat transfer coefficient because the higher fluid velocity produces smaller boundary layers that enhance heat-transfer rates. For fixed values of heat duty and temperature differential driving force, the area decreases as pressure drop increases. However, this higher pressure drop requires higher head pumps and/or compressors, which requires more work. Typical design heuristics set reasonable pressure drops of about 20 to 30 psi through heat exchangers handling liquids. For heat exchangers handling gases, the design heuristic is smaller (about 5 to10 psi) because compression is more expensive than pumping. Keep in mind also that gas heat-transfer coefficients are much lower than liquid heat-transfer coefficients. As a result of both of these effects, gas heat exchangers are usually much larger and more expensive than liquid heat exchangers.

Since the flow through most heat exchanges is turbulent, the pressure drop over the heat exchanger (and any other equipment with fixed resistance) will vary with the *square* of the flowrate. If the flowrate through the unit is doubled, the pressure drop over the unit will increase by a factor of four. These large changes in pressure drop have a strong effect on how control valves, pumps, and compressors should be sized.

The third item (control valve pressure drop) is by far the most controversial and the most important from the standpoint of dynamic controllability. The selection of control valve pressure drop is another example of an important engineering trade-off. This time the trade-off is between pump or compressor costs and dynamic controllability.

A process can be designed with small pressure drops over control valves. This minimizes energy costs. The valve is typically designed to be 50% open at design conditions. If the fail-safe condition is a wide open valve, an air-to-close (AC) valve is specified. Examples are cooling water and refrigeration valves. If the fail-safe condition is a completely shut valve, an air-to-open (AO) valve is specified. Examples are steam, fuel, and reactor-feed valves.

However, if the design control valve pressure drop is small, the increase in flow resulting from opening the valve will be small. This translates as poor controllability because only

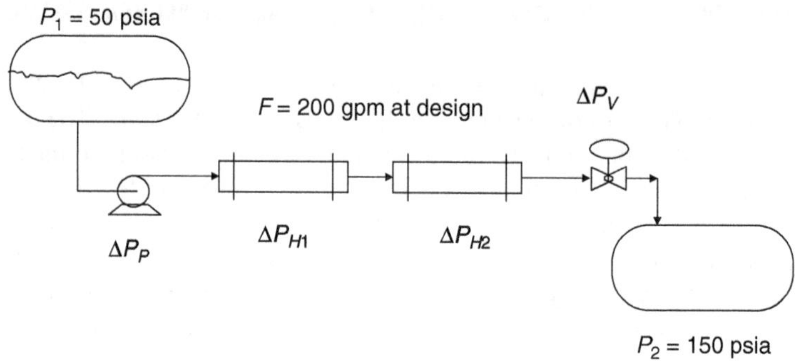


**Figure 4.12** Control valve sizing example pumping system.

small disturbances can be handled without the valve going wide open. On the other hand, if the design control valve pressure drop is large, opening the valve will produce a much larger increase in flow. So control valve pressure drop is a question of rangeability.

To illustrate this concept with a numerical example, consider the pumping system shown in Figure 4.12. Liquid is pumped from at tank at 50 psia through two heat exchangers and a control valve into a vessel at 150 psia. At the design flowrate of 200 gpm, the pressure drop through each of the heat exchangers is 20 psi. Obviously the pump must have a high enough head to overcome the static pressure difference between the vessels of 100 psi and the heat exchanger pressure drop of 40 psi. The pump must also put up enough pressure to overcome the pressure drop through the control valve.

The designer is free to select any pressure drop through the valve. Let us take two cases. In the first, a valve pressure drop of 20 psi at design conditions is specified. In the second, a valve pressure drop of 50 psi is used. Valve opening at design conditions is 50%. A flat pump curve is assumed, that is, the pump head is constant and does not change with flowrate.

*Case 1*: *$\Delta P_V = 20$ psi at design 200 gpm.* The required pump head is the sum of all the pressure drops plus the difference in pressures between the two vessels.

$$\Delta P_P = \Delta P_{H1} + \Delta P_{H2} + \Delta P_V + P_2 - P_1 = 20 + 20 + 20 + 150 - 50 = 160 \text{ psi}$$

The $C_V$ value of the control valve is calculated using Eq. (4.10).

$$F = f_{(x)} C_V \sqrt{\Delta P_V} \tag{4.10}$$

where $F$ is the flowrate and $f_{(x)}$ is the fraction that the valve is open. When the valve is wide open $f_{(x)} = 1$. The valve at design conditions is assumed to be half open, $f_{(x)} =$ 0.5. At design conditions, Eq. (4.10) can be used to size the valve.

$$\begin{aligned} F &= f_{(x)} C_V \sqrt{\Delta P_V} \\ 200 \text{ gpm} &= (0.5) C_V \sqrt{20} \\ C_V &= 89.44 \end{aligned} \tag{4.11}$$

Now let us calculate the flow through the system if the control valve is wide open ($f_{(x)} = 1$). This maximum flowrate is $F_{max}$.

The pump head is constant at 160 psi. The difference in static pressures between the vessels is constant at 100 psi. However, the pressure drop over the heat exchangers is not constant but varies with the square of the flowrate.

$$\Delta P_{H1} = \Delta P_{H2} = 20\left(\frac{F_{max}}{200}\right)^2 \tag{4.12}$$

Using Eqs. (4.10) and (4.12) gives:

$$\begin{gathered} \Delta P_{H1} + \Delta P_{H2} + \Delta P_V + 100 = \Delta P_P = 160 \\ 20\left(\frac{F_{max}}{200}\right)^2 + 20\left(\frac{F_{max}}{200}\right)^2 + \left(\frac{(F_{max})}{(1)(89.44)}\right)^2 + 100 = 160 \\ F_{max} = 231 \text{ gpm} \end{gathered} \tag{4.13}$$

*Case 2*: *$\Delta P_V = 50$ psi at design 200 gpm.* In this case the pump head is higher because of the larger valve pressure drop.

$$\begin{aligned} \Delta P_P &= \Delta P_{H1} + \Delta P_{H2} + \Delta P_V + P_2 - P_1 = 20 + 20 + 50 + 150 - 50 \\ &= 190 \text{ psi} \end{aligned} \tag{4.14}$$

The higher head pump will require a higher horsepower motor or turbine and consume more energy. The control valve will be smaller.

$$\begin{gathered} F = f_{(x)} C_V \sqrt{\Delta P_V} \\ 200 \text{ gpm} = (0.5) C_V \sqrt{50} \\ C_V = 56.57 \end{gathered} \tag{4.15}$$

Solving for the new maximum flowrate when the valve is wide open gives:

$$\begin{gathered} \Delta P_{H1} + \Delta P_{H2} + \Delta P_V + 100 = \Delta P_P = 190 \\ 20\left(\frac{F_{max}}{200}\right)^2 + 20\left(\frac{F_{max}}{200}\right)^2 + \left(\frac{(F_{max})}{(1)(56.57)}\right)^2 + 100 = 190 \\ F_{max} = 262 \text{ gpm} \end{gathered} \tag{4.16}$$

The two cases give different maximum flowrates. The higher the design pressure drop over the valve, the larger the maximum flowrate. This example illustrates that a control valve designed for a large pressure drop can provide larger increases in flowrates and thus avoid valve saturation and improve dynamic controllability and rangeability.

It is clear that a conflict exists between steady-state energy consumption and controllability. The question that needs answering is what is a reasonable valve pressure drop? The answer depends on the application. If only modest changes in flowrates are required, such as the flow of a feed stream to a unit, a fairly small control-valve pressure drop is adequate. The heuristic is to make the valve pressure drop equal to half the other pressure drops in the system. However, if large changes in flow are required, such as cooling water to the jacket of an exothermic chemical reactor, very large pressure drops are required. In some situations it may be more economical to use a variable speed motor or turbine to drive the pump instead of having a very high-pressure drop valve. Typical valve pressure drops are 2 to 4 atm.

If valve saturation occurs during a dynamic simulation in Aspen Dynamics, you can change the control valve size ($C_V$) or change the power to the pump to increase the pump head. Changing control valve size is achieved by clicking on the valve icon in the Aspen Dynamics process flow diagram (PFD), right clicking, selecting *Forms*, *All Variables*, and modifying the value of *COmax* (see Fig. 4.13). Changing the pump head is achieved by clicking on the pump icon, right clicking, selecting *Forms*, *Configure*, changing *Use performance curves* from *True* to *False*, and increasing the specified electrical power *EpowerR* (see Fig. 4.14).

Some mention should be made of valve characteristics and valve action. Real control valves have different characteristics (linear or equal-percentage trim) and can be AO or AC. In Aspen Plus simulations the default case is to use an AO valve with linear valve characteristics. In most cases this is adequate for dynamic simulations since flow controllers are used on many manipulated variables and the type of valve trim is not important. The action of the valve (AO or AC) can be changed once the file is exported to Aspen Dynamics if more reality is desirable in terms of the combined actions of the valve and the controller. For example, the cooling water valve on a reactor would be AC so that it fails wide open. The temperature controller would open the valve for an increase in reactor temperature. So the action of the controller would be "reverse" acting (*increasing* controller input temperature signal PV *decreases* the controller output signal OP to the valve, which increases water flowrate). If an AO valve had been used, the controller would be "direct" acting.

In a real plant and in a pressure-driven dynamic simulation, plumbing is important for stable operation of process. A few useful rules should be remembered when setting up the locations of pumps, compressors, and valves.

1. Never install two control valves in series in a single liquid-filled line because liquids are essentially incompressible and flow through the line should be determined by one

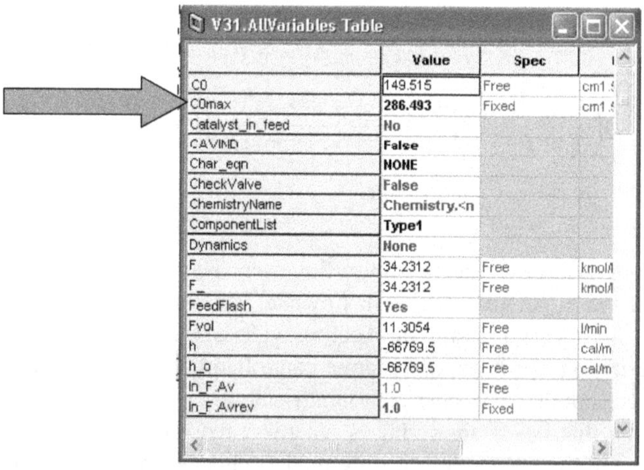


**Figure 4.13** Changing control valve size.

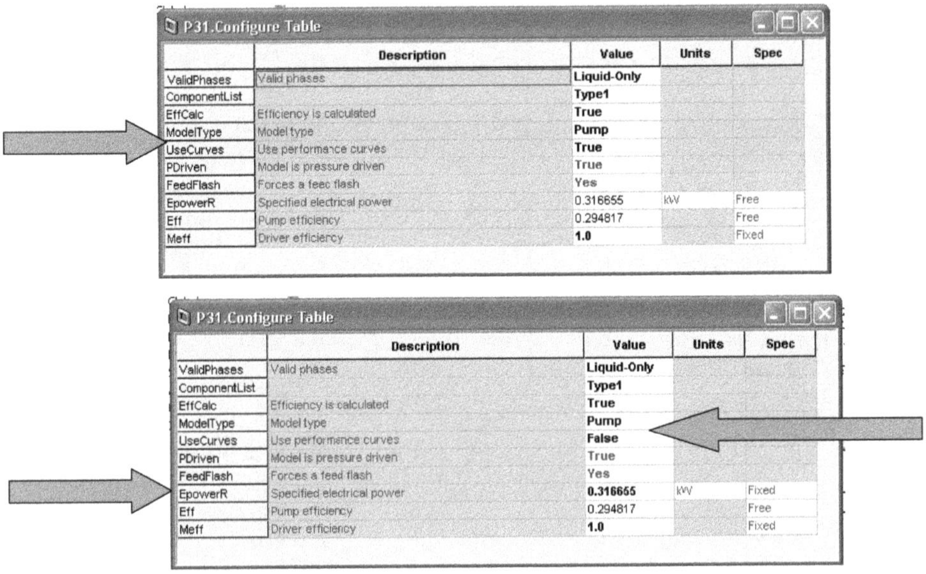


| | Description | Value | Units | Spec |
|---|---|---|---|---|
| ValidPhases | Valid phases | Liquid-Only | | |
| ComponentList | | Type1 | | |
| EffCalc | Efficiency is calculated | True | | |
| ModelType | Model type | Pump | | |
| UseCurves | Use performance curves | True | | |
| PDriven | Model is pressure driven | True | | |
| FeedFlash | Forces a feed flash | Yes | | |
| EpowerR | Specified electrical power | 0.316655 | kW | Free |
| Eff | Pump efficiency | 0.294817 | | Free |
| Meff | Driver efficiency | 1.0 | | Fixed |



| | Description | Value | Units | Spec |
|---|---|---|---|---|
| ValidPhases | Valid phases | Liquid-Only | | |
| ComponentList | | Type1 | | |
| EffCalc | Efficiency is calculated | True | | |
| ModelType | Model type | Pump | | |
| UseCurves | Use performance curves | False | | |
| PDriven | Model is pressure driven | True | | |
| FeedFlash | Forces a feed flash | Yes | | |
| EpowerR | Specified electrical power | 0.316655 | kW | Fixed |
| Eff | Pump efficiency | 0.294817 | | Free |
| Meff | Driver efficiency | 1.0 | | Fixed |

**Figure 4.14** Changing pump power.

valve. When the piping has parallel liquid-filled lines, which either join together into one line or one line that splits into parallel lines, one valve can be installed in each parallel line.

2. Place valves *after* pumps to avoid net-positive-suction-head (NPSH) problems.
3. Control compressors using by-passing (recycling some flow from the discharge back to the suction), suction throttling (put a valve in the compressor suction, which changes the gas density and affects the mass throughput), or changing the rotational speed of the turbine or motor drive.

A convenient way to achieve the variable speed option in a dynamic simulation is to manipulate the work to the compressor or pump.

## 4.3 CONTROLLERS AND DYNAMIC ELEMENTS

Once the file has been exported into Aspen Dynamics, controllers are installed to achieve the desired control structure and dynamic simulations are run to check the stability and performance of the control system. Various types of disturbances should be imposed on the system, such as throughput changes, feed composition changes, and changes in the set-points of the product-quality controllers (temperature and composition controllers). In this section we demonstrate, in a detailed step-by-step fashion, how these operations are performed in Aspen Dynamics.

We use a simple process as a numerical example to illustrate moving from a steady-state simulation in Aspen Plus to a dynamic simulation in Aspen Dynamics. Figure 4.15 shows the flowsheet and the control structure. The flash drum is the same as the one sized in Section 4.1.2. It is a vertical vessel 2 ft in diameter and 4 ft in height. Figure 4.16 shows the Aspen Plus process flow diagram.

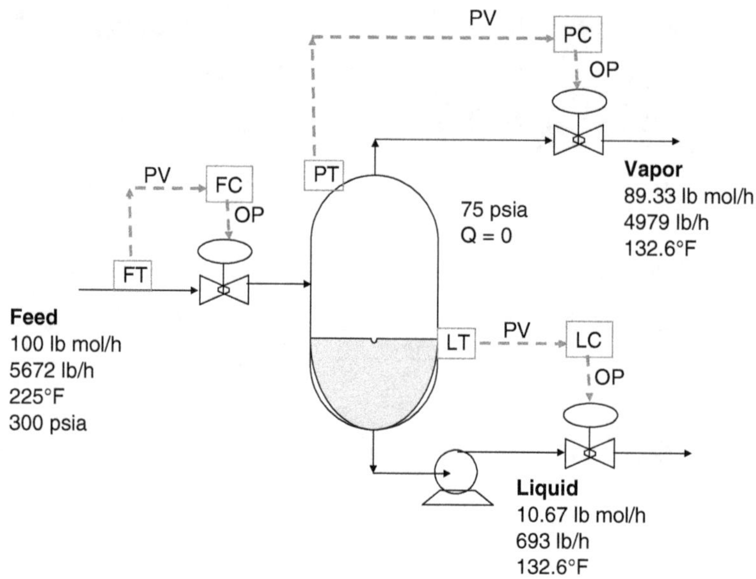


**Figure 4.15** Flash tank control.

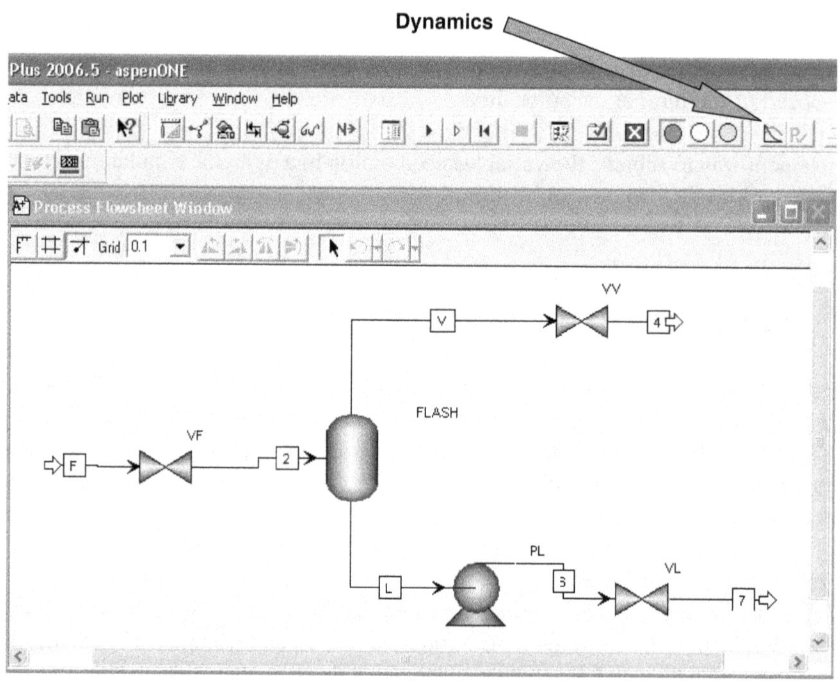


**Figure 4.16** Flash tank PFD in Aspen Plus.

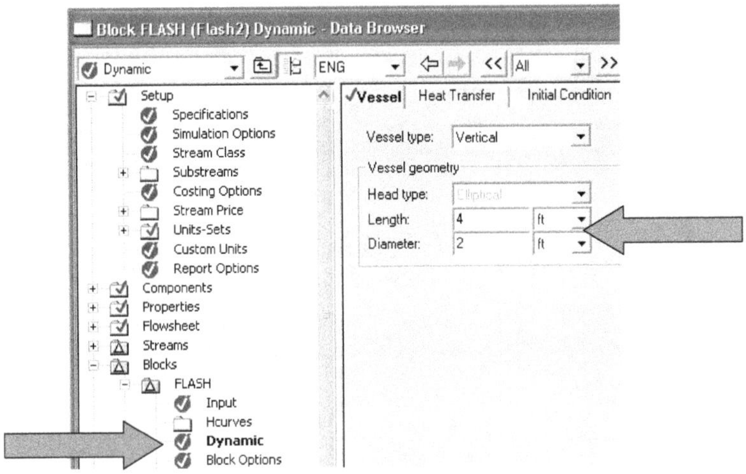


Figure 4.17 Specifying Flash tank size.

### 4.3.1 Equipment Sizing

The first thing to do in preparation for exporting into Aspen Dynamics is to click the "dynamics" button on the top toolbar, as shown in Figure 4.16. Then the *Dynamic* item under the block for the vessel (called *Flash*) is clicked, and the window shown in Figure 4.17 opens. This is where the size of the vessel is specified so that the dynamics of the process can be modeled. The *Vessel type* is selected to be *Vertical*, and the values of *Length* and *Diameter* are entered.

Next the pump head is specified to be 200 psi, as shown in Figure 4.18. The pressure drops over the two valves in the liquid and gas lines must be set. Figure 4.19 shows that a 50 psi pressure drop over the liquid valve is entered. Notice that the *Valid phases* is specified to be *Liquid-Only*. The gas valve is set up in a similar manner, using *Vapor-Only* and a pressure drop of 50 psi, which means the gas is flowing into a line at 25 psia since the

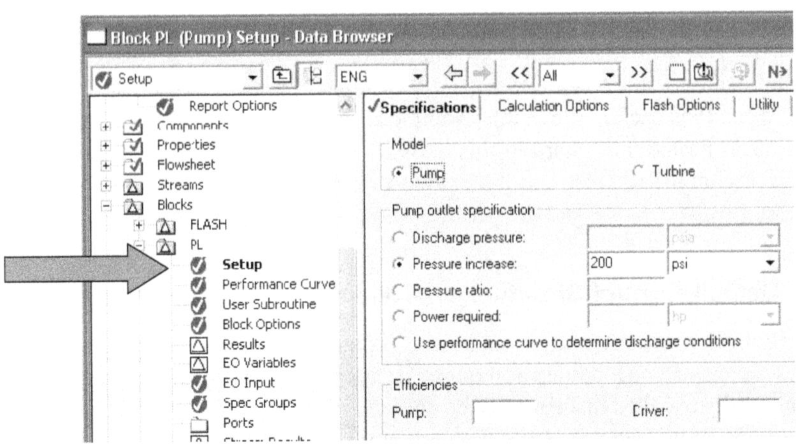


Figure 4.18 Specifying pump head.

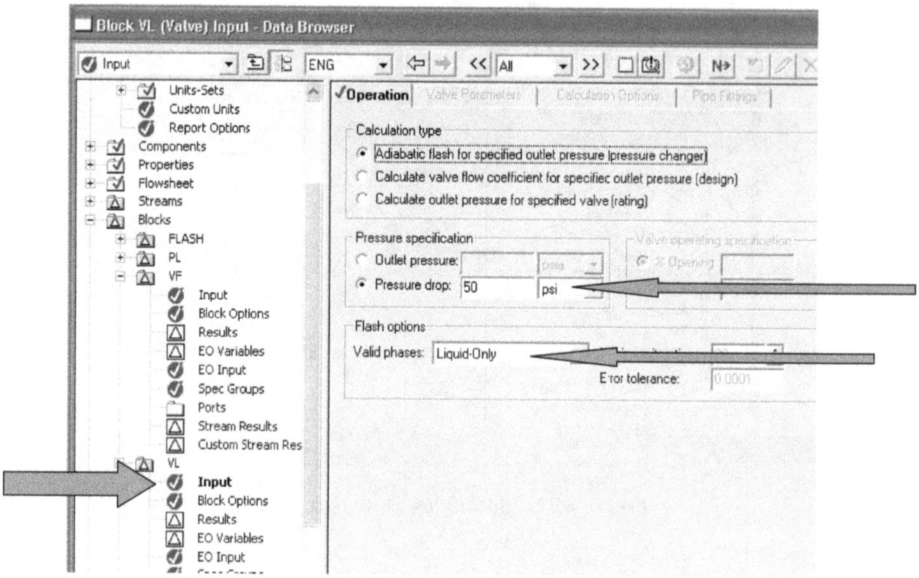


**Figure 4.19** Specifying valve pressure drop.

drum is at 75 psia. The valve on the feed line is set up by specifying that the *Outlet pressure* is 75 psia, and the *Valid phases* are *Vapor-Liquid* since there is flashing.

### 4.3.2 Pressure Checking and Exporting

After running the simulation (clicking the blue *N* button), the flowsheet is pressure checked to see if the plumbing has been correctly set up. This is done by clicking the "pressure check" button on the top toolbar, shown in Figure 4.20. If everything is in good shape, the message shown at the bottom of Figure 4.20 appears.

We are now ready to export the file. Go to *File* and select *Export*. The window shown in Figure 4.21 appears on which a *P Driven Dyn Simulation (*.dynf,*dyn.appdf)* is selected and the *Save* button is clicked. A message window appears, as shown in Figure 4.22, with some information about the need to install a liquid level controller and the use of a performance curve in the pump. No fatal errors are reported, so the export has been successful.

The Aspen Plus backup file in this example is called "flash.bkp" so the Aspen Dynamics file is called "flash.dynf." Starting this ".dynf" file in Aspen Dynamics opens the screen shown in Figure 4.23. There are three windows: *Process Flowsheet Window, Exploring,* and *Simulation Messages*. The flowsheet is shown with the vessel, valves, and pump.

### 4.3.3 Default Control Structure and Simplified Heat Transfer Models

Aspen Dynamics automatically installs some control loops on the flowsheet. This default control structure can be modified as required. In our flash-tank example, a pressure controller is automatically installed that controls the pressure in the vessel by manipulating the valve in the vapor line. If we had not placed a pump on the liquid line and only had a valve, Aspen Dynamics would automatically install a level controller that controls the liquid level in the tank by manipulating the valve in the liquid line.

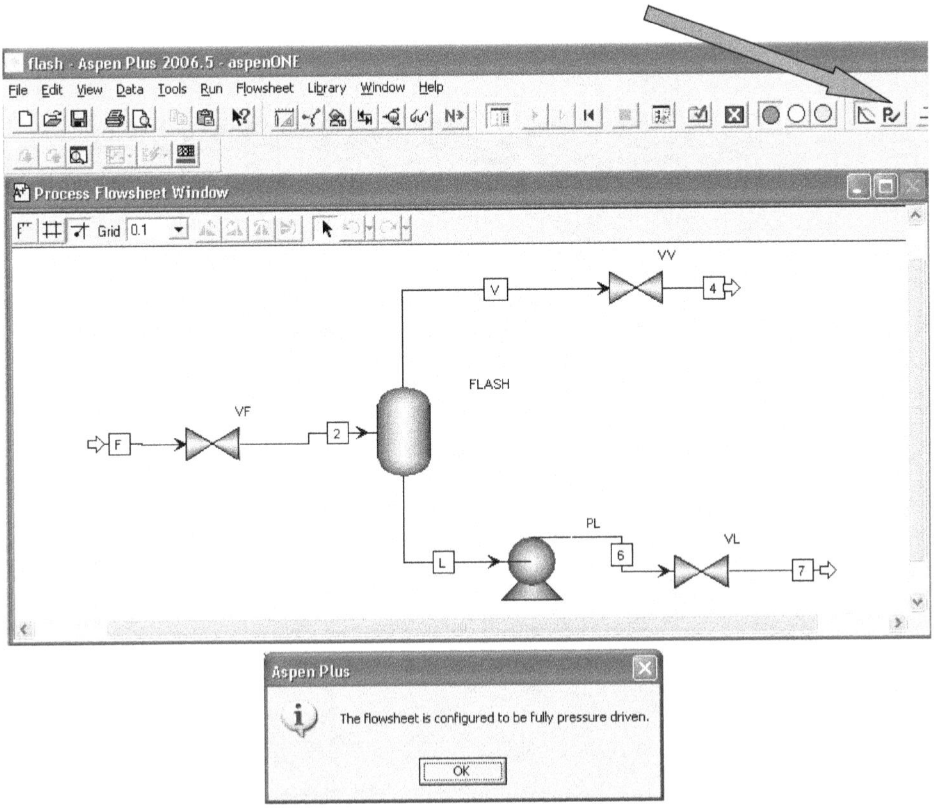


**Figure 4.20** Pressure check.

For a conventional distillation column with a total condenser, a pressure controller is automatically installed that controls the pressure in the condenser by manipulating heat removal from the condenser. This is an example of the situation where the simulation does not correspond precisely to reality. In terms of physical equipment, condenser heat removal is determined by the flowrate of cooling water to the condenser. So a rigorous simulation would manipulate a valve in the cooling water line and the heat-transfer relationships in the heat exchanger would be solved to get the heat removal for given process conditions and cooling water flowrate. Aspen Dynamics simplifies the model by specifying heat transfer directly.

A similar simplification is made for the heat input in the reboiler. If steam is the heating medium used, reboiler heat input is determined by position of a valve in the steam line feeding steam into the reboiler. Opening the steam valve puts more steam into the shell or tube side of the heat exchanger, which increases the steam pressure and increases the steam temperature, thus increasing heat transfer. Aspen Dynamics simplifies the model by specifying heat transfer directly.

### 4.3.4 Installing Controllers

Three controllers are required in the control structure shown in Figure 4.15 to control level, pressure, and feed flowrate. To place a controller on the flowsheet, go to the *Exploring Simulation* window and click on *Libraries, Dynamics*, and *ControlModels*. A list of dynamic

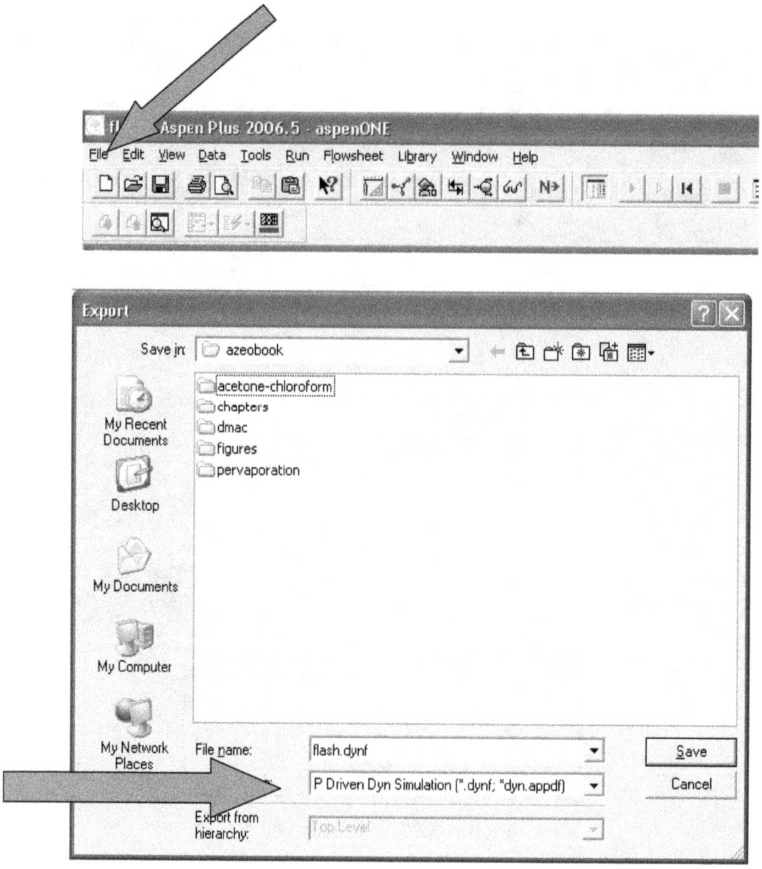


Figure 4.21 Export file.

elements is shown (see Fig. 4.24). These items are placed on the flowsheet by dragging and dropping.

A conventional proportional–integral–derivative (PID) controller is item *PIDIncr* on this list. To place a controller on the flowsheet, left click *PIDIncr* and, holding down the mouse button, drag it to the flowsheet. The little circle icon called *B*1 in Figure 4.24 appears

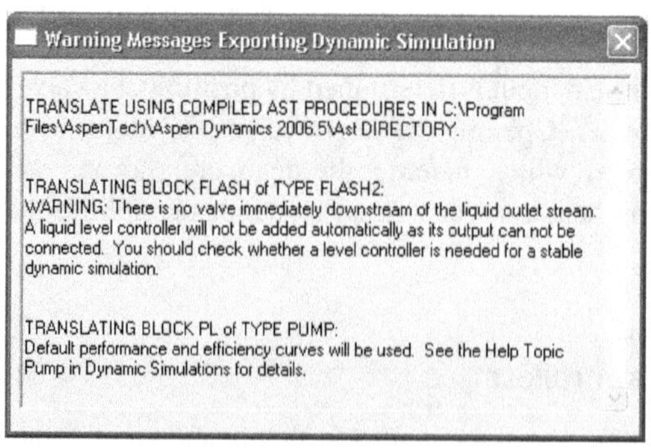


Figure 4.22 Export message.

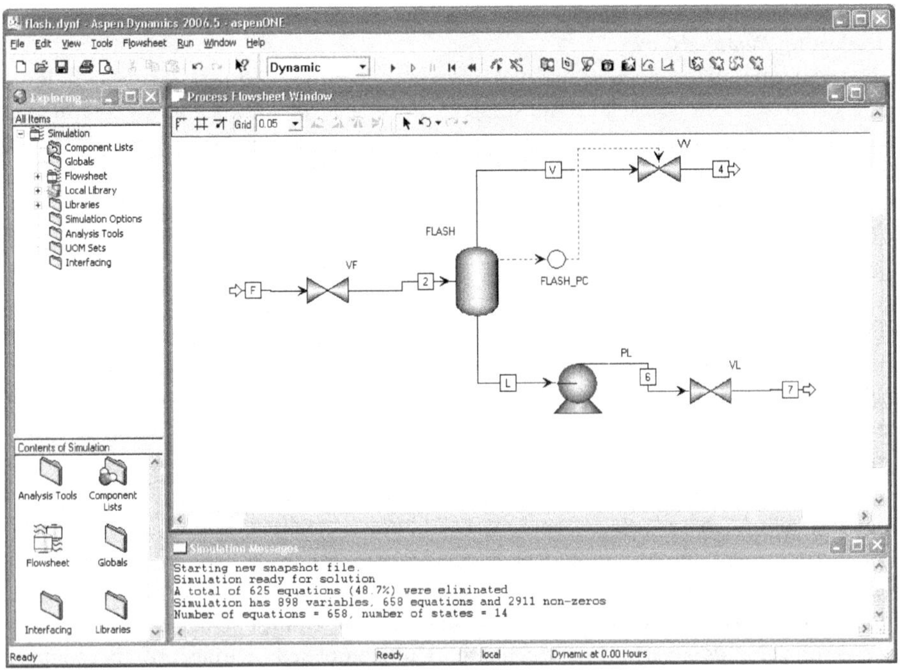


**Figure 4.23** Initial Aspen Dynamics window.

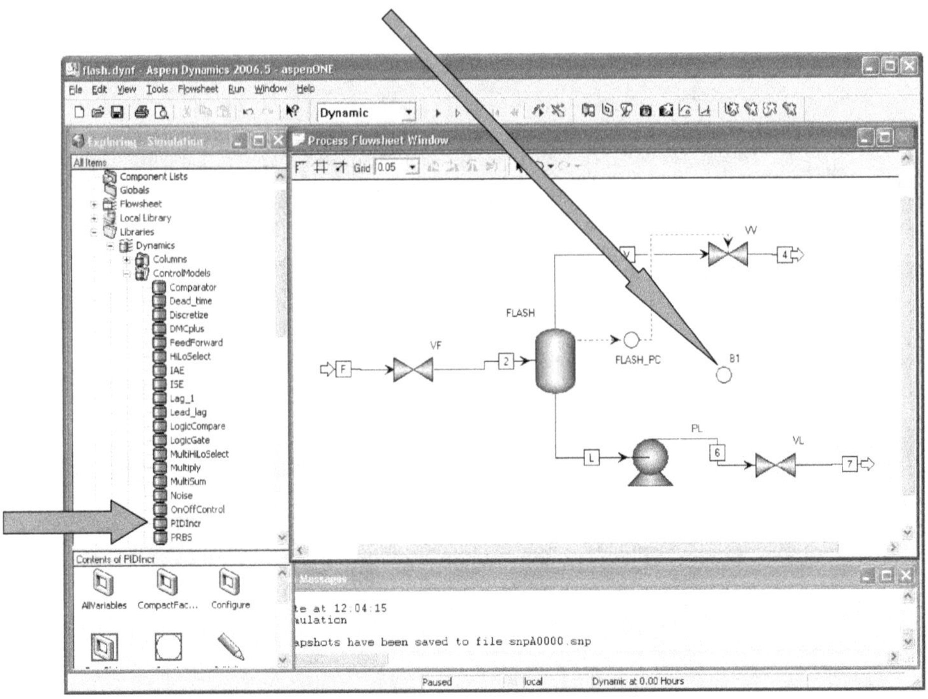


**Figure 4.24** Installing controller.

on the flowsheet. It can be renamed by left clicking on the icon to highlight it, right clicking and selecting *RenameBlock*.

The next job is to install control signals into and out of the controller. If the liquid level in the tank is to be controlled, a level transmitter senses the level and generates a control signal. Typically this is an electronic 4 to 20 milliamp signal that is fed into the controller. In Aspen Dynamics, the process variable signal to be controlled is called "PV." The output signal from the controller is called "OP." The setpoint signal to the controller is called "SP."

To insert these signals, select the *Stream Types* item under *Dynamics* as shown in Figure 4.25. Left click on *ControlSignal*, and holding down the left mouse button, drag the signal over to the flowsheet. By continuing to hold down the mouse button, a number of blue arrows will appear on the flowsheet. These are possible locations where the control signal can be attached. To control the liquid level in the tank, we want to select *Level* in the block called *Flash*. So the blue arrow coming out of the right-hand side of the vessel is clicked, which opens the window shown at the top of Figure 4.26. The item called *Blocks ("Flash"). Level* is selected and the *OK* button is clicked. This attaches the control signal to the tank. Then click the arrow pointing into the level controller. A window opens on which the item *LC.PV* is selected and the *OK* button is again clicked. The flowsheet now shows a control signal line running from the tank into the controller (see Fig. 4.25). A similar procedure is used to connect the controller OP to the control valve. The bottom window in Figure 4.26 shows that the item *LC.OP* is selected. The control signal is then attached to the control valve downstream of the pump. This completes "wiring up" the controller. Next, just as would be required in a real plant installation, we need to specify the action of controller, the range of the transmitter, the range of the output, and the tuning parameters of the controller (gain $K_C$ and integral time $\tau_I$ in minutes).

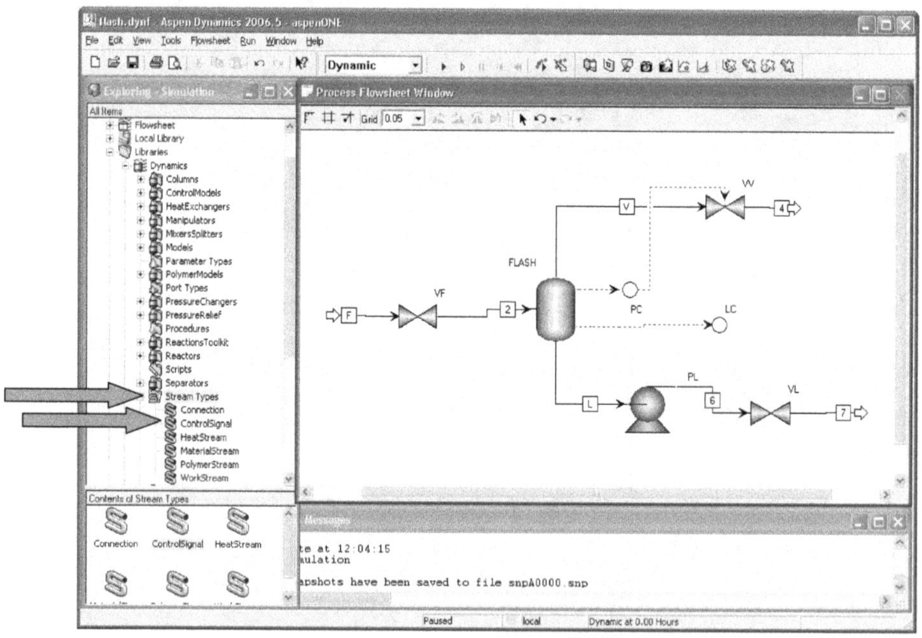


**Figure 4.25** Attaching control signals.

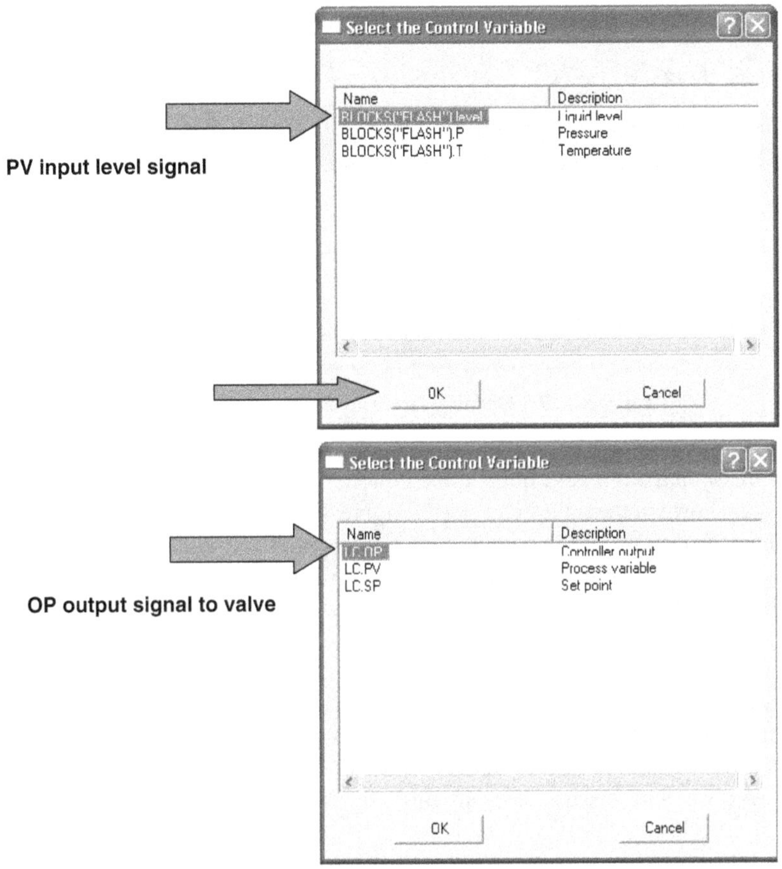


**Figure 4.26** Selecting PV and OP.

### 4.3.5 Controller Faceplates and Parameters

Double left clicking on the *LC* icon on the flowsheet opens a controller faceplate, as shown in Figure 4.27. The three important variables (*PV*, *OP*, and *SP*) are shown as both horizontal bar graphs and numerical values in the lower part of the faceplate. The toolbar at the top contains a number of items that permit the user to specify controller parameters, switch back and forth between manual and automatic, tune the controller, and generate a plot to see the dynamic responses of variables.

Selecting the left-most button places the controller on automatic. In this mode, a value for the SP of the controller can be entered in the box on the right by highlighting it, typing in the desired value, and pressing the *Enter* key. The second button from the left places the controller on manual. In this mode, a value for the OP signal can be entered to fix the controller output. The third button places the controller into cascade. In this mode, the controller receives its setpoint signal from some other source, such as the OP signal from another controller in a cascade control structure.

The first thing to do, after the controller is installed and the faceplate opened, is to click the third button from the right. This opens the window shown on the left-hand side of Figure 4.28. Notice that the values for PV, OP, and SP are meaningless at this point.

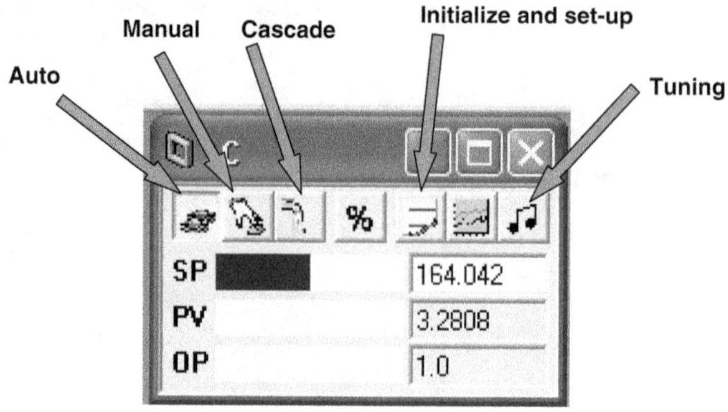


**Figure 4.27** Opening faceplate and initializing values.

Clicking the *Initialize Values* button at the bottom of the window brings in values of these variables, as shown in the right-hand side of Figure 4.28. Notice that the level is initialized at 2.5 ft. Since the drum is 4 ft in height, it should be initialized at 2 ft. You need to be aware that Aspen Dynamics sometimes give quirky initializations. In this case it does not matter that much. In other cases, particularly when initializing heat duties, the numbers can be way off. If this occurs, change the numbers to the correct initial values (and the correct ranges, as discuss below).

The action (direct or reverse) of the controller must be specified. The level controller should open the valve when the level increases, so direct action is selected. The controller tuning constants are initialized as default values of *Gain* ($K_C = 1$) and *Integral time* ($\tau_I = 20$ min). As we will discuss later in this chapter, most level controllers should use

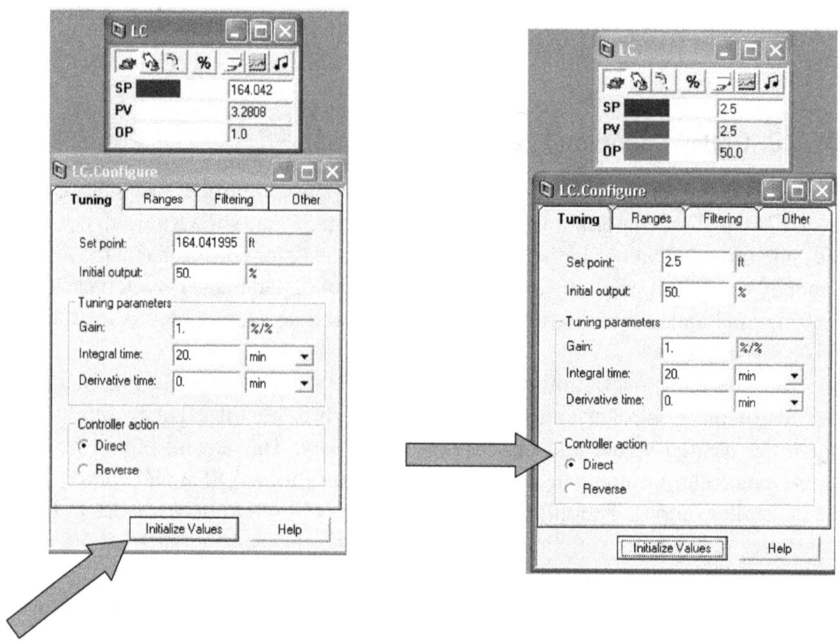

**Figure 4.28** Opening faceplate and initializing values.

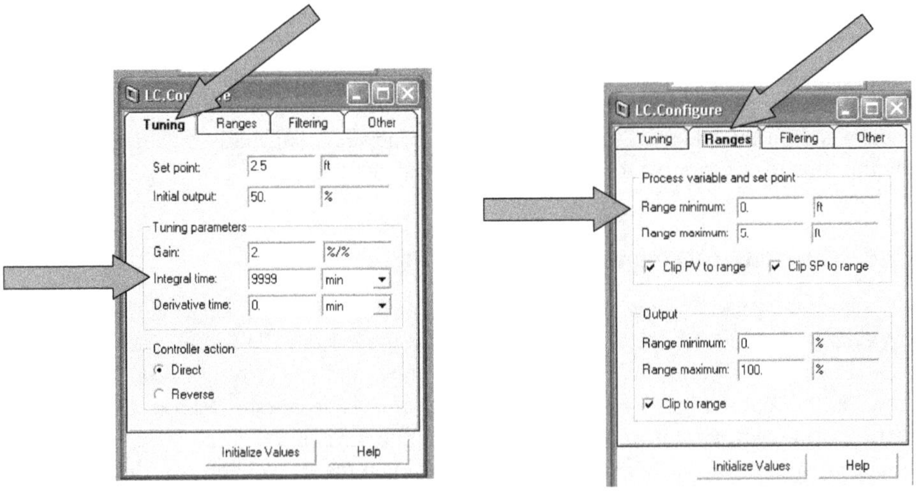


**Figure 4.29** Setting controller tuning constants and transmitter and valve ranges.

proportional-only control with a gain of 2. So the controller gain is changed to 2 and the integral time is changed to 9999 min, as shown on the left-hand side of Figure 4.29 where the *Tuning* page tab is open. Notice that the gain is dimensionless with units of %/%, where the % is percent of full scale. The gain depends on the ranges of the input and output variables.

Clicking the *Ranges* page tab opens the window shown on the right-hand side of Figure 4.29. The range of the level transmitter has been automatically set to twice the steady-state value, so the initial level (2.5 ft in the tank example) is at 50% of the maximum. The range of the process variable transmitter is from zero to 5 ft. Since the controller output is going to a valve, the range of the output is from zero to 100%, and the valve is half open at the initial steady state.

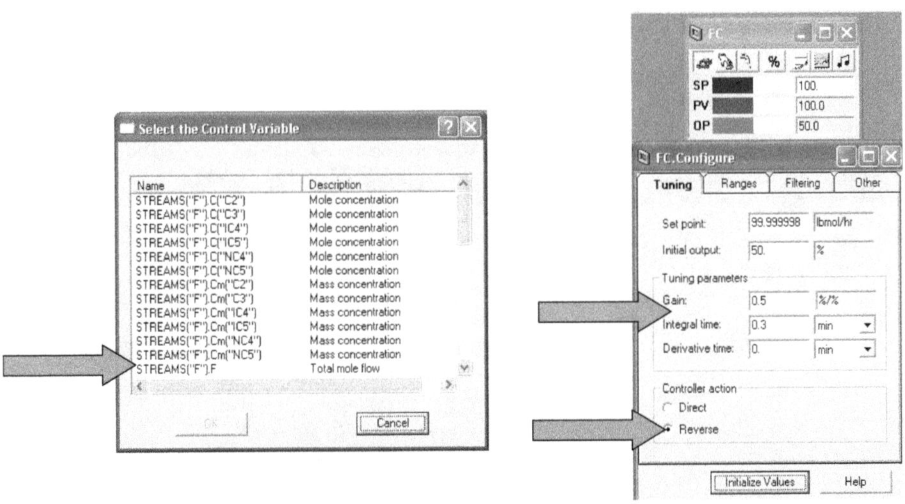


**Figure 4.30** Setting up flow controller.

The last controller to be set up is the flow controller on the feed. The left picture in Figure 4.30 shows how the molar flowrate is selected as the PV input signal. The OP signal goes to the valve. The right picture in Figure 4.30 shows that the action of the flow controller must be *Reverse* because an increase in flow should close the valve. As we will discuss later in this chapter, the tuning of flow controllers is usually a gain of 0.5 and an integral time of 0.3 min, so these values have been entered on the *Tuning* page tab.

### 4.3.6 Generating Dynamic Strip-Chart Plots

Strip charts, which show how important variables change with time, are needed to see what is happening when disturbances are introduced into the system. To install a strip chart plot, click on the *Tools* button at the top of the window, and select *New Form*. The window shown at the top of Figure 4.31 opens on which *Plot* is selected. Clicking the *OK* button opens a strip chart, as shown at the bottom of Figure 4.31 on the left.

The procedure for placing variables to be plotted on the strip chart is to "drag and drop." In Figure 4.31 the *Results Table* for the feed has been opened by double left clicking on the stream icon *F* on the flowsheet. The top line in this table is the total molar flowrate of the feed. Place the cursor at the left-hand side of this row in the table, click the left mouse button, then press the left mouse button again and hold it down as you drag it to the plot window. Releasing the mouse button places this variable on the plot.

Figure 4.32 illustrates how the various features of the strip chart can be set up. Place the cursor on the plot, right click, and select *Properties*. A number of page tabs can be selected to select colors, label variables, specify the numerical values of ordinate axis for each variable, set the time abscissa axis, and so on.

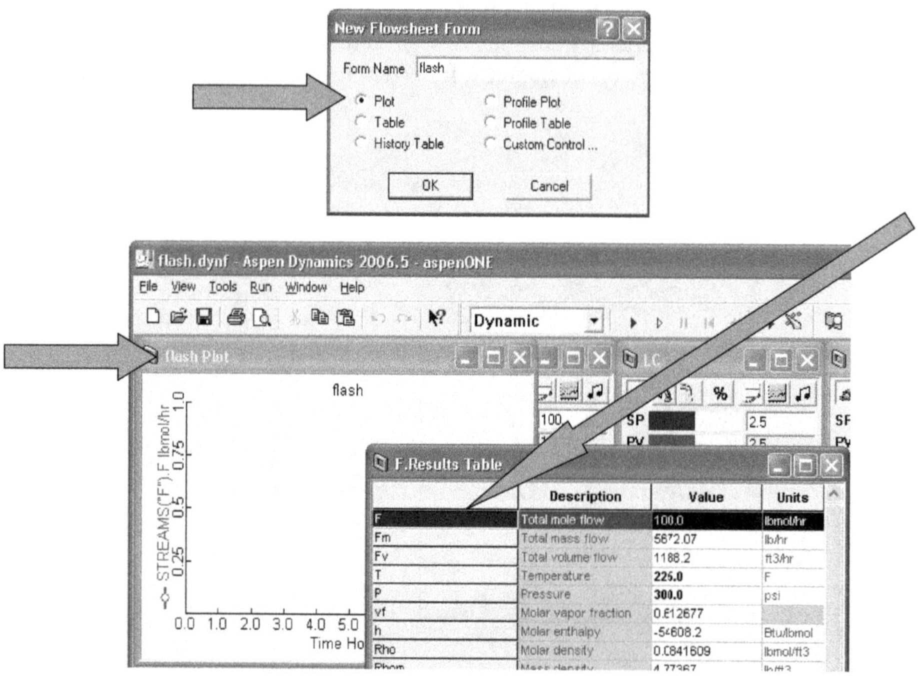


**Figure 4.31** Installing strip chart plot.

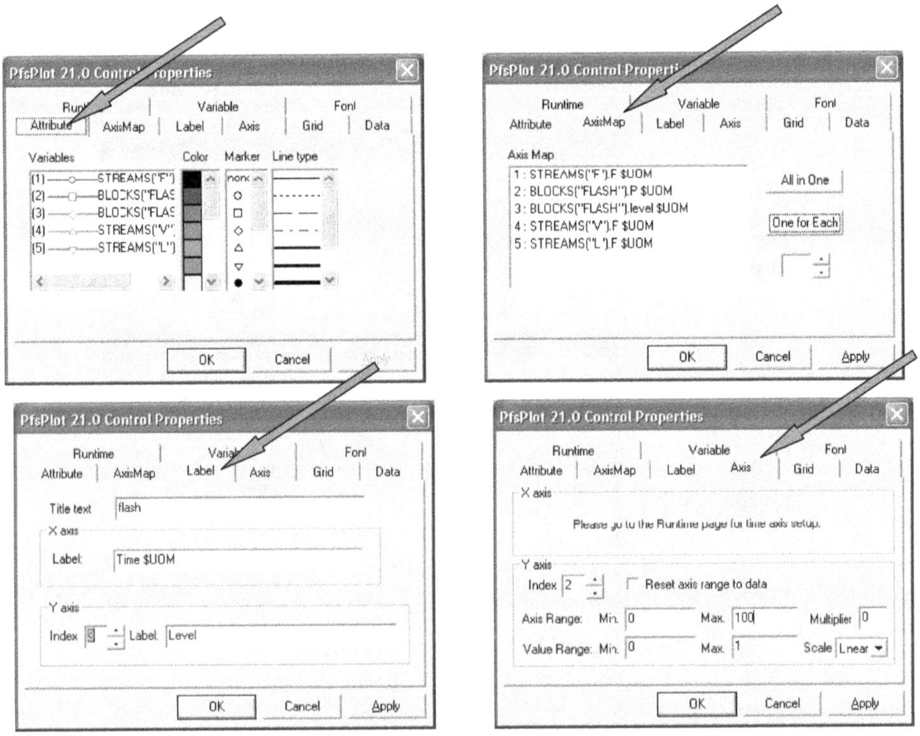


**Figure 4.32** Setting up properties of plot.

The plotting capability in Aspen Dynamics is fairly limited, so it is often convenient to copy all the dynamic information shown in the strip-chart plots and use a more flexible plotting software. For example, the data from Aspen Dynamics can be loaded into Matlab and multiple plots can be generated in whatever form you want. To get a file of the data in Aspen Dynamics, right click the plot and select *Show as History*. A window opens that contains a table of all the variables, including time in the first column, which can be copied. Many examples of these kinds of plots are presented in subsequent chapters.

### 4.3.7 Running the Simulation

We are now ready to run the simulation. As shown at the top of Figure 4.33, there are a number of buttons that control the simulation. Clicking the first button to the right of the window showing *Dynamic* starts the simulation running in time. The run can be paused (time stops changing and all variables are held at their current values) by clicking the fourth button.

The normal procedure is to let the simulation run out in time until variables stop changing. This is the steady state, and we want to save this so that we can start subsequent runs from the same condition. Pause the run, use the dropdown arrow on the *Dynamic/Initialization* window to select *Initialization* and click the run button. The window shown at the bottom left of Figure 4.33 will tell you the initialization run is successful. Now select *Dynamic* and click the *Rewind* button (see Fig. 4.33). The window shown at the top of Figure 4.34 opens. Select the top row and click the *Rewind* button on this window. All variables will

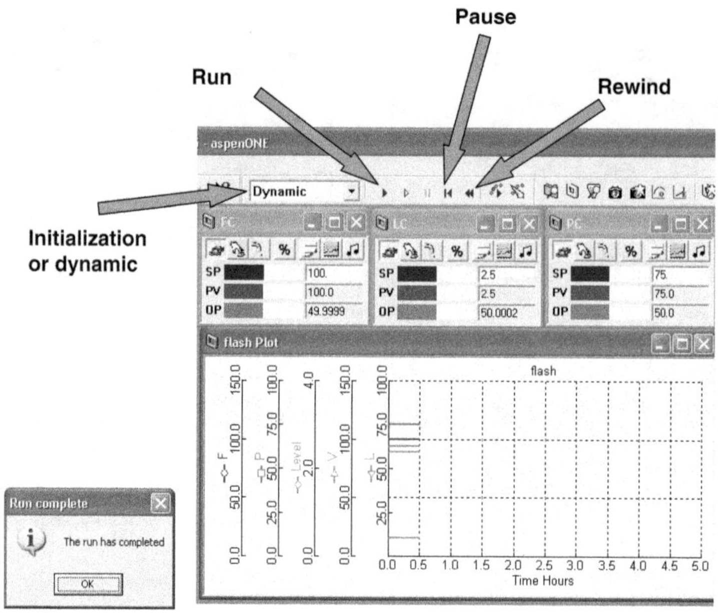


Figure 4.33 Running simulation.

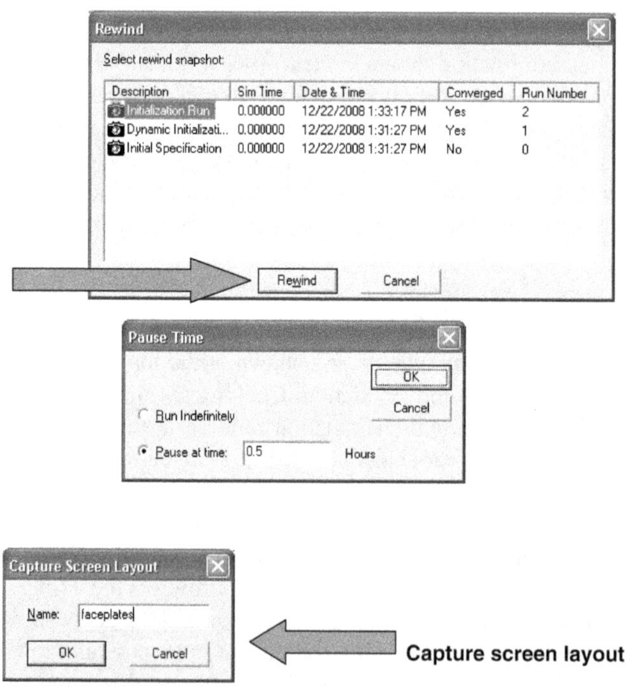


Figure 4.34 Rewinding and pausing.

be set at these conditions and time will be set to zero. These are initial conditions for the ordinary differential equations that model the process in Aspen Dynamics. Be sure to save the file.

If you want to run the simulation for a fixed period of time (specify the time to stop the simulation), go to the very top line above the toolbar and click *Run* (to the left of *Window*). Then select *Pause Time* and enter the time you want to stop the simulation (see the middle of Fig. 4.34).

For a large simulation with many controllers, it is convenient to arrange the faceplates and strip-chart plots in a convenient way on your monitor. To avoid having to do all this arranging every time you start up the simulation, the layout of the screen can be saved by selecting *Tools* and *Capture Screen Layout.* The window at the bottom of Figure 4.34 opens on which a label is provided; save the file. When you start up the simulation again, click the *Flowsheet* item in the *Explore Simulation* window and double click the icon with the label you selected.

To illustrate making a run, let us start from a steady state where the flowrate is 100 lb mol/h, the pressure is 75 psia, liquid level is at 2.5 ft, and all three control valves are 50% open. Figure 4.35 shows the three controller faceplates with these PV and OP values. The simulation is started and paused at 0.5 h. Notice that, up to this point in time, all variables shown on the plot are drawing straight lines (are constant with time). The setpoint of the flow controller is now changed from 100 to 120 lb mol/h, as shown in Figure 4.35. The pause time is changed to 5 h and the run is restarted. Figure 4.36 shows the response. The feed flowrate changes very quickly to the new setpoint. Pressure is held almost constant by increasing the vapor flow; however, the liquid level increases slowly, eventually opening the liquid valve completely (the OP signal of the level controller goes to 100%). The liquid flowrate increases only a very small amount, which is not enough to maintain the level. This is an example of designing a control valve with too small a pressure drop. The valve has saturated.

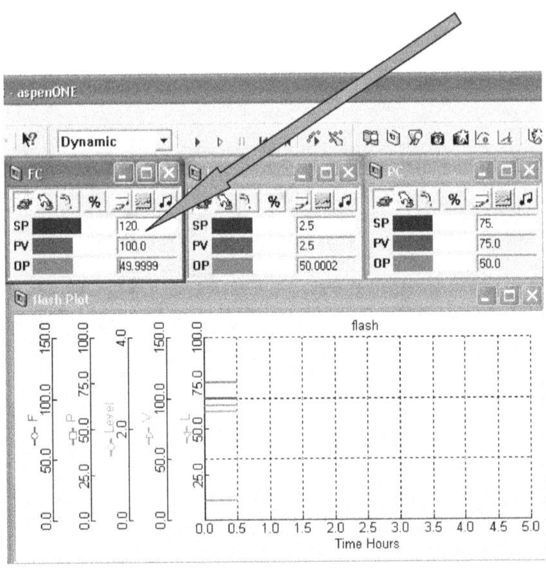


**Figure 4.35** Flow setpoint disturbance.

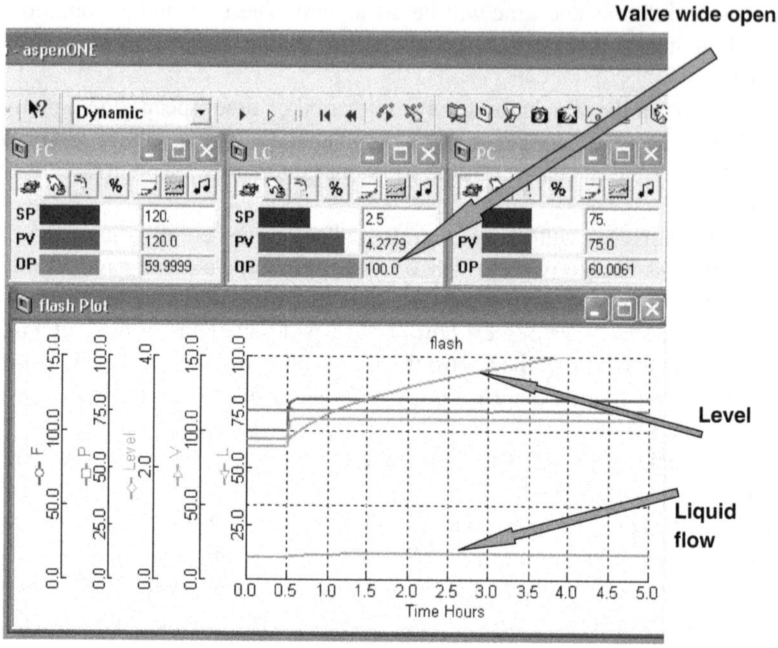


**Figure 4.36** Response.

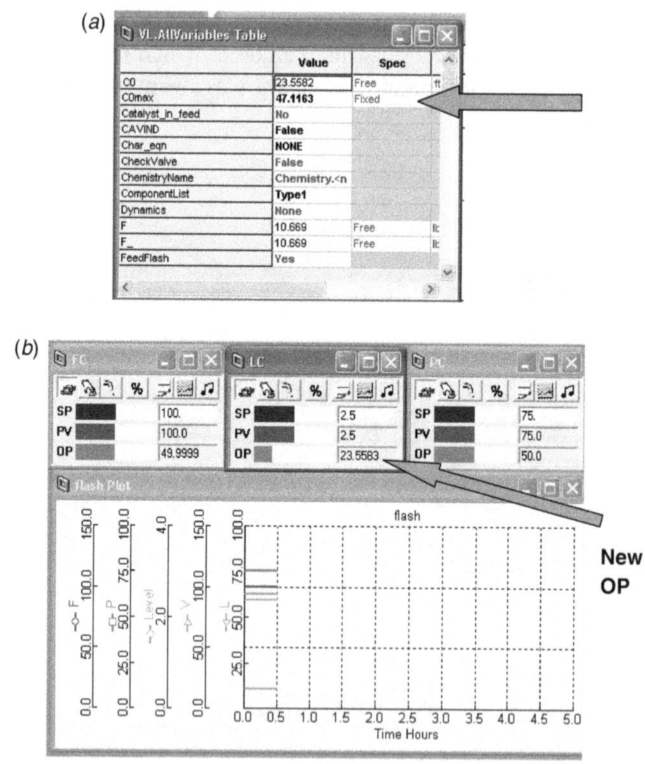


**Figure 4.37** (*a*) Increasing valve size. (*b*) New steady state with bigger liquid valve.

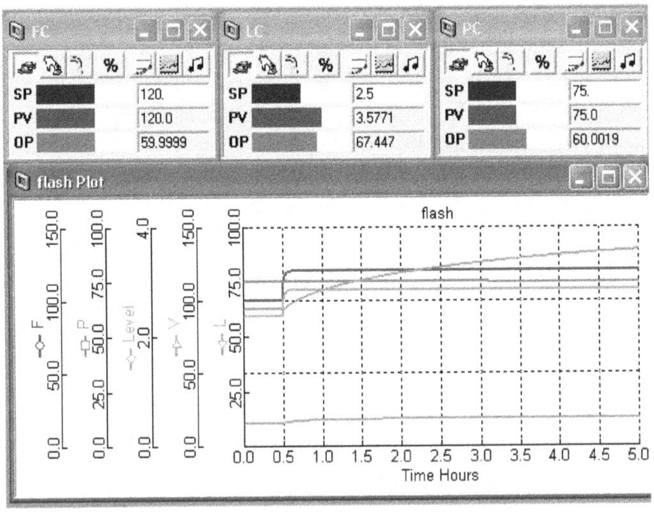


**Figure 4.38** Response with bigger liquid valve.

So what do we do if this occurs? There are two choices: increase the size of the valve or increase the pump head by increasing the power. Figure 4.37 illustrates how the valve size is increased by changing the *COmax* value from 23 to 47.This window is opened by left clicking the valve icon on the flowsheet, right clicking, selecting Forms, and *All* Variables.

With the feed flow at the original 100 lb mol/h, the program is run out to a new steady state. The only thing that is different from the original steady state is the new value of OP signal of the level controller, which changes from the original 50% to 23.5% because the valve is bigger. The response of the system with the larger valve is shown in Figure 4.38. Now the liquid flowrate increases enough to maintain the level in the tank. The level controller OP increases to 67%.

## 4.4 CONTROLLER TUNING

Each of the controllers in a process must be tuned. Some of the loops are easily tuned using process control heuristics and conventional wisdom. Others require some kind of dynamic testing and tuning procedure to obtain reasonable tuning constants. The tuning should be done in a well-defined and consistent way, so that alternative control structures and alternative flowsheets can be fairly compared for dynamic performance.

In the discussion below, controllers with proportional and integral action are considered. Derivative action is only rarely used in process control applications because of the problem of noise and the additional complexity. In simulation studies it is conservative to only use PI controllers.

### 4.4.1 Level Controllers

We have already indicated that most level controllers should use only proportional action with no integral action. A proportional level control smoothes out (filters) disturbances in flowrates, which lessens the impact of these disturbances on downstream units. Proportional controllers do not maintain the level at the setpoint, but in most surge volumes a constant level is not required. The exception is a chemical reactor where the liquid level is important because it affects reaction rates. In liquid reactor applications a proportional controller with a higher gain ($K_C = 10$) is often used.

### 4.4.2 Flow Controllers

Most flow control loops use an orifice plate and a differential pressure transmitter to generate the PV flow signal fed to the controller. The controller output signal goes to a valve. Both the transmitter and the valve have fairly small time constants, so the controller can use a small integral time. A value of 0.3 min is typical.

The pressure-drop signal can be noisy (high-frequency oscillations) owing to the turbulent flow through the orifice, so a small value of controller gain is used to reduce the oscillation of the valve as a result of the fluctuations of the noise. A value of 0.5 for the gain is typical.

There are flow loops that are different from this standard configuration and require different tuning. An important example is a flooded reboiler in which the control valve is on the liquid condensate leaving the shell side of the reboiler. Steam enters the reboiler from the steam supply line. There is no valve in the steam line. The steam flowrate is measured and its flowrate is controlled by manipulating the condensate valve. In this configuration, there is a significant dynamic lag between changing the position of the condensate valve and a change in the steam flowrate because the liquid level of condensate on the steam side of the reboiler must change to affect the heat transfer rate and subsequently the incoming steam flowrate. Changes in heat transfer in this system are governed by changes in the heat transfer area, not changes in the differential temperature driving force. This kind of flow loop requires the tuning procedure discussed in the next section.

### 4.4.3 Temperature and Composition Controllers

***Deadtime and Lags.*** Most temperature and composition controllers need to be tuned because the dynamics lags in these loops cannot be neglected. Deadtimes and lags degrade dynamic performance, so not including realistic dynamic lags in the simulation of these loops can lead to a prediction of dynamic performance that is unrealistically better than what will actually be seen in the plant.

Temperature measurements typically involve a thermocouple inserted in a thermowell, which is installed in the pipe or vessel. A change in the process temperature must first change the temperature of the thermowell by transferring heat from the process to the thermowell as a result of the temperature difference. The heat-transfer coefficient depends on the thickness of the boundary layer of fluid around the thermowell, which depends on the velocity of the fluid flowing past the thermowell and the physical properties of the fluid. Therefore, temperature measurements are not instantaneous. Time lags and deadtimes of 30–60 s are typically experienced. Using a deadtime element of 1 min in the simulation gives fairly conservative controller tuning.

Composition measurements often involve the use of a chromatographic column, and these devices exhibit significant deadtimes because of both sample-line and column-cycle time delays. These deadtimes can range from 3 to 30 min depending on the application. Using a deadtime element of at least 5 min in the simulation is recommended.

***Relay-Feedback Testing.*** A relay-feedback test is an easy and accurate method for experimentally determining important dynamic parameters of a process. The test can be used on a real process or on a simulation. It is widely used in both industrial and academic applications. Relay-feedback tests can be easily performed in Aspen Dynamics simulations.

This dynamic test gives very accurate estimates of the ultimate gain and ultimate period of the loop, which can then be used for controller tuning. The method inserts an on–off relay in the feedback loop that positions the controller output signal at a specified percent higher or lower than the initial steady-state value. For example, if the OP signal to the valve is 50% at steady state and the specified displacement is 5%, the controller output will fluctuate back and forth between 45 and 55%.

The height of this displacement is called "$h$." The loop will settle into a periodic oscillation because the relay fires up or down every time the PV signal crosses the SP signal (see Fig. 4.39). The period of the oscillation is the ultimate period $P_U$. The amplitude of the PV signal is called "$a$" and is read off the strip chart plot. Knowing $h$ and $a$, the ultimate gain can be calculated from Eq. (4.17).

$$K_U = \frac{4h}{a\pi} \tag{4.17}$$

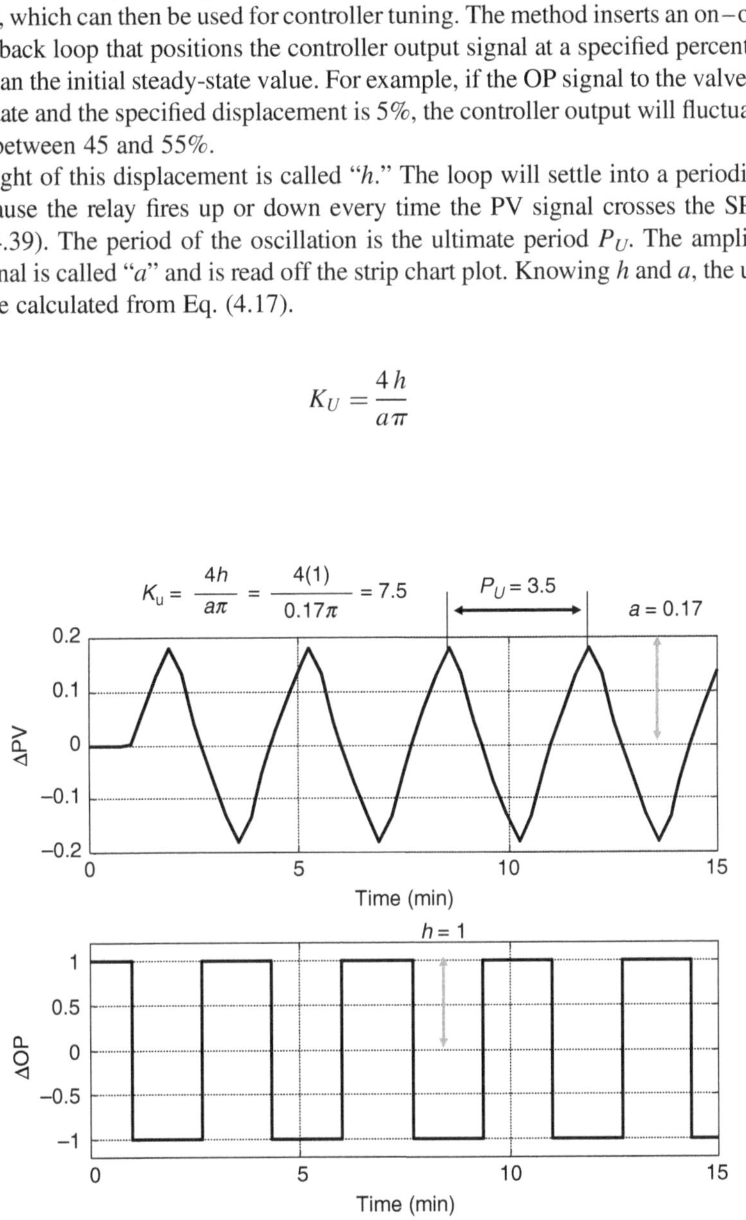


**Figure 4.39** Relay-feedback test.

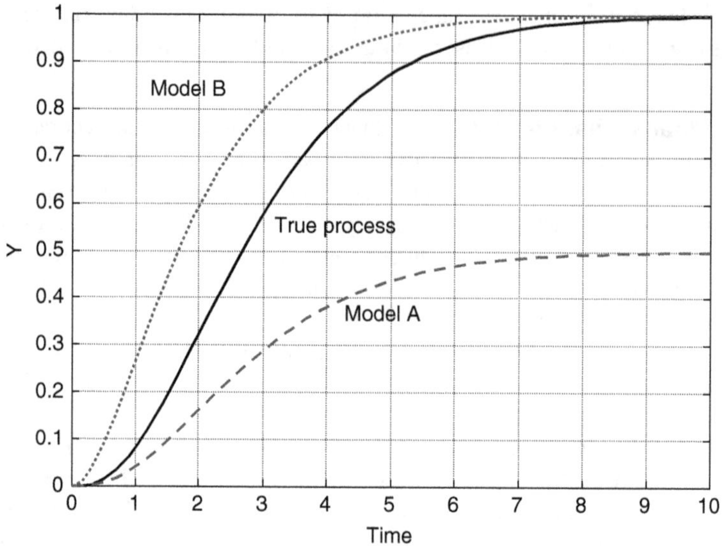


**Figure 4.40** Step responses of a process and two alternative models.

The ultimate frequency (radians/time) is related to the ultimate period.

$$\omega_U = \frac{2\pi}{P_U} \tag{4.18}$$

Knowing the ultimate gain and ultimate period provides important information about the dynamics of the process around the critical $(-1, 0)$ point in the complex plane. The shape and location of the plot of the real and imaginary parts of the open-loop transfer

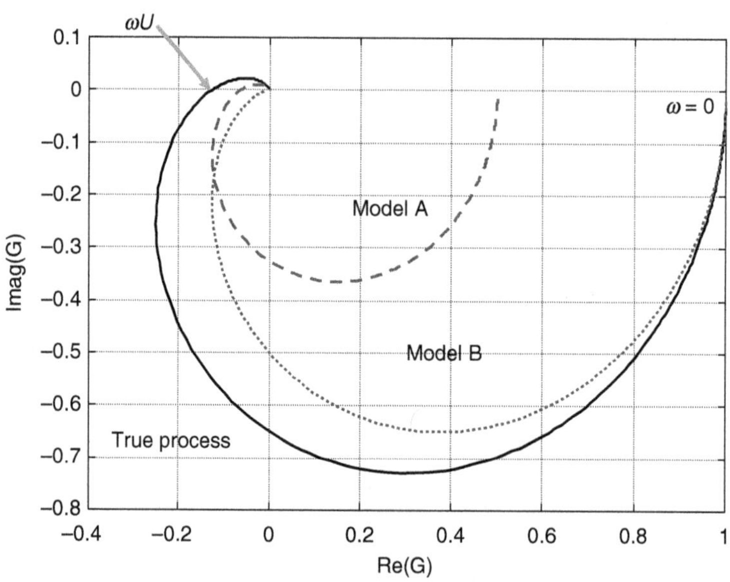


**Figure 4.41** Nyquist plots of a process and two alternative models.

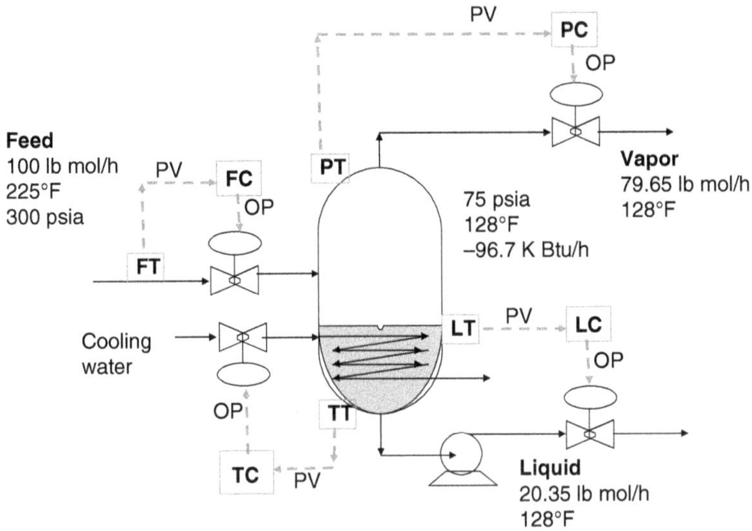


**Figure 4.42** Flash tank with temperature control.

function of the process (Nyquist plot) near this critical point governs closed-loop stability. To emphasize this point, consider the three different step responses shown in Figure 4.40. The solid line is the step response of the real process. The dotted and dashed lines are the step responses of two alternative models. Model A matches the real process well during the initial part of the transient but the match at longer times is poor. The reverse situation

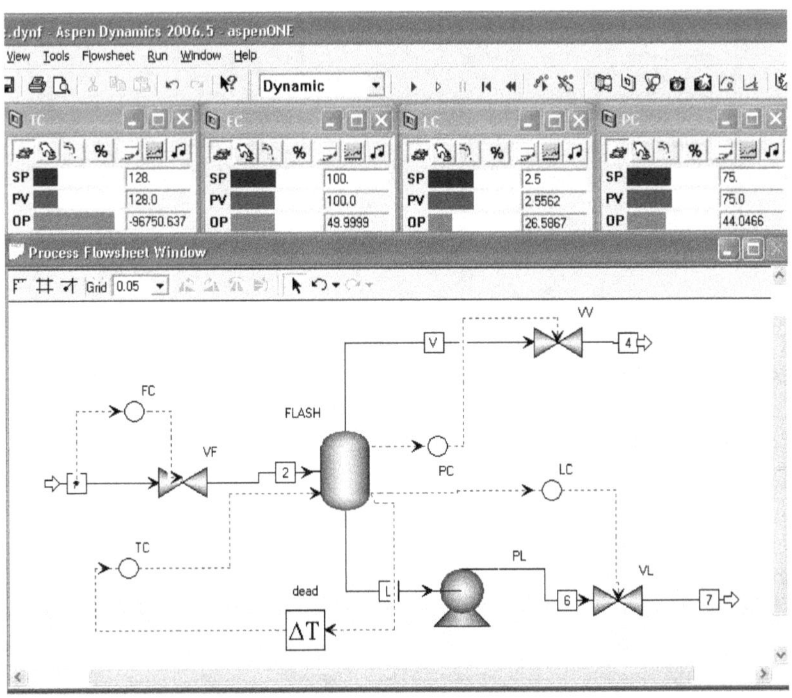


**Figure 4.43** Flash tank with temperature control.

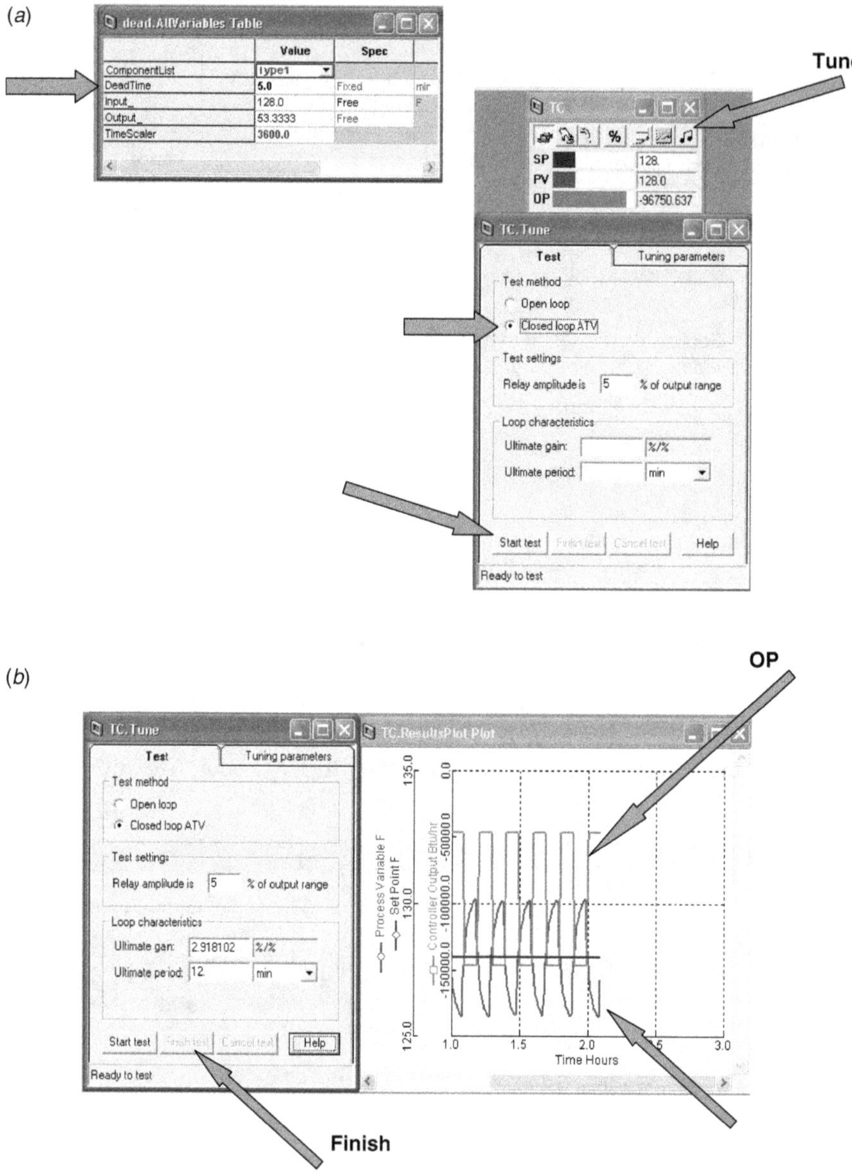


**Figure 4.44** (*a*) Deadtime and relay-feedback testing. (*b*) Relay-feedback testing results.

applies for Model B. Which of the models is better for the purpose of designing a feedback controller?

Figure 4.41 gives Nyquist plots of the three systems. It is clear that Model A does a much better job of matching the real process curve near the critical $(-1, 0)$ point. So using Model A to design a feedback controller will produce more reliable controller tuning constants than using Model B. The relay-feedback test provides accurate information at the $(-1, 0)$ point.

***Tuning Methods.*** There are many controller tuning methods. Some only require values of $K_U$ and $P_U$ to calculated the controller gain $K_C$ and integral time $\tau_I$. The two methods incorporated into Aspen Dynamics are Ziegler–Nichols and Tyreus–Luyben. The former is more aggressive than the latter and should be used when tight control is required and rapid changes in the manipulated variable are permissible. Chemical reactor temperature control is an example. The latter method is more conservative and is suitable for many chemical engineering applications. Distillation column temperature control is a common example.

Ziegler–Nichols:

$$\begin{aligned} K_C &= \frac{K_U}{2.2} \\ \tau_I &= \frac{P_U}{1.2} \end{aligned} \tag{4.19}$$

Tyreus–Luyben:

$$\begin{aligned} K_C &= \frac{K_U}{3.2} \\ \tau_I &= 2.2 P_U \end{aligned} \tag{4.20}$$

***Example.*** To illustrate the application of the relay-feedback test, we modify the flash tank example, considered previously, by adding heat removal to the tank and a temperature controller. Figure 4.42 shows the modified process. The heat-removal rate at steady state is $96.7 \times 10^3$ Btu/h, which reduces the tank temperature to 128°F and produces more liquid (20.35 lb mol/h) and less vapor (79.65 lb mol/h) than the adiabatic flash case shown in Figure 4.15.

Figure 4.43 shows the new Aspen Dynamics process flow diagram with a temperature controller and a deadtime element installed on the temperature measurement. The controller

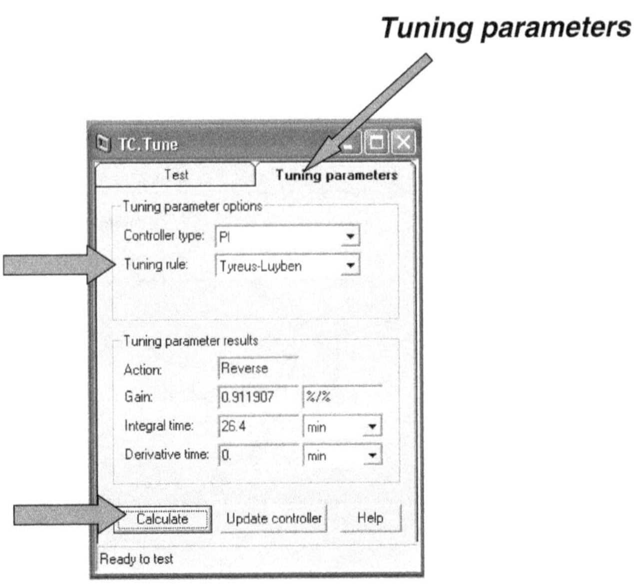


**Figure 4.45** Controller tuning constants.

faceplates give steady-state values of all variables and controller output signals. Notice that the OP signal of the temperature controller is −96,750. The maximum range of the output is zero and the minimum is set up to be −1,000,000. This is an example of a loop in which the controller output signal does not position a valve but is a rate of energy transfer.

The controller is reverse acting (an increase in temperature reduces the controller output, i.e., more heat removal). The span of the temperature transmitter is 100–200°F, and the temperature controller setpoint is 128°F. The value of the deadtime is inserted by highlighting the deadtime icon on the flowsheet, right clicking, selecting *Forms* and *AllVariables*.

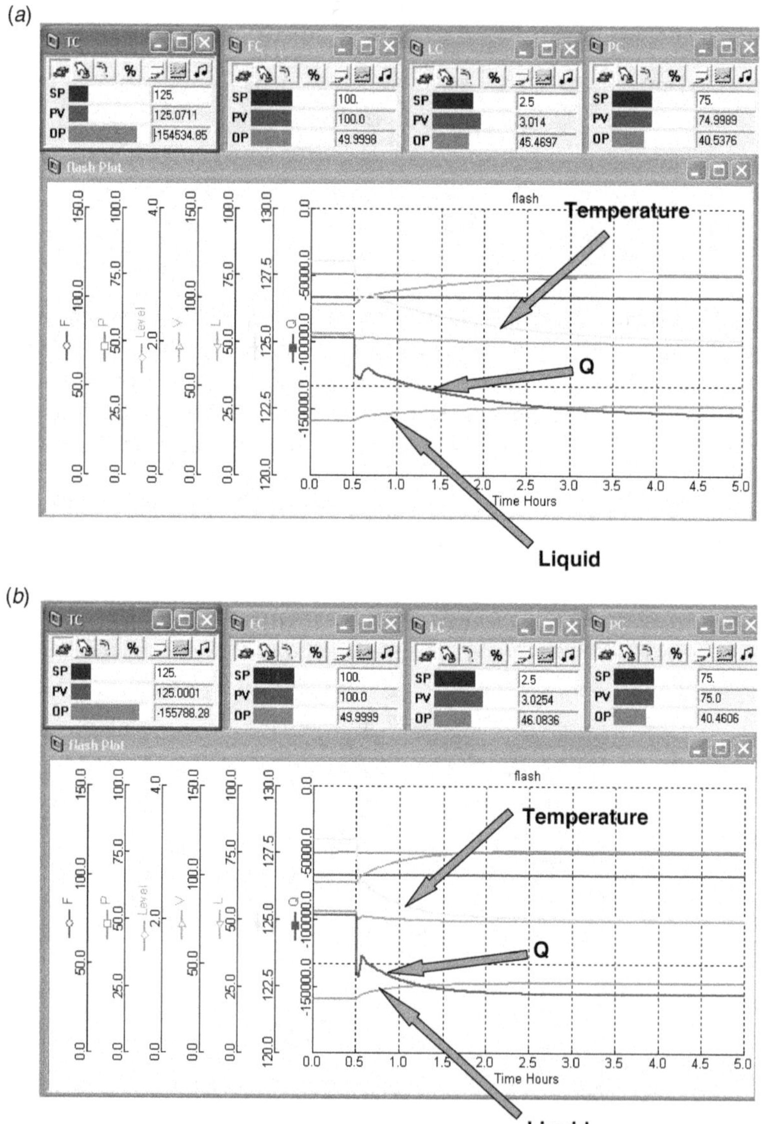


**Figure 4.46** (*a*) Response with D = 5. (*b*) Response with D = 2.

The window shown on the left-hand side of Figure 4.44 opens and a value of the deadtime is entered. We will look at the effect of deadtime by using values of 2 and 5 min.

The relay-feedback test is initiated by clicking the *Tune* button at the top right of the controller faceplate as shown on the right-hand side of Figure 4.44*a*. The *Closed loop ATV* method is selected, the amplitude is set at 5%, and the *Start test* button is clicked. Figure 4.44*b* shows the resulting dynamic response. Notice the periodic step functions in the controller output OP and the resulting responses in the process temperature PV. Remember that the only thing we do to run this test is to specify a 5% displacement of the OP signal from its steady-state value. The loop automatically switched the OP signal up and down.

After several cycles, the simulation is paused and the *Finish test* button on the bottom of the *Tune* window is clicked. The ultimate gain and ultimate period are displayed. For a 5 min deadtime, they are $K_U = 2.9$ and $P_U = 12$ mins. For a 2 min deadtime, they are $K_U = 4.3$ and $P_U = 5.4$ min.

Controller tuning parameters are calculated by clicking the *Tuning parameters* page tab, selecting the desired method (*Tyreus-Luyben* shown in Fig. 4.45), the desired type of controller (*PI*), and clicking the *Calculate* button. The resulting tuning constants are $K_C = 0.91$ and $\tau_I = 26$ min for a deadtime of 5 min. For a 2 min deadtime, the results are $K_C = 1.34$ and $\tau_I = 12$ min. The tuning constants can be inserted into the controller by clicking the *Update controller* button.

Figure 4.46 shows the responses of the process to a step change in the temperature controller setpoint signal from 128 to 125°F at time equal 0.5 h. In Figure 4.46*a* the deadtime is 5 min. The heat removal gets larger (more negative), and it takes almost 4 h to arrive at the new steady state. In Figure 4.46*b* the deadtime is 2 min, and it takes only 2 h to arrive at the new steady state. The two steady states are identical, but the time trajectories are different.

## 4.5 RATIO AND CASCADE CONTROL

Advanced control structures can be easily implemented in Aspen Dynamics. In this section we illustrate two of the more important methods. A simple distillation column is used to illustrate the installation of ratio elements (multipliers) and the use of cascade control.

### 4.5.1 Ratio Control

There are many situations when the flowrate of one stream needs to be changed when the flowrate of another stream changes. The blending of two or more streams is a common example. In distillation column control, it is often desirable to control a reflux ratio or to control a reflux-to-feed ratio. We may also want to control a reboiler heat input-to-feed ratio.

The distillation column used as a numerical example is shown in Figure 4.47. A binary mixture of methanol and water are separated in a 16-stage column operating a 14.7 psia in the reflux drum. The feed is 40 mol% methanol. Distillate purity is 99 mol% methanol. Bottoms purity is 99.5 mol% water. A reflux ratio of 1.27 is required to achieve these specifications. Reboiler heat input is $152 \times 10^3$ Btu/h. Column diameter is 0.474 ft. The reflux drum and column base are sized for 5 min holdup when half full. Valve pressure drops are 30 psi.

The Aspen Plus file is pressure checked and exported into Aspen Dynamics. Controllers are installed to achieve the following control structure (see Fig. 4.48).

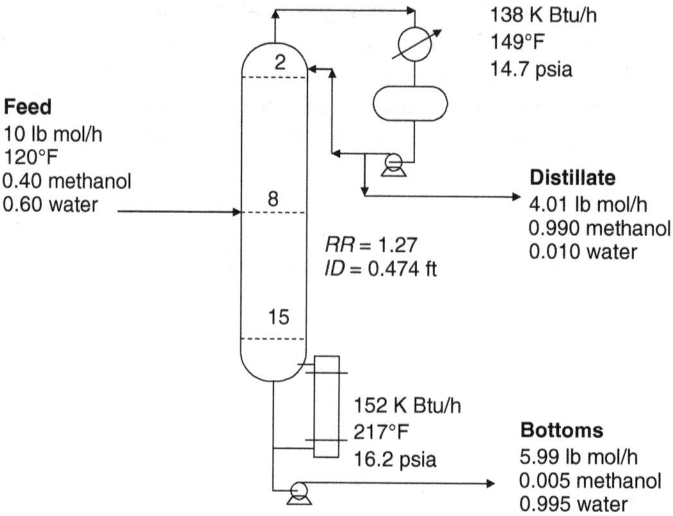


**Figure 4.47** Distillation flowsheet.

1. Feed is flow controlled.
2. Reflux drum level is controlled by manipulating distillate flowrate.
3. Base level is controlled by manipulating bottoms flowrate.
4. Pressure is controlled by manipulating condenser heat removal.
5. Stage 12 temperature is controlled by manipulating reboiler heat input. A 1 min deadtime is inserted in this loop.
6. Reflux flowrate is fixed.

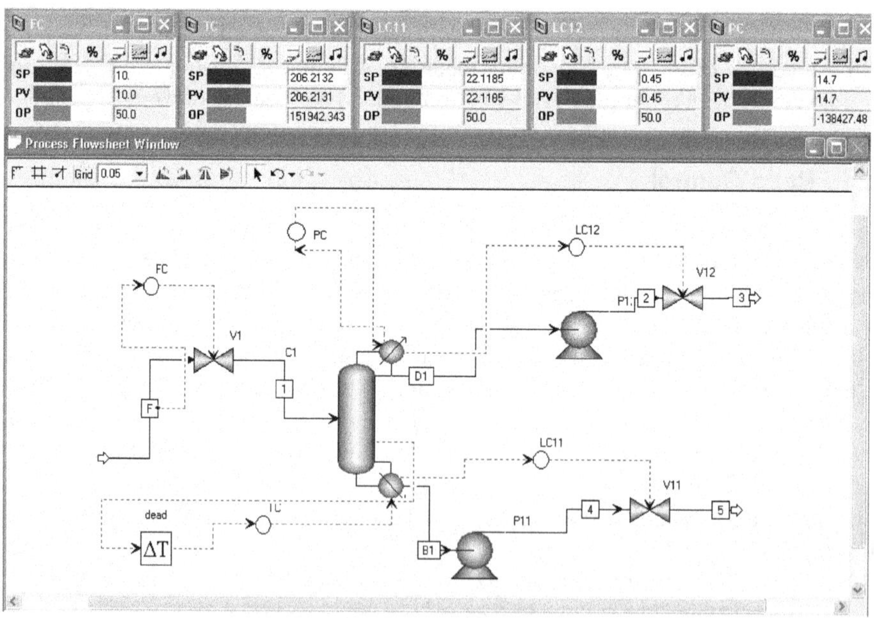


**Figure 4.48** Flowsheet with fixed reflux.

The last item is not shown in Figure 4.48 because the default setup in Aspen Dynamics is to fix the *mass* flowrate of the reflux. We will compare the performance of a fixed reflux structure with the performances of both a fixed reflux ratio structure and a fixed reflux-to-feed ratio control structure.

With the reflux flowrate fixed, running a relay-feedback test and using Tyreus–Luyben tuning give temperature controller constants $K_C = 0.73$ and $\tau_I = 16$ min for a deadtime of 1 min. The controller is reverse acting. The temperature transmitter range is 150–250°F with a 206.2°F setpoint. The temperature controller output is 0 to $300 \times 10^3$ Btu/h.

***Reflux Ratio.*** To set up a ratio, a *Multiplier* is dropped onto the flowsheet from the list of *ControlModels* in the *Dynamic* Library. The *Multiplier* block has two inputs. For reflux ratio control, the first input is connected from the mass flowrate of the distillate using a *ControlSignal* stream type. Remember that the flowrate of the distillate is set by the reflux drum level controller. The output of the multiplier is connected to the column (*Block C*1), using a second *ControlSignal*, and selecting the *mass* flowrate of reflux (see Fig. 4.49). The second input to the multiplier is the desired reflux ratio ($RR = 1.27$), and it is entered into the window shown in Figure 4.50, which appears when the multiplier block label "RR" is highlighted, right clicked, and *Forms* and *All Variables* are clicked. In other control structures this ratio signal could come from another source. For example, the output signal from a temperature or composition controller could change the reflux ratio. The flowsheet with the reflux ratio control structure is shown in Figure 4.51.

The temperature controller must be retuned with the new control structure. With the reflux ratio fixed, running a relay-feedback test and using Tyreus–Luyben tuning gives temperature controller constants $K_C = 1.16$ and $\tau_I = 11.9$ min for a deadtime of 1 min. Notice that the temperature controller integral time is smaller when a reflux ratio structure is in place. Changes in reboiler heat input produces changes in the flowrate of the distillate because of the level controller. This then changes the flowrate of the reflux, so both the vapor boilup and the reflux are affecting the controlled temperature on Stage 12 in the column.

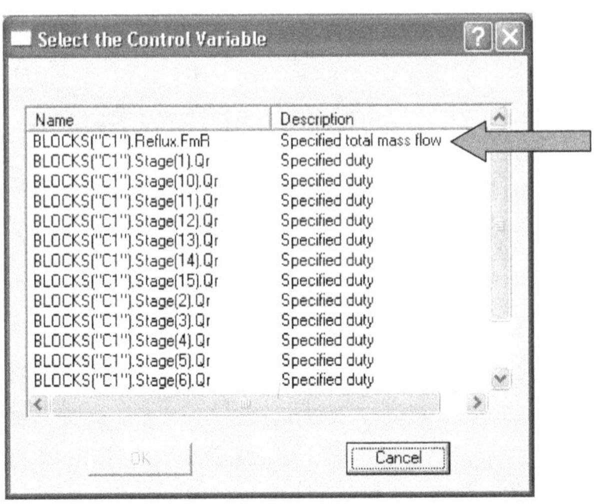


**Figure 4.49** Selecting mass flowrate of reflux.

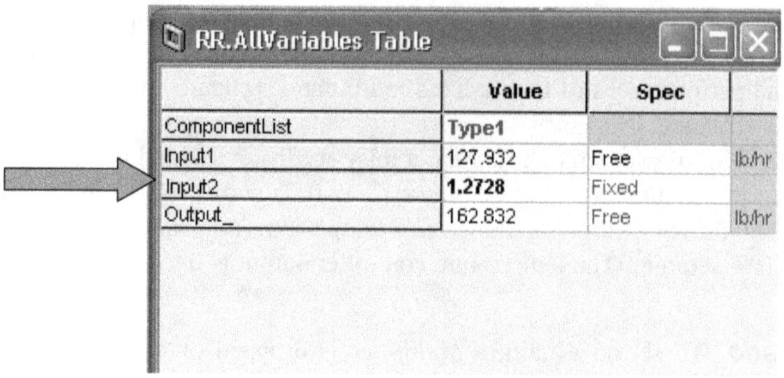


**Figure 4.50** Fixing reflux ratio.

***Reflux-to-Feed Ratio.*** Achieving this structure uses a multiplier with its first input being the mass flowrate of the feed. The output is the mass flowrate of the reflux. The second input is the ratio of the mass flowrates.

$$R/F = \frac{R_W}{F_W} = \frac{162.8\ \text{lb/h}}{236.3\ \text{lb/h}} = 0.6891 \tag{4.21}$$

Figure 4.52 shows this fixed reflux to feed control structure. The tuning of the temperature controller is the same as the fixed reflux case ($K_C = 0.73$ and $\tau_I = 16$ min for a deadtime of 1 min).

***Results.*** Disturbances in feed flowrate and feed composition are imposed on the three alternative systems. Figure 4.53 gives results for a step change in the setpoint of the feed flow controller from 10 to 12 lb mol/h at time equal 0.5 h. All of the structures give stable regulatory control; however, as we would expect, the fixed reflux flowrate structure

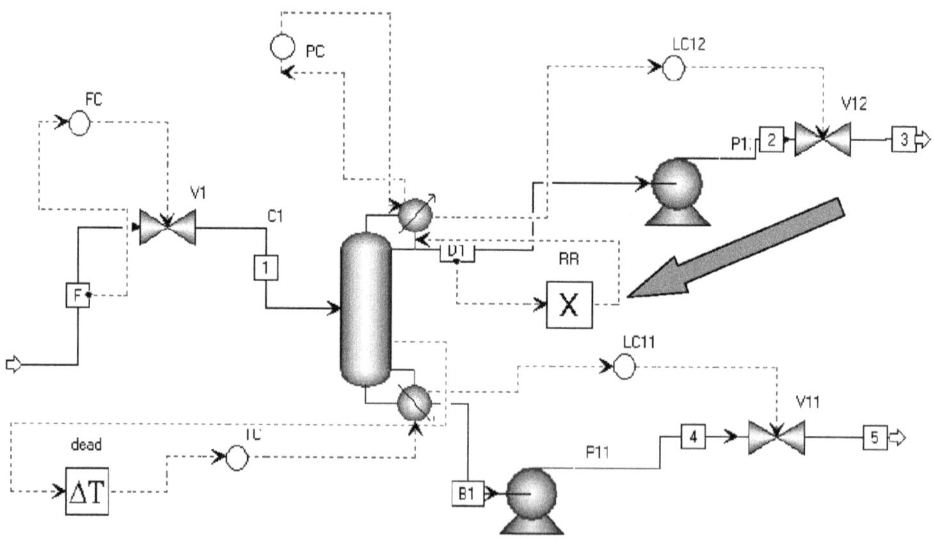


**Figure 4.51** Flowsheet with fixed reflux ratio.

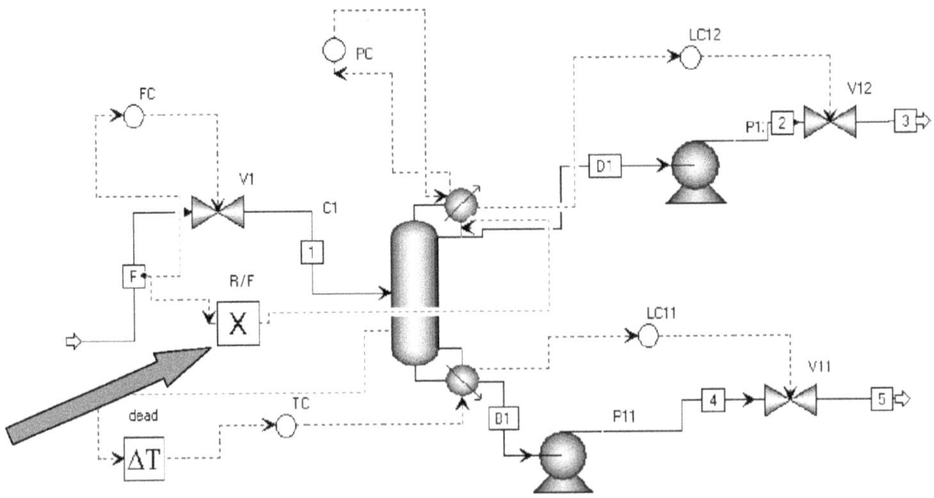


**Figure 4.52** Flowsheet with fixed reflux-to-feed ratio.

does not maintain product compositions when the feed flowrate is changed. The upper right graph shows that the impurity of water in the distillate climbs from its desired value of 1 to over 3 mol% water when reflux is fixed. The other two structures return the product purities back essentially to the original desired values. This is what we would expect from these ratio structures because all flows change in direct proportion to the feed flow. There are small

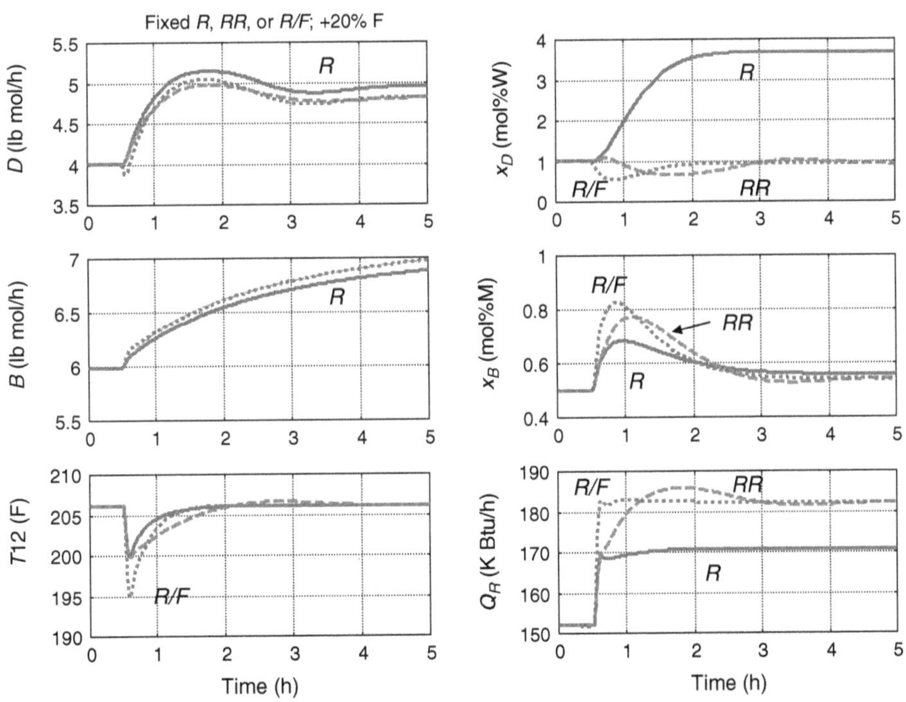


**Figure 4.53** Feed flowrate disturbance; fixed reflux, reflux ratio, or reflux-to-feed ratio.

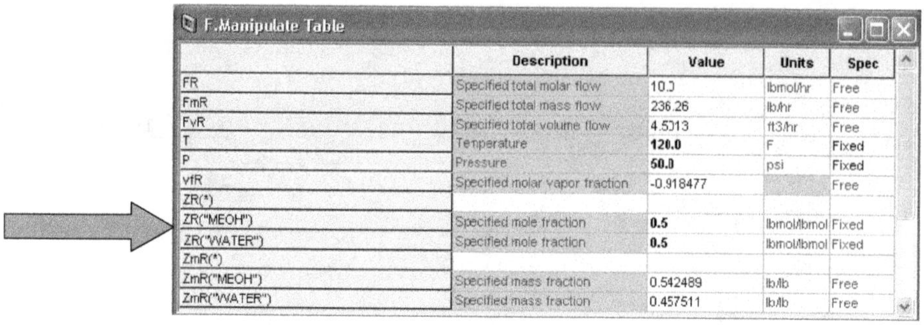

F.Manipulate Table

| | Description | Value | Units | Spec |
|---|---|---|---|---|
| FR | Specified total molar flow | 10.0 | lbmol/hr | Free |
| FmR | Specified total mass flow | 236.26 | lb/hr | Free |
| FvR | Specified total volume flow | 4.5013 | ft3/hr | Free |
| T | Temperature | **120.0** | F | Fixed |
| P | Pressure | **50.0** | psi | Fixed |
| vfR | Specified molar vapor fraction | -0.918477 | | Free |
| ZR(*) | | | | |
| ZR("MEOH") | Specified mole fraction | **0.5** | lbmol/lbmol | Fixed |
| ZR("WATER") | Specified mole fraction | **0.5** | lbmol/lbmol | Fixed |
| ZmR(*) | | | | |
| ZmR("MEOH") | Specified mass fraction | 0.542489 | lb/lb | Free |
| ZmR("WATER") | Specified mass fraction | 0.457511 | lb/lb | Free |

**Figure 4.54** Changing feed composition.

changes in pressure at different flowrates, and this affects temperatures, but the effect in this column is small. So we cannot select between the reflux ratio and reflux-to-feed ratio control structures by only looking at feed flowrate changes.

Subjecting the column to feed compositions disturbances will give us more information. Figure 4.54 gives the window on which this disturbance is specified. The window is obtained by left clicking the feed stream icon *F*, right clicking, selecting *Forms*, and selecting *Manipulate*. The disturbance is a change in feed composition from 40 to 50 mol% methanol with the corresponding reduction in water composition in the feed.

The solid lines in Figure 4.55 are for the fixed reflux ratio structure. The dashed lines are for the fixed reflux-to-feed ratio structure. Both give stable regulatory control, with the temperature on Stage 12 well controlled. The composition of the bottoms is also held close to the desired value. However, the response of composition of the distillate is different in the two

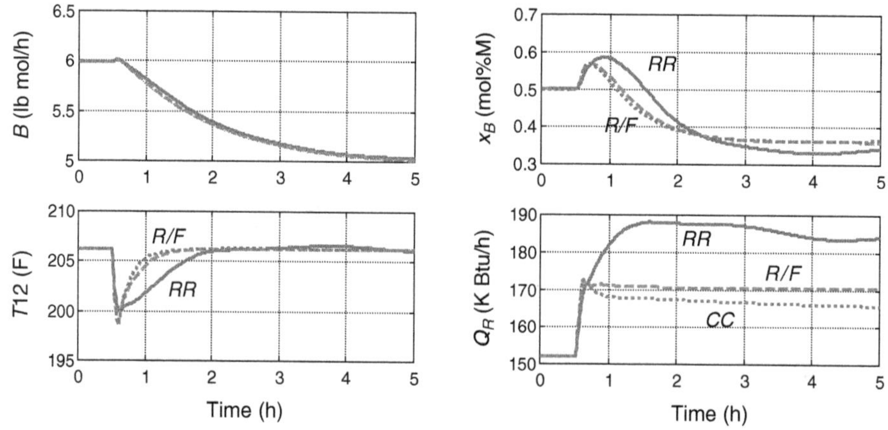


**Figure 4.55** Feed composition disturbance.

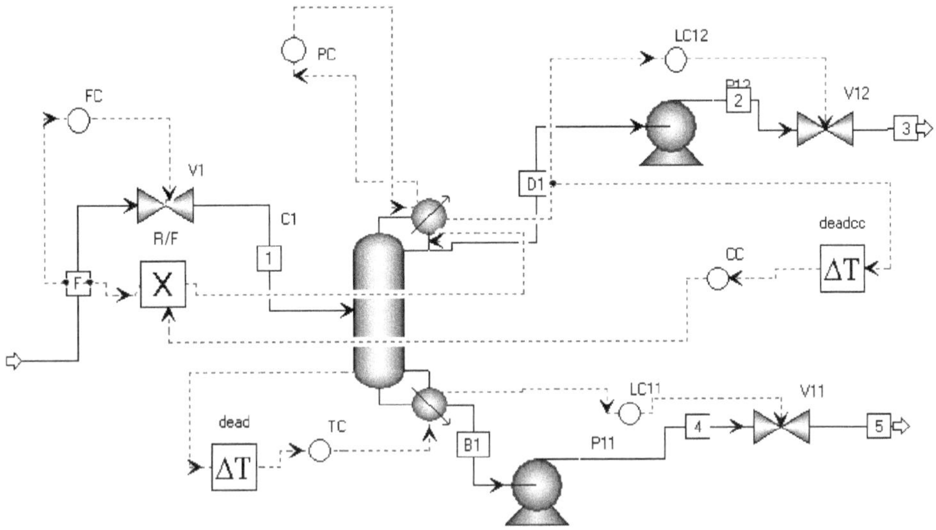


**Figure 4.56** Reflux-to-feed ratio reset by composition controller.

cases. The reflux-to-feed ratio structure holds the distillate closer to the desired value, so it is the preferred structure when only a single temperature is controlled in the column.

Another alternative is to use a composition controller on the distillate to adjust the R/F ratio. Figure 4.56 shows this structure. The composition controller controls the *impurity* of water in the distillate by adjusting the R/F ratio. With the Stage 12 temperature on automatic, running a relay-feedback test and using Tyreus–Luyben tuning give composition controller constants $K_C = 0.77$ and $\tau_I = 80$ min. for a deadtime of 5 min. The controller is direct acting. The composition transmitter range is 0–0.05 mf water with a 0.01 mf water setpoint. The controller output range is 0.4–0.8 with a steady-state value of R/F = 0.689. The dashed lines in Figure 4.55 give results for the composition control structure. Notice that the distillate impurity $x_{D(W)}$ is slowly returned to the setpoint, but it takes over 5 h.

### 4.5.2 Cascade Control

A cascade control structure features a primary controller and a secondary controller. The OP signal of the primary controller is the SP signal of the secondary controller.

To illustrate cascade control, we add a second composition controller to the distillation column considered above; see Figure 4.57. The controller maintains the *impurity* of the methanol in the bottoms by adjusting the setpoint of the Stage 12 temperature controller. Thus, the secondary controller is the temperature control, and its tuning does not change. The primary controller is the "*CCxB*" controller, and it must be tuned. The other composition controller on the distillate ("*CCxD*") and the temperature controller (TC) are placed on automatic. Running a relay-feedback test and using Tyreus–Luyben tuning give bottoms composition controller constants $K_C = 0.62$ and $\tau_I = 59$ min. for a deadtime of 5 min. The controller is direct acting (more methanol impurity increases the setpoint of the temperature controller). The composition transmitter range is 0–0.02 mf methanol with a 0.005 mf methanol setpoint. The controller output range is 150–250°F with a steady-state value of 206.2°F.

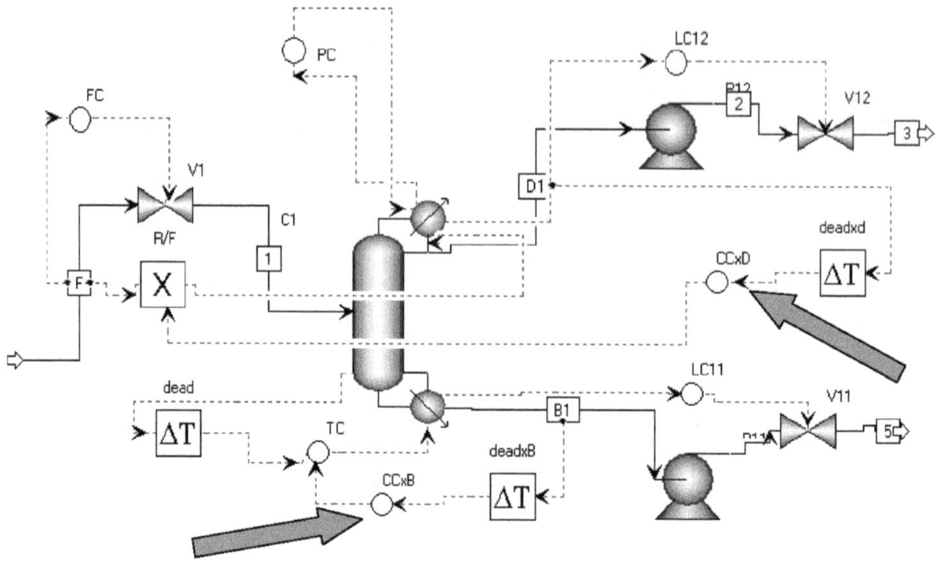


**Figure 4.57** Dual composition control structure.

Figure 4.58 shows the controller faceplates of this dual composition control structure. There are two composition controllers (*CCxB* and *CCxD*) and one temperature controller (*TC*). Note that the temperature controller is on cascade, receiving its setpoint from the *OP* signal for the *CCxB* composition controller.

The performance of this structure is shown in Figure 4.59 for increases and decreases in feed composition. The solid lines are for changes from 40 to 50 mol% methanol in the feed. The dashed lines are for changes from 40 to 30 mol% methanol. This dual composition control structure holds the compositions of both product streams at their desired values. Notice that the SP of the temperature controller changes for feed composition disturbances. A higher Stage 12 temperature is required when the feed contains less methanol. For more methanol in the feed, the Stage 12 temperature must be lower. Of course the reflux flowrate is also changing because the top composition controller is changing the reflux-to-feed ratio.

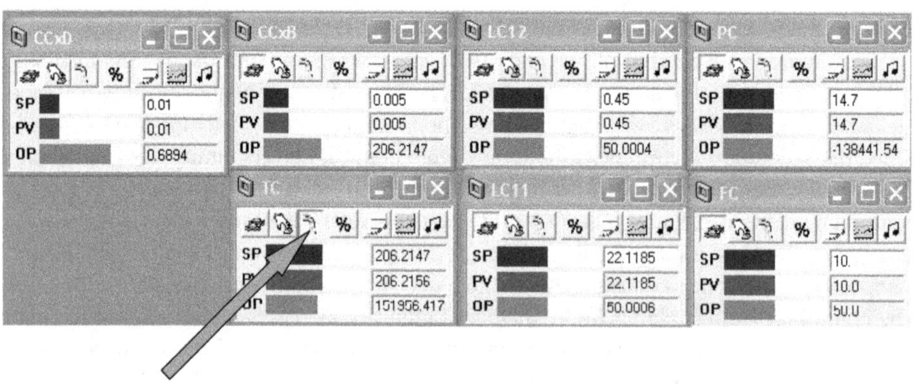


**Figure 4.58** Controller faceplates.

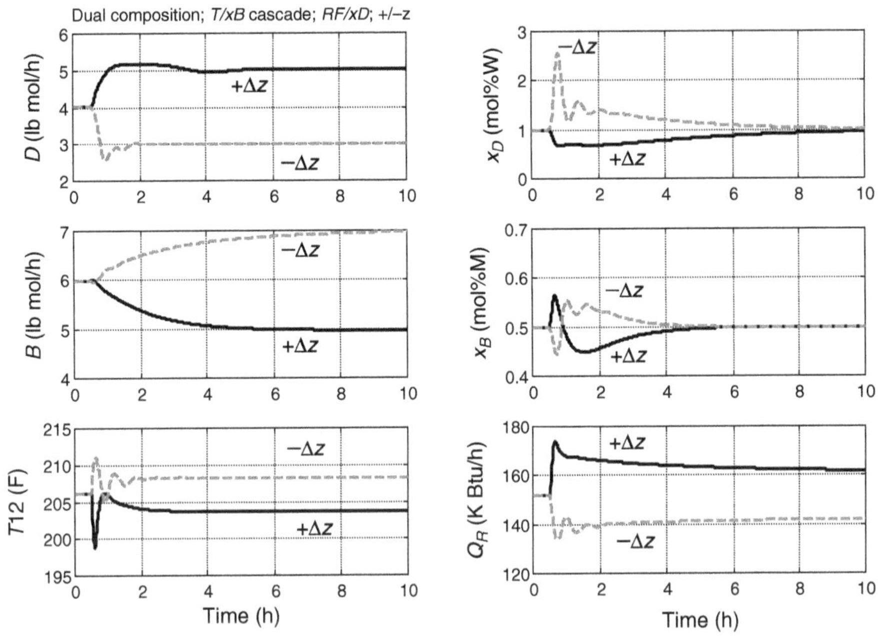


**Figure 4.59** Dual composition control; feed composition disturbances.

## 4.6 CONCLUSION

In this chapter we have attempted to present many of the useful features and tools available in Aspen Dynamics. Detailed step-by-step procedures have been given for performing the various tasks required to run a dynamic simulation. These methods will be applied in the dynamic simulations reported in subsequent chapters.

# PART 2

# SEPARATIONS WITHOUT ADDING OTHER COMPONENTS

Although it is not possible to separate an azeotrope in a single distillation column, it is sometimes possible to achieve the separation using two distillation columns without adding other components (light or heavy entrainers). Pressure swing can be used in some systems for homogeneous binary azeotropes whose azeotropic compositions change significantly with pressure. Chapter 5 gives some examples of this type of system. Chapter 6 extends the treatment to consider the use of heat integration in pressure-swing systems. Chapter 7 demonstrates that no other components must be added when the binary azeotrope is *heterogeneous*. A two-column system with a decanter achieves the separation by taking advantage of the liquid–liquid phase equilibrium in the decanter.

CHAPTER 5

# PRESSURE-SWING AZEOTROPIC DISTILLATION

Some azeotropic systems exhibit the desirable property that pressure has a strong effect on the composition of the azeotrope. When this occurs, a two-column system can be used to achieve the desired separation. The basic idea is to operate one column at low pressure and a second column at high pressure. One of the components comes out from one end of the low-pressure column. If the azeotrope is minimum boiling, the azeotrope will come out of the top and the product stream out the bottom. The composition of this distillate stream will be close to that of the azeotrope at the low pressure. If the azeotrope is maximum boiling, the azeotrope will come out the bottom and the product stream out the top. The azeotrope is then fed to the second high-pressure column in which a similar separation occurs, except now the other component is removed from one end of the column and a stream with composition close to the high-pressure azeotrope is removed from the other end. This azeotropic stream is fed back to the low-pressure column.

This chapter gives several examples of pressure-swing azeotropic distillation systems. All of the examples involve minimum-boiling azeotropes, so the azeotropes come overhead in the distillate streams and the bottom streams from the two columns are the high-purity products.

## 5.1 TETRAHYDROFURAN–WATER SYSTEM

The first example is the THF–water separation, which was one of the first studied[1] in terms of both design and control. As shown in Figures 5.1 and 5.2, the composition of the azeotrope in this system changes from 82.3 mol% THF at 1.01 bar to 63.9 mol% THF at 7.9 bar. This significant shift makes it economically feasible to separate the two components by using two distillation columns operating at two different pressures.

*Design and Control of Distillation Systems for Separating Azeotropes.* By William L. Luyben and I-Lung Chien

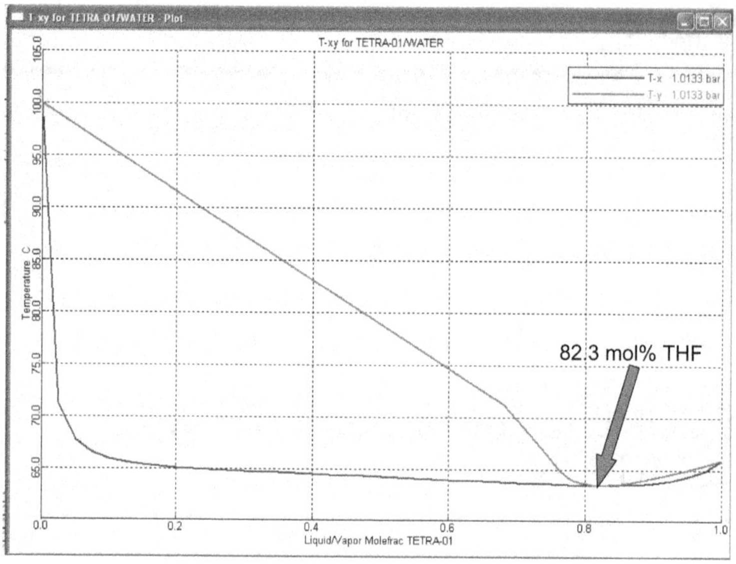


**Figure 5.1** THF/water azeotrope at 1.01 bar.

Figure 5.3 illustrates the Aspen Plus flowsheet using two independent columns, each with its own reboiler and condenser. The feed mixture is fed to the low-pressure column C1. The bottoms product is high-purity water (1 ppm THF). The overhead distillate has a composition that is slightly lower than the low-pressure azeotrope. It is fed to the high-pressure column C2, which produces a bottoms that is high-purity THF (10 ppm water) and a distillate with a composition that is slightly higher than the high-pressure azeotrope. The distillate from C2 is recycled back to the low-pressure column. Heat exchangers (economizers) are used on both columns to recover energy from the hot bottoms by preheating the feed streams.

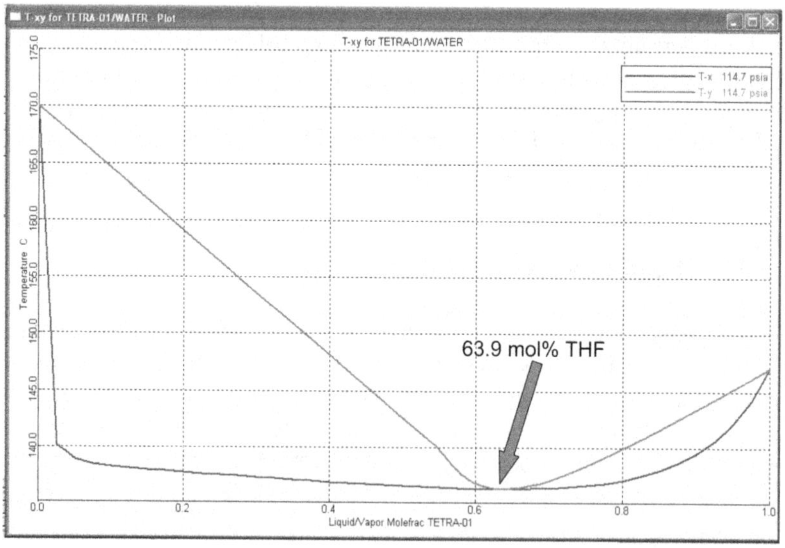


**Figure 5.2** THF/water azeotrope at 7.9 bar.

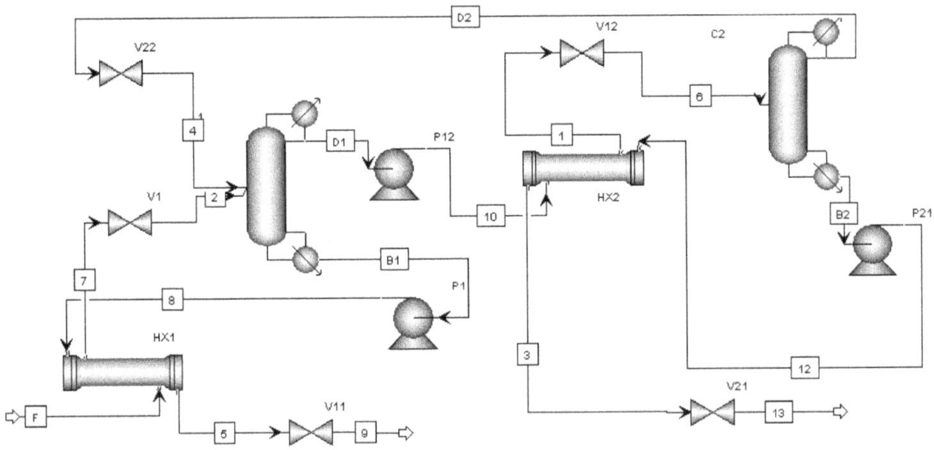


**Figure 5.3** Flowsheet for pressure-swing azeotropic separation.

The flowsheet in Figure 5.3 shows independent reboilers and condensers on each column. However, the reflux-drum temperature of the high-pressure column (136°C at 7.9 bar) is higher than the base temperature of the low-pressure column (106°C at the pressure in the base of this column 1.1 bar), so heat integration could be attractive in terms of energy consumption.

We will postpone a detailed discussion of the THF–water system until Chapter 6 where quantitative comparison of the steady-state designs and dynamic controllability of heat-integrated and nonheat-integrated systems are presented.

## 5.2 ACETONE–METHANOL SYSTEM

Another example of the use of pressure-swing distillation is for overcoming the binary minimum-boiling azeotrope between acetone and methanol. As shown in Figure 5.4, the minimum-boiling azeotrope at 1 atm contains 77.6 mol% acetone at 328 K. At 10 atm the

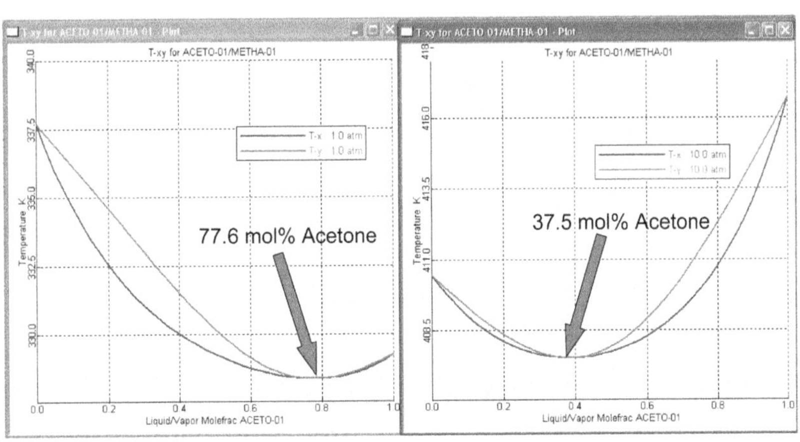


**Figure 5.4** *Txy* diagrams for acetone–methanol at 1 and 10 atm.

azeotropic composition is 37.5 mol% acetone at 408 K, so pressure-swing separation is feasible. Extractive distillation is also feasible in this system, using water as the solvent. This alternative is discussed in Chapter 11 where comparisons are made with the pressure-swing design. Both systems require two distillation columns.

Knapp and Doherty[2] studied heat-integration of binary homogeneous azeotropic systems using extractive distillation methods. One of their examples considered the acetone–methanol system with water as the solvent. They did not consider pressure-swing distillation, nor did they consider dynamics and control.

The phase equilibrium results shown in Figure 5.4 use the UNIQUAC physical property package. Knapp and Doherty[2] used the van Laar method, but the azeotropic search in Aspen Plus using van Laar predicts that there are no azeotropes. Therefore, the UNIQUAC method is used in this work. Notice that Figure 5.4 shows an unusual behavior in the pure component boiling points. At low pressure, acetone is the lower-boiling component. At high pressure, methanol becomes the lower-boiling component.

### 5.2.1 Steady-State Design

Figure 5.5 shows the detailed flowsheet with stream information, heat duties, equipment sizes, and operating conditions. The design is heuristically optimized by running the rigorous simulation with an increasing number of stages until the reflux ratio stopped decreasing, which gives an approximation of the minimum reflux ratio. These runs are made with the distillate and bottoms compositions held at their specified values. In the low-pressure column, the bottoms composition specification is 99.5 mol% methanol, and the distillate composition specification is 75 mol% acetone (2.6 mol% lower than the azeotrope at 1 atm). In the high-pressure column, the bottoms composition specification is 99.4 mol%

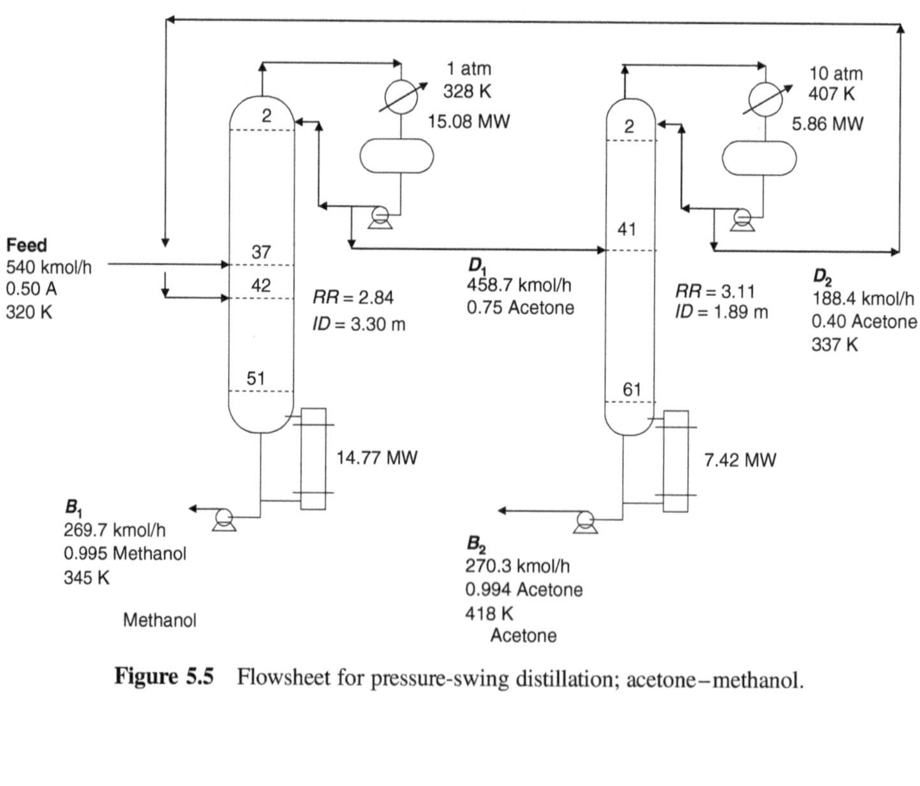


**Figure 5.5** Flowsheet for pressure-swing distillation; acetone–methanol.

acetone, and the distillate composition specification is 40 mol% acetone (2.5 mol% higher than the azeotrope at 10 atm).

The minimum reflux ratios are found with 103 stages to be 2.3 and 2.6 in the low- and high-pressure columns, respectively. The number of stages required for reflux ratios about 1.2 times these minima are 52 in the low-pressure column and 62 in the high-pressure

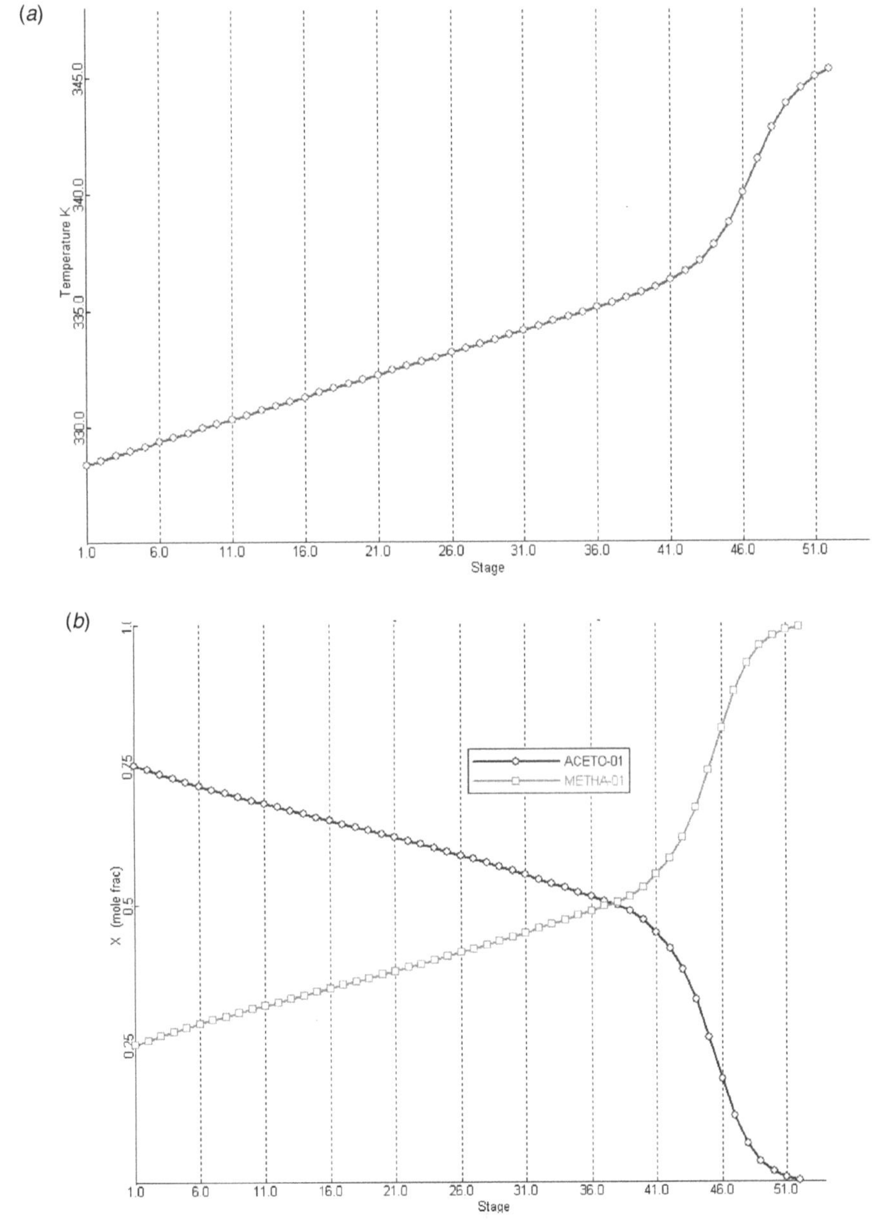


**Figure 5.6** (*a*) Low-pressure column temperature profile. (*b*) Low-pressure column composition profiles.

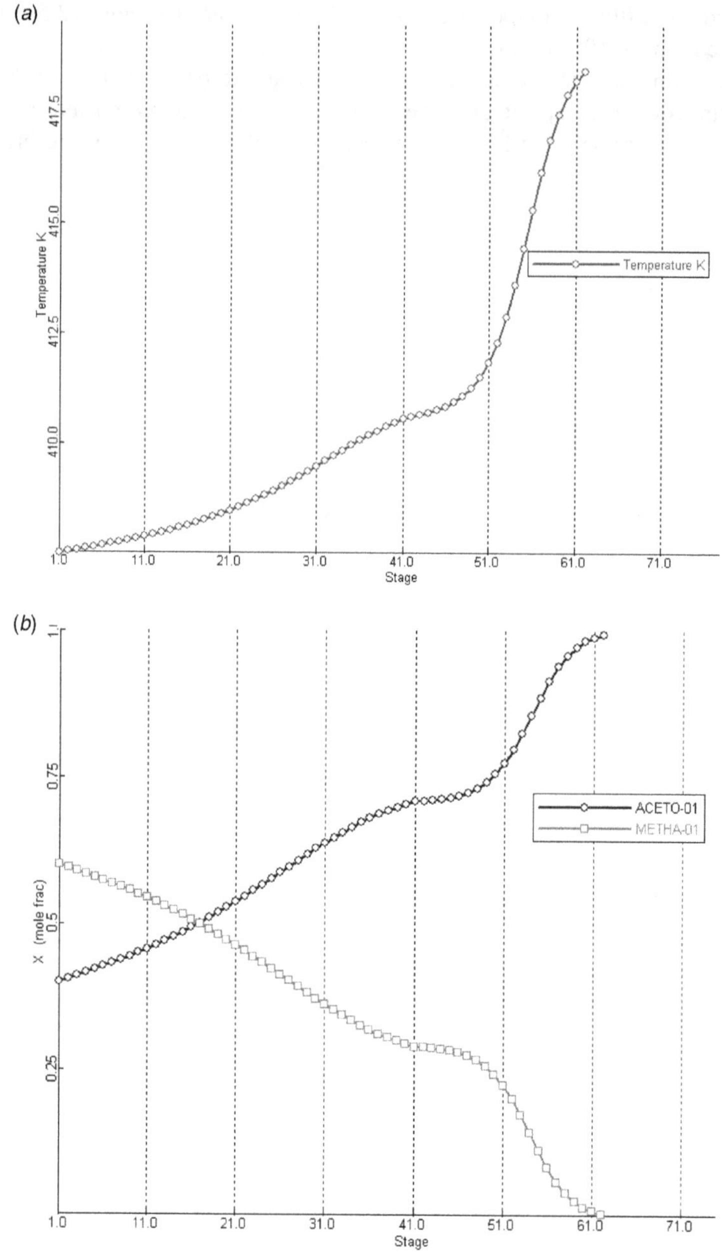


**Figure 5.7** (*a*) High-pressure column temperature profile. (*b*) High-pressure column composition profiles.

column. The optimum feed stage locations are determined by finding the feed trays that minimized reboiler heat input. Figures 5.6 and 5.7 give temperature and composition profiles in the two columns.

The first column in Table 5.1 gives design parameters and economic data for this pressure-swing distillation system. Column diameters are found using Aspen Tray Sizing.

**TABLE 5.1 Design Results Acetone–Methanol.**

| | | Pressure Swing | Extractive |
|---|---|---|---|
| C1 | *N* stage | 52 | 57 |
| | *ID* (m) | 3.30 | 2.79 |
| | $Q_R$ (MW) | 15.8 | 11.4 |
| | $Q_C$ (MW) | 15.1 | 9.87 |
| | $Q_{HX}$ (MW) | — | 1.35 |
| | $T_{reflux}$ (K) | 328 | 329 |
| | $A_C$ (m$^2$) | 996 | 609 |
| | $A_R$ (m$^2$) | 758 | 576 |
| | $A_{HX}$ (m$^2$) | — | 52.7 |
| | Shell (10$^6$ $) | 1.13 | 1.03 |
| | HX (10$^6$ $) | 1.19 | 1.02 |
| | Total capital (10$^6$ $) | 2.32 | 2.06 |
| | Energy ($10$^6$/y) | 2.22 | 1.69 |
| | TAC ($10$^6$/y) | 3.00 | 2.38 |
| C2 | *N* stage | 62 | 26 |
| | *ID* (m) | 1.89 | 1.89 |
| | $Q_R$ (MW) | 7.42 | 7.10 |
| | $Q_C$ (MW) | 5.86 | 6.89 |
| | $T_{reflux}$ (K) | 407 | 338 |
| | $A_C$ (m$^2$) | 72.2 | 288 |
| | $A_R$ (m$^2$) | 383 | 350 |
| | Shell (10$^6$ $) | 0.722 | 0.346 |
| | HX (10$^6$ $) | 0.466 | 0.624 |
| | Total capital (10$^6$ $) | 1.19 | 0.970 |
| | Energy ($10$^6$/y) | 1.13 | 1.03 |
| | TAC ($10$^6$/y) | 1.52 | 1.38 |
| Total system | Total capital (10$^6$ $) | 3.51 | 3.03 |
| | Total energy ($10$^6$/y) | 3.35 | 2.72 |
| | TAC | 4.52 | 3.75 |

The relationships given in Luyben[3] are used for sizing condensers and reboilers and for calculating capital cost of equipment. Temperature differentials are 34.8 K in reboilers. In condensers, the temperature differential is the reflux drum temperature minus 310 K cooling water. Notice that the reflux drum temperature in the high-pressure column (407 K) is much higher than that in the low-pressure column (328 K). This results in a very small area condenser. These temperatures suggest that heat integration would be attractive in this pressure-swing system, and this design is explored in Chapter 6.

Energy cost is assumed to be $4.7 per million kJ, and a payback period of three years is used to calculate annual capital cost. The total annual cost (TAC) of this system is $4,520,000 per year.

The last column in Table 5.1 gives results for an extractive distillation system for this same separation. We will consider this alternative in Chapter 11.

### 5.2.2 Control System Design

The Aspen Plus file is exported to Aspen Dynamics as a pressure-driven simulation after reflux-drum and base volumes are specified to provide 5 min of holdup when at 50%

liquid level. Pumps and valves are sized to give adequate pressure drops to handle changes in flowrates.

Figure 5.8 shows the control structure developed for this system. Conventional PI controllers are used for all flows, pressures, and temperatures. Proportional controllers are used for all liquid levels. Relay-feedback tests are run on the two temperature controllers to determine ultimate gains and periods, and Tyreus–Luyben tuning is used. Each temperature controller has a 1-min deadtime in the loop. Reflux ratios are held constant in each column (2.84 in the low-pressure column and 3.11 in the high-pressure column).

The temperature control tray location is selected by choosing the tray where there is a large temperature change from stage to stage. In the low-pressure column, Stage 48 is selected with a SP of 342.8 K (see Fig. 5.6*a*). In the high-pressure column, Stage 53 is selected with a SP of 413 K (see Fig 5.7*a*). Controller faceplates are shown at the bottom of Figure 5.8. The first column in Table 5.2 gives controller parameters for the two temperature controllers.

Figures 5.9 and 5.10 demonstrate the effectiveness of this control structure. Stable regulatory control is achieved for several large disturbances. The responses of the system to large 20% increases and decreases in feed flowrate are shown in Figure 5.9. Responses for feed composition disturbances from 50 to 60 mol% acetone and from 50 to 40 mol% acetone

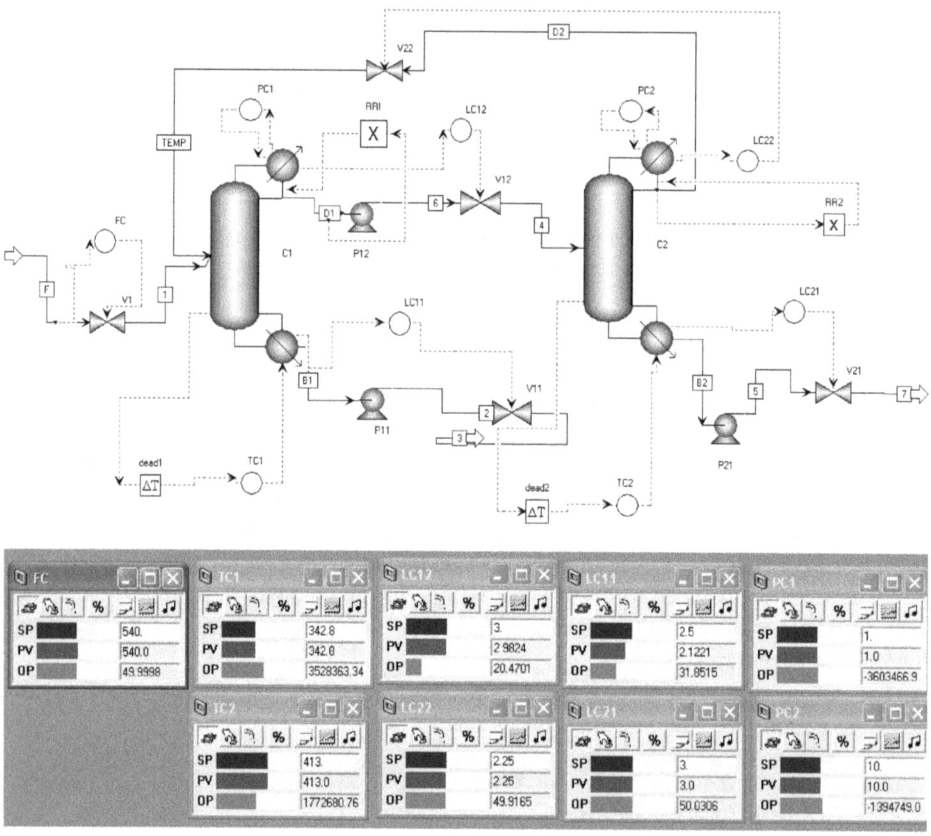


**Figure 5.8** Control structure for pressure-swing distillation system.

TABLE 5.2 **Controller Tuning Parameters Acetone–Methanol.**

| | | Pressure Swing | Extractive | Pressure Swing with Heat Integration | Extractive with Heat Integration |
|---|---|---|---|---|---|
| TC1 | Controlled variable | $T_{1,48} = 342.8$ K | $T_{1,50} = 347.6$ K | $T_{1,48} = 342.8$ K | $T_{1,50} = 347.6$ K |
| | Manipulated variable | $Q_{R1}$ | $Q_{R1}$ | $Q_{AUX}$ | $Q_{AUX}$ |
| | Transmitter range | 300–400 K | 300–400 K | 300–400 K | 300–400 K |
| | Controller output range | $0–6.6 \times 10^6$ cal/s | $0–6 \times 10^6$ cal/s | 0–40 GJ/h$^{(1)}$ | −30–30 GJ/h$^{(2)}$ |
| | Ultimate gain | 26.1 | 7.15 | 44 | 10.3 |
| | Ultimate period | 2.4 min | 4.2 min | 3.6 min | 4.8 min |
| | $K_C$ | 8.16 | 2.23 | 13.7 | 3.2 |
| | $\tau_I$ | 5.28 min | 9.24 min | 7.9 min | 10.6 min |
| TC2 | Controlled variable | $T_{2,53} = 413$ K | $T_{2,22} = 371.7$ K | $T_{2,53} = 140°C$ | $T_{2,21} = 136.3°C$ |
| | Manipulated variable | $Q_{R2}$ | $Q_{R2}$ | $Q_{R2}$ | $Q_{R2}$ |
| | Transmitter range | 350–450 K | 325–425 K | 350–450 K | 300–400 K |
| | Controller output range | $0–3.6 \times 10^6$ cal/s | $0–3.6 \times 10^6$ cal/s | $0–3.6 \times 10^6$ cal/s | $0–6 \times 10^6$ cal/s |
| | Ultimate gain | 13.1 | 4.23 | 24.9 | 1.20 |
| | Ultimate period | 3.6 min | 4.8 min | 4.8 min | 6.0 min |
| | $K_C$ | 4.11 | 1.32 | 7.8 | 0.375 |
| | $\tau_I$ | 7.92 min | 10.6 min | 10.6 min | 13.2 min |

*Notes:* (1) Faceplate shows heat input in metric (GJ/h).
(2) Auxiliary condenser needs to handle an increase in methanol in feed.

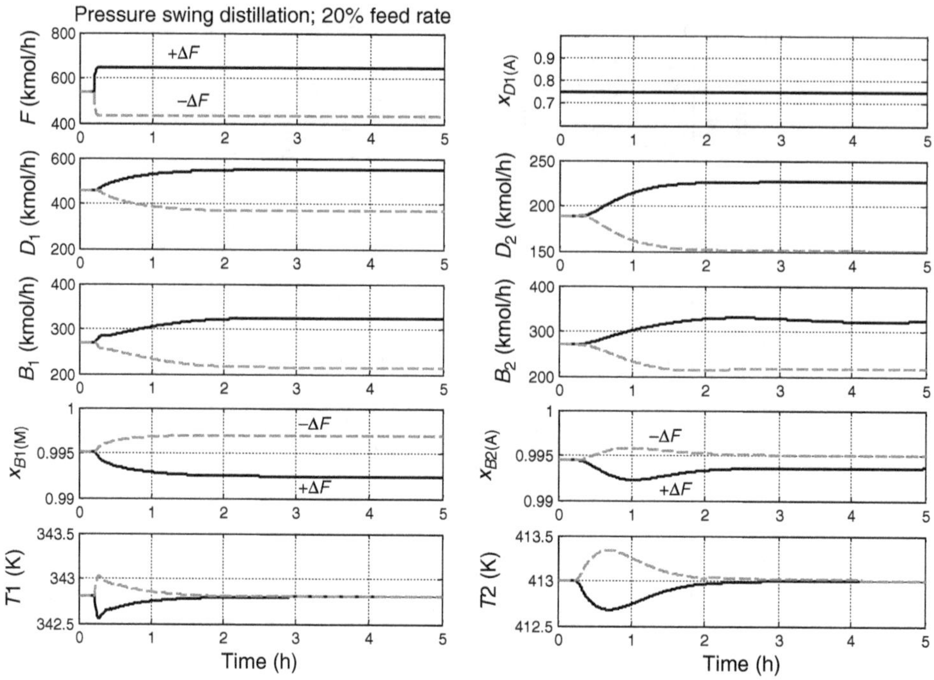


**Figure 5.9** Pressure-swing system; 20% feed flow disturbances.

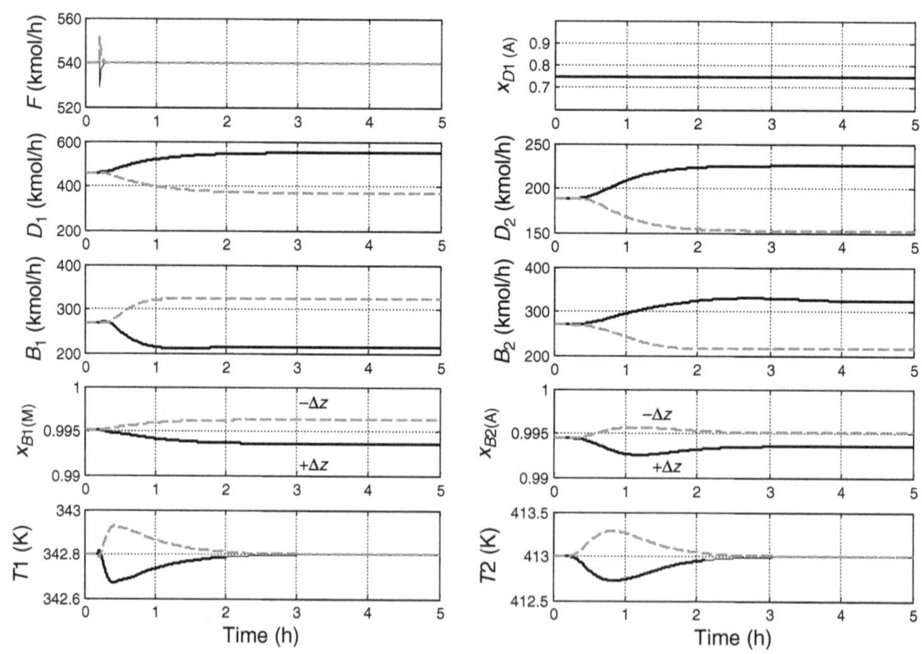


**Figure 5.10** Pressure-swing system; feed composition disturbances.

are shown in Figure 5.10. Product purities ($x_{B1(M)}$ and $x_{B2(A)}$) are held quite close to their specifications. The system takes about 2 h to come to a new steady state.

The pressure-swing distillation system is well suited for heat integration because the temperatures in the high-pressure column are much higher than those in the low-pressure column. Therefore, it is often possible to use the hot vapor from the top of the high-pressure column to provide heat in the reboiler of the low-pressure column. Heat-integrated pressure-swing systems are discussed in Chapter 6.

## 5.3 PENTANE-METHANOL SYSTEM

The final pressure-swing azeotropic distillation system presented in this chapter is the separation of pentanes from methanol. This application arises in the production of tert-amyl methyl ether (TAME), which is used as a high-octane gasoline blending component. TAME is produced in a reactive distillation column by the reaction of methanol with the unsaturated five-carbon iso-amylenes (2-methyl–1-butene and 2-methyl–2-butene).

The process flowsheet has a prereactor, a reactive distillation column, and a methanol-recovery section. A methanol-recovery section is required because the inert C5 components coming in with the reactive iso-amylenes in the C5 fresh feed form azeotropes with methanol. The result is that a significant amount of methanol is present in the distillate from the reactive column.

### 5.3.1 Phase Equilibrium

The chemically inert components in the C5 feed stream include isopentane, *n*-pentane, 1-pentene, and 2-pentene. These components are not inert from the standpoint of phase equilibrium. Essentially all of these chemically inert components go out the top of the reactive distillation column as azeotropes with methanol. Isopentane ($iC_5$) is the major C5 component. We illustrate the phase equilibrium by looking at the binary system $iC_5$ and methanol.

The normal boiling points of $iC_5$ and methanol are 301 and 338 K, respectively. The reactive column operates at 4 bar, at which the boiling points are 348 and 377 K, respectively. Using UNIFAC physical properties, the binary azeotrope is 25.01 mol% methanol at 4 bar and 339 K. This means that the distillate from the reactive distillation column is a mixture of C5 components and methanol.

Figure 5.11 gives *Txy* diagrams for $iC_5$ and methanol at the pressures used in the two columns: 2 atm and 10 atm. The corresponding azeotropic compositions are 21.4 and 33.4 mol% methanol at 317.2 and 372.8 K, respectively.

### 5.3.2 Steady-State Design

Figure 5.12 gives the flowsheet for the entire process. The distillate from the reactive distillation column C1 has a methanol concentration of 28.1 mol%. It is fed into the two-column pressure-swing methanol recovery system. Equipment sizes and economic data are given in Table 5.3.

The methanol-recovery column C2 operates at a pressure of 2 bar so that the reflux-drum temperature (318 K) is high enough for the use of cooling water in the condenser. The separation is a fairly easy one, so using only 10 stages and a reflux ratio of 0.4 yield a

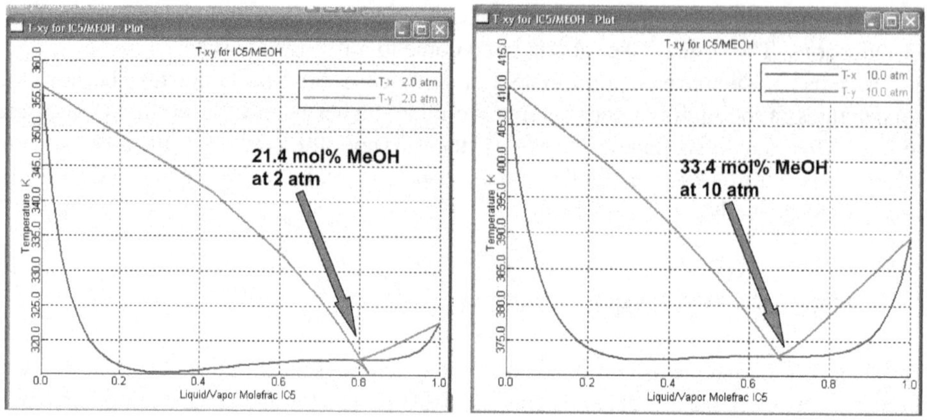


**Figure 5.11** *Txy* diagrams for $iC_5$–methanol at 2 and 10 atm.

bottoms purity of 99.9 mol% methanol. The bottoms (318 kmol/h) is mixed with a fresh feed of pure methanol (232 kmol/h), and the total is split between the methanol fed to the prereactor and the methanol fed to the reactive column.

The distillate $D_2$ composition (22.8 mol% methanol) is near the azeotropic composition at the 2 bar pressure. Reboiler heat input and condenser heat removal are 19.5 MW and 24.5 MW, respectively. The column diameter is 4.2 m.

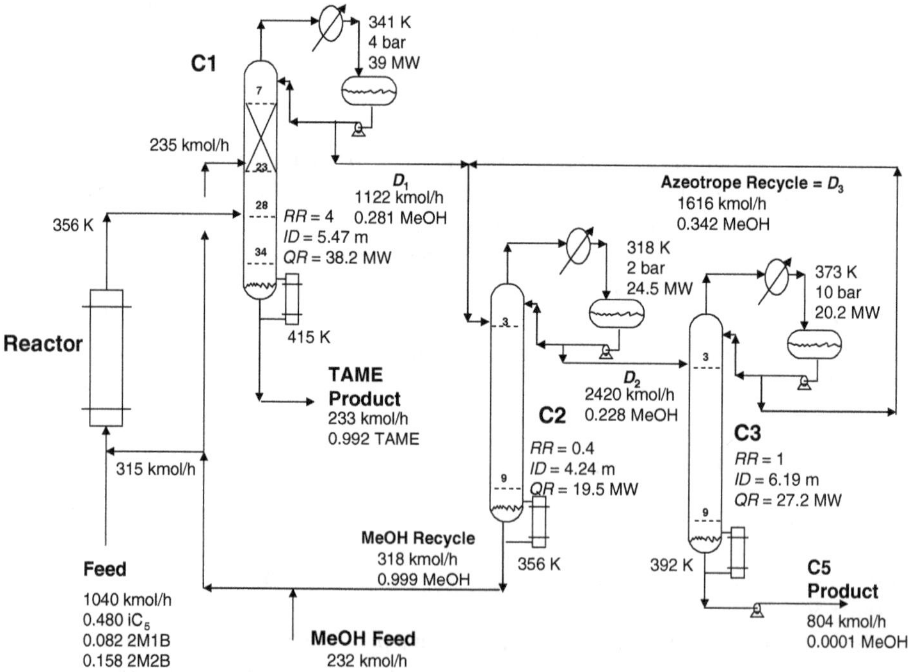


**Figure 5.12** TAME process with pressure-swing methanol recovery.

**TABLE 5.3 Design Parameters and Economics Pressure-Swing Columns C2 and C3 in TAME Process.**

| | | Pressure-Swing Process |
|---|---|---|
| C2 | Diameter (m) | 4.2 |
| | Stages | 10 |
| | $Q_C$ (MW) | 24.5 |
| | $A_C$ (m$^2$) | 2065 |
| | $Q_R$ (MW) | 19.5 |
| | $A_R$ (m$^2$) | 985 |
| | $Q_{HX}$ (MW) | — |
| | $A_{HX}$ (m$^2$) | — |
| | Shell cost ($10^6$ \$) | 0.339 |
| | HX cost ($10^6$ \$) | 1.69 |
| | Energy cost ($10^6$ \$/y) | 2.89 |
| | Capital ($10^6$ \$) | 2.02 |
| C3 | Diameter (m) | 6.2 |
| | Stages | 10 |
| | $Q_C$ (MW) | 20.2 |
| | $A_C$ (m$^2$) | 1703 |
| | $Q_R$ (MW) | 27.2 |
| | $A_R$ (m$^2$) | 1374 |
| | Shell cost ($10^6$ \$) | 0.508 |
| | HX cost ($10^6$ \$) | 0.304 |
| | Energy cost ($10^6$ \$/y) | 4.03 |
| | Capital ($10^6$ \$) | 2.23 |
| Total capital | ($10^6$ \$) | 4.25 |
| Total energy | ($10^6$ \$/y) | 6.92 |
| TAC | ($10^6$ \$/y) | 8.34 |

The distillate $D_2$ is fed to the C5-recovery column C3. This column operates at a pressure of 10 bar, which shifts the azeotropic composition so that the distillate stream from this column $D_3$ has a composition of 34.2 mol% methanol. Note that this is slightly above the composition of the $iC_5$–methanol binary azeotrope, but remember that there is a mixture of a number of C5 components in this system. Higher and lower pressures were explored to see their effect on the economics. The 10 bar pressure seems to be about the optimum since going above this pressure does not shift the azeotrope significantly and raises the base temperature, which would require higher-temperature energy input.

The separation is a fairly easy one, so using only 10 stages and a reflux ratio of 1 yield a bottoms impurity of 0.01 mol% methanol. This bottoms stream $B_3$ is the C5 product stream containing $iC_5$ and the rest of the C5 inerts. The reflux drum temperature is 373 K at this high pressure, which means that some heat integration between C2 and C3 may be economical (the base of C2 is at 356 K and the reflux drum of C3 is at 373 K).

The distillate is recycled back to C2 at a flowrate of 1616 kmol/h. Reboiler heat input and condenser heat removal are 27.2 MW and 20.2 MW, respectively. The column diameter is 6.2 m.

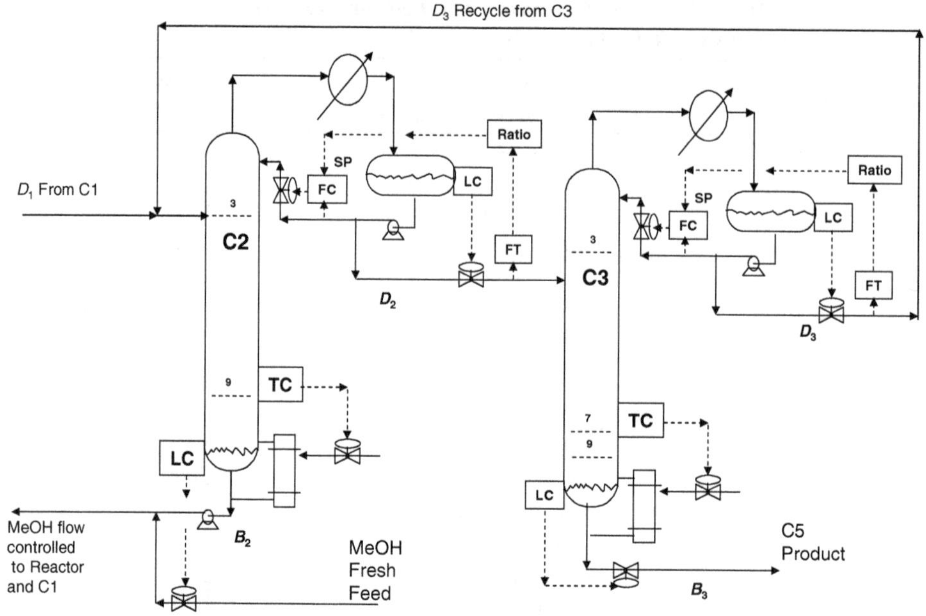


**Figure 5.13** Control structure for pressure-swing process: C2 and C3.

### 5.3.3 Dynamics and Control

In preparation for exporting the steady-state flowsheet into Aspen Dynamics, all equipment is sized. Column diameters are calculated by Aspen Tray Sizing. Reflux drums and column bases are sized to provide 5 min of holdup when 50% full, based on the total liquid entering the surge capacity. Pumps and control valves are specified to give adequate dynamic rangeability. Typical valve pressure drops are 2 atm.

The control structure for the methanol recovery section of the process (C2 and C3) is shown in Figure 5.13. A tray temperature is controlled in each column by manipulating reboiler heat input. The trays are selected by finding the location where the temperature profile is steep: Stage 9 in column C2 and Stage 7 in column C3. All temperature controllers have 1 min deadtimes. The PI controllers are tuned by running a relay-feedback test and using the Tyreus–Luyben settings. Table 5.4 gives controller constants. All liquid levels are controlled by proportional controllers with gains of 2 for all level loops. Liquid levels in reflux drums are controlled by manipulating distillate flowrates. The reflux ratios in all

**TABLE 5.4 Controller Tuning Constants for Pressure-Swing Columns C2 and C3 for TAME Process.**

| Column | Controller | $K_C$ | $\tau_I$ (min) | Transmitter Range | Set Point |
|---|---|---|---|---|---|
| Pressure-swing | | | | | |
| | TC2 | 0.41 | 11 | 300–400 K | 346 K |
| | TC3 | 1.8 | 8.4 | 350–450 K | 387 K |

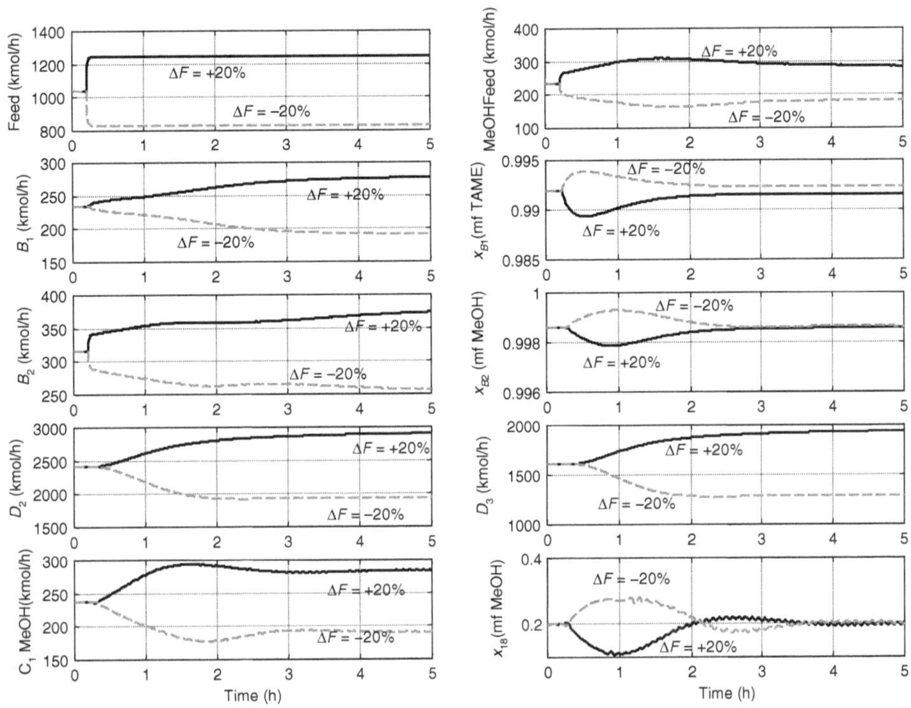


**Figure 5.14** Feed rate disturbance with pressure-swing TAME process.

columns are controlled by manipulating reflux. Column pressure controllers use default controller settings and manipulate condenser heat removal.

The liquid level in the base of column C3 is controlled by manipulating bottoms flowrates. In column C2, the base level is controlled by manipulating the fresh feed of methanol, as shown in Figure 5.13. This exactly balances the stoichiometry in the entire plantwide system by appropriately adjusting the addition of fresh methanol as it is consumed in the TAME reaction. Remember that the downstream methanol is set by two flow controllers (methanol flow to the reactor and to C1). This means that increasing methanol fresh feed flow produces an *immediate* decrease in the flowrate of bottoms $B_2$ from C2. Thus there is an instantaneous effect on the base level.

This control structure effectively handles quite large disturbances. Figure 5.14 gives the responses to positive and negative 20% changes in the feed to the process occurring at 0.2 hours. The control structure maintains TAME purity very close to its specification. The system reaches a new steady state in about 3 h.

Feed composition disturbances are also tested. The responses of the system to two types of composition disturbances are shown in Figure 5.15. The curves labeled $+\Delta z$ correspond to 20% increases in the molar flowrates of the two reactants in the feed (2M1B and 2M2B), with an appropriate reduction in the $iC_5$ molar flowrate such that the total feed flowrate remains constant. The curves labeled $-\Delta z$ correspond to 20% decreases in the molar flowrates of the two reactants, with an appropriate increase in the $iC_5$ molar flowrate. The control structure provides effective control of the process. The compositions of the C5 product stream and the methanol recycle stream remain close to their desired values.

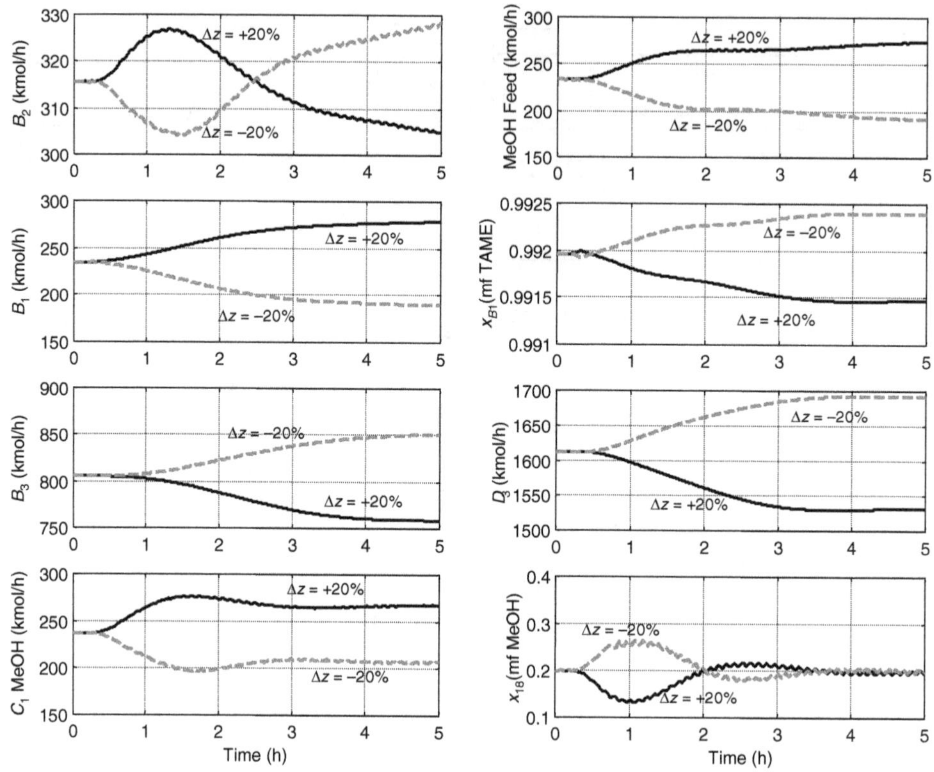


**Figure 5.15** Feed composition disturbance with pressure-swing TAME process.

## 5.4 CONCLUSION

The application of pressure swing as a method for separating a homogeneous binary azeotrope has been illustrated in this chapter. Three chemical systems have been explored. Steady-state designs have been developed, and effective dynamic control structures have been tested for significant disturbances in throughput and feed composition.

Comparison with the alternative method of extractive distillation for the acetone–methanol will be presented in a subsequent chapter.

## REFERENCES

1. Abu-Eishah S. I. and W. L. Luyben, Design and control of a two-column azeotropic system, *Ind. Eng. Chem. Process Des. Dev.*, **24**, 132–140 (1985).
2. Knapp J. P. and M. F. Doherty, Thermal integration of homogeneous azeotropic distillation sequences, *AIChE J.*, **36**, 969–984 (1990).
3. Luyben W. L., *Distillation Design and Control Using Aspen Simulation*, 2006, Wiley, Hoboken, NJ, p. 88.

CHAPTER 6

# PRESSURE SWING WITH HEAT INTEGRATION

In Chapter 5 we illustrated the use of pressure-swing distillation to separate homogeneous azeotropic mixtures whose azeotropic compositions changed significantly with pressure. We noted that operating two distillation columns at two different pressures produces low temperatures in one and high temperatures in the other. Therefore, the use of heat integration is often economically attractive in pressure-swing systems. In this chapter we illustrate the steady-state economic effects of using heat integration. We also show how heat integration affects the dynamic control of the system.

The heat integrated systems can be designed for *complete* heat integration in which there is only one steam-heated reboiler (in the high-pressure column) and one water-cooled condenser (in the low-pressure column). There is another heat exchanger, with process fluid on both the condensing and boiling sides, which serves as the condenser for the high-pressure column and the reboiler for the low-pressure column. Thus the heat removal in the condenser of the high-pressure column is exactly equal to the heat input in the reboiler of the low-pressure column.

Alternatively, heat integrated systems can be designed for *partial* heat integration in which the high-pressure condenser heat removal is not equal to the low-pressure reboiler heat input. The use of auxiliary reboilers and/or auxiliary condensers is required. As we will demonstrate in this chapter, there is an economic/controllability trade-off. Complete heat integration usually is more economical from the standpoint of capital investment and energy cost. Partial heat integration usually results in better dynamic control because there are more control degrees of freedom.

*Design and Control of Distillation Systems for Separating Azeotropes.* By William L. Luyben and I-Lung Chien

## 6.1 THF–WATER SYSTEM STEADY-STATE DESIGN WITH HEAT INTEGRATION

We begin with the THF–water system. A comparison of systems with no heat integration, with partial heat integration and with complete heat integration will be presented. The phase equilibria for this system and a nonheat-integrated system have been discussed in Chapter 5.

### 6.1.1 Partial Heat Integration

The 1985 paper by Abu-Eishah and Luyben[1] discussing the design and control of the THF–water system used partial heat integration. The optimum design of a heat-integrated two-column system was determined by minimizing the total energy consumption (the sum of the two reboiler duties). They then used the heat removed in the condenser of the high-pressure column as a portion (about 60%) of the heat input to the base of the low-pressure column. The remaining heat was provided by an auxiliary reboiler, which was heated by low-pressure steam. The reboiler in the high-pressure column was driven by high-pressure steam (base temperature was 148°C at 7.9 bar, requiring 10 bar steam). Thus, a partially heat-integrated system was studied. The auxiliary reboiler provided an additional control degree of freedom, so both a tray temperature and a reflux ratio could be independently manipulated in each column.

The flowsheet is shown in Figure 6.1. The base case has 2000 kmol/h of feed with a composition of 6 mol% THF. Operating pressures in the two columns are 1.1 and 7.9 bar. The numbers of stages in each column are the same as those used by Abu-Eishah and

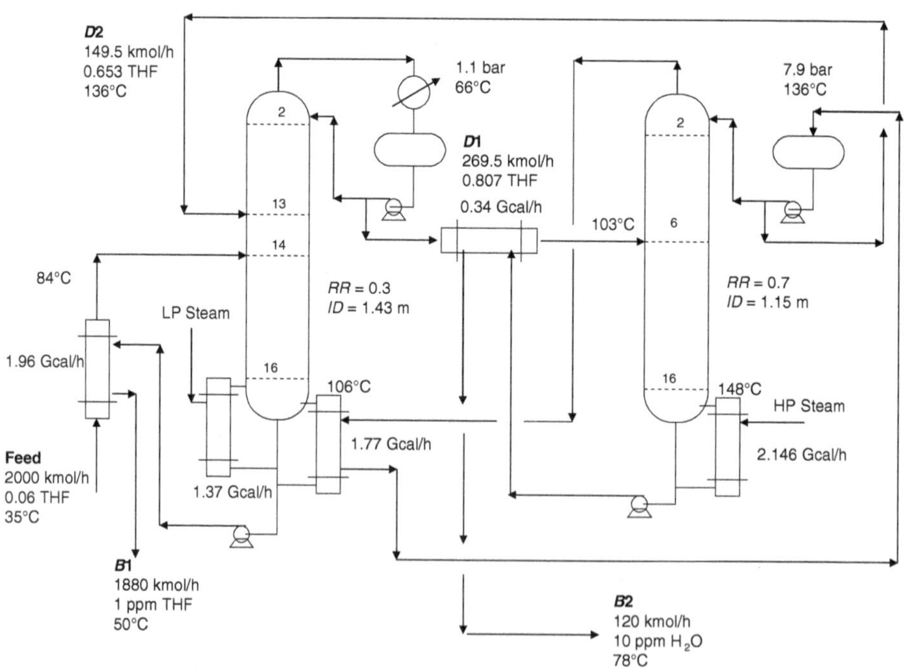


**Figure 6.1** Flowsheet conditions with partial heat integration.

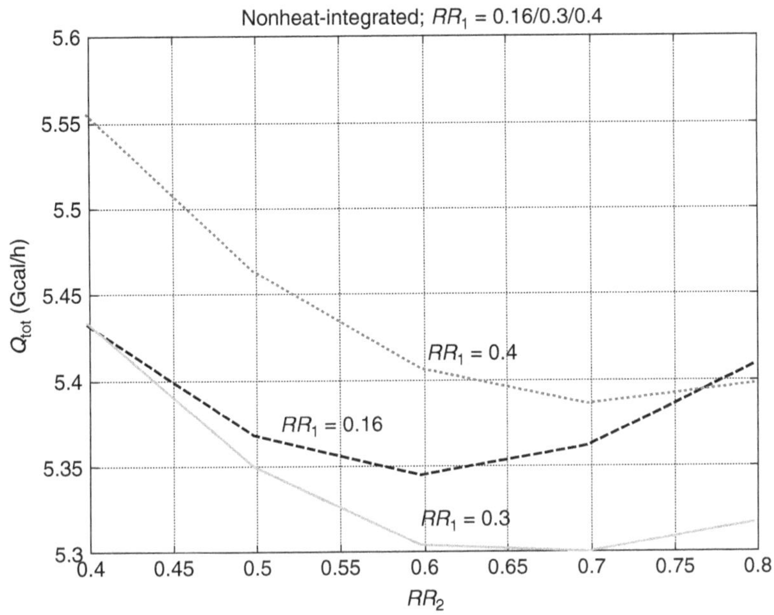


**Figure 6.2** Minimum total energy.

Luyben[1] (17 stages using Aspen notation with the condenser as Stage 1). Product purities are 1 ppm THF in the water product $B_1$ from the bottom of the low-pressure column C1 and 10 ppm water in the THF product $B_2$ from the bottom of the high-pressure column C2.

The Aspen *Design Spec* and *Vary* capability is used to hold the two product purities by manipulating the flowrates of the two bottoms streams. The reflux ratios in the two columns are the design optimization variables. They are varied until the minimum reboiler energy consumption is determined (the sum of $Q_{R1}$ and $Q_{R2}$). As shown in Figure 6.2, the optimum values of the reflux ratios are $RR_1 = 0.3$ and $RR_2 = 0.7$. The corresponding total energy consumption is 5.29 Gcal/h ($Q_{R1} = 3.14$ and $Q_{R2} = 2.15$ Gcal/h).

Under these conditions the heat removal in the condenser of the high-pressure column is 1.77 Gcal/h. This suggests that a partially heat-integrated flowsheet could be used. A condenser/reboiler could transfer 1.77 Gcal/h of energy into the low-pressure column. The temperature differential between the condensing vapor from the high-pressure column (136°C) and the boiling liquid at the bottom of the low-pressure column (106°C) is adequate for heat integration with a reasonable area heat exchanger. The remaining energy requirement in the low-pressure column can be satisfied by a steam-heated auxiliary reboiler supplying 3.14−1.77 = 1.37 Gcal/h. The heat-transfer area of the condenser/reboiler is 79.9 $m^2$. Note that the pressure in the high-pressure column is not controlled but floats with operating conditions. The effect of this feature on the use of temperature control in the high-pressure column will be discussed later in this chapter.

### 6.1.2 Complete Heat Integration

Next we study the case where the two columns are fully heat integrated and no auxiliary reboiler or condenser is used. The only energy input is the heat-transfer duty in the high-pressure column $Q_{R2}$. This "neat" configuration minimizes the total energy consumed. However, a control degree of freedom is lost since the heat input to the low-pressure column $Q_{R1}$ cannot be independently set. It must be equal to the heat removal rate $Q_{C2}$ in the condenser of the high-pressure column.

The *Design Spec* and *Vary* feature in each of the column blocks in Aspen Plus is used to adjust the bottoms flowrate to achieve the desired product purity in each column. The specifications for the product purities are the same in all cases. The water product from the base

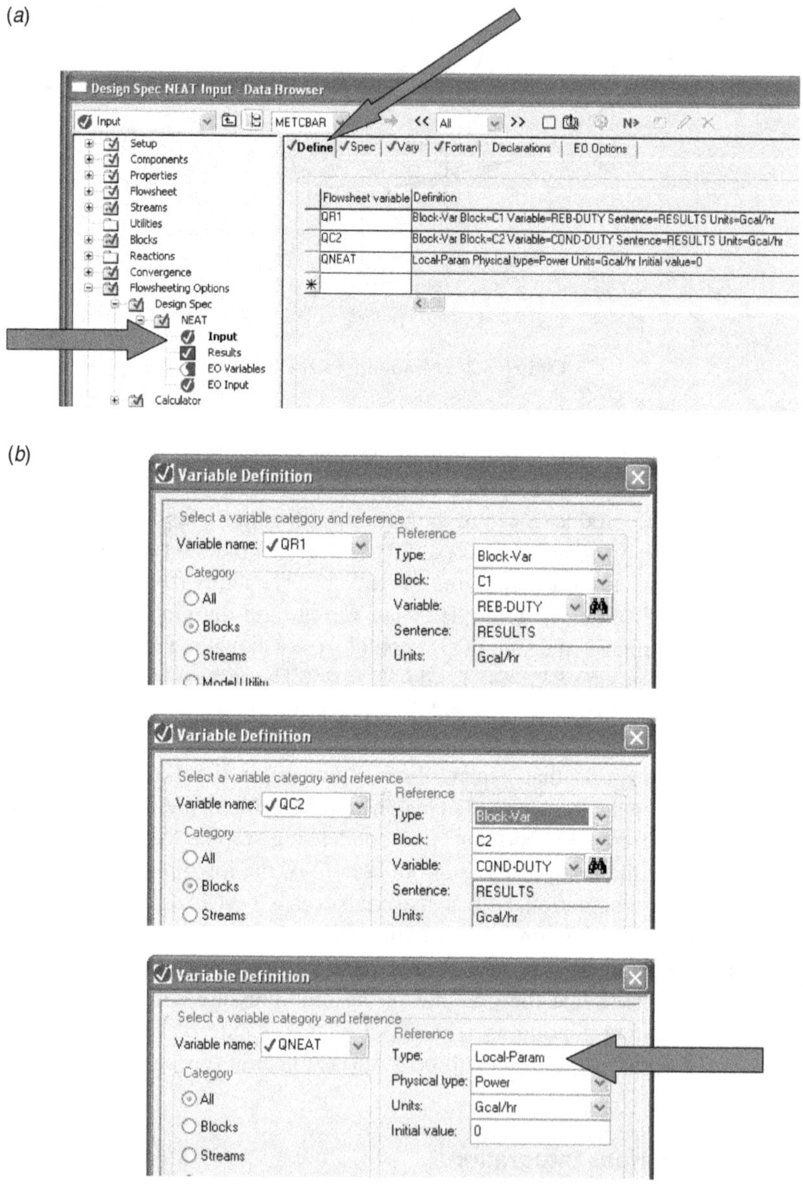


**Figure 6.3** (*a*) Setting up flowsheet design spec. (*b*) Defining variables. (*c*) Specifying QNEAT and specification. (*d*) Varying reflux ratio in high-pressure column.

(c)

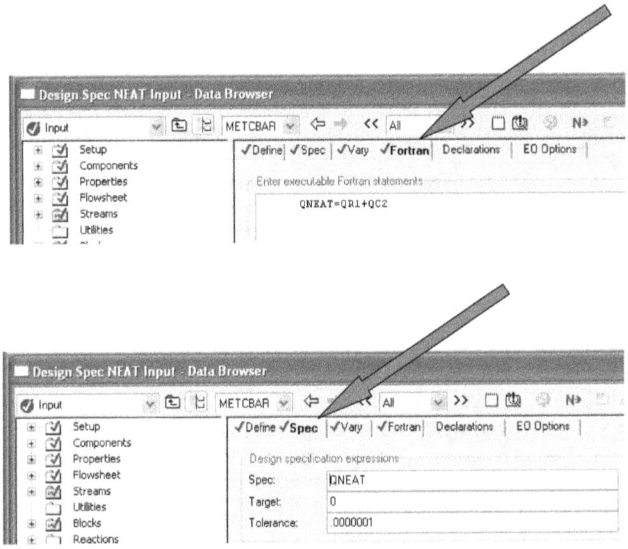


(d)

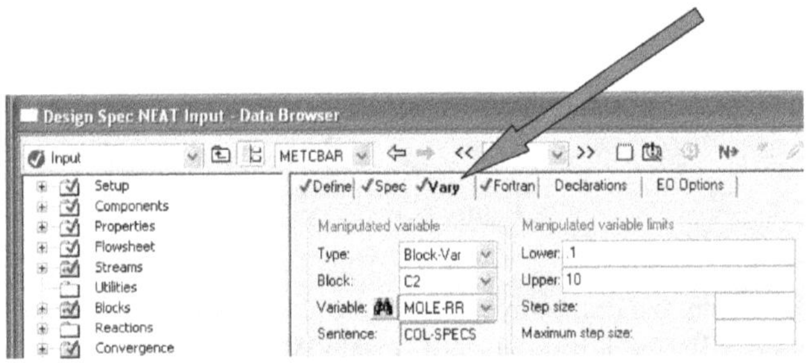


**Figure 6.3** *Continued.*

of the low-pressure column $B_1$ has a composition $x_{B1(\mathrm{THF})}$ of 1 ppm (molar) THF. The THF product from the base of the high-pressure column $B_2$ has a composition $x_{B2(\mathrm{water})}$ of 10 ppm (molar) water.

With complete heat-integration, the heat removal in the condenser of the high-pressure column $Q_{C2}$ must be equal to the heat input to the reboiler of the low-pressure column. Therefore, a degree of freedom is lost, and we can only set the reflux ratio on one column, not both as is the case with partial heat integration or no heat integration.

Achieving steady-state simulations in Aspen Plus that rigorously capture the neat heat integration requires the use of a "flowsheet design spec" to make $Q_{R1} = -Q_{C2}$. Figure 6.3 shows how this is set up. Variables and parameters are defined (Figs. 6.3*a* and *b*), specifications are given (Fig. 6.3*c*), and the manipulated variable is selected to be the reflux ratio in the high-pressure column $RR_2$ (Fig. 6.3*d*). Thus the reflux ratio of the high-pressure column is varied by the flowsheet design spec to make $Q_{R1} = -Q_{C2}$.

With the reflux ratio fixed in the low-pressure column, there are three variables being used ($B_1$, $B_2$, and $RR_2$) to drive the two product compositions to their desired specifications and to make the heat duties equal in magnitude but opposite in sign.

The next step in the design is to vary the reflux ratio in the low-pressure column $RR_1$ to find the value that minimizes the heat input to the reboiler of the high-pressure column $Q_{R2}$.

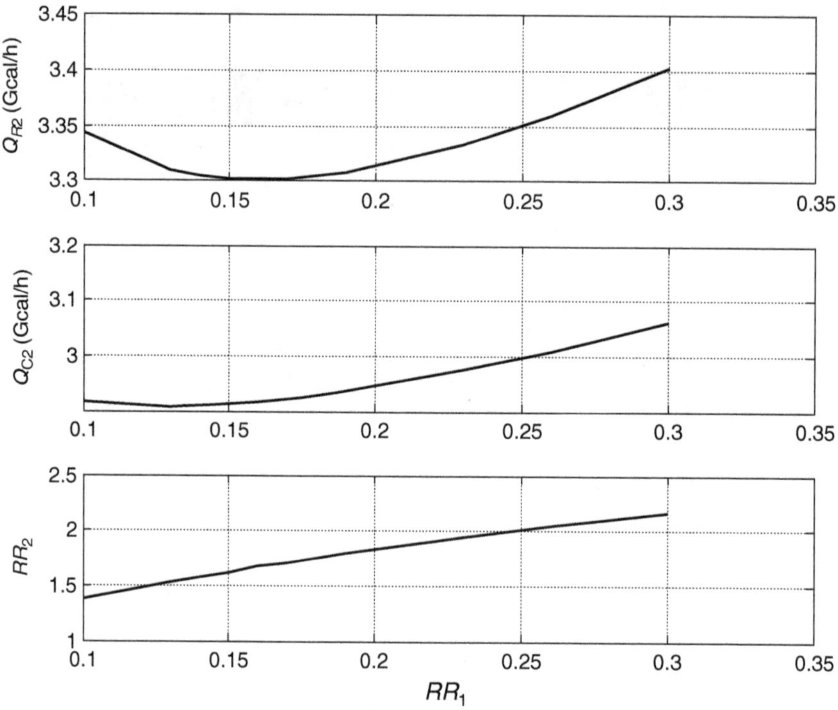


**Figure 6.4** Effect of $RR_1$.

Figure 6.4 shows that the optimum reflux ratio $RR_1$ is 0.16. At this value, the corresponding heat input $Q_{R2}$ is 3.3016 GJ/h and the reflux ratio $RR_2$ is 1.669. Figure 6.5 gives stream information at these steady-state design conditions. Figure 6.6 gives temperature and composition profiles in the two columns.

| | F | B1 | D1 | B2 | D2 |
|---|---|---|---|---|---|
| Temperature C | 35.0 | 106.2 | 66.0 | 148.1 | 136.3 |
| Pressure bar | 5.000 | 1.260 | 1.100 | 8.068 | 7.908 |
| Vapor Frac | 0.000 | 0.000 | 0.000 | 0.000 | 0.000 |
| Mole Flow kmol/hr | 2000.000 | 1880.002 | 275.986 | 120.003 | 155.988 |
| Mass Flow kg/hr | 42521.557 | 33868.858 | 16860.379 | 8652.953 | 8207.680 |
| Volume Flow cum/hr | 43.516 | 37.151 | 19.676 | 11.920 | 10.535 |
| Enthalpy MMkcal/hr | -134.358 | -125.537 | -14.816 | -5.648 | -8.432 |
| Mole Flow kmol/hr | | | | | |
| TETRA-01 | 120.000 | 0.002 | 219.783 | 120.001 | 99.785 |
| WATER | 1880.000 | 1880.000 | 56.204 | 0.001 | 56.203 |
| Mole Frac | | | | | |
| TETRA-01 | 0.060 | 1 PPM | 0.796 | 1.000 | 0.640 |
| WATER | 0.940 | 1.000 | 0.204 | 10 PPM | 0.360 |

**Figure 6.5** Stream conditions.

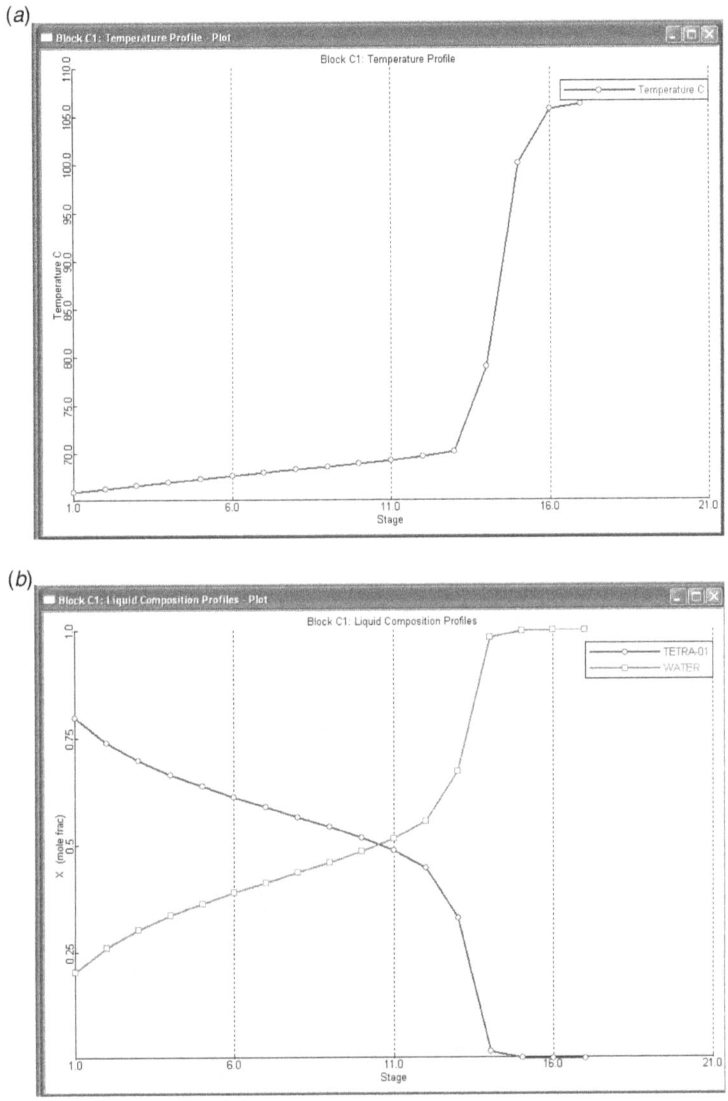


**Figure 6.6** (*a*) Low-pressure column; temperature profile. (*b*) Low-pressure column; composition profiles. (*c*) High-pressure column; temperature profile. (*d*) High-pressure column; composition profiles.

Figure 6.7 gives flowsheet conditions and equipment sizes. Based on a feed flowrate of 2000 kmol/h and a feed composition of 6 mol% THF, the column diameters calculated by Aspen's *Tray Sizing* are 1.36 and 1.43 m. The total number of stages in each column is set at 17 since the separations are fairly easy. Feed tray locations are shown in the figure. The economizers are sized using minimum approach temperatures of 10–15°C and overall heat-transfer coefficients of $U = 0.00306\ \mathrm{GJ\ h^{-1}\ m^{-2}\ ^{\circ}C^{-1}}$ ($150\ \mathrm{Btu\ h^{-1}\ ft^{-2}\ ^{\circ}F^{-1}}$).

At this point, the heat duty in the condenser/reboiler is known. All the temperatures throughout both columns are also known. Therefore, the heat-transfer area of this heat exchanger can be determined by using the difference between the reflux drum temperature

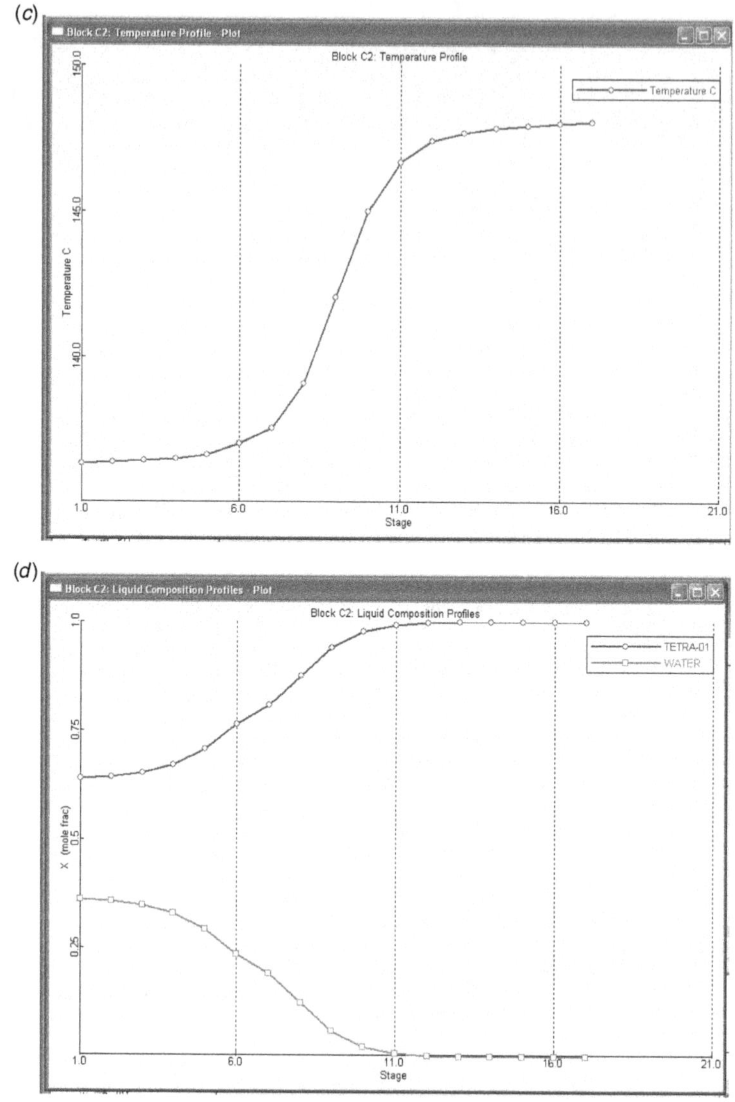


**Figure 6.6** *Continued.*

of the high-pressure column, the base temperature of the low-pressure column, and an appropriate overall heat-transfer coefficient. In the numerical example a value of $U = 0.00306$ GJ $h^{-1}\,m^{-2}\,{}^{\circ}C^{-1}$ (150 Btu $h^{-1}\,ft^{-2}\,{}^{\circ}F^{-1}$) is assumed. The resulting area is 133 $m^2$.

### 6.1.3 No Heat Integration

If the two columns have independent reboilers and condensers, both a temperature and a reflux ratio can be manipulated in each column. In addition, the pressures in both columns can be controlled. Therefore, the pressure in the high-pressure column does not float with operating conditions. As we will see, this is advantageous for temperature control. The flowsheet for this design is given in Figure 6.8. The columns are exactly the same, but there is a

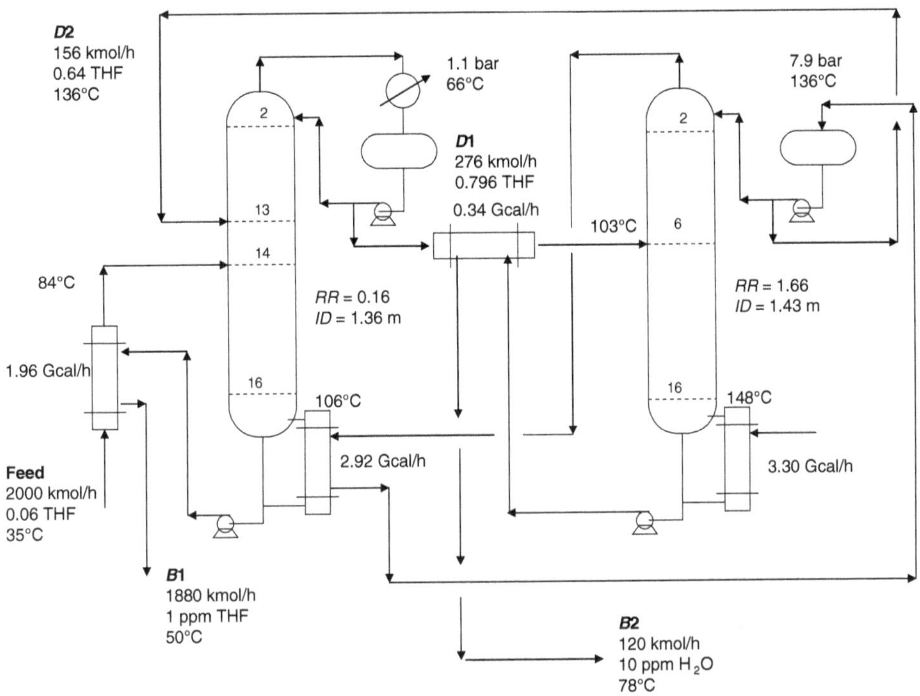


**Figure 6.7** Flowsheet conditions; fully heat integrated.

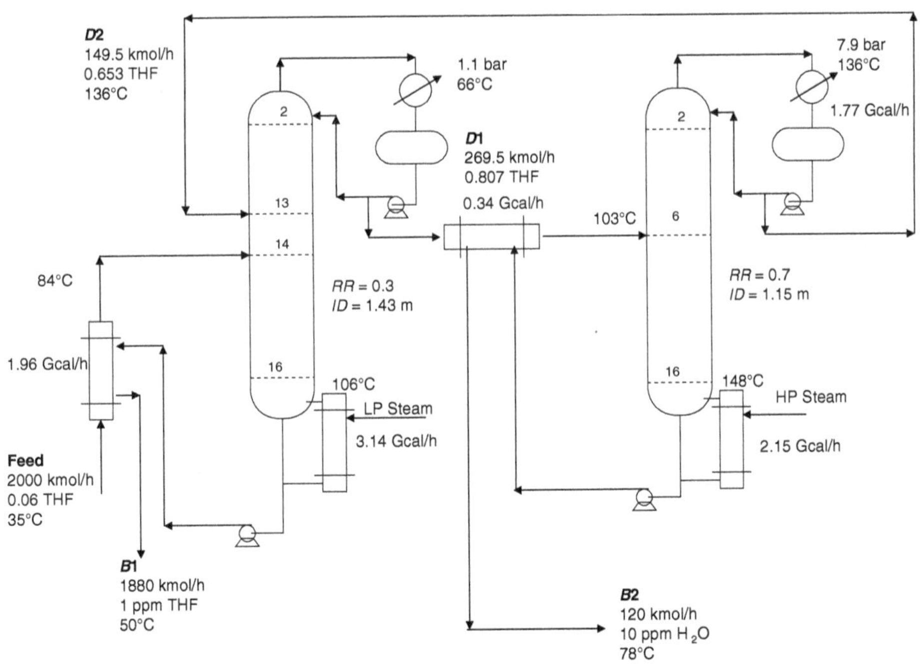


**Figure 6.8** Flowsheet conditions with no heat integration.

condenser on the high-pressure column and only one reboiler on the low-pressure column. The total energy consumption in this design (5.29 Gcal/h) is the sum of $Q_{R1}$ (3.14 Gcal/h) and $Q_{R2}$ (2.15 Gcal/h). This should be compared with the energy consumption of the totally heat-integrated system (3.30 Gcal/h) and the energy consumption of the partially heat-integrated system (3.52 Gcal/h).

These results clearly demonstrate that heat integration can result in very significant reductions in energy consumption. However, as we demonstrate in the next section, the dynamic controllability of the completely heat-integrated system is not as good as that of the nonheat-integrated system.

## 6.2 THF–WATER SYSTEM DYNAMICS AND CONTROL

The dynamic controllabilities of the three designs considered above are compared in this section. Implementing heat integration in Aspen Dynamics requires the use of *Flowsheet Equations*. First we look at the case with complete heat integration.

### 6.2.1 Complete Heat Integration

The flowsheet is exported to Aspen Dynamics after all the parameters required for a dynamic simulation are specified (equipment sizes), and the flowsheet is pressure checked so that a realistic "pressure driven" dynamic simulation can be used. Reflux drums and column bases are sized to provide 5 min of holdup when half full. Pump heads and control valve pressure drops are specified to give reasonable plumbing. Typical design control valve pressure drops are 3 bar with the valve half open at design flowrates.

Achieving dynamic simulations that rigorously capture the "neat" heat integration require the use of *Flowsheet Equations* in Aspen Dynamics. Two conditions must exist at each point in time during the dynamic simulation. First, the heat transfer in the condenser/reboiler must be equal to the product of the area, the overall heat-transfer coefficient, and the current temperature difference between the reflux drum of the high-pressure column and the base of the low-pressure column. These two temperatures both change dynamically as compositions and pressures vary. The pressure in the high-pressure column is not controlled but floats.

The second condition is that $Q_{R1}$ must equal the negative of $Q_{C2}$. From a fundamental perspective, these two conditions make sure that the first and second laws of thermodynamics are satisfied.

With heat integration, there is no pressure controller on the high-pressure column. The pressure $P_2$ in this column is not controlled and varies with operating conditions. In addition, the reboiler heat input in the low-pressure column cannot be specified. Therefore, as shown in Figure 6.9, there is no pressure controller on the high-pressure column.

The next step is to set up the "flowsheet equations." As shown in Figure 6.10*a*, the *Flowsheet* item is selected in the *Exploring Simulation* window, and *Flowsheet* in the lower window (two parallel blue bars) is double clicked. A text editor window opens (Fig. 6.10*b*) on which the appropriate equations are entered.

The first line calculates the heat-transfer rate from the fixed area (132.4 $m^2$), the fixed overall heat-transfer coefficient (0.00306 GJ $h^{-1}$ $m^{-2}$ °$C^{-1}$), and the current difference in temperature between the reflux drum of the high-pressure column and the reboiler of low-pressure column. The second line equates the two heat duties with the appropriate change in sign. Right clicking the window and selecting *Compile* complete the operation.

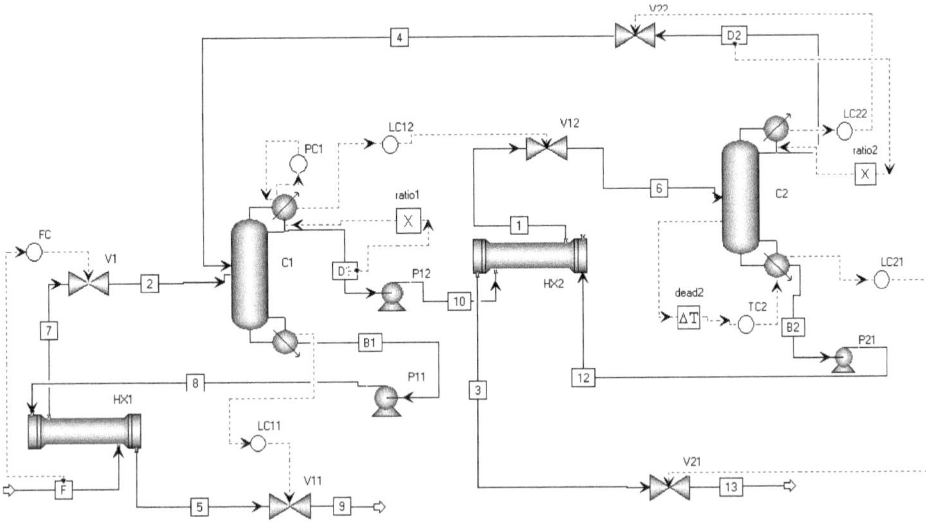


**Figure 6.9** Control structure with heat integration.

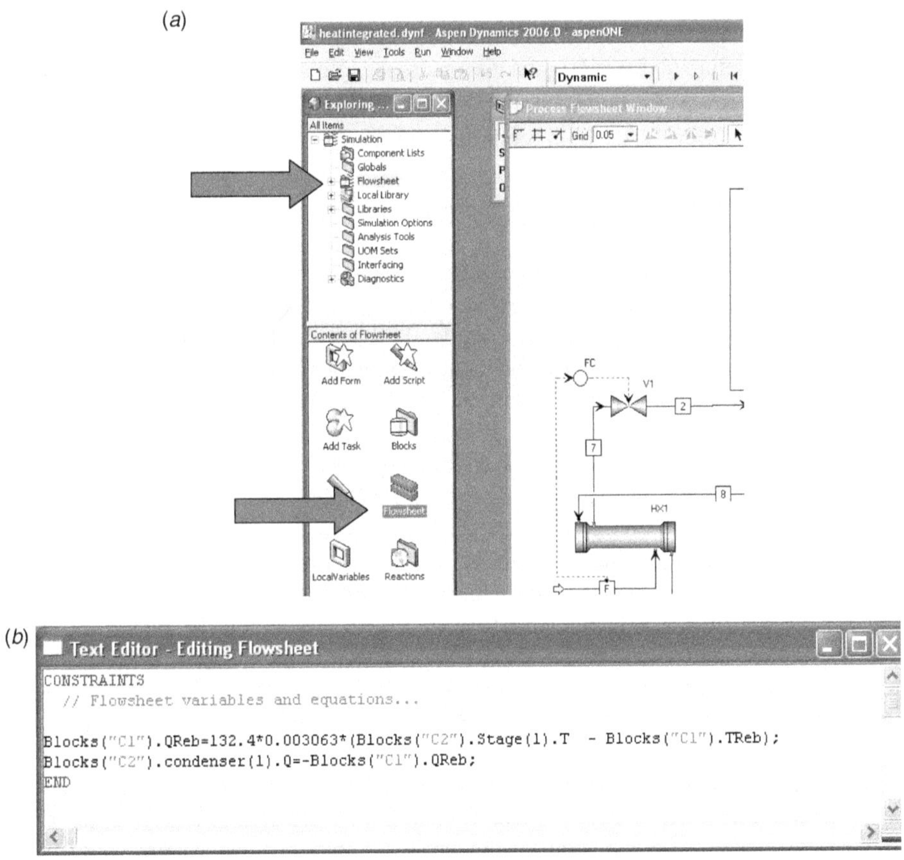


**Figure 6.10** (*a*) Setting up flowsheet equations. (*b*) Equations for condenser/reboiler coupling and heat transfer.

The little status arrow at the bottom of the Aspen Dynamics window changes from green to red, however, indicating that the simulation is not ready to run. Clicking the arrow opens the window shown in Figure 6.11*a*, which reveals that the problem is caused by having too many fixed variables. The two variables that must be changed from *Fixed* to *Free* are the heat duty variable in the condenser of the high-pressure column and the heat duty variable in the reboiler of the low-pressure column. These changes are achieved by clicking on the appropriate column, selecting *Forms* and *All Variables* and scrolling down to the appropriate variable. Figure 6.11*b* shows the operation for the high-pressure column C2. The *Condenser(1).QR* variable is changed from *Fixed* to *Free.* Figure 6.11*c* shows the operation for the low-pressure column C1. The *QRebR* variable is changed from *Fixed* to *Free.* The flowsheet is now correctly specified.

The basic control structure evaluated is shown in Figure 6.12. There are 14 controllers in this system.

1. Feed is flow controlled.
2. The temperature on Stage 14 in the low-pressure column is controlled by manipulating the reboiler heat input in the high-pressure column.
3. The reflux ratio in the low-pressure column is fixed.
4. The pressure in the low-pressure column is controlled by manipulating the rate of heat removal in the condenser of the low-pressure column.
5. The temperature on Stage 9 in the high-pressure column is controlled by manipulating the reflux ratio in the high-pressure column.
6. Reflux-drum levels in both columns are controlled by manipulating distillate flowrates.
7. Base levels in both columns are controlled by manipulating bottoms flowrates.

The controller output signal from the low-pressure column temperature controller TC1 is the heat-input to the reboiler of the high-pressure column $Q_{R2}$. The controller output signal from the high-pressure column temperature controller TC2 is the reflux ratio of the high-pressure column $RR_2$. Note that there is no pressure controller on the high-pressure column. The lower right faceplate shown at the bottom of Figure 6.12 is used as a pressure indicator to show what $P_2$ is doing during the dynamic simulations.

The alternative control structure of controlling the temperature in the low-pressure column with reflux ratio in the low-pressure column might seem a more logical choice. However, the very low reflux ratio in this column makes this structure ineffective. Temperatures are affected much more by vapor boilup (which results from changes in the high-pressure column) than by reflux ratio. Simulations of this alternative structure confirmed that it was unworkable with shutdowns occurring for extremely small disturbances.

Deadtimes of 1 min are inserted in the two temperature loops. Table 6.1 gives temperature transmitter ranges, controller output ranges, and several tuning parameters. Relay-feedback tests are run, and Tyreus–Luyben tuning is used in both controllers. All level controllers are proportional only with $K_C = 2$.

Figure 6.13 shows the response of the system to a 5% step increase in the setpoint of the feed flow controller at time equals 0.2 h. The solid lines are for the case with the control structure discussed above and shown in Figure 6.12. Both product flowrates ($B_1$ and $B_2$) and the distillate flowrates ($D_1$ and $D_2$) increase. The temperature on Stage 14 in the low-pressure column drops quickly from 79°C to about 73°C, but the controller eventually

(a)

(b)

(c)

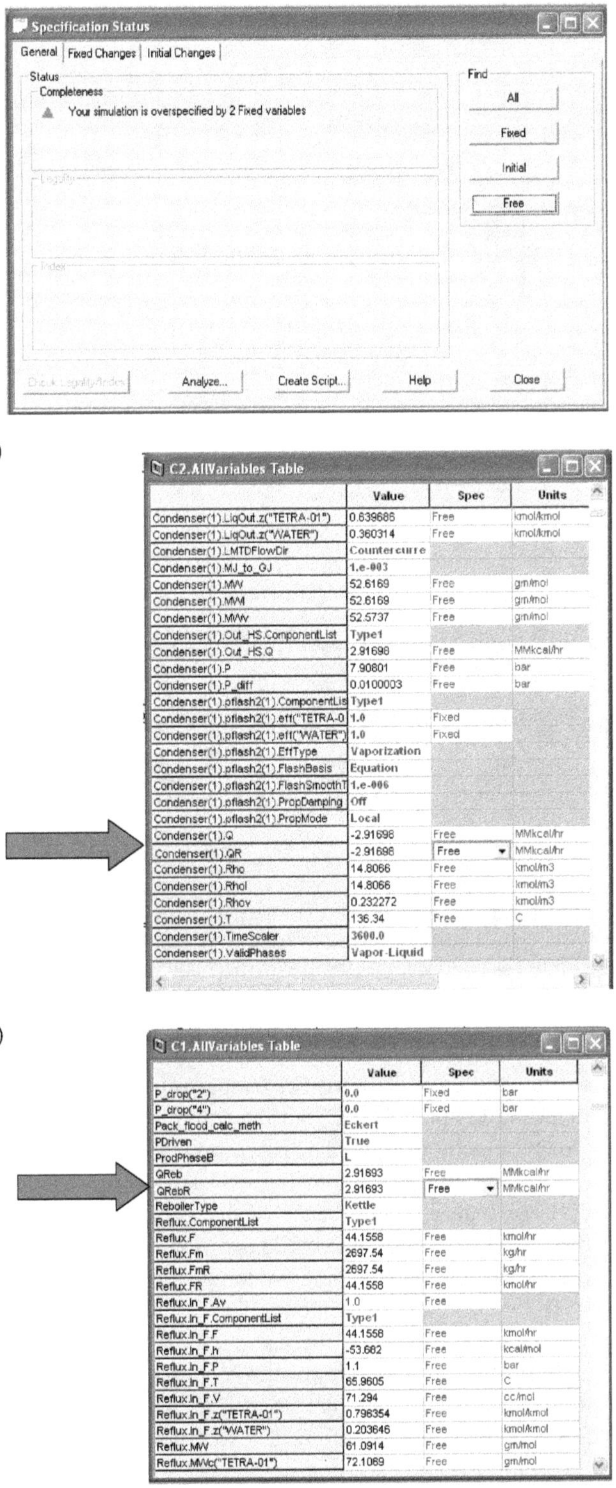


Figure 6.11 (*a*) Over specified simulation. (*b*) Changing $Q_{C2}$ to free. (*c*) Changing $Q_{R1}$ to free.

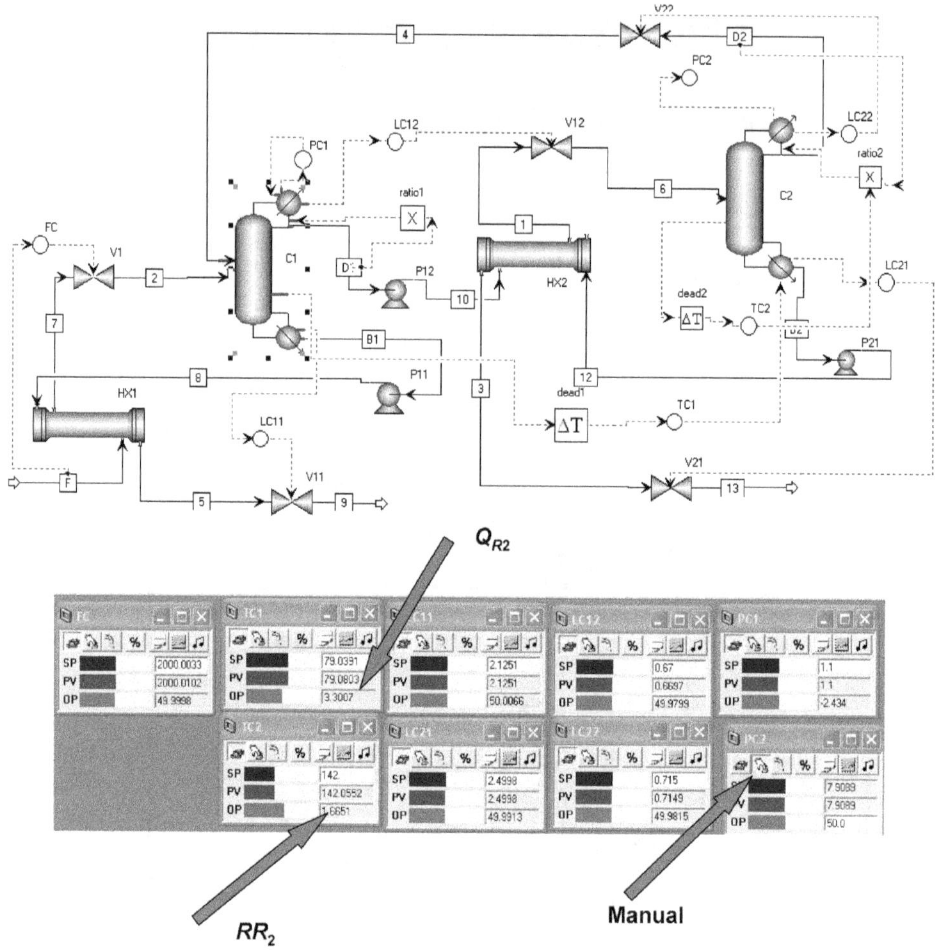


**Figure 6.12** Heat-integrated flowsheet and control structure.

increases $Q_{R2}$ and brings the temperature back to the setpoint in about 2 h. This loop is slow because of the large integral time (38 min). The temperature on Stage 9 in the high-pressure column rises from 142°C to about 146°C, but the controller increases $RR_2$ and brings the temperature back to the setpoint in about 2 h. This loop is slow because of the large integral time (41 min).

There is a very large transient disturbance of the THF impurity in the water product $x_{B1(\text{THF})}$, which rises from 1 ppm to almost 200 ppm for about 1 h. This is caused by the increase in the feed to the column dropping the temperatures in the stripping section and permitting more THF to escape out the bottom.

This problem can be greatly improved by using a feedforward ratio control structure in which the reboiler heat input to the high-pressure column is ratioed to the feed flowrate, with the ratio adjusted by the temperature controller. The dashed lines in Figure 6.13 shows the responses with this control structure. The transient disturbance in $x_{B1}$ is drastically reduced to about 20 ppm instead of 200 ppm.

To set up the ratio scheme, a multiplier block is inserted. One input is the molar flowrate of the feed (kmol/h). The second input is the ratio of $Q_{R2}$ to $F$. Since Aspen Dynamics uses

**TABLE 6.1 Controller Tuning Parameters Full Heat Integration; THF–Water.**

| | | Without Ratio | With Ratio |
|---|---|---|---|
| TC1 | Controlled variable | $T_{1,14}$ | $T_{1,14}$ |
| | Manipulated variable | $Q_{R2}$ | $Q_{R2}/F$ |
| | Transmitter range | 50–100°C | 50–100°C |
| | Controller output range | 0–6 Gcal/h | 0–0.01 Gcal/h |
| | Ultimate gain | 0.375 | 0.41 |
| | Ultimate period | 17.4 min | 19.2 min |
| | $K_C$ | 0.117 | 0.128 |
| | $\tau_I$ | 38 min | 42 min |
| TC2 | Controlled variable | $T_{2,9}$ | $T_{2,9}$ |
| | Manipulated variable | $RR_2$ | $RR_2$ |
| | Transmitter range | 100–200°C | 100–200°C |
| | Controller output range | 0–3 Gcal/h | 0–3 Gcal/h |
| | Ultimate gain | 1.57 | 1.57 |
| | Ultimate period | 18.6 min | 18.6 min |
| | $K_C$ | 0.49 | 0.49 |
| | $\tau_I$ | 41 min | 41 min |

metric units, the flow must be in kmol/h and the energy must be in GJ/h. The steady-state value of this ratio is (3.30 Gcal/h)(4.184 GJ/Gcal)/(2000 kmol/h) = 0.00691. Figure 6.14 shows the parameter values. Figure 6.15 gives the revised control structure with the ratio scheme. Note that the output signal of the TC1 temperature controller is the $Q_{R2}/F$ ratio.

The effectiveness of this structure is illustrated in Figure 6.16 for several disturbances in feed flowrate. The dashed lines are for a 5% increase. The solid lines are for a 10% increase. The dotted lines are for a 10% decrease in feed flowrate. The larger the disturbance, the larger the deviations in product purities. The impurity $x_{B2(\text{water})}$ of the THF product stream $B_2$ shows a large transient increase for the decrease in feed flowrate. This occurs because of the rapid decrease in $Q_{R2}$ caused by the ratio, which lets more water drop out the bottom of the high-pressure column.

Notice that the purity of THF product stream $x_{B2(\text{water})}$ is not maintained at the 10 ppm specification at the new steady-state conditions. The impurity level ends up higher than 10 ppm for increases in feed flowrate and lower than 10 ppm for decreases. The larger the change in feed flowrate, the larger the deviation. This occurs because the pressure in the high-pressure column is changing and we are controlling a fixed temperature in this column. An increase in feed produces a higher pressure, which will give a higher water concentration on the control tray for the same temperature. Remember, water is the light component on the other side of the azeotrope. A pressure-compensated temperature scheme can eliminate this problem. Implementation of this scheme will be discussed in a later section.

Larger changes in feed flowrate (>10%) result in a system shutdown. The recycle of material between the two columns increases to the point that the reflux drum levels exceed 100%.

Figure 6.17 gives responses to feed composition disturbances. The solid lines correspond to changing the feed composition from 6 to 6.5 mol% THF. The dashed lines correspond to changing the feed composition from 6 to 5.5 mol% THF. More THF in the feed produces an increase in bottoms product $B_2$ from the high-pressure column, as expected. Circulation rates

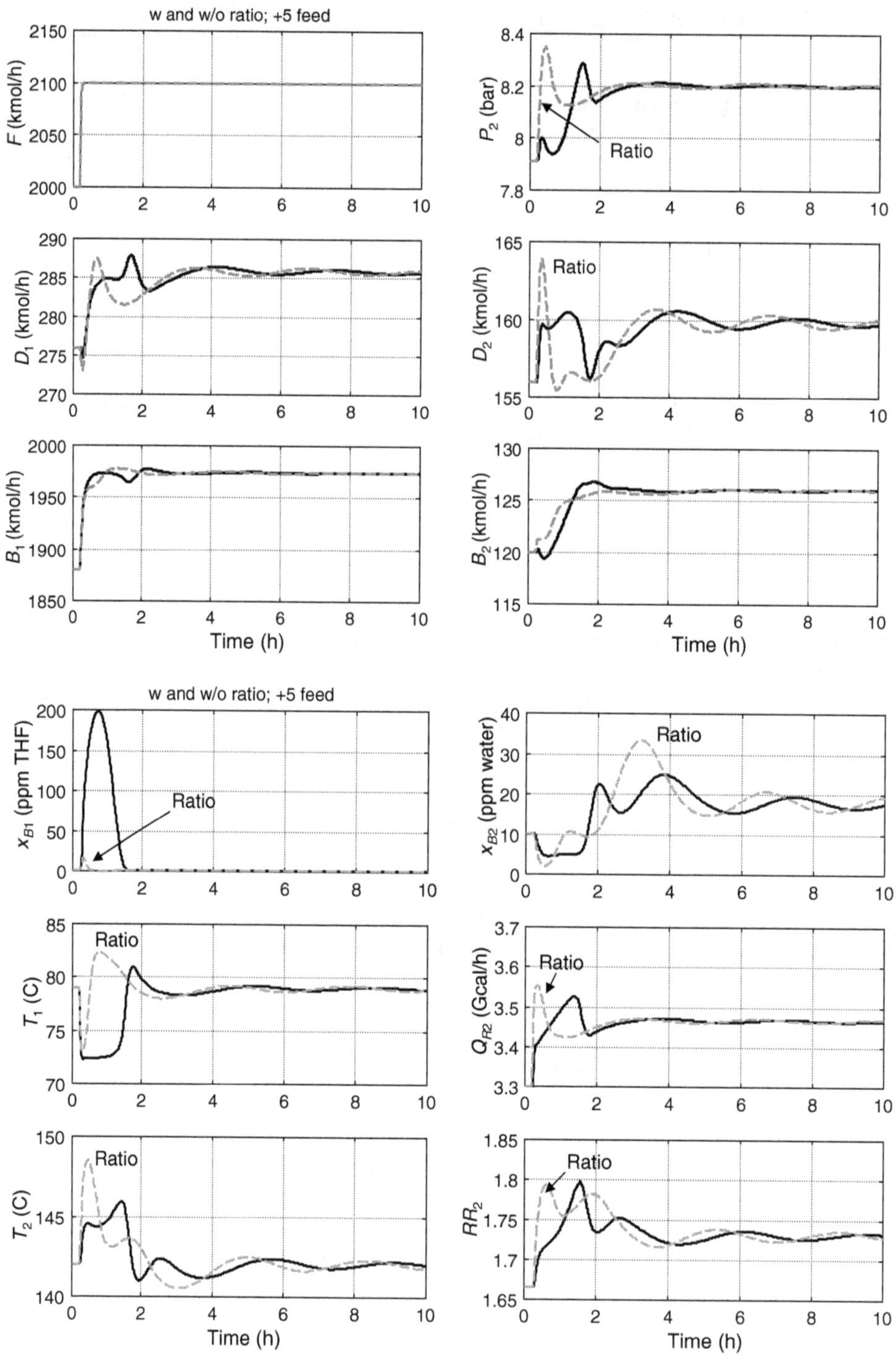


**Figure 6.13** Feed flowrate disturbance $+5\%$; with and without ratio $Q_{R2}/F$.

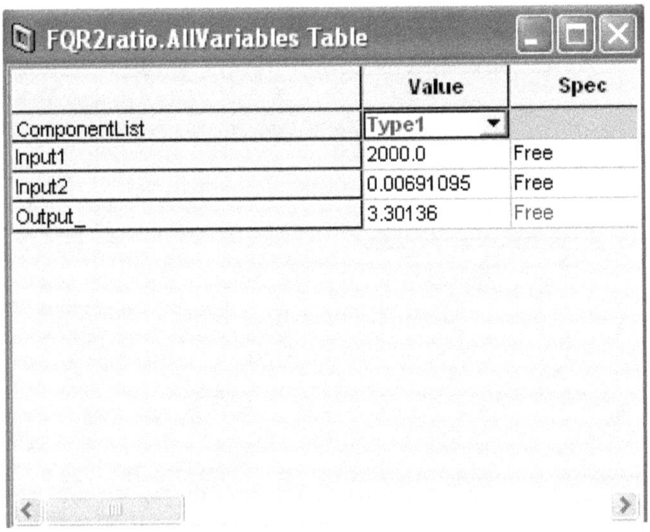


**Figure 6.14** Setting up ratio.

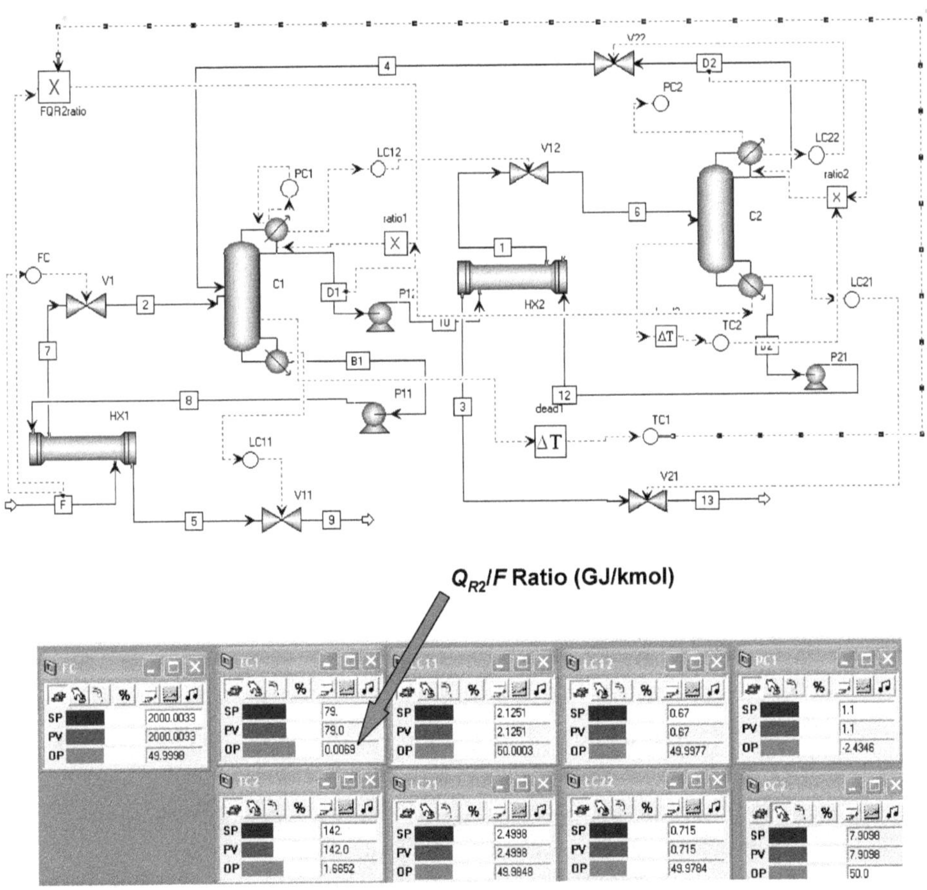


**Figure 6.15** Control structure with ratio.

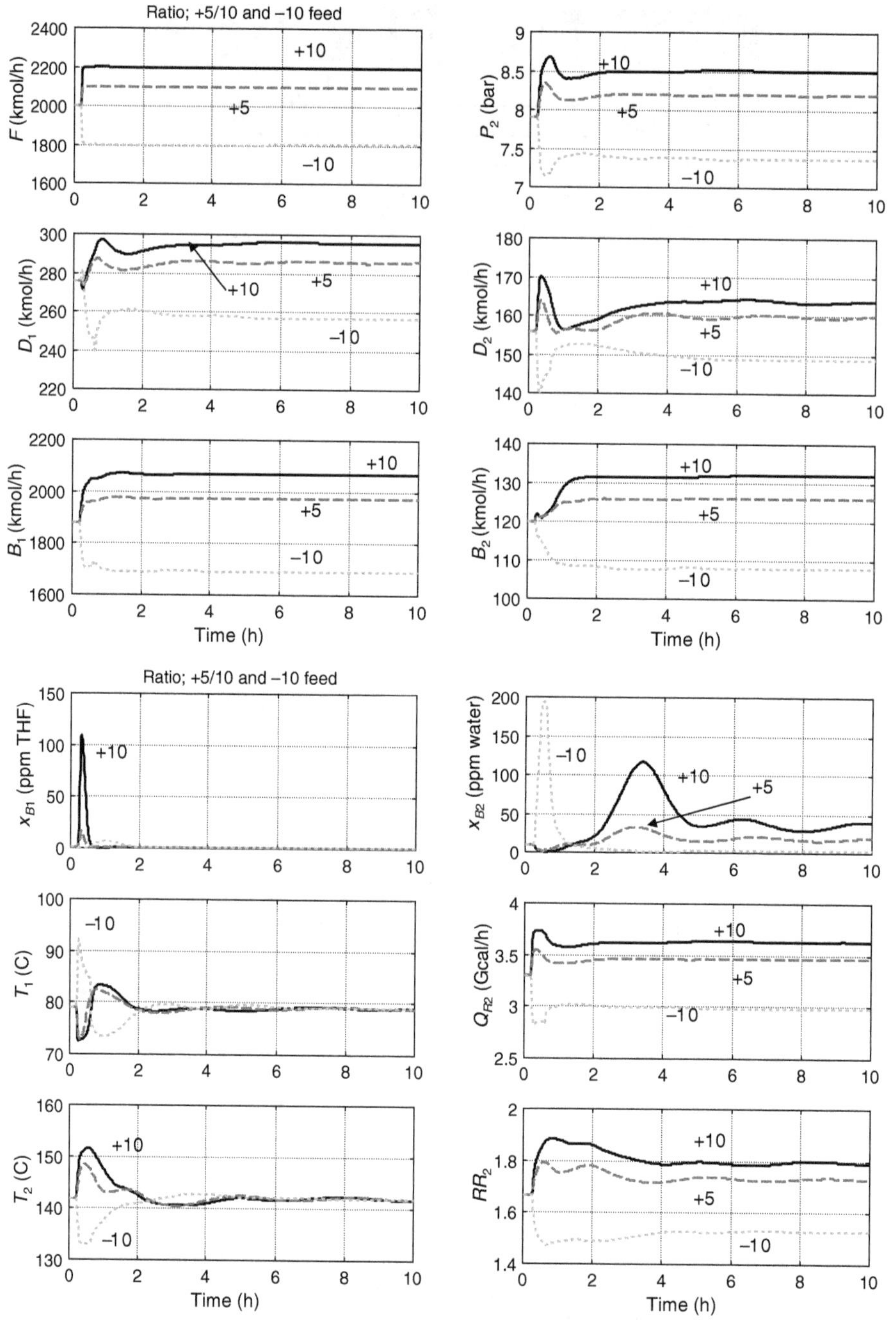


**Figure 6.16** Feed flowrate disturbances with ratio.

between the columns $D_1$ and $D_2$ also increase, as does the pressure in the high-pressure column $P_2$. The reverse is true for less THF in the feed.

However, the purity of the water product $B_1$ from the high-pressure column undergoes a very large transient disturbance (up to over 300 ppm). This occurs because of the drop in

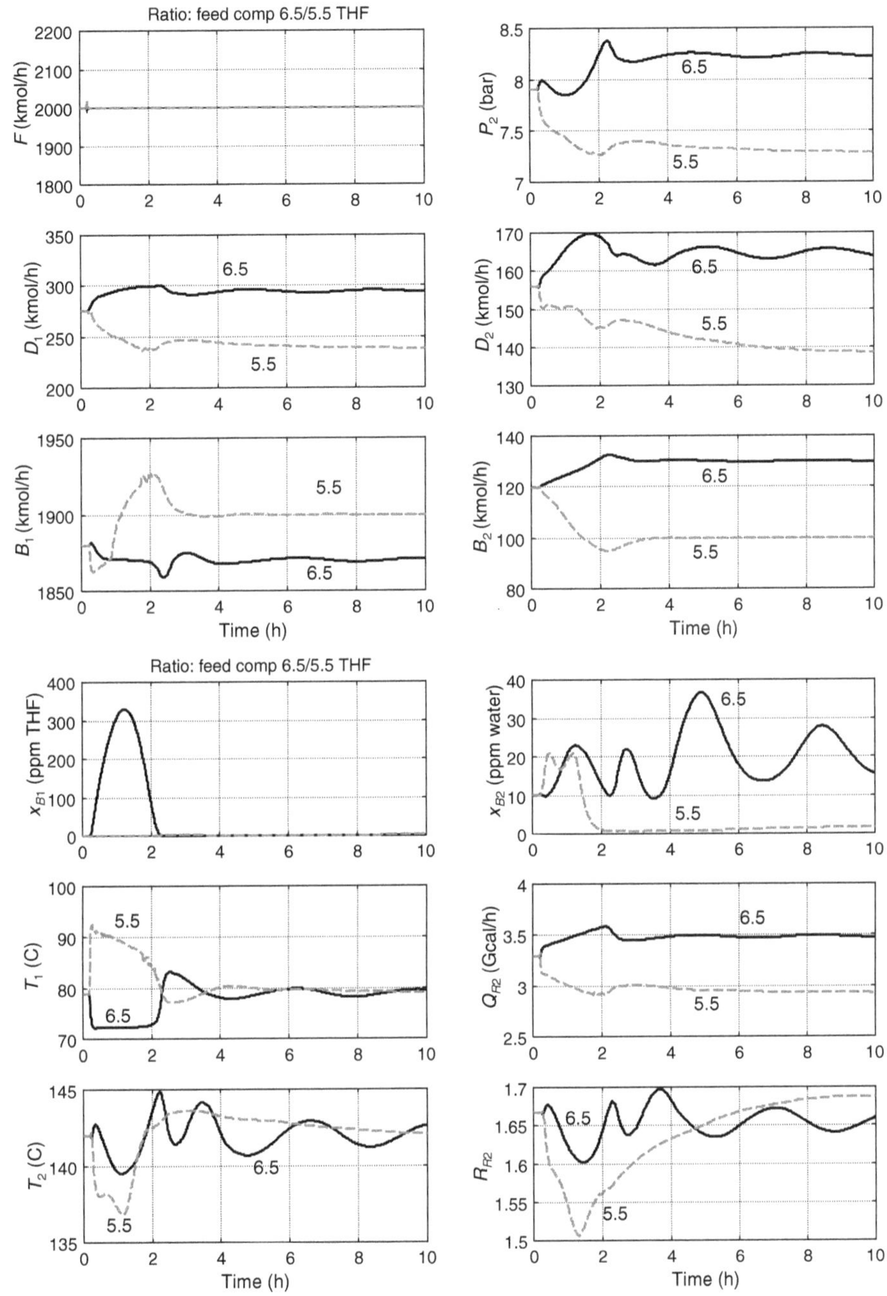


**Figure 6.17** Feed composition disturbances with ratio.

temperature in the high-pressure column ($T_1$), and the fact that the temperature controller TC1 takes over 2 h to ramp up the heat input in $Q_{R2}$ because of the large integral time and small controller gain (see Table 6.1). The dynamics of changing the heat input in the second column to affect the temperature in the first column are inherently somewhat slow.

The pressure in the high-pressure column must increase in order to raise the temperature in its reflux drum before an increase in heat transfer can occur.

These results demonstrate that the fully heat-integrated system can only handle fairly small disturbances without having purity specifications violated.

In the following sections the fully heat-integrated system is compared with a completely nonheated-integrated system and with a partially heat-integrated system.

### 6.2.2 No Heat Integration

If the two columns are not heat integrated and have separate condensers and reboilers, a temperature in each column can be controlled by manipulating the corresponding reboiler heat input, and the reflux ratio in each column could also be controlled. This structure is shown in Figure 6.18. Each column has a pressure controller manipulating condenser heat removal. The output signals of the two temperature controllers are the two reboiler duties.

If the two columns have independent reboilers and condensers, both a temperature and a reflux ratio can be manipulated in each column. A change in reboiler heat input has an immediate effect on temperature in the column; therefore, tighter temperature control is expected. In addition, the pressures in both columns can be controlled; therefore the pressure in the high-pressure column does not float with operating conditions. The $Q_{R1}/F$ ratio is used in the low-pressure column with the ratio reset by the TC1 temperature controller.

If we assume that the design conditions are the same as those used in the previous section, nothing changes except the tuning of the two temperature controllers. The first column in Table 6.2 shows that both temperature controllers can have higher gains and smaller integral time than those of the heat-integrated system (see Table 6.1). Gains are twice as large and integral times are three times smaller.

Figure 6.19 demonstrates the drastic improvement in dynamic control performance. Large 20% disturbances in feed flowrate are made. Product purities are held very close to their desired values during the transient. The deviations in $x_{B1(\mathrm{THF})}$ are less than 4 ppm for these large disturbances in feed flowrate.

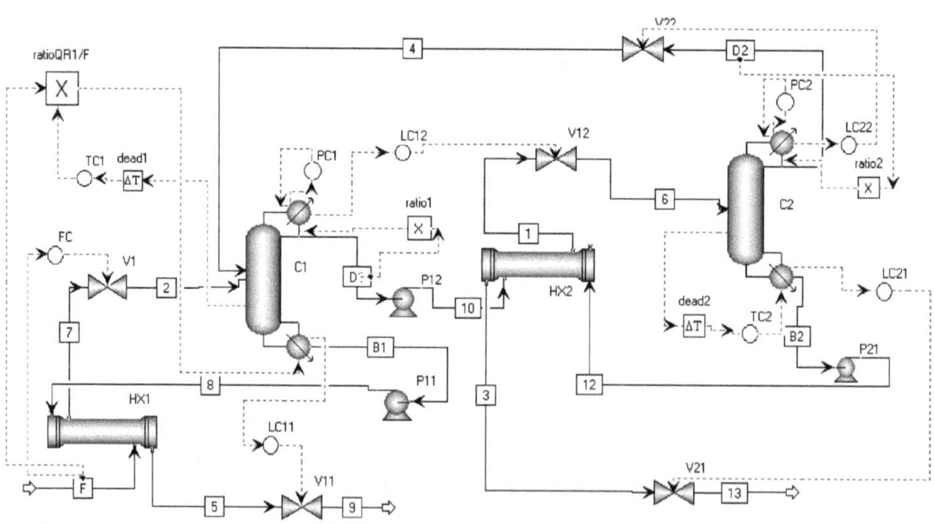


**Figure 6.18** Nonheat-integrated control structure.

**TABLE 6.2 Controller Tuning Parameters; No Heat Integration and Partial Heat Integration: THF–Water.**

| | | No Integration | Partial Integration |
|---|---|---|---|
| TC1 | Controlled variable | $T_{1,14} = 79°C$ | $T_{1,14} = 77.8°C$ |
| | Manipulated variable | $Q_{R1}/F$ Ratio | $Q_{aux}$ |
| | Transmitter range | 50–100°C | 50–100°C |
| | Controller output range | 0°0.1 Gcal/h | 0–20 Gcal/h |
| | Ultimate gain | 0.609 | 1.18 |
| | Ultimate period | 5.4 min | 5.4 min |
| | $K_C$ | 0.19 | 0.37 |
| | $\tau_I$ | 12 min | 12 min |
| TC2 | Controlled variable | $T_{2,9} = 142°C$ | $T_{2,9} = 144°C$ |
| | Manipulated variable | $Q_{R2}$ | $Q_{R2}$ |
| | Transmitter range | 100–200°C | 100–200°C |
| | Controller output range | 0–10 Gcal/h | 0–6 Gcal/h |
| | Ultimate gain | 2.93 | 3.64 |
| | Ultimate period | 4.2 min | 4.8 min |
| | $K_C$ | 0.92 | 1.14 |
| | $\tau_I$ | 9.2 min | 10.6 min |

Figure 6.20 gives results for changes in feed composition from 6 to 7 mol% THF (solid lines) and from 6 to 5 mol% THF (dashed lines). The transient deviation in $x_{B1(THF)}$ is about 70 ppm, which is much smaller than that seen in the heat-integrated system for a small feed composition disturbance (6–6.5 mol% THF).

These results demonstrate that there is a significant dynamic controllability penalty associated with the fully heat-integrated system.

### 6.2.3 Partial Heat Integration

A design using an auxiliary reboiler to supply some of the heat required in the low-pressure column was discussed in Section 6.1.1. A condenser/reboiler transfers 1.77 Gcal/h of energy into the low-pressure column. The remaining energy requirements are satisfied by a steam-heated auxiliary reboiler supplying 3.14–1.77 = 1.37 Gcal/h. The heat-transfer area of the condenser/reboiler is 79.9 m$^2$. Figure 6.1 gives the flowsheet of this system. Note that the pressure in the high-pressure column is not controlled but floats with operating conditions.

Implementing this structure in Aspen Dynamics requires some modification of the *Flowsheet Equations* as shown in Figure 6.21. The first equation calculates the heat-transfer rate in the reboiler condenser $Q_{C2}$. The second equation takes the negative of this value and adds the heat input from the auxiliary reboiler, which is specified to be the output signal from the TC1 controller. The control structure is shown in Figure 6.22. Note that the output of the TC1 is not connected to anything on the diagram. This output signal is used in the second flowsheet equation.

Adding a ratio block between feed and heat input to the low-pressure column improves the response of the system. Figure 6.23 shows one way to implement this structure using an additive feedforward scheme. The output signal of the multiplier block "ratioQtot/F" is the calculated total heat input to the reboiler of the low-pressure column. This is fed

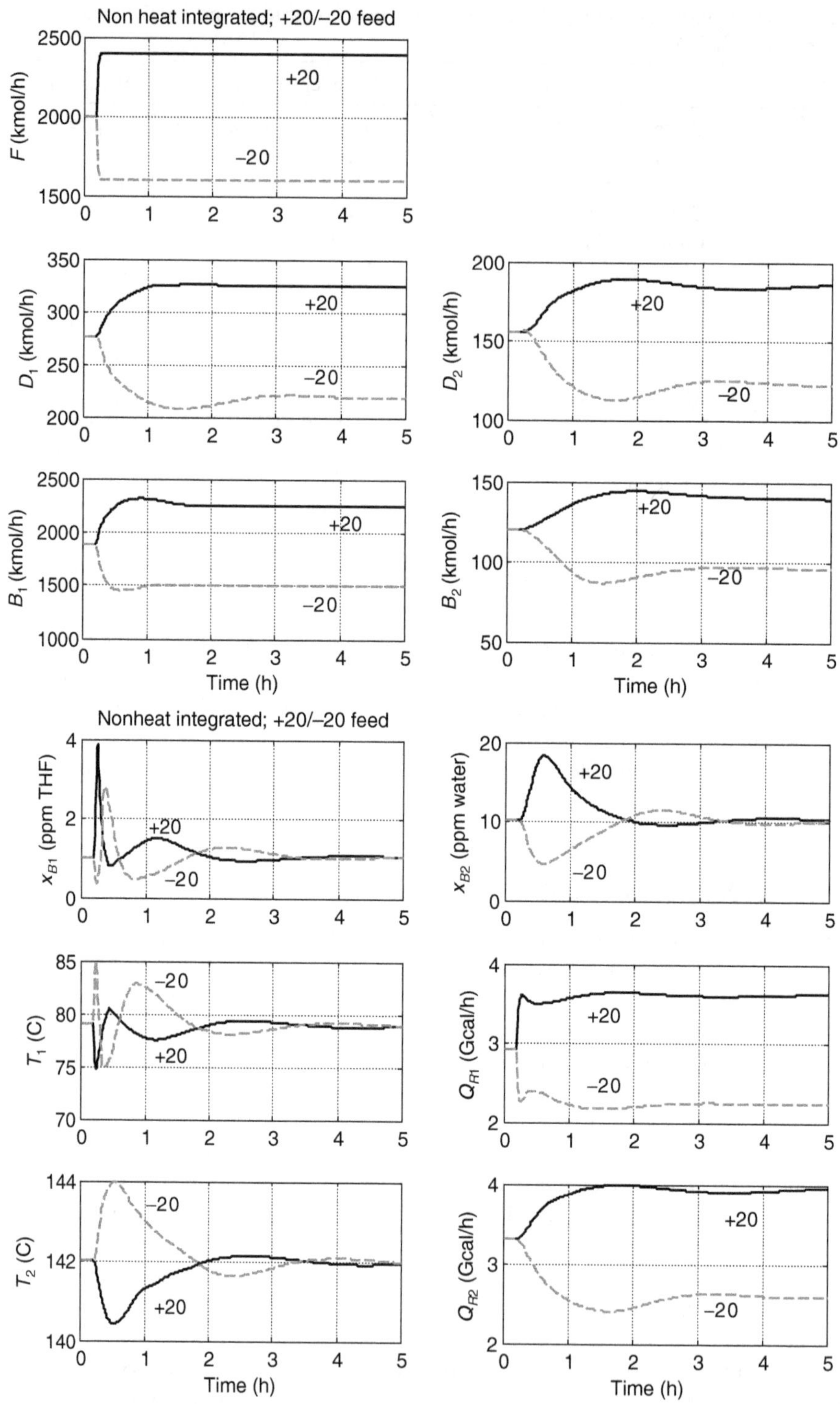


**Figure 6.19** Nonheat-integrated process: feed flowrate changes.

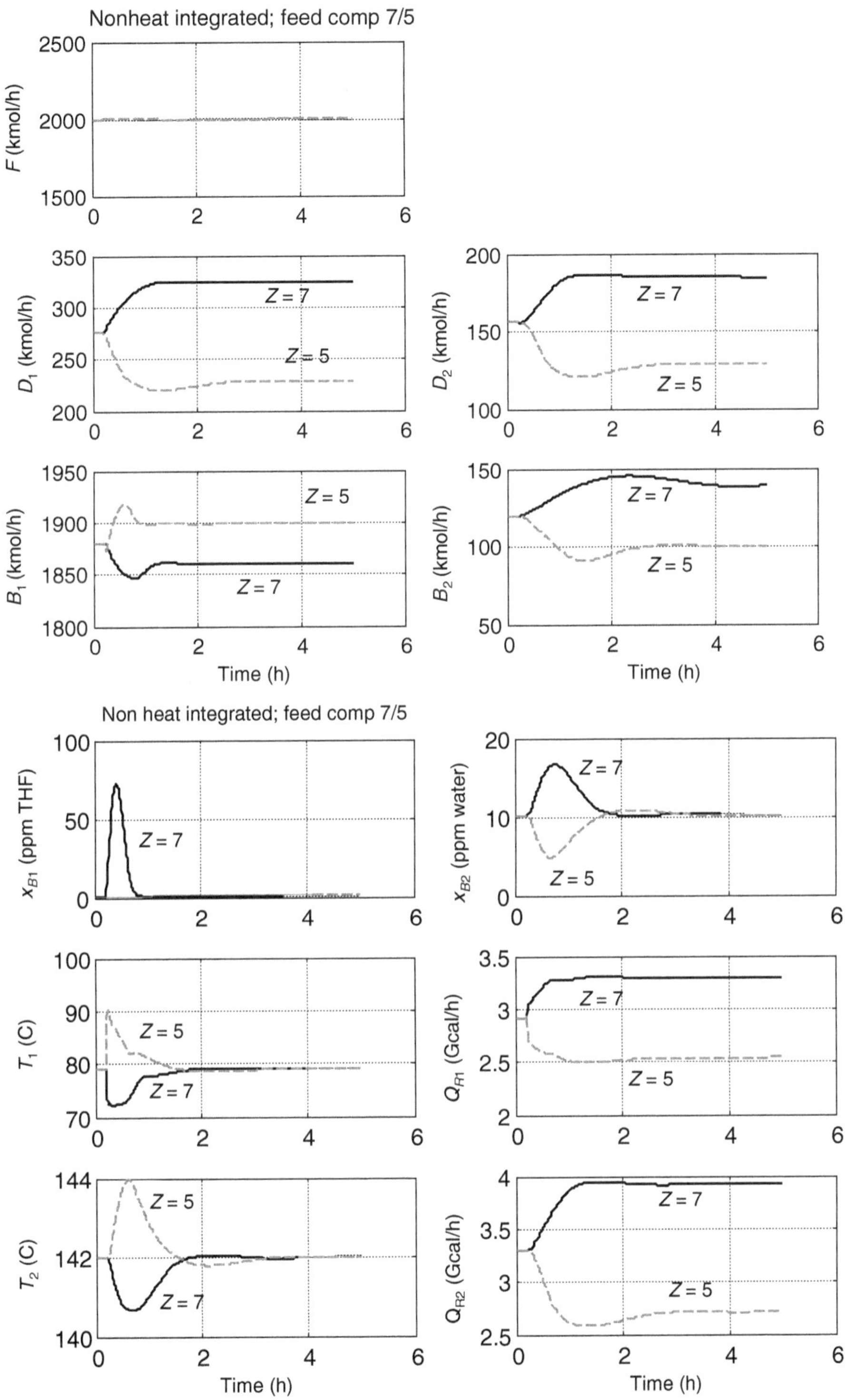


**Figure 6.20** Nonheat-integrated process: feed composition changes.

```
Text Editor - Editing Flowsheet
CONSTRAINTS
  // Flowsheet variables and equations...

Blocks("C2").condenser(1).Q= - 79.93*0.003063*(Blocks("C2").Stage(1).T  - Blocks("C1").TReb);
Blocks("C1").QReb=Blocks("TC1").OP - Blocks("C2").condenser(1).Q;
END
```

**Figure 6.21** Flowsheet equations with auxiliary reboiler.

into a summation block whose output is the required heat duty of the auxiliary reboiler $Q_{aux}$. The other input to the summation block comes from the TC1 temperature controller after a sign change in the multiplier block "negative." This structure produces an immediate change in the reboiler heat-input if the feed flowrate changes. The temperature controller trims up the heat input by adjusting $Q_{aux}$. The total heat input to the low-pressure column reboiler(s) is $Q_{R1}$, which is the sum of $Q_{aux}$ and the negative of $Q_{C2}$. The second equation of the flowsheet equations is changed, as shown in Figure 6.24.

All of the energy terms must have units of GJ/h in the blocks. The ratio of $Q$tot to $F$ is (3.15 Gcal/h)(4.184 GJ/Gcal)/(2000 kmol/h) = 0.00658. The output signal from the TC1 temperature controller is the auxiliary reboiler duty (1.37 Gcal/h)(4.184 GJ/Gcal) = 5.76 GJ/h. The output signal of the sum is the total heat input $Q_{R1}$, which is the sum of $Q_{aux}$ and the negative of $Q_{C2}$. All these block variables are shown in Figure 6.25.

Notice in the faceplates shown at the bottom of Figure 6.23 that the setpoints of the two temperature controllers are different in the partially heat-integrated case from those used in the fully heat-integrated system because the temperature profiles are slightly different. Of course the TC1 temperature controller must be retuned since its output signal is a ratio.

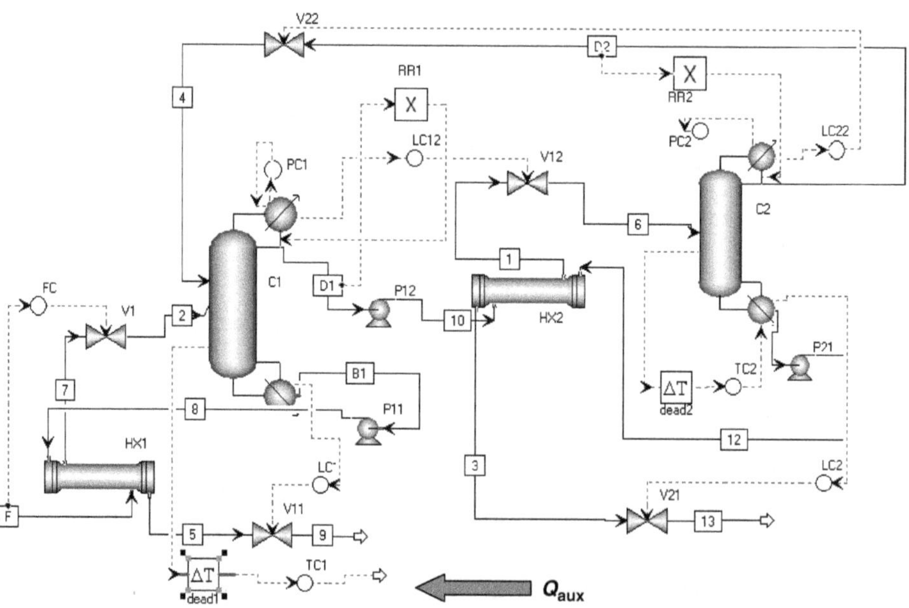


**Figure 6.22** Control structure with auxiliary reboiler.

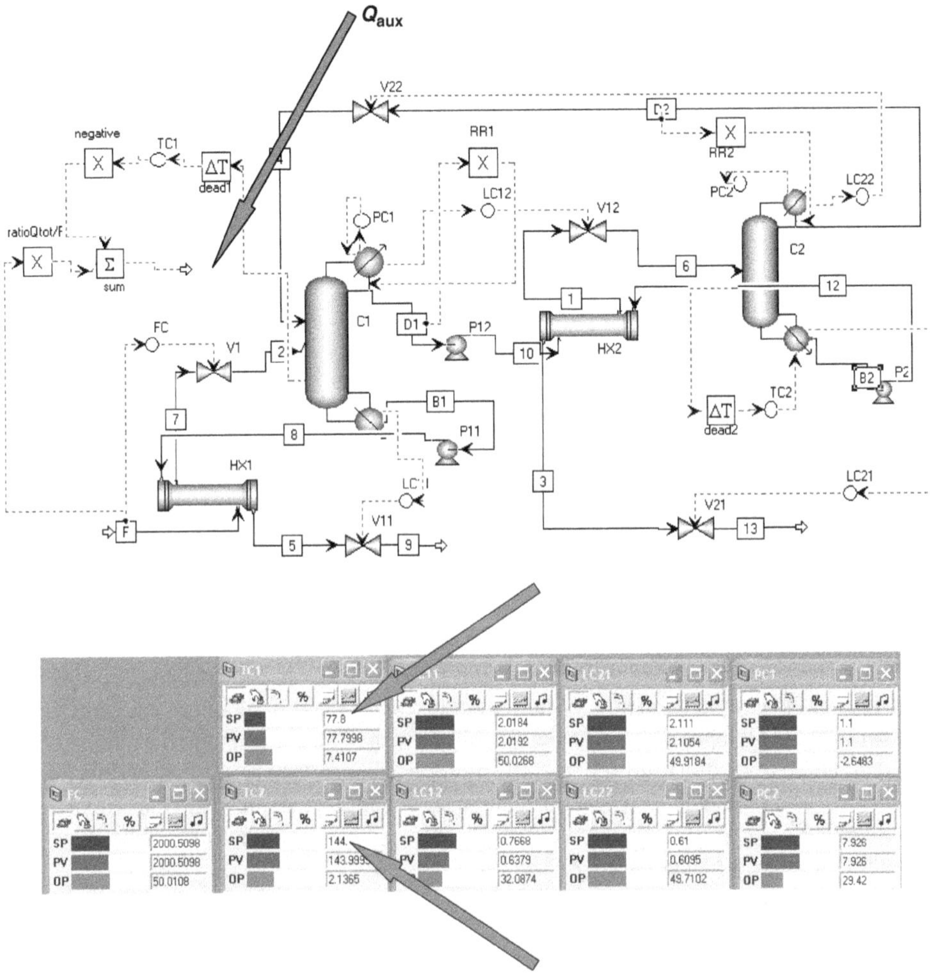


Figure 6.23 Control structure with auxiliary reboiler and ratio.

The second column in Table 6.2 gives tuning parameters. The TC2 temperature controller is also retuned because of the somewhat small heat duty in the high-pressure column with partial heat integration. These tuning parameters are very similar to those found in the completely nonheat-integrated system, so we anticipate that control should be good.

```
Text Editor - Editing Flowsheet
CONSTRAINTS
  // Flowsheet variables and equations...


Blocks("C2").condenser(1).Q= - 79.93*0.003063*(Blocks("C2").Stage(1).T  - Blocks("C1").TReb);
Blocks("C1").QReb= - Blocks("C2").condenser(1).Q + Blocks("sum").Output;
END
```

Figure 6.24 Flowsheet equations with ratio.

ratioQtot/F.Results Table

| | Description | Value | Units |
|---|---|---|---|
| Input1 | Input signal 1 | 2000.51 | kmol/hr |
| Input2 | Input signal 2 | 0.006586 | |
| Output_ | Output signal | 13.1754 | |

sum.Results Table

| | Description | Value | Units |
|---|---|---|---|
| Input1 | Input signal 1 | -7.41237 | |
| Input2 | Input signal 2 | 13.1754 | |
| Output_ | Output signal | 5.76299 | |

**Figure 6.25** Ratio and sum blocks.

It is interesting to note that the action of the TC1 controller is "direct acting" in this ratio scheme. An increase in temperature should produce a decrease in reboiler heat input. In the normal setup, the temperature controller is "reverse acting"; however, with the ratio structure using an additive feedforward scheme, the multiplier block "negative" changes the sign of the signal that is adjusting the $Q_{aux}$. So an increase in temperature increases the output signal from the direct acting TC1, which reduces the output signal from the summation block and provides the correct reduction in $Q_{aux}$.

An alternative control structure would be to use a "Total Q" controller. In this scheme the heat transfer in the condenser/reboiler and the heat input in the auxiliary reboiler are added together and become the process variable signal to a Total Q controller. The setpoint of this controller is the output signal of the TC1 temperature controller. The output signal of the Total Q controller goes to the control valve in the steam line feeding the auxiliary reboiler.

The response of the system with partial heat integration for 20% changes in feed flowrate is shown in Figure 6.26. The solid lines are +20% step changes. The dashed lines are −20% changes. Stable regulatory control is achieved for these large disturbances. The $Q_{C2}$ variable is negative, indicating heat removal. Increasing feed flowrate produces increases in $Q_{aux}$, $Q_{R1}$, and $Q_{R2}$. However, the water impurity in the THF product $x_{B2(water)}$ increases from the desired value of 10 ppm to about 50 ppm water, despite holding the temperature on Stage 9. This occurs because of the change in the pressure in the high-pressure column, which is not controlled but floats with operating conditions in this partially heat-integrated system. As previously discussed, a pressure-compensated temperature control scheme should be able to correct this problem.

### 6.2.4 Pressure-Compensated Temperature Control

The design pressure is 7.9 bar in the high-pressure column. As shown in Figure 6.26 (upper right figure), the pressure $P_2$ increases to about 8.6 bar for a 20% increase in feed flowrate

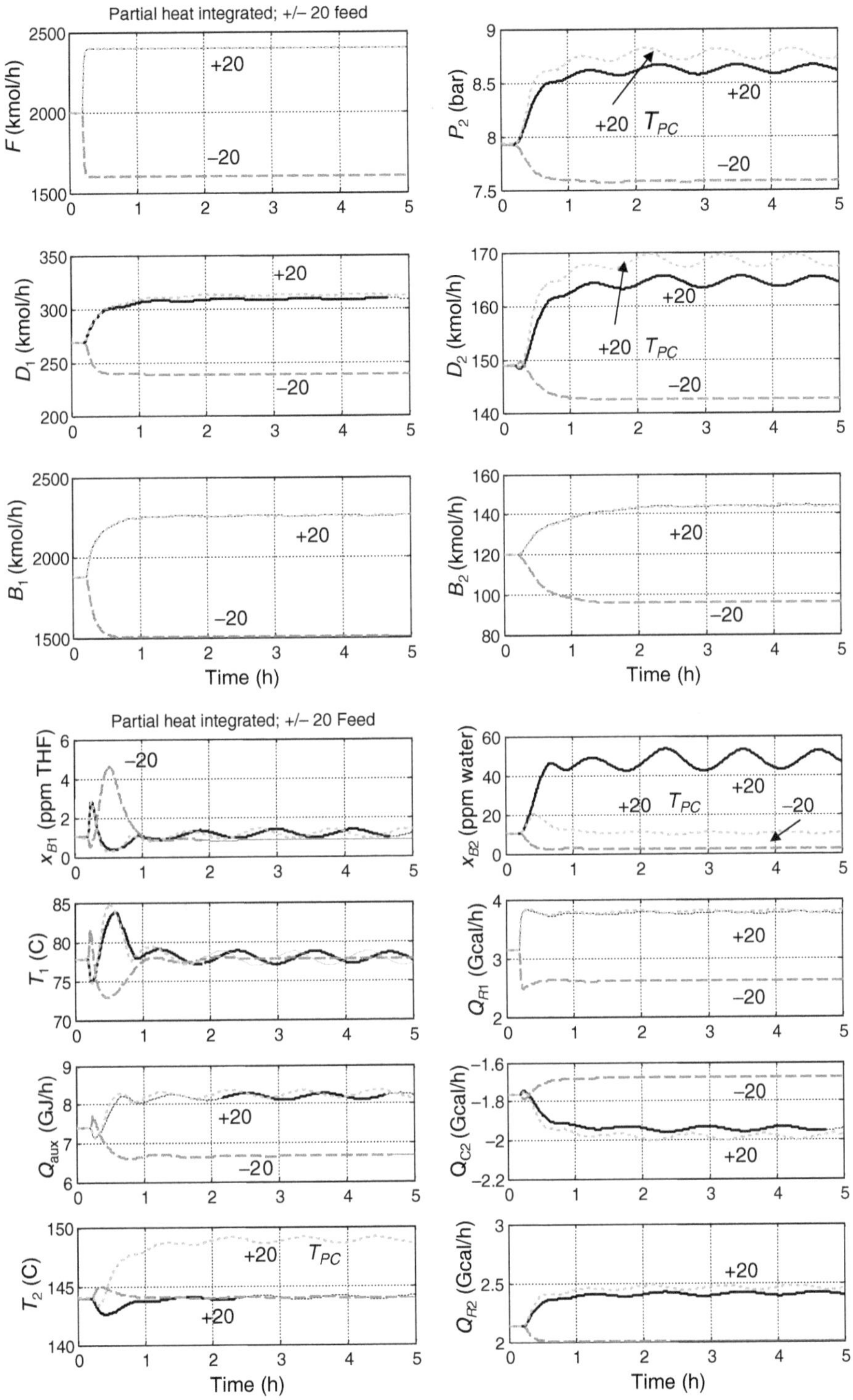


**Figure 6.26** Feed flowrate disturbances; partial heat integration.

because the higher heat-transfer rate in the condenser/reboiler requires a high differential temperature driving force. With the same Stage 9 temperature and a higher pressure, the water composition on Stage 9 is higher since water is lower boiling than the azeotrope in this column. The result has larger water impurity levels in the THF bottoms product.

To set up a pressure-compensated temperature scheme, VLE data are calculated at two different pressures for the same liquid composition. For a liquid mole fraction of 0.90 THF, the temperature is 139.499°C at 7.91 bar and 142.858°C at 8.5 bar. Therefore, the temperature signal is adjusted as given below:

$$T_{PC} = T - (P - 7.91)\frac{(142.858 - 139.499)}{(8.5 - 7.91)}$$

Both the temperature on Stage 9 and the column pressure are used to calculate $T_{PC}$, which is the process variable signal fed to the TC2 temperature controller. In Aspen Dynamics, this is easily achieved by using flowsheet equations, as shown in Figure 6.27. The last equation calculates the signal fed to the deadtime element in the TC2 loop. Figure 6.28 gives the control structure.

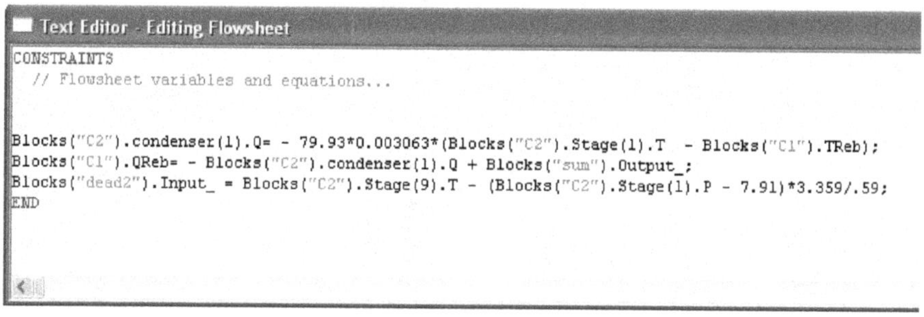


```
CONSTRAINTS
  // Flowsheet variables and equations...

Blocks("C2").condenser(1).Q= - 79.93*0.003063*(Blocks("C2").Stage(1).T  - Blocks("C1").TReb);
Blocks("C1").QReb= - Blocks("C2").condenser(1).Q + Blocks("sum").Output_;
Blocks("dead2").Input_ = Blocks("C2").Stage(9).T - (Blocks("C2").Stage(1).P - 7.91)*3.359/.59;
END
```

**Figure 6.27** Flowsheet equations with pressure-compensated temperature.

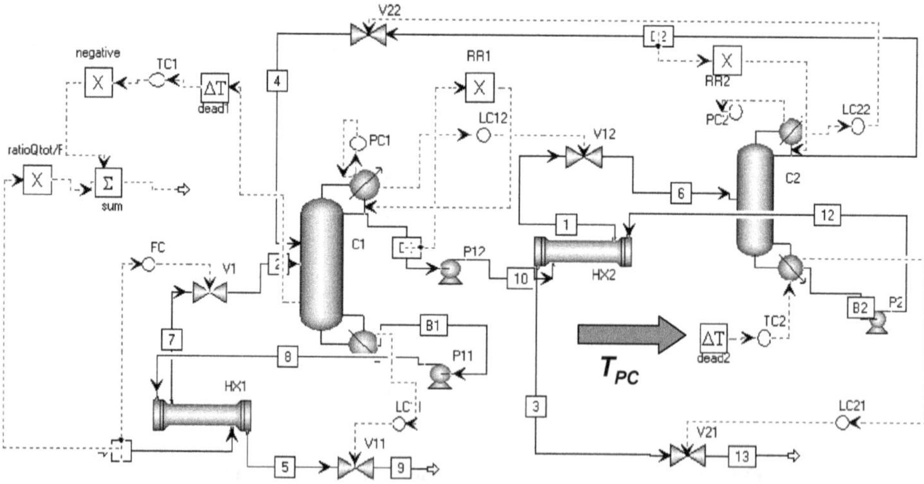


**Figure 6.28** Control structure with pressure-compensated temperature.

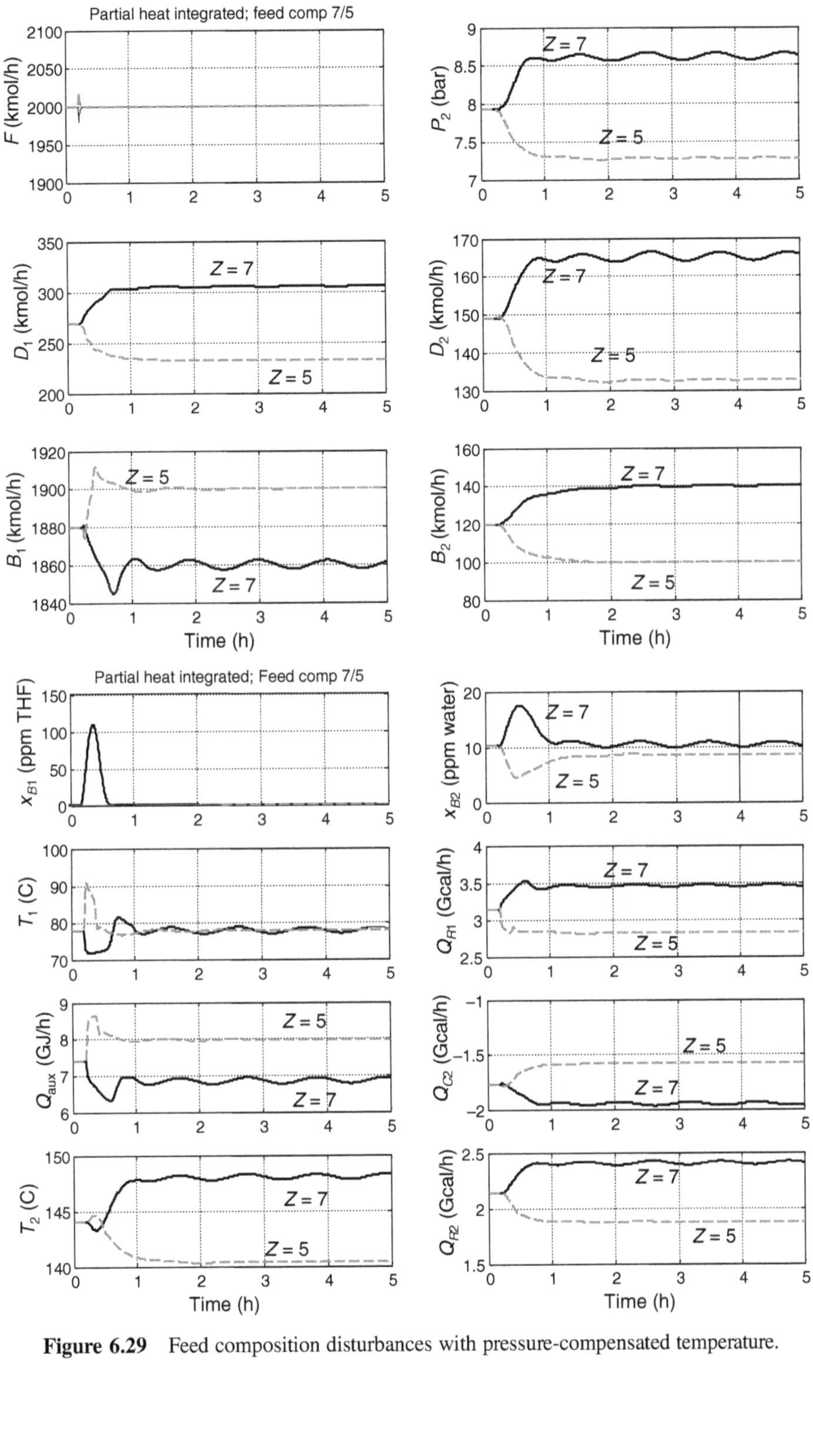


**Figure 6.29** Feed composition disturbances with pressure-compensated temperature.

The dotted lines in Figure 6.26 (labeled "+20 $T_{PC}$") give results for a +20% change in feed flowrate using the pressure-compensated temperature scheme. The temperature on Stage 9 in the high-pressure column is not held at 144°C, but increases to about 149°C at the higher pressure. The water impurity $x_{B2(\text{water})}$ in the THF product stream is held very close to the desired 10 ppm level.

Figure 6.29 shows responses for feed composition disturbances using the pressure-compensated scheme. Feed composition is changed from 6 to 7 mol% THF or from 6 to 5 mol% THF. Notice that the Stage 9 temperature ($T_2$) changes as the pressure in the high-pressure column $P_2$ changes. Except for the large transient increase in $x_{B1(\text{THF})}$, product

**TABLE 6.3 Design Results for Heat-Integrated Systems; Acetone–Methanol.**

| | | Pressure Swing with Heat Integration | Extractive with Heat Integration |
|---|---|---|---|
| C1 | *Nstage* | 52 | 57 |
| | *ID* (m) | 3.30 | 2.79 |
| | $Q_{CR}$ (MW) | 5.86 | 9.31 |
| | $Q_{AUX}$ (MW) | 8.93 | 2.10 |
| | $Q_C$ (MW) | 15.1 | 9.87 |
| | $Q_{HX}$ (MW) | — | 1.35 |
| | $T_{reflux}$ (K) | 328 | 329 |
| | $A_{RC}$ ($m^2$) | 110 | 514 |
| | $A_C$ ($m^2$) | 982 | 609 |
| | $A_{HX}$ ($m^2$) | — | 52.7 |
| | $A_{AUX}$ ($m^2$) | 450 | 106 |
| | Shell ($10^6$ \$) | 1.13 | 1.02 |
| | Condenser ($10^6$ \$) | 0.643 | 0.567 |
| | Condenser/reboiler ($10^6$ \$) | 0.155 | 0.421 |
| | Aux. reboiler ($10^6$ \$) | 0.388 | 0.151 |
| | Cooler HX ($10^6$ \$) | — | 0.096 |
| | Aux. condenser ($10^6$ \$) | — | ? |
| | Total capital ($10^6$ \$) | 1.58 | 2.26 |
| | Energy (\$$10^6$/y) | 1.32 | 0.311 |
| C2 | *Nstage* | 62 | 26 |
| | *ID* (m) | 1.89 | 1.61 |
| | $Q_R$ (MW) | 7.59 | 11.3 |
| | $A_C$ ($m^2$) | — | — |
| | $A_R$ ($m^2$) | 383 | 570 |
| | Shell ($10^6$ \$) | 0.722 | 0.292 |
| | Condenser ($10^6$ \$) | — | — |
| | Reboiler ($10^6$ \$) | 0.349 | 0.451 |
| | Total capital ($10^6$ \$) | 1.07 | 0.743 |
| | Energy (\$$10^6$/y) | 1.13 | 1.67 |
| Total system | Total capital ($10^6$ \$) | 2.65 | 3.00 |
| | Total energy (\$$10^6$/y) | 2.45 | 1.98 |
| | TAC | 3.33 | 2.98 |

purities are maintained close to their specifications. These responses are quite similar to those found in the nonheat-integrated system, so partial heat-integration also provides good dynamic control.

### 6.2.5 Conclusion for a THF–Water System

Several types of heat-integrated systems have been explored, ranging from full-heat integration to no integration, in a two-column pressure-swing azeotropic distillation system. Significant dynamic penalties have been demonstrated for fully heat-integrated systems. Only small disturbances can be handled, and product purities are not held close to specifications. The basic problem is the smaller number of control degrees of freedom, which does not permit the independent setting of both reflux ratio and reboiler heat input in both columns. Partial heat integration, using both a condenser/reboiler and an auxiliary reboiler on the low-pressure column, provides much more robust control.

Since the pressure in the high-pressure column changes with operating conditions, a pressure-compensated temperature control structure is required to maintain product specification in this column. The detailed steps in implementing the various flowsheets and control structures in Aspen Plus and Aspen Dynamics have been discussed.

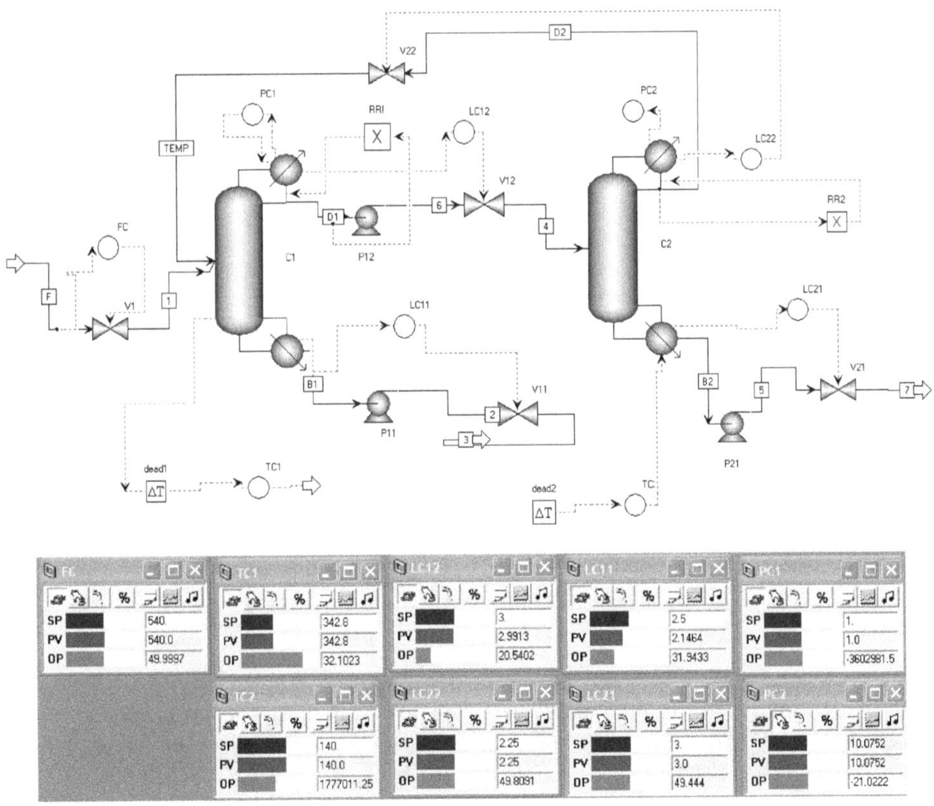


**Figure 6.30** Control structure for acetone–methanol with partial heat integration.

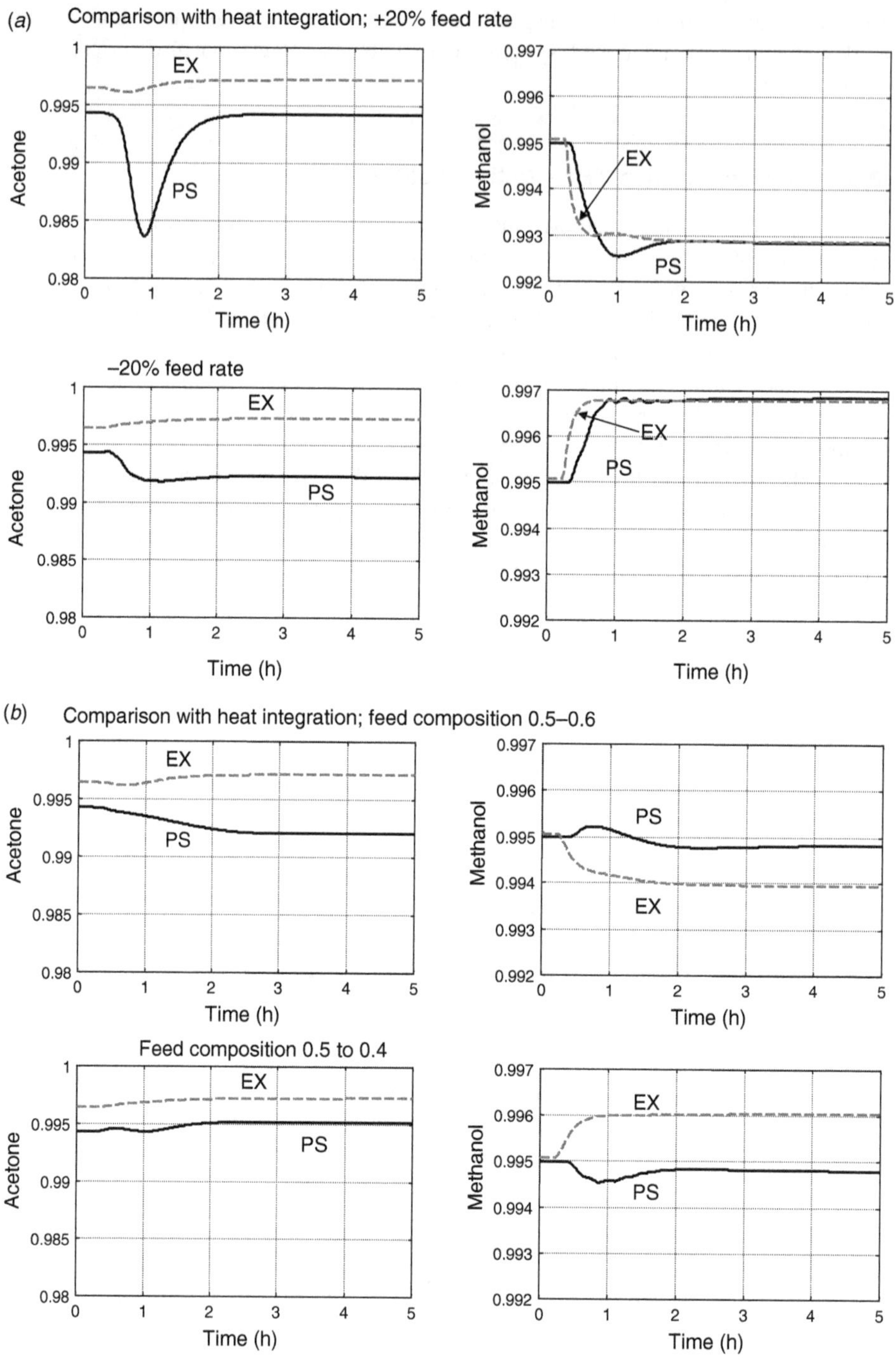


**Figure 6.31** (*a*) Comparison of extractive and pressure-swing distillation with heat integration: feed rate disturbances. (*b*) Comparison of extractive and pressure-swing distillation with heat integration: feed composition disturbances.

## 6.3 HEAT INTEGRATION IN AN ACETONE–METHANOL SYSTEM

For the acetone–methanol pressure-swing system, we consider a partially heat integrated flowsheet. This partially heat-integrated flowsheet is essentially the same as without heat integration, described in Chapter 5 (see Table 5.1). The only change is that there is a reboiler/condenser and an auxiliary steam-heated reboiler on the low-pressure column. The temperature difference between the condenser of the high-pressure column and the base of the low-pressure column is quite large (407.5–345.3 = 62.2 K). The heat duty in the condenser of the high-pressure column is 5.86 MW, which requires a heat-transfer area of 110 $m^2$.

The first column in Table 6.3 gives process and economic data for this process. The TAC of the heat-integrated pressure-swing system is $3,330,000/y. This should be compared with the base-case TAC of $4,520,000/y. Total capital investment is unexpectedly smaller than for the base design ($2,650,000 instead of $3,510,000). This is the result of the small reboiler/condenser owing to the large temperature driving force. The energy cost is reduced from $3,350,000 to $2,450,000 per year.

The energy requirement in the low-pressure column is 14.79 MW, so the auxiliary reboiler must provide 14.79–5.86 = 8.93 MW (32 GJ/h, as shown in the TC1 faceplate in Fig. 6.30). Flowsheet equations similar to those described in the THF–water system are needed in this system. The pressure-compensated temperature measurement uses the temperature (411.5 K) and pressure (10.354 atm) on Stage 53.

$$T_{PC} = T_{53} - (P_{53} - 10.354)\left(\frac{414.26 - 411.51}{11 - 10.354}\right)$$

Figure 6.30 gives the control structure developed for the heat-integrated pressure-swing process. Notice that the temperatures shown in the faceplates for TC2 are in degrees Centigrade.

Figure 6.31 gives a direct comparison of pressure-swing and extractive distillation (discussed in Chapter 11) designs with heat-integration for feed flowrate and feed composition disturbances. The solid lines are for the pressure-swing system. The dashed lines are for the extractive system. Both systems provide stable base-level regulatory control. The results are quite similar to those found in the nonheat-integrated systems.

These results show that there is no control penalty for going to the partial heat-integrated systems when auxiliary reboilers or condensers are added. This result is to be expected because there is no loss of control degrees of freedom.

## 6.4 CONCLUSION

Pressure-swing systems are very amenable to heat integration because of the inherent temperature differences in the two columns. Complete heat integration is the most economical from a steady-state economic point of view. However, a considerable dynamic controllability penalty can occur, which can be greatly reduced by using partial heat integration.

## REFERENCES

1. Abu-Eishah S. I. and W. L. Luyben, Design and control of a two-column azeotropic system, *Ind. Eng. Chem. Process Des. Dev.*, **24**, 132–140 (1985).

# CHAPTER 7

# HETEROGENEOUS BINARY AZEOTROPES

The binary azeotropes considered in Chapters 5 and 6 were homogeneous, that is, a single liquid phase. The interaction between molecules that have similar structures and elements is usually small, and the liquid–liquid behavior is ideal. The molecular interaction increases as the molecules become more dissimilar. In many systems the molecules repel each other and result in nonideal behavior with activity coefficients greater than unity. For example, methanol and water are somewhat dissimilar, so their activity coefficients are modestly greater than unity. The methanol–water system does not have an azeotrope because the nonideality is modest. Ethanol and water are more dissimilar, so their activity coefficients are fairly large. The result is a homogeneous azeotrope.

Butanol and water are very dissimilar with very large activity coefficients. This produces a heterogeneous azeotrope, as shown in the *Txy* diagram given in Figure 7.1. The pressure in this figure is 0.5 atm at which the boiling points of pure *n*-butanol and water are 372.5 and 354.8 K, respectively. At this pressure the azeotropic composition is 76.1 mol% water with a temperature of 348 K. Note that the temperature of the azeotrope is lower than the boiling points of both components. The composition of the *aqueous* liquid phase is 97 mol% water, and the composition of the *organic* liquid phase is 58 mol% *n*-butanol.

The separation of a binary heterogeneous azeotrope is often much easier than the separation of a binary homogeneous azeotrope because the liquid–liquid phase equilibrium in a decanter can be used to facilitate the separation. No third component has to be introduced, which avoids many problems of trace amounts of the added component in the product streams.

We use the *n*-butanol–water system in this chapter as an example of this type of binary heterogeneous azeotropic system. A straight-forward two-column distillation system can be used to easily achieve high-purity products. A simple control structure is developed that is capable of handling very large disturbances in throughput and feed composition. The control

*Design and Control of Distillation Systems for Separating Azeotropes.* By William L. Luyben and I-Lung Chien

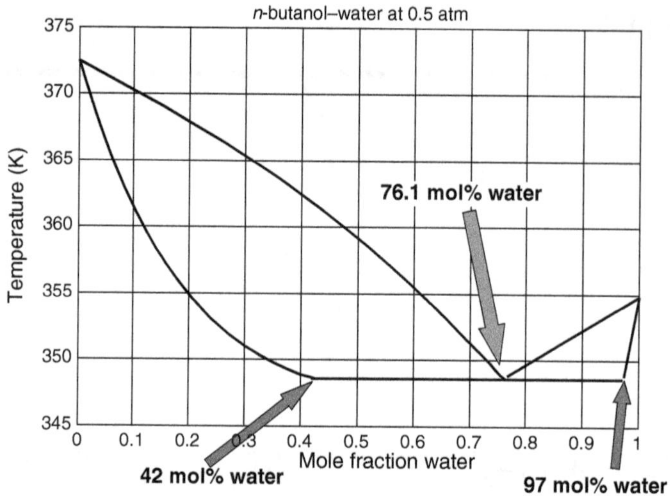


**Figure 7.1** *Txy* diagram for *n*-butanol–water at 0.5 atm.

system is very robust and is demonstrated to be applicable for systems that are designed for a wide range of feed compositions.

## 7.1 *n*-BUTANOL–WATER SYSTEM

The interest in *n*-butanol as a biofuel has increased in recent years owing to its superior fuel qualities compared to ethanol. These include a higher octane number, lower heat of vaporization, higher energy density (energy/volume), and lower vapor pressure. However, in the traditional ABE (acetone-butanol-ethanol) fermentation process, the concentration of *n*-butanol coming from the fermenter is lower than that achieved in ethanol fermentation. In addition, acetone and ethanol are also produced. Recent studies to improve yield and increase *n*-butanol concentration have explored fed-batch systems with stripping, adsorption, liquid–liquid extraction, distillation, and/or pervaporation to recover products.

The purification of *n*-butanol involves removing the acetone and ethanol, and separating the *n*-butanol from the water. This separation cannot be achieved in a single distillation column because of the presence of an azeotrope.

Industrial production of *n*-butanol from fermentation was developed almost a century ago (the ABE process) and was widely used in Europe and the United States until inexpensive petroleum made its production from petrochemical raw materials a less expensive process. The recent surge in crude-oil prices and issues of energy independence and carbon dioxide emissions have generated new interest in biofuels. Ethanol is now being produced on a fairly large scale for blending into gasoline.

Qurishi et al.[1] discuss the advantages of *n*-butanol over ethanol: lower heat of vaporization, lower vapor pressure, higher octane number, and higher energy density (energy/volume). The last characteristic translates into more range for a given volume fuel tank on the vehicle. However, the concentration of *n*-butanol produced in the batch fermenter is quite low (20 g/L) because of toxicity problems. In addition, acetone and ethanol are also produced, which must be removed from the mixture coming from the fermenter.

There are four isomers of butanol: *n*-butanol, isobutanol, 2-butanol, and tert butyl alcohol. The butanol produced in fementation is *n*-butanol. Water and *n*-butanol form a heterogeneous azeotrope at atmospheric pressure and 364.6 K with a composition of 76.33 mol% water.

Qurishi et al.,[1] Ladisch,[2] and Phillips and Humphrey[3] discuss some of the issues concerning the separation of the fermenter products. Doherty and Malone[4] indicate that a two-column distillation system in conjunction with a decanter can be used to separate the heterogeneous binary *n*-butanol–water azeotrope. Sticklmair and Fair[5] discuss the separation of the ternary acetone–water–*n*-butanol system in a two-column system. Pucci et al.[6] suggest that the ternary separation can be achieved in a single column in which a side decanter is used to remove the water.

## 7.2 PHASE EQUILIBRIUM

Figure 7.1 gives the *Txy* diagram for *n*-butanol–water at 0.5 atm. This pressure is selected because it gives a decanter temperature that is high enough (343 K) to permit the use of inexpensive cooling water in the condenser. The UNIQUAC physical properties from Aspen Plus are used to generate this plot and in the simulations of the columns.

In setting up the Aspen simulation, the selection of *Vapor-Liquid-Liquid* must be made, as shown in Figure 7.2. To generate this diagram, go to *Tools* and select *Analysis* and *Binary*. The pressure and physical property package are specified. Clicking the *Go* button at the bottom of the window gives the *Txy* diagram. Figures 7.3*a* and *b* give results at pressures of 1 and 0.5 atm, respectively.

The composition of the azeotrope is found by going to *Tools*, selecting *Conceptual Design* and clicking *Azeotropic Search*. The window shown in Figure 7.4*a* opens on

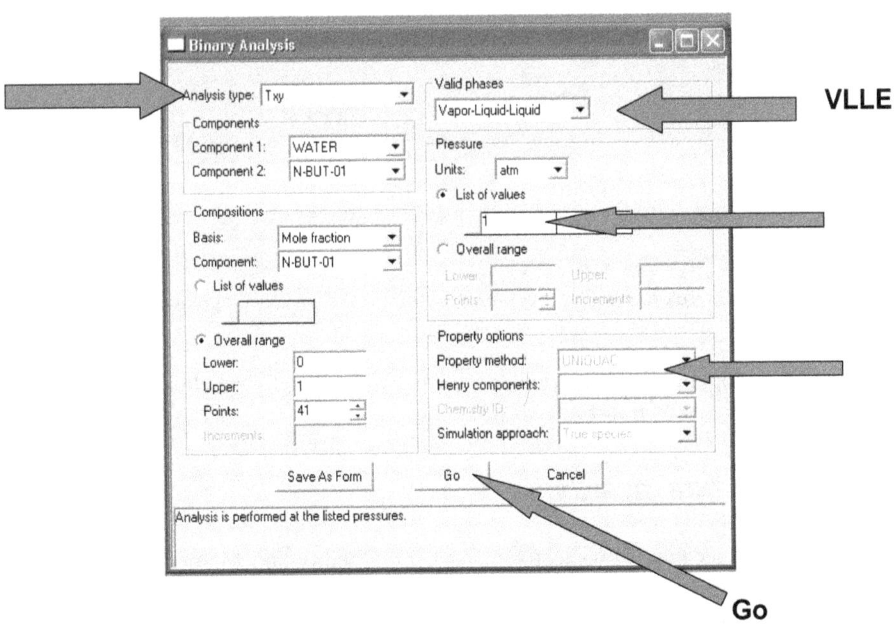


**Figure 7.2** Aspen setup for *Txy* diagram.

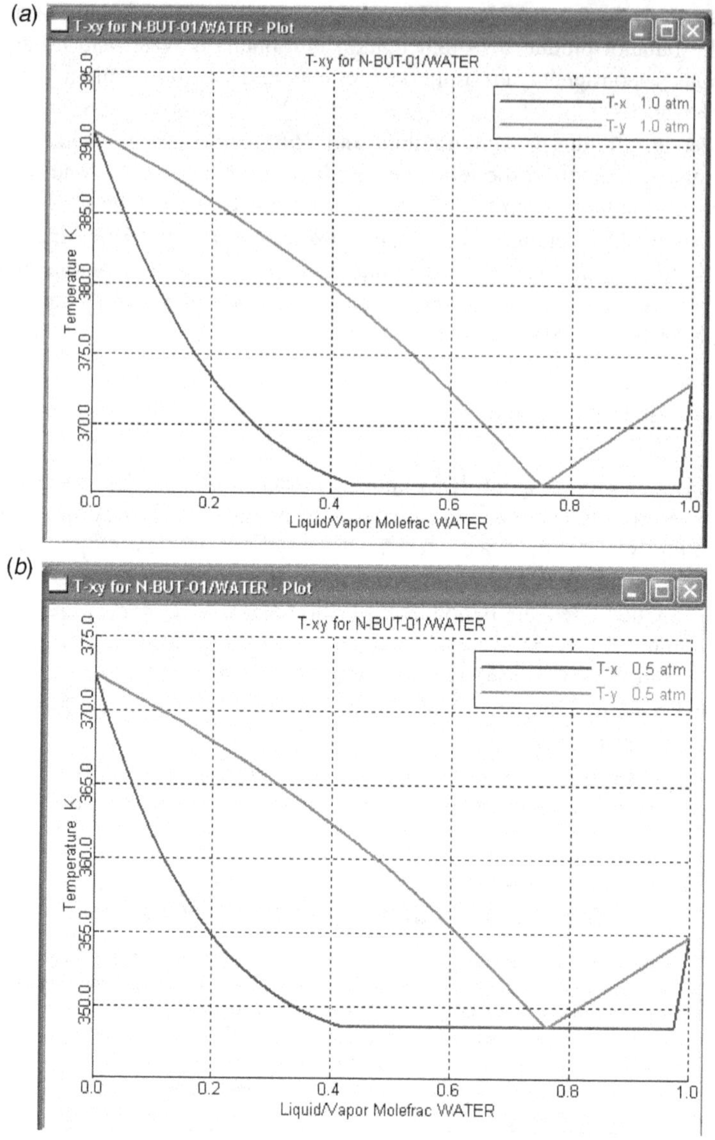


**Figure 7.3** (*a*) Aspen *Txy* diagram at 1 atm. (*b*) Aspen *Txy* diagram at 0.5 atm.

which the two components are selected, as well as *Vap-Liq-Liq*. Then clicking on *Azeotropes* generates the results shown in Figure 7.4*b*.

## 7.3 STEADY-STATE DESIGN

The two-column/decanter flowsheet suggested by Doherty and Malone[4] is used. Several different designs are generated for feed compositions ranging from 2 mol% *n*-butanol to 40 mol% *n*-butanol. Aspen Technology simulation software is used for steady-state and dynamic studies. The Aspen stage numbering convention is used in which stages are

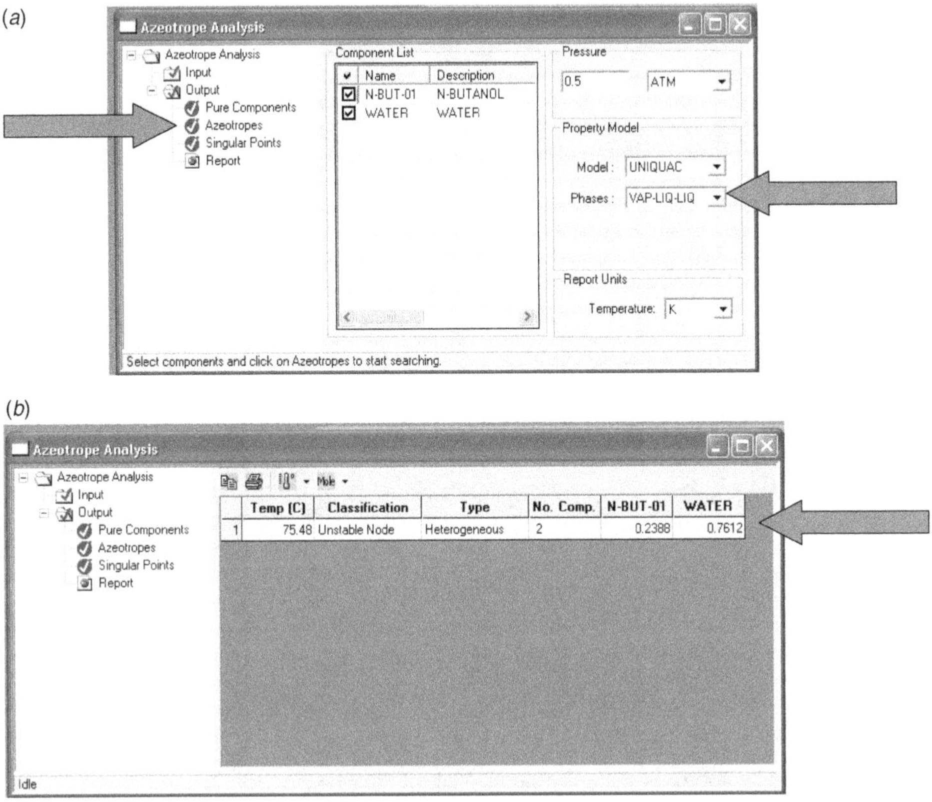


**Figure 7.4** (*a*) Azeotrope at 0.5 atm. (*b*) Azeotrope results at 0.5 atm.

numbered from the top down (reflux drum is Stage 1). The design feed flowrate is 1000 kmol/h; product purities are 99.9 mol%.

Figure 7.5 shows the Aspen Plus process flowsheet of the two-column system. Each column is a stripper with its individual reboiler. The bottoms of the first column C1 is 99.9 mol% water. The bottoms of the second column C2 is 99.9 mol% *n*-butanol. The overhead vapors from the two strippers are fed to a single condenser HX1. The compositions of the two vapor streams are almost the same, with values that are on opposite sides of the azeotropic composition.

Liquid from the condenser flows to a decanter in which the aqueous and organic liquid phases separate. The aqueous phase is returned to the top of the first column C1. The organic phase is fed to the top of the second column C2.

The point where the feed is introduced depends on its composition. For low butanol feed compositions (2 mol% *n*-butanol), feed is added on an intermediate tray in the first column. For higher butanol feed compositions (10 mol% or greater), the feed is fed directly to the decanter, since it is in the two-liquid phase region.

The economics are explored by evaluating capital and energy costs. The design optimization variables are the number of trays in each column, pressure, and feed location. As the *Txy* diagram given in Figure 7.1 shows, the separation between the saturated-liquid and saturated-vapor curves are quite similar on both sides of the azeotrope. This means that the difficulty of separation is similar and suggests that the number of trays in both

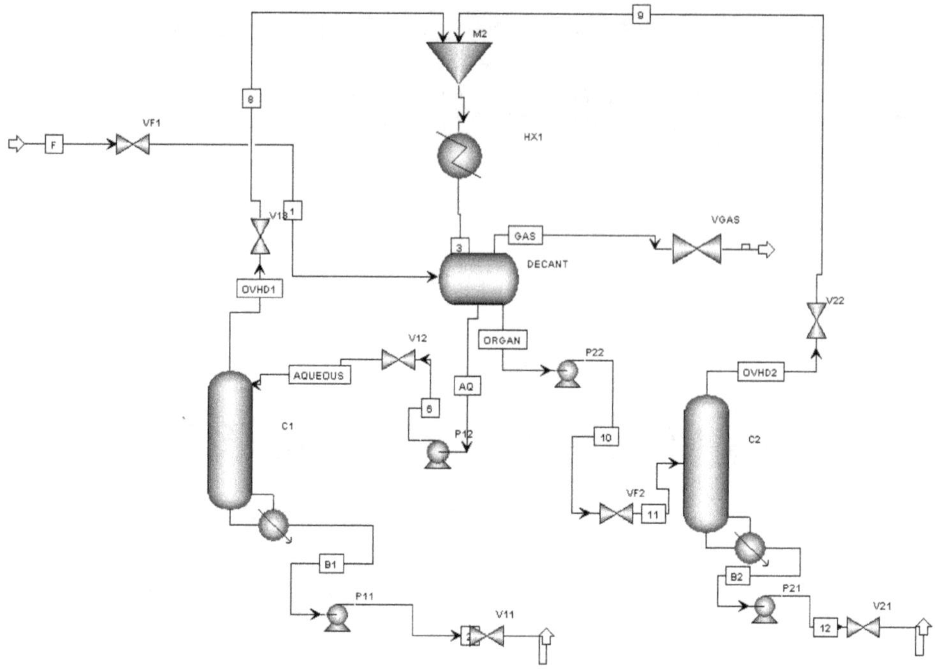


**Figure 7.5** Aspen Plus process diagram.

columns should be approximately the same. Making this assumption, the number of trays is varied from 20 down to 10. A ten-tray column is assumed to be a reasonable practical minimum size. The reboiler heat inputs actually decrease slightly as the number of trays is reduced, which is counterintuitive. For example, in the design for a 2 mol% butanol feed at an operating pressure of 0.5 atm, the reboiler energy requirements are $Q_{R1} = 2.237$ MW and $Q_{R2} = 0.3222$ MW when each column has 20 trays. When each column has 10 trays, the reboiler energy requirements are $Q_{R1} = 2.203$ MW and $Q_{R2} = 0.3183$ MW. On the other hand, the removal requirement in the overhead condenser $Q_C$ increases from 1.215 MW with 20 trays to 1.243 MW with 10 trays. These small changes have little effect on the economics. Since the capital cost of the 10-tray column is lower, this design is selected.

Operating the columns at a pressure of 1 atm gives higher energy consumption: $Q_{R1} =$ 2.431 MW and $Q_{R2} = 0.3257$ MW. The operation at 0.5 atm still permits the use of cooling water in the condenser since the decanter temperature of 343 K is sufficiently higher than the available cooling water temperature.

The final designs for the different feed compositions are shown in Figures 7.6, 7.7, and 7.8. As the *n*-butanol feed composition increases, the diameter of the first column decreases, the diameter of the second column increases, and the size of the decanter increases (the decanter is sized to provide 20 min of holdup). Also, as the feed *n*-butanol composition increases, the reboiler energy in the first column decreases, while the reboiler energy in the second column increases. Temperature and composition profiles in the two columns are quite similar in all cases. Figure 7.9 gives these profiles in both columns for a feed composition design of 40 mol% *n*-butanol. The products streams are the bottoms of each column.

The compositions of the two bottoms products are fixed at 99.9 mol% purity of their respective component. The design spec/vary feature of Aspen Plus is used to hold these

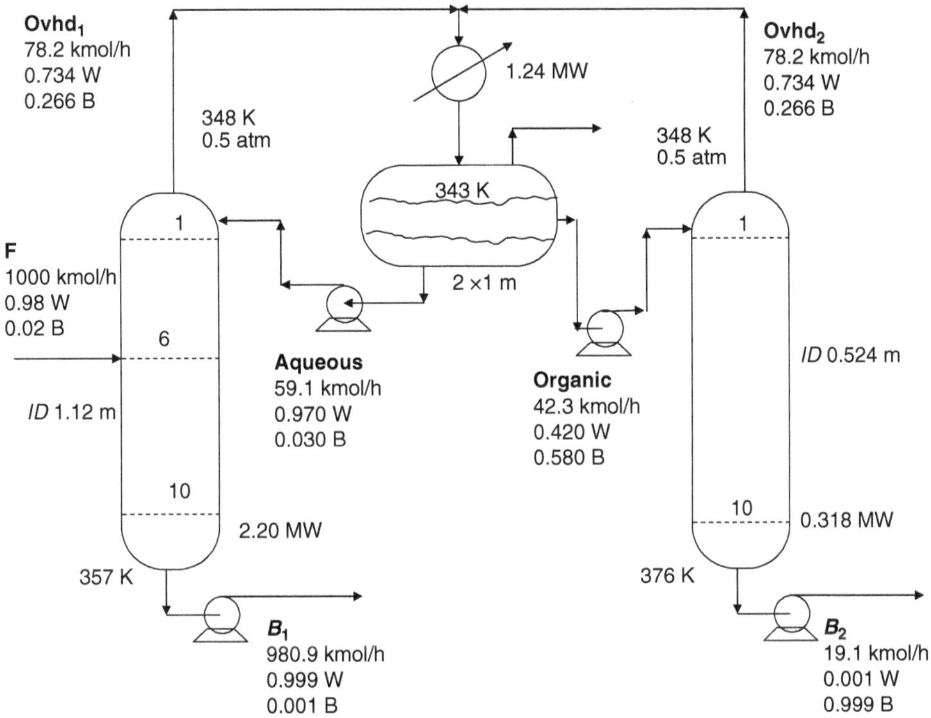


**Figure 7.6** Butanol–water flowsheet for 2 mol% butanol feed.

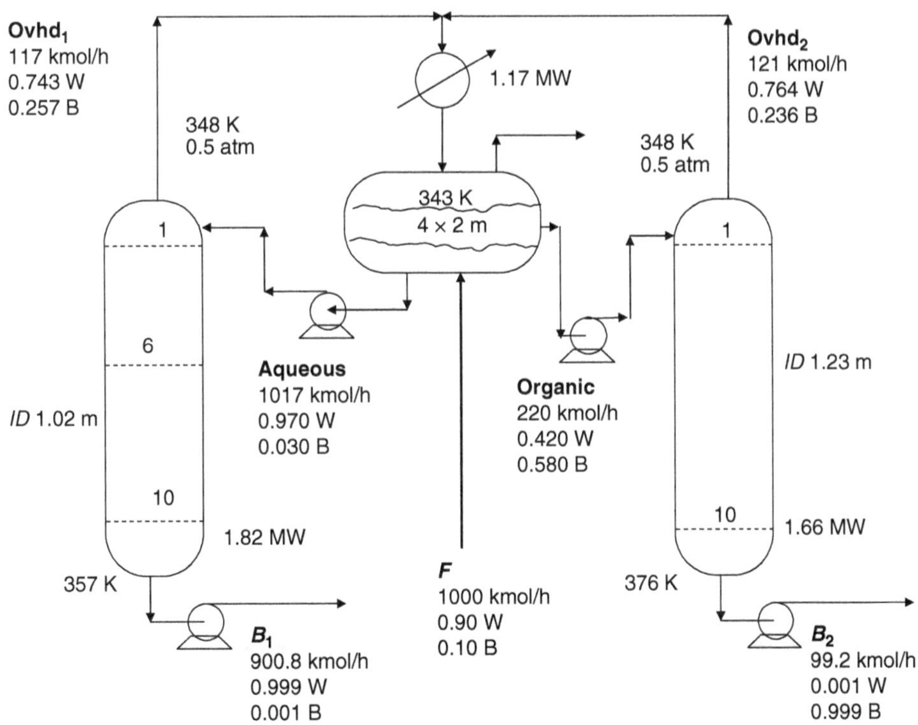


**Figure 7.7** Butanol–water flowsheet for 10 mol% butanol feed.

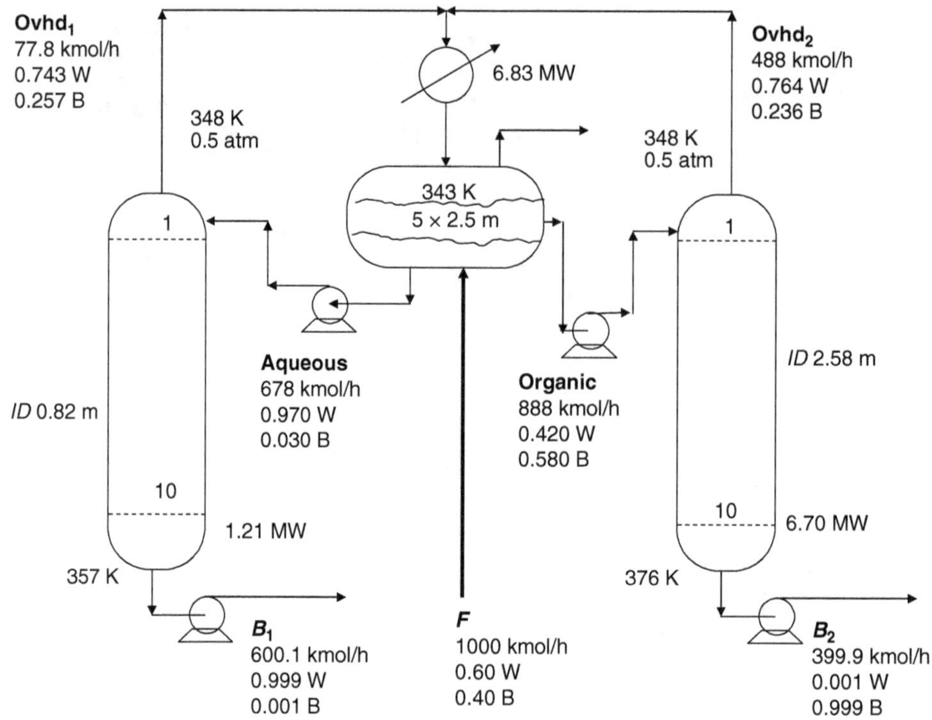


**Figure 7.8** Butanol–water flowsheet for 40 mol% butanol feed.

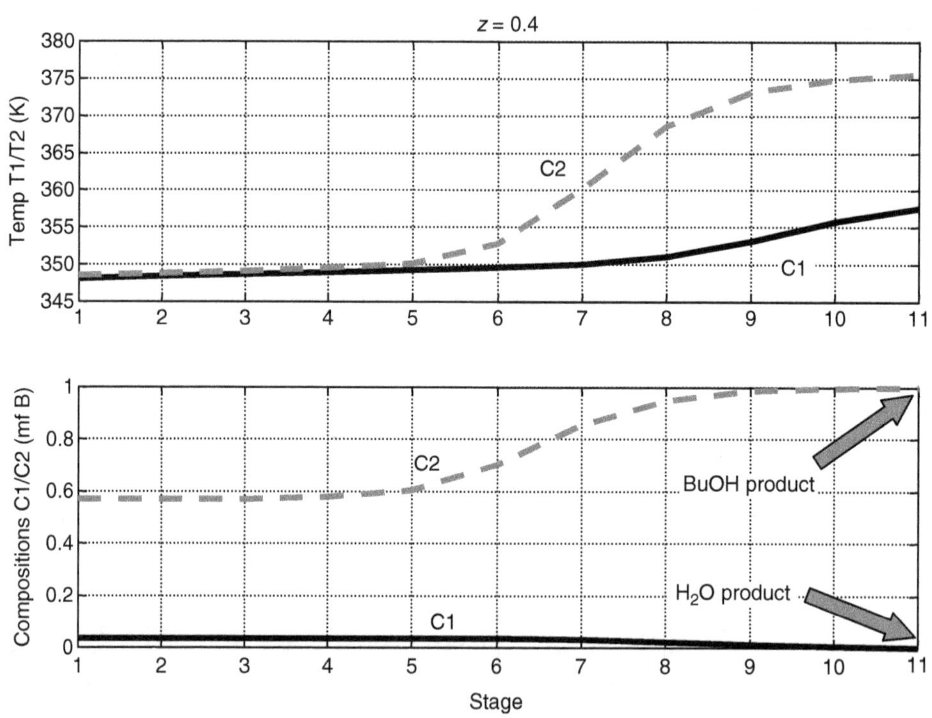


**Figure 7.9** Temperature and composition profiles for 40% case.

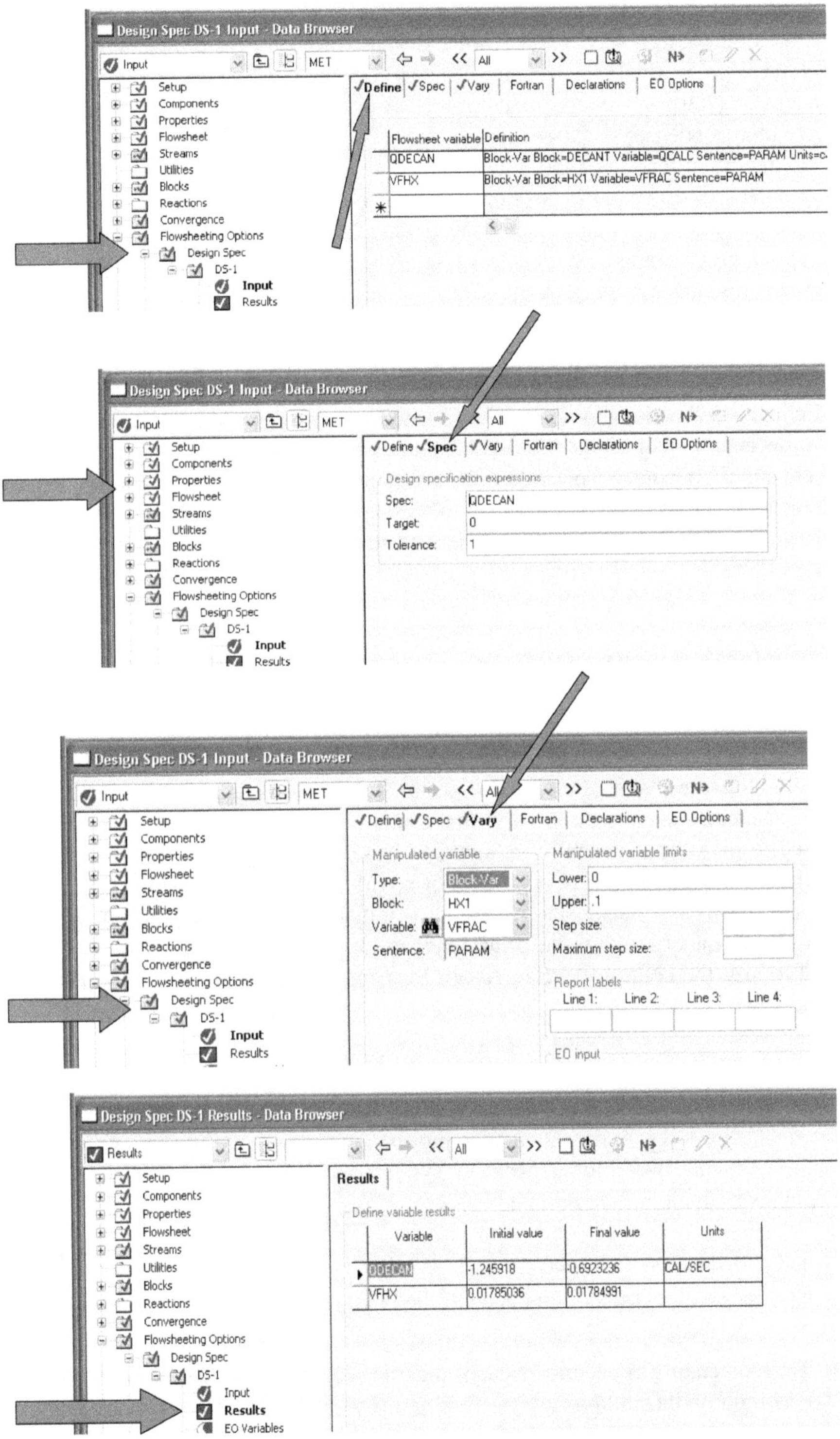


**Figure 7.10** Design spec to achieve an adiabatic decanter.

compositions by varying the flowrates of the two bottom product streams. Each column block uses one design spec/vary. A third design spec/vary (in *Flowsheeting Options*) is used to achieve an adiabatic decanter. This is illustrated in Figure 7.10. The heat input in the decanter *QDECAN* is specified to be driven to zero by varying the vapor fraction *VFHX* in the condenser (~0.018).

One important feature of the flowsheet shown in Figure 7.5 is a vapor stream leaving the decanter. As discussed in Section 4.1.3, there are two liquid–liquid decanter models available in Aspen Plus and Aspen Dynamics. The first is "DECANTER," which has only two liquid streams leaving the vessel. The second is "FLASH3," which has two liquid streams and a vapor stream. Both of the models worked for this system in the steady-state simulations conducted in Aspen Plus. However, in the dynamic simulations conducted in Aspen Dynamics, we found that the "FLASH3" model worked well for the butanol–water system, but the "DECANTER" model predicted that the organic level steadily increased and could not be controlled. The reason for this problem is unclear.

The decanter model using "FLASH3" requires two specifications. The first is selected as pressure and the second is specified to be a very small vapor fraction (0.0001). During the dynamic simulations, the control valve on the vapor stream is completely shut by putting the decanter pressure controller on manual and setting its output signal at 0%.

## 7.4 DYNAMICS AND CONTROL

Each of the three designs is exported to Aspen Dynamics after the volumes of the column bases and the decanter have been specified. A holdup time of 10 min in the bottom of each column is used. A holdup time of 20 min is used in the decanter. The effect of using larger decanter holdups is explored in the results given below. Control valves and pumps are sized to give the rangeability required to handle the large disturbance tested later in the dynamic simulations. If a valve becomes wide open, the appropriate pump work is increased and a new steady-state is used for initial conditions. The resulting controller output signals under steady-state conditions are lower than the default value of 50% that is used to size the control valve in Aspen Plus for the given flowrate and pressure drop over the valve.

### 7.4.1 Control Structure

Figure 7.11 shows the Aspen Dynamics process flow diagram with the control structure developed for this two-column/decanter system. Conventional PI controllers are used. The features of the loops are outlined below.

1. The feed is flow controlled.
2. The pressure in each column is controlled by a valve in its overhead vapor line.
3. The base level of each column is controlled by manipulating the bottoms flowrate.
4. The temperature of one tray in each column is controlled by manipulating the reboiler heat input in that column.
5. The aqueous interface in the decanter is controlled by manipulating the flowrate of the aqueous stream to the top tray of the first column.
6. The organic level in the decanter is controlled by manipulating the flowrate of the organic stream to the top tray of the second column.

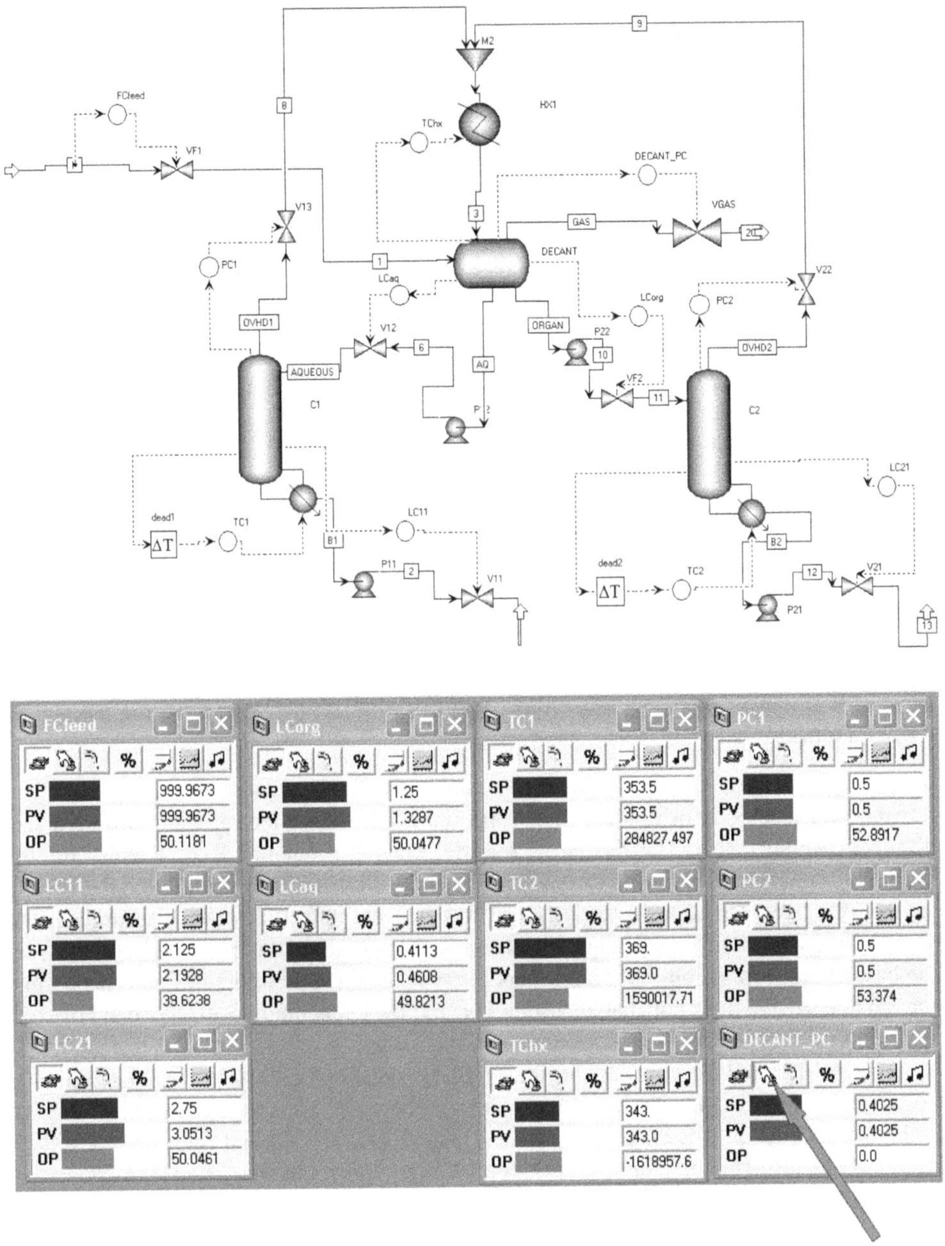


**Figure 7.11** Control structure and controller faceplates.

7. The temperature of the decanter is controlled by manipulating the heat removal in the condenser.
8. The pressure controller on the decanter is on manual with the vapor valve shut.

All level controllers are proportional only with a gain of 2. The trays selected for temperature control are located near the bottom of the columns, where the temperatures are changing the most rapidly from tray to tray (see Figure 7.9). A 1-min deadtime is inserted in each temperature loop. Relay-feedback tests and Tyreus–Luyben tuning are used to obtaining

**TABLE 7.1 Controller Tuning Parameters.**

| | | 2 mol% Butanol Feed | 10 mol% Butanol Feed | 40 mol% Butanol Feed |
|---|---|---|---|---|
| TC1 | Controlled variable (*K*) | $T_9 = 352.8$ | $T_9 = 353.2$ | $T_9 = 353.2$ |
| | Manipulated variable | $Q_{R1}$ | $Q_{R1}$ | $Q_{R1}$ |
| | Transmitter range (*K*) | 300–400 | 300 to 400 | 300 to 400 |
| | Controller output range (cal/s) | $0–1.52 \times 10^6$ | $0–1.5 \times 10^6$ | $0–0.6 \times 10^6$ |
| | $K_C$ | 0.55 | 1.35 | 2.28 |
| | $\tau_I$ (min) | 10.6 | 6.6 | 10.6 |
| TC2 | Controlled variable (*K*) | $T_8 = 386.7$ | $T_8 = 368.7$ | $T_8 = 368.7$ |
| | Manipulated variable | $Q_{R2}$ | $Q_{R2}$ | $Q_{R2}$ |
| | Transmitter range (*K*) | 300–400 | 300–400 | 300–400 |
| | Controller output range (cal/s) | $0–1.26 \times 10^6$ | $0–0.80 \times 10^6$ | $0–3.0 \times 10^6$ |
| | $K_C$ | 1.60 | 0.76 | 1.19 |
| | $\tau_I$ (min) | 9.2 | 10.6 | 7.9 |

controller tuning parameters in the column temperature loops. Table 7.1 gives tuning constants for the three design cases. Controller faceplates are shown at the bottom of Figure 7.11.

### 7.4.2 Results

To test the effectiveness of this control structure for each of the cases, disturbances in feed flowrate and feed composition are made. In Figure 7.12, the effect of using different size decanters is illustrated. The design is for the 2 mol% *n*-butanol feed. Results for the normal 20-min decanter are the dashed lines; results for a decanter with twice the holdup are the solid lines. The disturbance is a large 30% increase in feed flowrate. The large decanter reduces the transient peak change in the composition of the *n*-butanol product $x_{B2(B)}$ from the base of the second column. This occurs because the flowrate of the feed to this column, the organic stream, does not change as quickly, and the temperature controller can hold tray temperature closer to the desired value.

Figures 7.13*a* and *b* give results for the 2 mol% *n*-butanol feed case with a 20 min holdup decanter. In Figure 7.13*a*, the disturbances are very large 30% changes in feed flowrate. The solid lines are for a 30% increase at time equals 0.2 h. The dashed lines are for a 30% decrease. Stable regulatory control is achieved. The compositions of the two product streams ($x_{B1(W)}$ and $x_{B2(B)}$) are held quite close to the desired values. Notice that the composition of the stream leaving the bottom of the first column undergoes a larger transient deviation than the deviation seen in the bottoms of the second column. This occurs because the first column sees a step change in feed flowrate since the feed is introduced into this column for the 2 mol% feed composition case. The second column sees a more gradual change in the flowrate of the stream entering the column (the organic). The use of a steam-to-feed ratio system in the first column would help to reduce this transient deviation. Notice that it takes almost 10 h for this system to come to its new steady state.

Figure 7.13*b* shows results for very large 50% (on a relative basis) feed composition disturbances. The solid lines are for a step change at 0.2 h from 2 mol% up

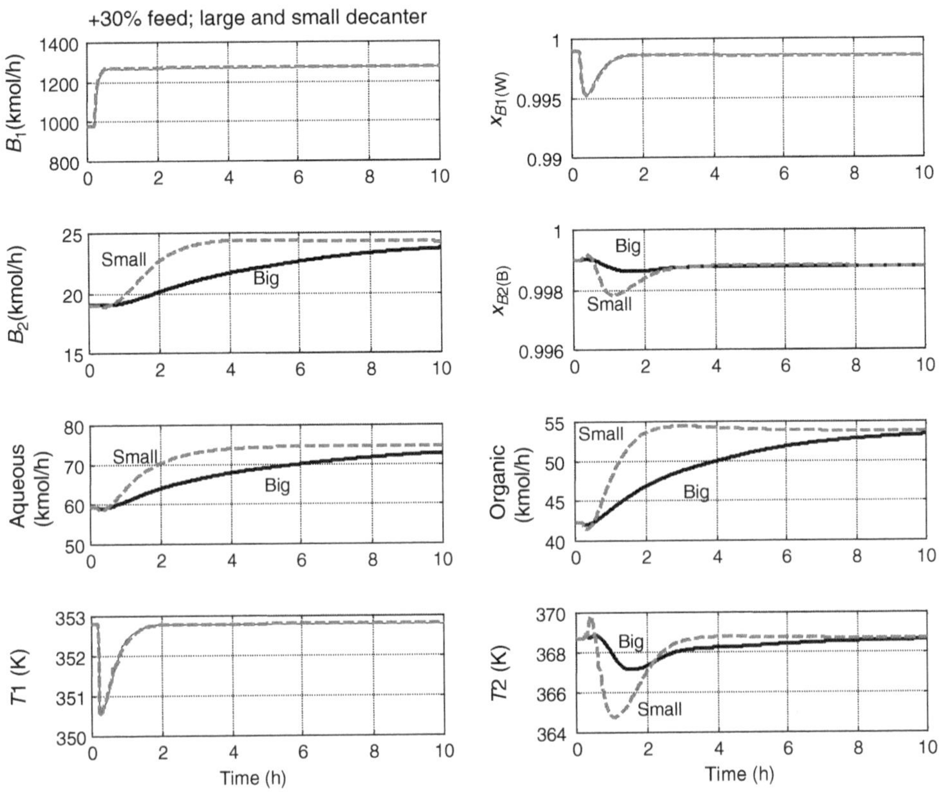


**Figure 7.12** Effect of decanter size; $z = 2\%$.

to 3 mol%. The dashed lines are for a step change from 2 mol% down to 1 mol%. Stable regulatory control is achieved, and product purities are held close to the desired value. The increase in $n$-butanol in the feed causes a transient drop in the purity of the water product stream from the bottom of the first column $x_{B1(W)}$ down to about 99.6 mol%, but the composition returns to 99.9 mol% in about 2 h. There is a small offset in the purity of the $n$-butanol stream from the bottom of the second column because we are controlling a tray temperature, not the bottoms composition. A composition–temperature cascade control structure could be used to achieve the desired purity, but a composition analyzer would be required.

Similar dynamic results are observed in Figures 7.14 and 7.15 for the other feed composition cases. Figure 17.14 gives results for the 10 mol% case, and Figure 7.15 gives results for the 40 mol% case. Notice that the time scale is smaller for these cases. Transients die out in about 2 h. Notice also that the transient deviations in $x_{B1(W)}$ for feed flowrate disturbances are smaller than those seen in the 2 mol% case. This occurs because the feed is entering the decanter, not the column. The column sees a more gradual disturbance in the aqueous stream instead of a step change.

The 40 mol% case has somewhat slower dynamics than the 10 mol% case and exhibits more sensitivity to disturbances in feed composition. The peak deviations in product purities are larger.

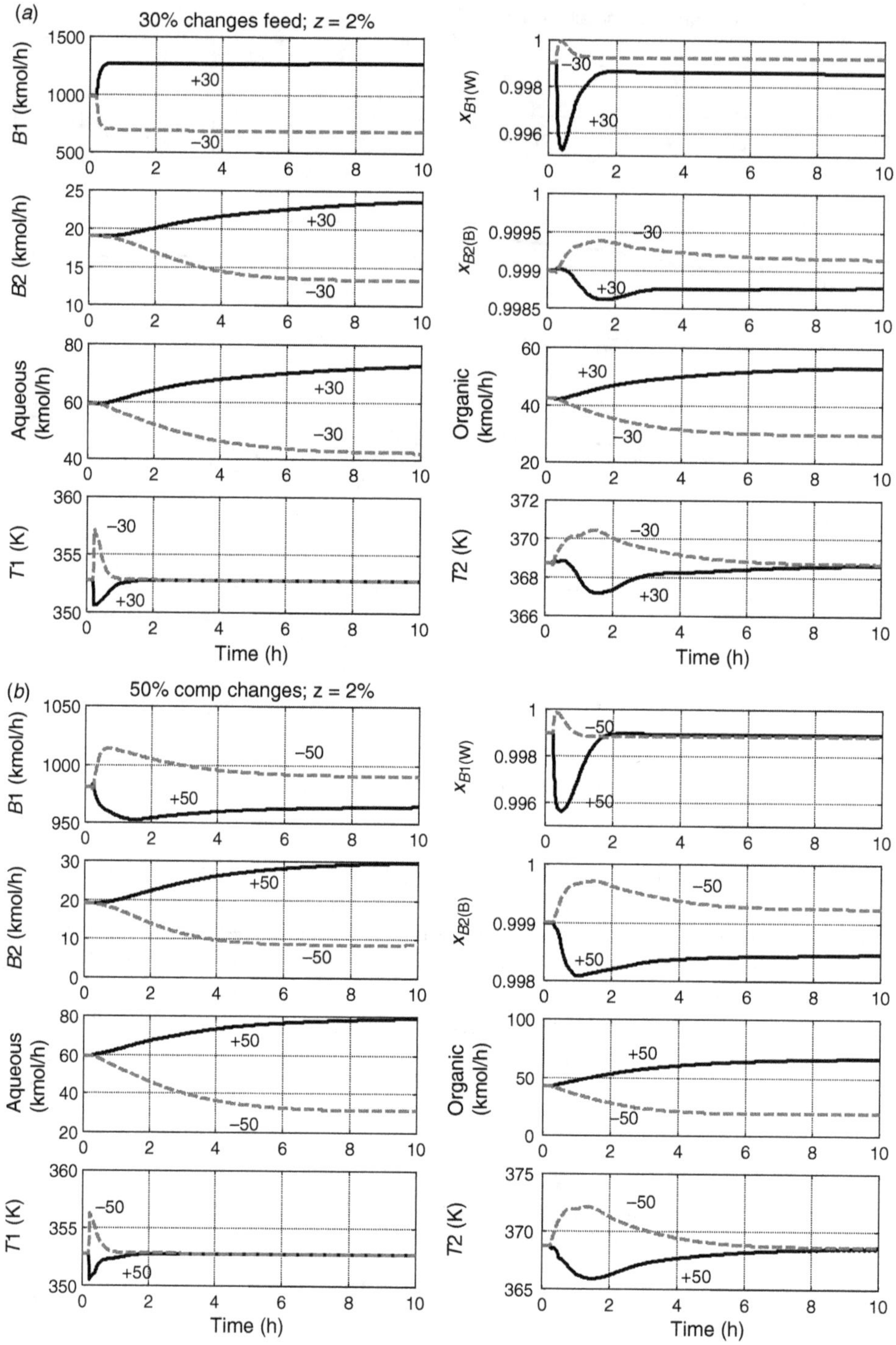


**Figure 7.13** (*a*) 30% changes in feed flowrate; $z = 2\%$ case. (*b*) 50% changes in feed composition; $z = 2\%$ case.

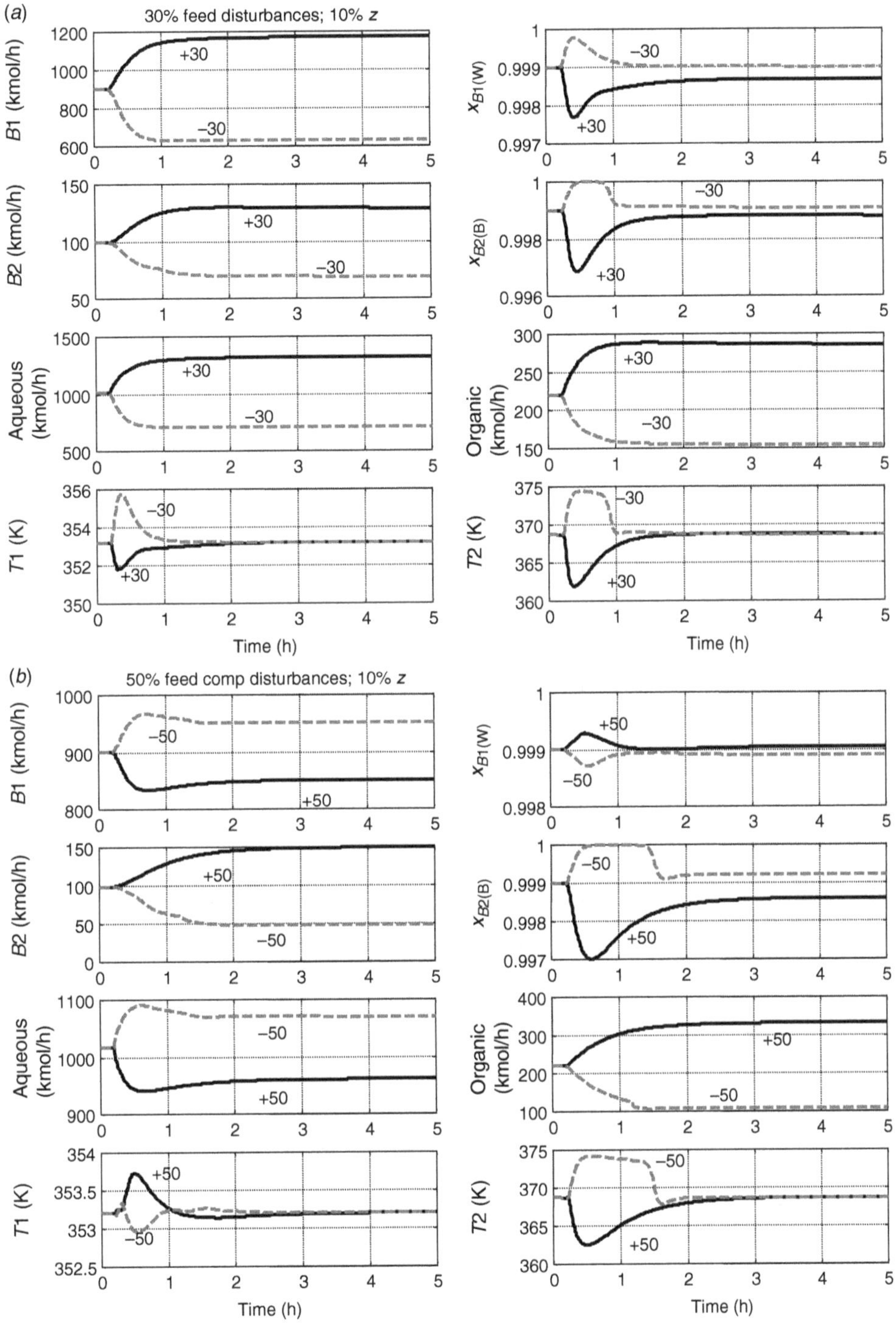


**Figure 7.14** (*a*) 30% changes in feed flowrate; $z = 10\%$ case. (*b*) 50% changes in feed composition; $z = 10\%$ case.

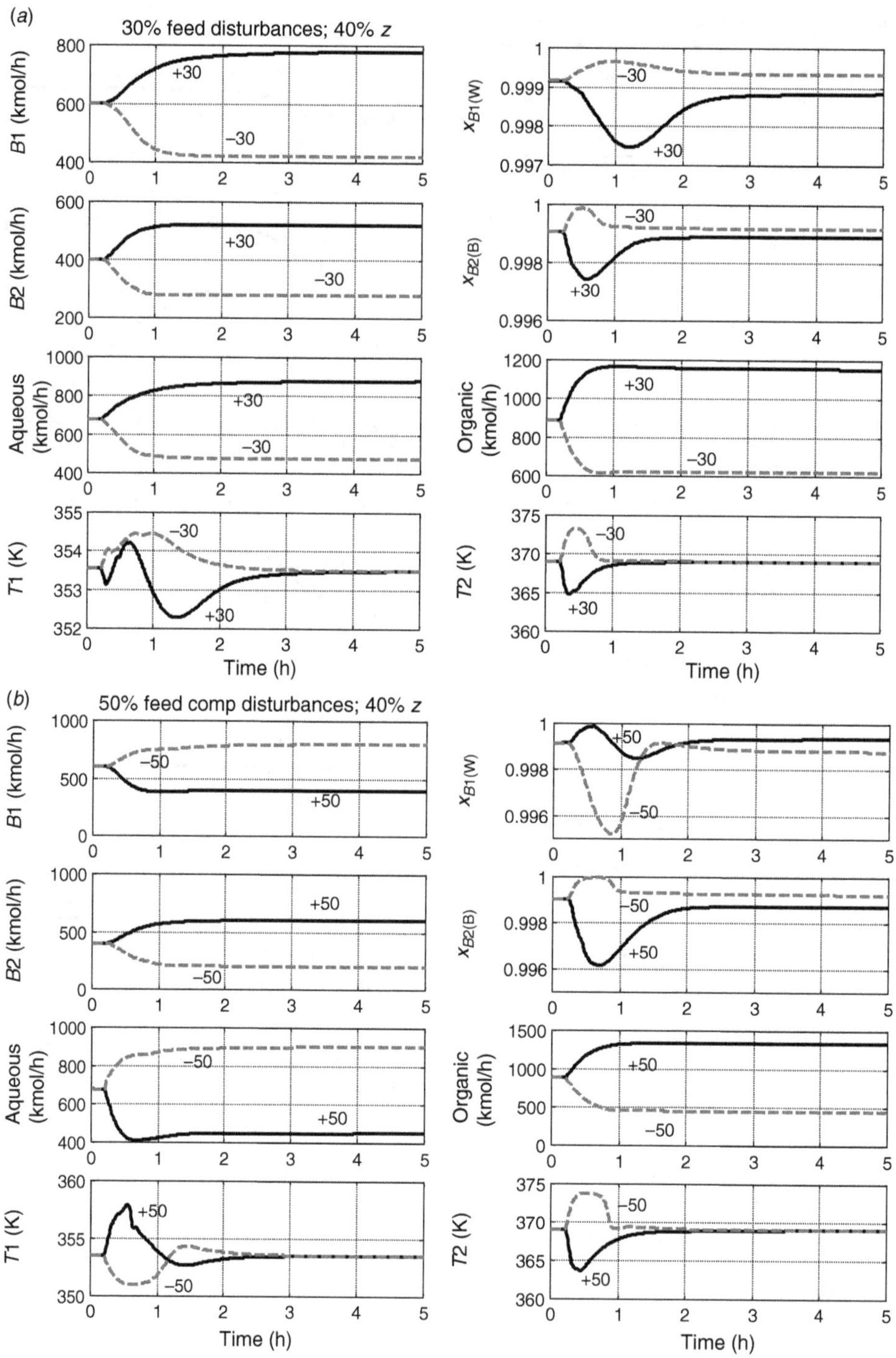


**Figure 7.15** (*a*) 30% changes in feed flowrate; $z = 40\%$ case. (*b*) 50% changes in feed composition; $z = 40\%$ case.

## 7.5 CONCLUSION

A simple control structure has been developed and demonstrated to be remarkably robust for the separation of the binary heterogeneous azeotropic *n*-butanol–water system. The process configuration consists of two stripping columns, a decanter, and a single condenser. A tray temperature is controlled in each column. Very large disturbances in feed flowrate and feed composition are effectively handled with product compositions maintained close to the desired values.

## REFERENCES


1. Qureshi N. and H. P. Blaschek, Butanol production from agricultural biomass, In *Food Engineering*, Second Edition, CRC Taylor & Francis, Boca Raton, FL, 2006, pp. 525–549.
2. Ladisch M. R., Fermentation-derived butanol and scenarios for its use in energy-related applications, *Enzyme Micro. Technol.*, **13**, 280–283 (1991).
3. Phillips J. A. and A. E. Humphrey, An overview of process technology for the production of liquid fuels and chemical feedstocks via fermentation. In *Organic Chemicals from Biomass*, Ed. Wise, D. L., Benjamin Cummings, Menlo Park, CA, 1983, pp. 249–304.
4. Doherty M. F. and M. F. Malone, *Conceptual Design of Distillation Systems*, McGraw-Hill, New York, 2001, p. 383.
5. Stichlmair J. G. and J. R. Fair, *Distillation Principles and Practices*, Wiley-VCH, New York, 1998, pp. 222–342.
6. Pucci A., P. Mikitendo, and L. Asselineau, Three-phase distillation simulation and application to the separation of fermentation products, *Chem. Eng. Sci.*, **41**, 485–494 (1986).

## 7.2 CONCLUSIONS

# PART 3

# SEPARATIONS USING A LIGHT ENTRAINER (HETEROGENEOUS AZEOTROPIC DISTILLATION)

One way to separate a mixture containing an azeotrope is to add a light entrainer into the system so that an additional azeotrope(s) can be formed that helps in the separation. In order to make the separation feasible, there are two important characteristics of this additional azeotrope (or one of the additional azeotropes). First, the azeotropic temperature of one additional azeotrope should be the minimum temperature of the whole ternary system. Second, this azeotrope should be heterogeneous so that natural liquid–liquid separation, without energy requirement, can be performed in a decanter at the top of the column.

Depending on the number of additional azeotropes, quite a few different RCMs could be generated. Chapter 8 will show the feasible design of three different systems. The first system is with only one additional heterogeneous binary azeotrope by adding an entrainer (separating propylene glycol monomethyl ether and water with isopropyl acetate as the entrainer). The second system is with two additional binary azeotropes. One is heterogeneous and the other is homogeneous. An example is the separation of pyridine and water with toluene as the entrainer. The third system is with three additional azeotropes: one binary homogeneous azeotrope, one binary heterogeneous azeotrope, and one ternary heterogeneous azeotrope. An example is the separation of isopropanol and water with cyclohexane as the entrainer. A second example is the separation of ethanol and water with benzene as the entrainer. The third system is the most complex one of all the above systems, so we will study the design and control of this system in detail in Chapter 8.

Another application of an entrainer addition is for the separation of a close-boiling mixture using heterogeneous azeotropic distillation. One common industrial example is the separation of acetic acid and water with an ester as the entrainer. The reason for adding the

*Design and Control of Distillation Systems for Separating Azeotropes.* By William L. Luyben and I-Lung Chien

entrainer into such a system is because there is a tangent pinch for the binary mixture of acetic acid and water near the pure water end. Thus many stages are required to obtain high-purity products via a regular column. Chapter 9 will study the design and control of this system using isobutyl acetate as the entrainer. Special attention will also be given to study the common industrial situation when the fresh feed contains small amounts of another component that is an impurity.

# CHAPTER 8

# ISOPROPANOL–WATER (CYCLOHEXANE AS THE ENTRAINER)

We will start this chapter by explaining how the heterogeneous azeotropic distillation works in separating a mixture with an azeotrope. Three systems with different RCMs will be used as examples to illustrate the column sequence for the separation. After that, we will focus on the detailed design and control of the isopropanol–water system, which is a system that exhibits the most complex RCM.

## 8.1 FEASIBLE COLUMN SEQUENCE FOR THE SEPARATION

Depending on the number of additional azeotropes generated by adding an entrainer, quite a few different RCMs could be produced. In the following, we will show the feasible column sequence of three different systems.

### 8.1.1 Propylene Glycol Monomethyl Ether–Water System

As shown in Figure 8.1, the propylene glycol monomethyl ether (PM)–water mixture has an azeotrope with composition around 81 mol% $H_2O$ and azeotropic temperature around 97°C. By adding isopropyl acetate into the system, as shown in Figure 8.2, an additional azeotrope is formed that has a minimum azeotropic temperature of 76.46°C. This azeotrope is also inside the LLE envelope. The RCM at 1 atm and LLE envelope at 40°C easily can be generated in Aspen Plus. With the aid of the information in Figure 8.2, the feasible column sequence can be designed. As shown in Figure 8.3, the flowsheet features a preconcentrator column and a heterogeneous azeotropic distillation column.

The purpose of the preconcentrator column is to remove part of the water in the fresh feed (assume the fresh feed composition is 90 mol% water). The distillate composition of this

*Design and Control of Distillation Systems for Separating Azeotropes.* By William L. Luyben and I-Lung Chien

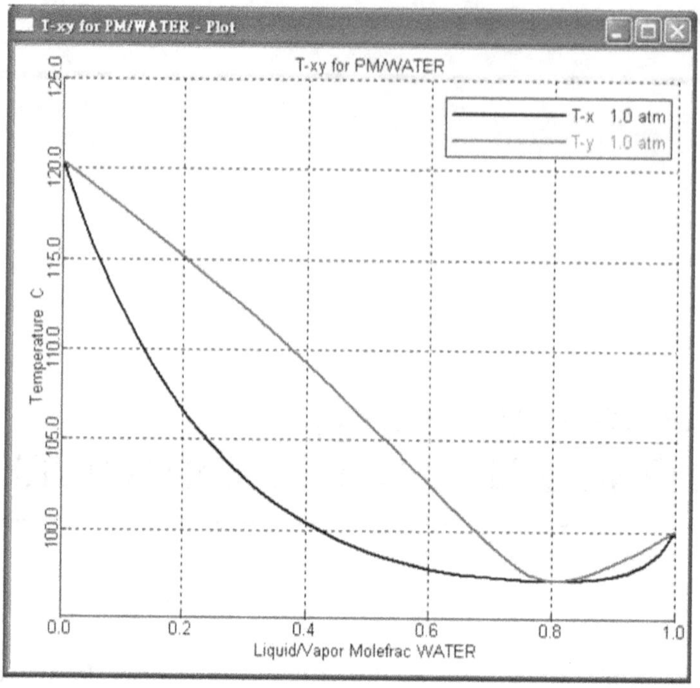


**Figure 8.1** *Txy* plot of PM–water system.

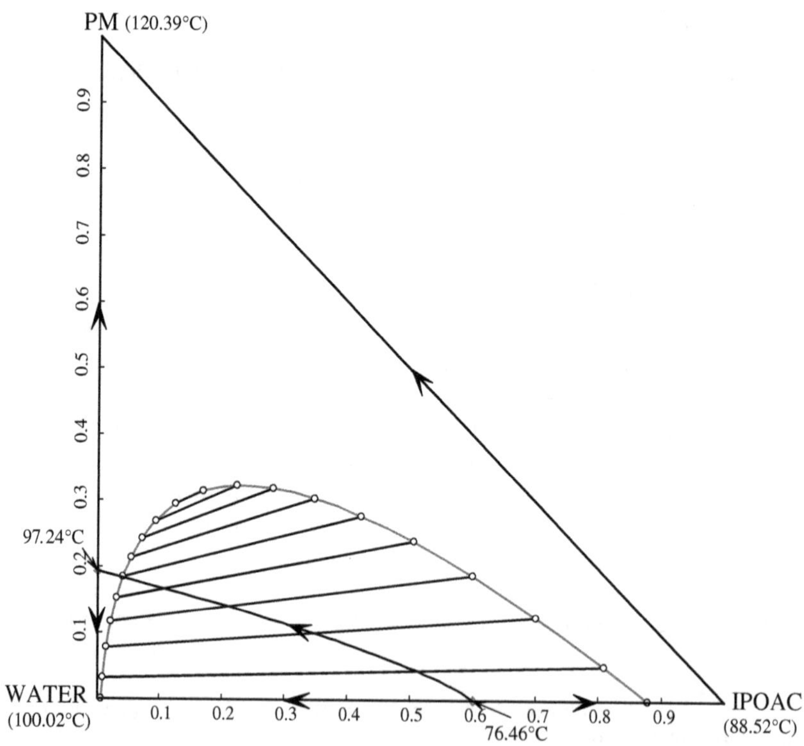


**Figure 8.2** RCM with LLE envelope of PM–water system with isopropyl acetate as the entrainer.

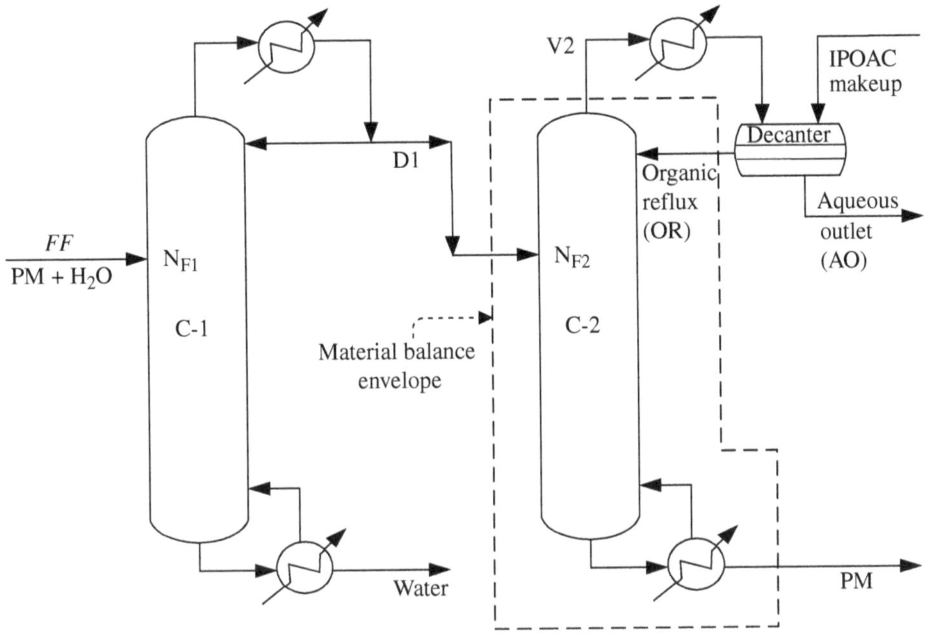


**Figure 8.3** Desired column sequence for PM–water separation.

preconcentrator column will approach azeotropic composition of PM–water. This distillate stream is fed into a heterogeneous azeotropic column. With the aid of adding another component (isopropyl acetate) into the system through an OR stream, the bottom composition will approach pure PM composition and the column top vapor will approach the lowest temperature of the ternary system, which is the azeotropic composition of isopropyl acetate–water.

This azeotrope is heterogeneous; therefore, by condensing and cooling the top vapor to 40°C and feeding the condensate to a decanter, a natural liquid–liquid separation into an organic phase and an aqueous phase will occur. The organic phase, which is rich in entrainer, is recycled back to the heterogeneous azeotropic column. The composition of the aqueous phase is pure enough that it can be drawn out of the system as an aqueous product. A small entrainer makeup stream is needed to feed into the system at the decanter to balance the small entrainer loss in the AO stream. The material balance lines for each of the two columns (shown in Fig. 8.4) clearly explain the conceptual design of this separation flowsheet.

Another point worth mentioning is that the products of the column sequence (PM and water) are in different distillation regions (see Fig. 8.4). This normally cannot happen for a column sequence with relatively linear distillation boundaries, as is the case for this system. The reason that the distillation boundary can be crossed in this system is the liquid–liquid separation occurring in the decanter.

### 8.1.2 Pyridine–Water System

As shown in Fig. 8.5, a pyridine–water mixture has an azeotrope with composition of around 77 mol% $H_2O$ and azeotropic temperature of around 95°C. By adding toluene

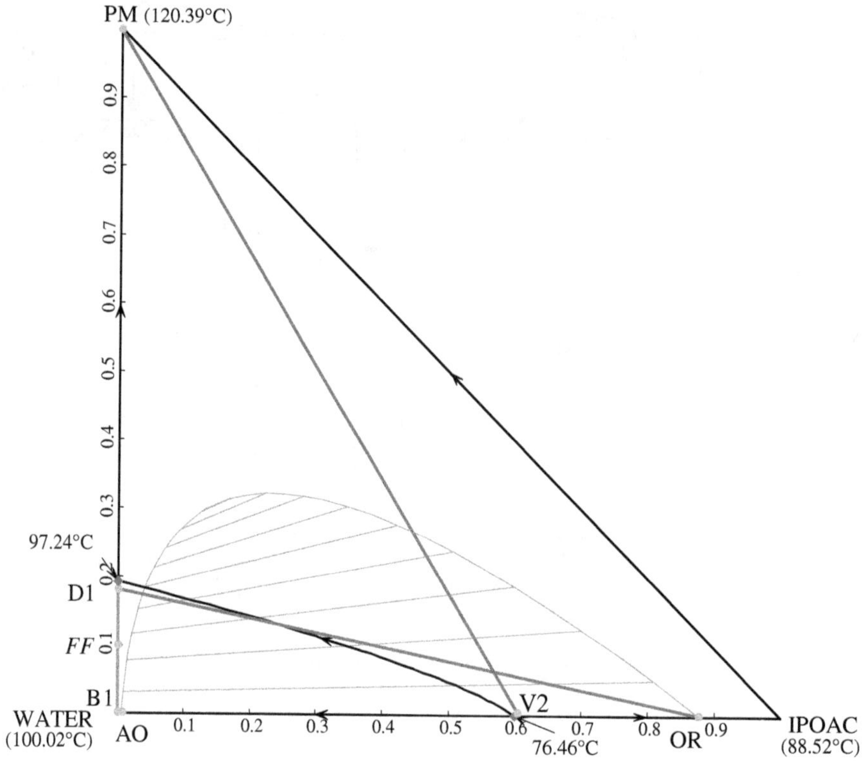


**Figure 8.4** Material balance lines for PM–water separation.

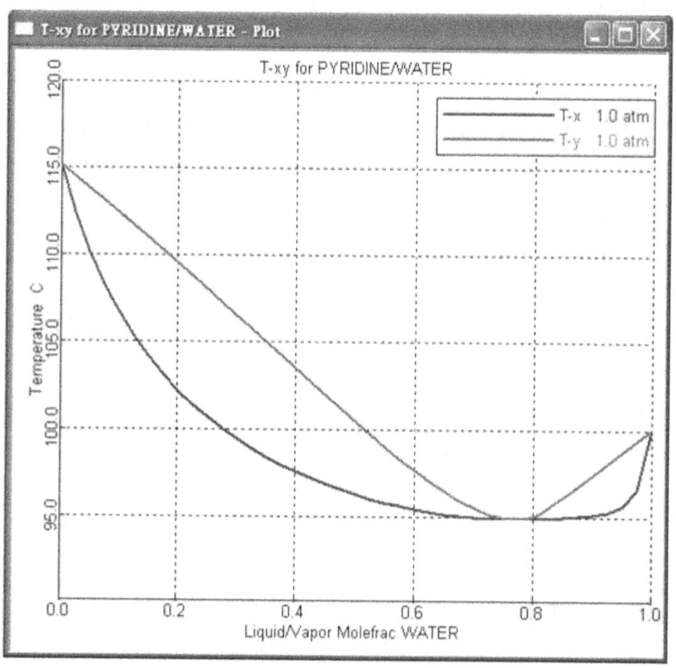


**Figure 8.5** *Txy* plot of pyridine–water system.

into the system, as shown in Figure 8.6, two additional azeotropes are formed. One desirable azeotrope (toluene–water) is heterogeneous with an azeotropic temperature of 84.53°C, which is the minimum for the entire ternary system. Another azeotrope between pyridine and toluene is also formed with a higher azeotropic temperature of 110.15°C. As shown in Figure 8.6, the RCM divides the ternary system into three distillation regions. The feasible column sequence is the same as the previous system. As shown in Figure 8.7, the flowsheet features a preconcentrator column and a heterogeneous azeotropic distillation column.

The purpose of the preconcentrator column is again to remove part of the water in the fresh feed (assume the fresh feed composition is 90 mol% water). The lowest temperature in this ternary system is the toluene–water azeotrope, and the highest temperature is pyridine; therefore, the heterogeneous azeotropic column top vapor is designed to approach a toluene–water azeotrope and the bottom product is pure pyridine. The material balance lines of the column sequence can be seen in Figure 8.8.

One difficulty in the design flowsheet in Figure 8.7 is that reaching the toluene–water azeotrope needs to go through a very narrow funnel (see Fig. 8.8). Hence, this heterogeneous column either requires a large number of total stages or even proves practically impossible to reach this point. Failure to reach this correct toluene–heterogeneous azeotrope will result in some pyridine being lost from the AO stream. Another worry is that the operability of this system may be very fragile owing to the same problem of a narrow operating funnel. An alternative design flowsheet will be proposed in Section 8.2 that is more economical and also enhances the operability of the separation system.

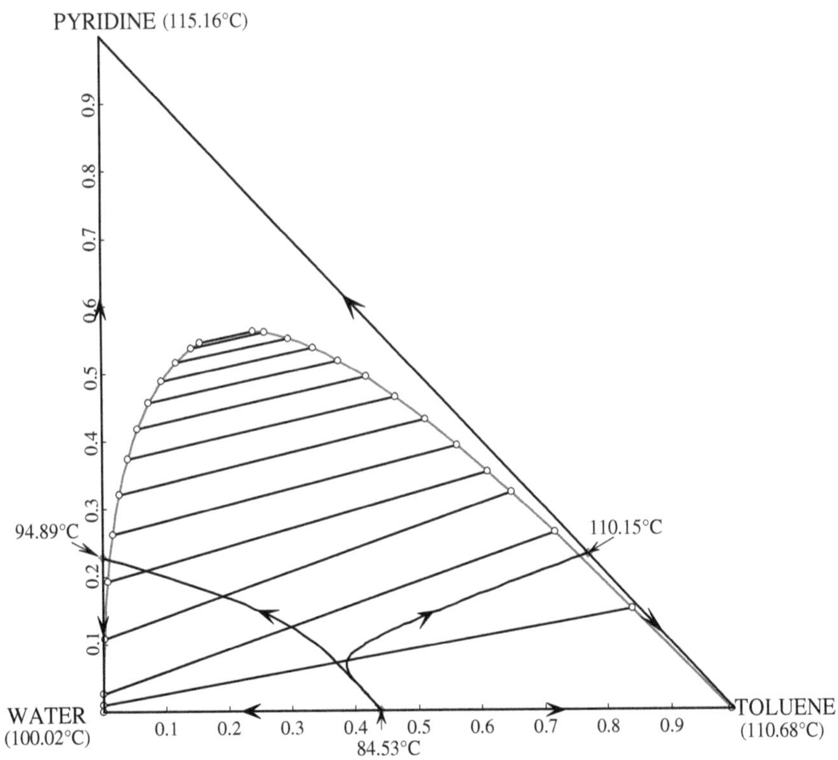


**Figure 8.6** RCM with LLE envelope of pyridine–water system with toluene as the entrainer.

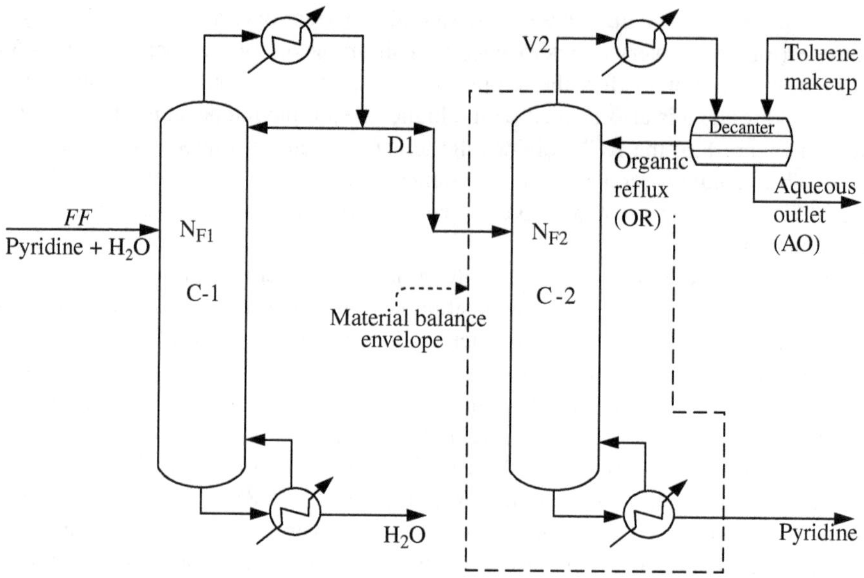


**Figure 8.7** Column sequence for pyridine–water separation.

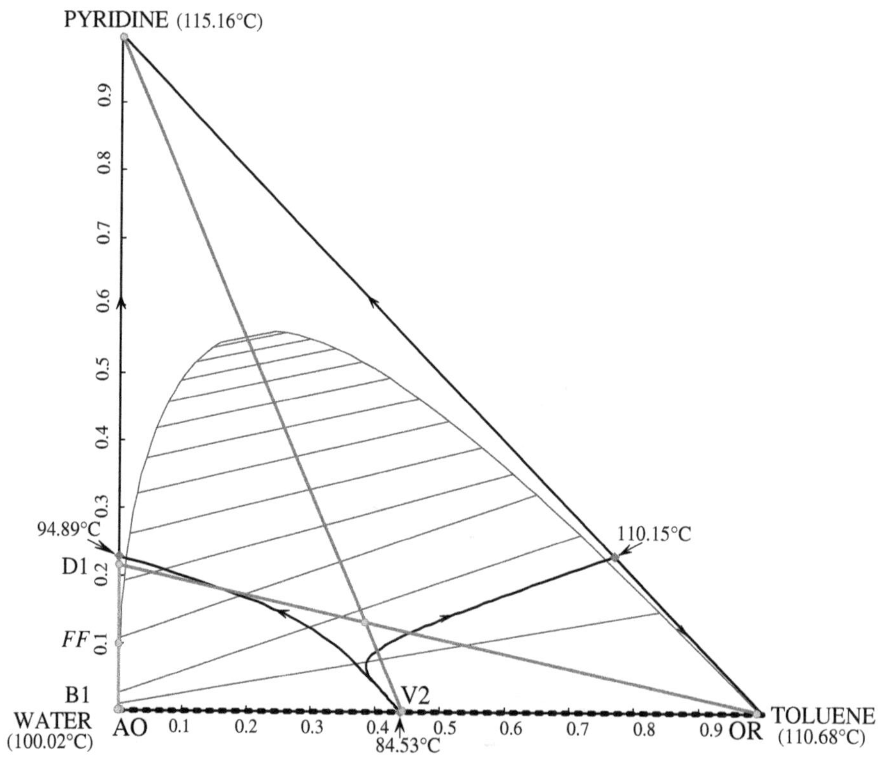


**Figure 8.8** Material balance lines for pyridine–water separation.

### 8.1.3 Isopropanol–Water System

As shown in Figure 8.9, the IPA–water mixture has an azeotrope with composition of around 31 mol% $H_2O$ and azeotropic temperature of around 80°C. By adding cyclohexane into the system, as shown in Figure 8.10, three additional azeotropes are formed. One desirable ternary azeotrope (IPA–water–cyclohexane) is heterogeneous with an azeotropic temperature of 63.77°C that is the minimum temperature for the entire ternary system. Another two azeotropes (one heterogeneous and another one homogeneous) both have higher azeotropic temperatures. As shown in Figure 8.10, the RCM divides the ternary system into three distillation regions. The feasible column sequence shown in Figure 8.11 contains a preconcentrator column, a heterogeneous azeotropic distillation column, and a recovery column.

The purpose of the preconcentrator column is again to remove part of the water in the fresh feed (assume the fresh feed composition is 50 mol% IPA to agree with a true application of a waste IPA stream in the semi-conductor industry). The distillate composition of this preconcentrator column will approach the azeotropic composition of IPA–water. This distillate stream is fed into a heterogeneous azeotropic column. With the aid of adding another component (cyclohexane) into the system through an OR stream, the bottom composition will approach a pure IPA composition and the column top vapor will approach the lowest temperature of the ternary system, which is the ternary azeotrope. Note that this column is operated inside the upper distillation region (see Fig. 8.12).

This azeotrope is heterogeneous, so by condensing and cooling the top vapor and feeding the condensate at 40°C to a decanter, natural liquid–liquid separation into an organic phase and an aqueous phase will occur. The organic phase, which is rich in entrainer, is designed

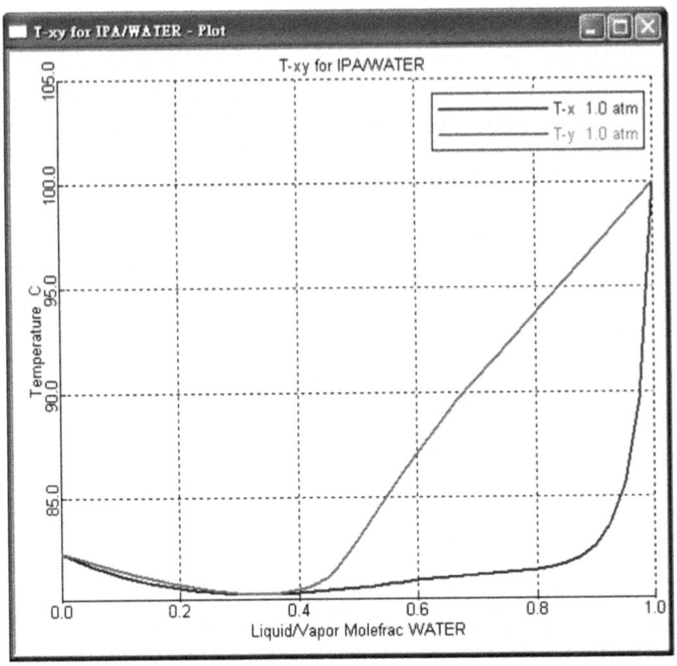


**Figure 8.9** *Txy* plot of IPA–water system.

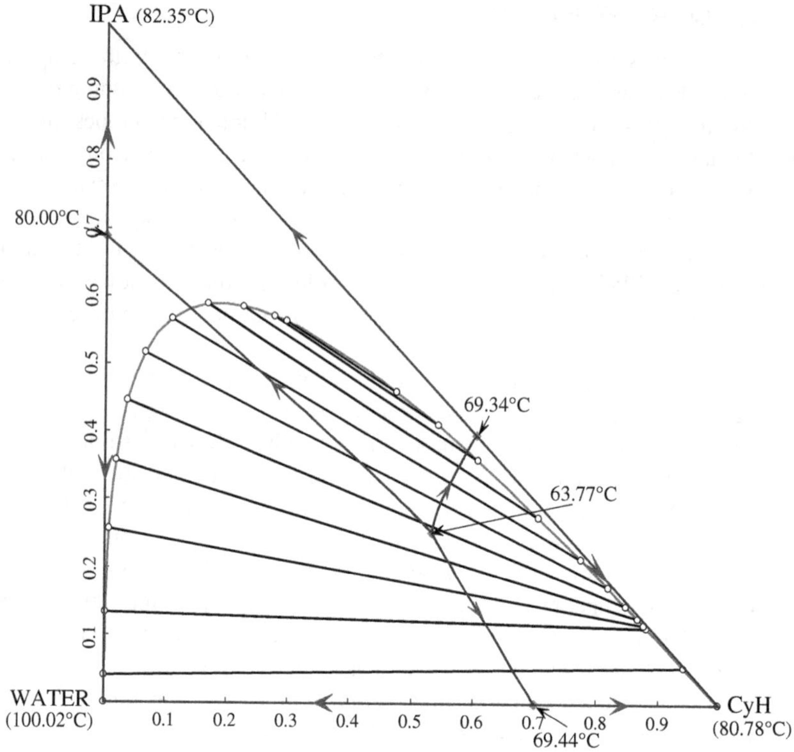


**Figure 8.10** RCM with a LLE envelope of IPA–water system with cyclohexane as entrainer.

to go back to the heterogeneous azeotropic column. The composition of the aqueous phase still contains significant amounts of the IPA product, so it is fed into a third recovery column. In this third column, pure water can be obtained out the bottom. The composition of the distillate product can extend into the lower left distillation region. Since the composition of this distillate product is close to the distillate product of the preconcentrator

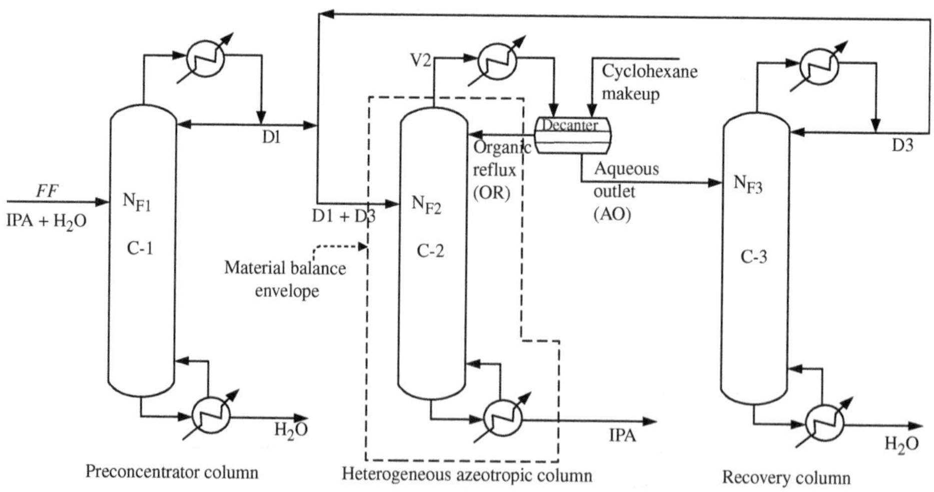


**Figure 8.11** Desired column sequence for IPA–water separation.

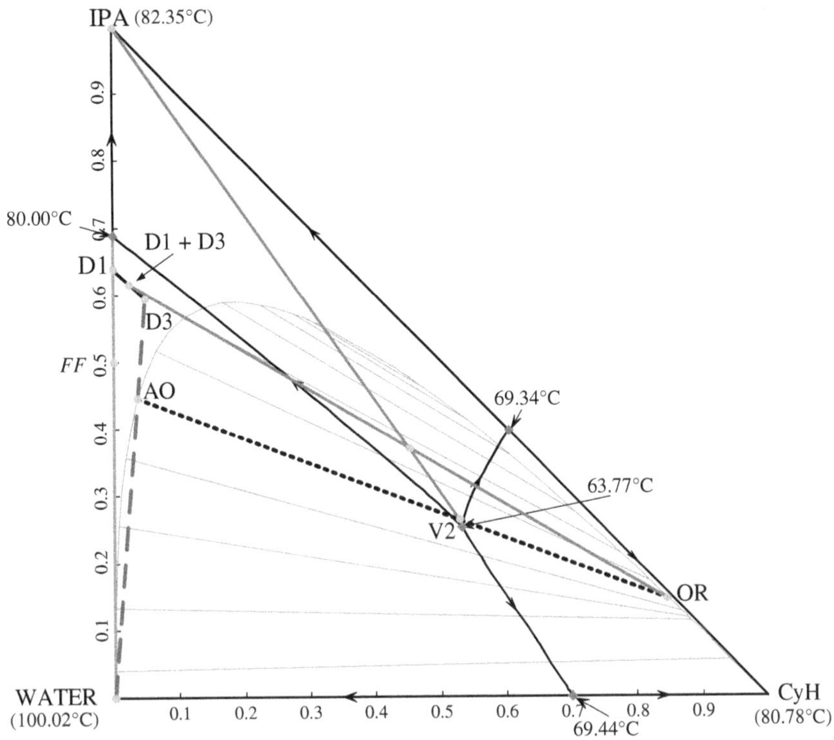


**Figure 8.12** Material balance lines for IPA–water separation.

column, this stream is recycled back to the heterogeneous azeotropic column. The material balance lines in Figure 8.12 clearly explain the conceptual design of this separation flowsheet.

In the next section, a detailed study will be conducted to see if this three-column sequence can be reduced to a two-column sequence.

## 8.2 STEADY-STATE DESIGN OF AN ISOPROPANOL–WATER SYSTEM

A waste stream containing IPA and water typically occurs in the semi-conductor industry where IPA is used as a cleaning agent. Taking a typical waste IPA stream with equal molar amounts of IPA and water as an example, design and control of the overall IPA dehydration process will be investigated. The feed rate of the fresh feed stream is chosen to be at an industrial scale of 100 kmol/h. For the following study, the NRTL property method is selected with the binary parameters from Wang et al.[1] For the vapor pressure calculations, Aspen Plus built-in parameters were used. The NRTL parameters for calculating the liquid activity coefficients and the coefficients of the Antoine equation for calculating the vapor pressure are listed in Tables 8.1 and 8.2, respectively.

### 8.2.1 Alternative Design 1 for a Two-Column System

We would like to investigate several design alternatives to the one in Figure 8.11 so that the number of total columns can be reduced. An alternative design with only two columns is

**TABLE 8.1 NRTL Parameters for IPA–Water–Cyclohexane System.**

| Comp. $i$<br>Comp. $j$ | IPA<br>Water | IPA<br>CyH | CyH<br>Water |
|---|---|---|---|
| $a_{ij}$ | 0 | 0 | 0 |
| $a_{ji}$ | 0 | 0 | 0 |
| $b_{ij}$ $(K)$ | 185.4 | 294.6 | 1629 |
| $b_{ji}$ $(K)$ | 777.3 | 622.5 | 2328 |
| $\alpha_{ij}$ | 0.5 | 0.5 | 0.242 |

Aspen Plus NRTL:

$$\ln \gamma_i = \frac{\sum_j x_j \tau_{ji} G_{ji}}{\sum_k x_k G_{ki}} + \sum_j \frac{x_j G_{ij}}{\sum_k x_k G_{kj}} \left[ \tau_{ij} - \frac{\sum_m x_m \tau_{mj} G_{mj}}{\sum_k x_k G_{kj}} \right]$$

where $G_{ij} = \exp(-\alpha_{ij}\tau_{ij})$

$$\tau_{ij} = a_{ij} + \frac{b_{ij}}{T}$$

$$\alpha_{ij} = c_{ij},\ \tau_{ii} = 0,\ G_{ii} = 1$$

shown in Figure 8.13. The material balance lines in the ternary diagram (see Fig. 8.14) can be used to explain the conceptual design of this design flowsheet.

In Figure 8.14, the fresh feed (FF) is combined with the distillate from the recovery column (D2) to become the point $(FF + \mathrm{D2})$. This point is combined with the OR, and the mixture is separated into pure IPA at the bottom of the heterogeneous azeotropic column and the top vapor V1 close to the ternary azeotrope. This top vapor stream, after subcooling to 40°C, will naturally split into two liquid phases. The organic phase, which contains mostly entrainer (CyH), is designed to be totally refluxed back to the heterogeneous azeotropic column while the aqueous phase, still containing a significant amount of IPA, is fed into the recovery column. The bottom stream of the recovery column (B2) is pure water, and the distillate (D2) is recycled back to the heterogeneous azeotropic column.

**TABLE 8.2 Coefficients of Antoine Equation for IPA–Water–Cyclohexane System.**

| Components | IPA | Water | CyH |
|---|---|---|---|
| Temperature units | °C | °C | °C |
| Property units | atm | atm | atm |
| $C_1$ | 64.904 | 62.12 | 39.56 |
| $C_2$ | −7607 | −7258 | −5226 |
| $C_3$ | 0 | 0 | 0 |
| $C_4$ | 0 | 0 | 0 |
| $C_5$ | −7.4086 | −7.3037 | −7.4086 |
| $C_6$ | 4.399E−18 | 4.1659E−6 | 4.399E−18 |
| $C_7$ | 6 | 2 | 6 |

Aspen Plus extended Antoine vapor pressure model is:

$$\ln P_i^* = C_{1i} + \frac{C_{2i}}{T + C_{3i}} + C_{4i}T + C_{5i}\ln T + C_{6i}T^{C_{7i}}, \quad i = \text{component}$$

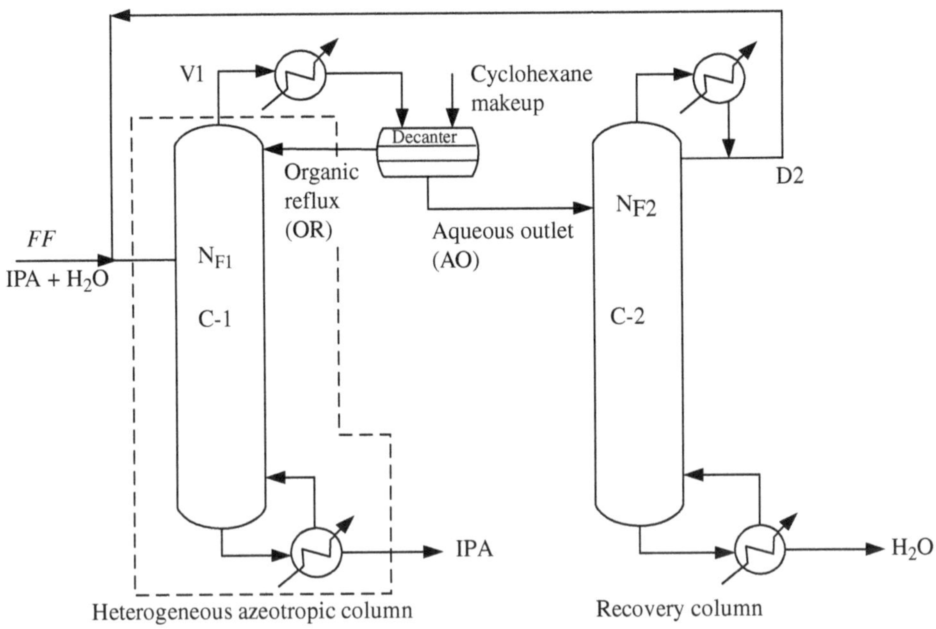


**Figure 8.13** Alternative design 1 with two-column system.

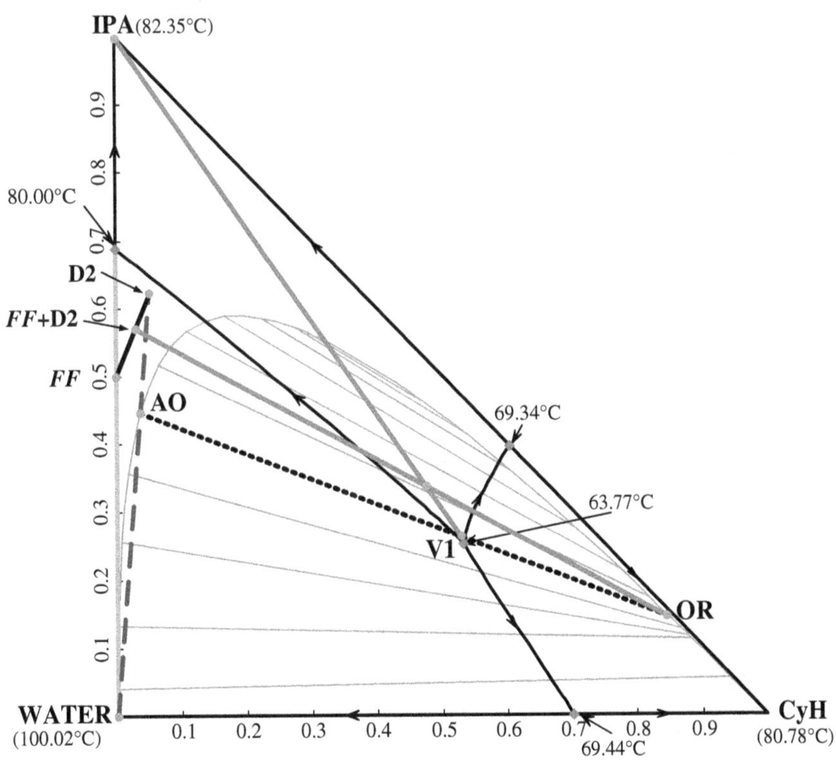


**Figure 8.14** Material balance lines for alternative design 1.

From Figure 8.14 one would immediately see that, because the fresh feed is more diluted, the material balance line drawn from the point of $FF + \mathrm{D2}$ to the point of OR is very close to the point of ternary azeotrope. From the lever rule, the top vapor flowrate can be estimated to be much greater than the flowrate of the bottom IPA product, thus requiring a large recycle flowrate in the overall system (large OR and large AO as feed to the recovery column).

The economic optimal process flowsheet was obtained by minimization of the total annual cost (TAC) with five design variables: IPA distillate composition (XD2) in the recovery column, total number of stages for the heterogeneous azeotropic column and the recovery column ($N_1$ and $N_2$), and the two feed stages ($N_{F1}$ and $N_{F2}$). The product specification of IPA is set to be ultrapure (99.9999 mol%) to be used in the semi-conductor industry. The product specification of water is set to be 99.9 mol%. In each simulation run, the IPA product specification is achieved by varying the reboiler duty of the heterogeneous azeotropic column, and the water product specification is achieved by varying the reboiler duty of the recovery column. The entrainer makeup flowrate will be very small to balance the entrainer loss from the two bottom streams.

The TAC of the overall system includes the annualized capital costs and the operating costs. The annualized capital costs include the column shells of the two columns, internal trays of the two columns, and also the reboilers and condensers of the two columns. The payback period was assumed to be 3 years in the calculation. The operating costs include the steam and cooling water required for operating these two columns and also a very small entrainer makeup cost.

A sequential iterative procedure was used to obtain the optimal design flowsheet with the procedure summarized in Figure 8.15. The optimization method is intuitive; however, it is rather time-consuming when the numbers of design variables becomes large in a plantwide optimization. Figure 8.16 shows some of the TAC plots within the optimization search.

The top plot is a summarized plot at each XD2 and shows that the optimal distillation composition is roughly at 0.62 (or can be said to be between 0.615 and 0.625). Of course a more precise optimal XD2 can be calculated, which would require longer simulation time. However, because the plot is flat at low TAC, not much can be gained in being more precise. This IPA distillate composition represents a tradeoff in the overall process design. The costs associated with the recovery column will be increased when this distillate composition is purer in IPA. However, setting XD2 to be more pure in IPA will make the material balance line connecting points $FF + \mathrm{D2}$ and OR (see Fig. 8.14) further away from the ternary azeotropic point, thus reducing the recycle flowrate in the overall system.

The optimal design and operating conditions can be seen in Figure 8.17. Notice that the ratio of AO flow (2653.2 mol/min) to the bottom IPA product flow (832.5 mol/min) is 3.19; thus, large flowrate is going from the heterogeneous azeotropic column to the recovery column and recycled back. The TAC and the operating cost will be compared to two other designs later.

### 8.2.2 Proposed Design for a Two-Column System

The above design flowsheet results in a large flow circulated between the heterogeneous azeotropic column and the recovery column. The main reason (as can be seen in

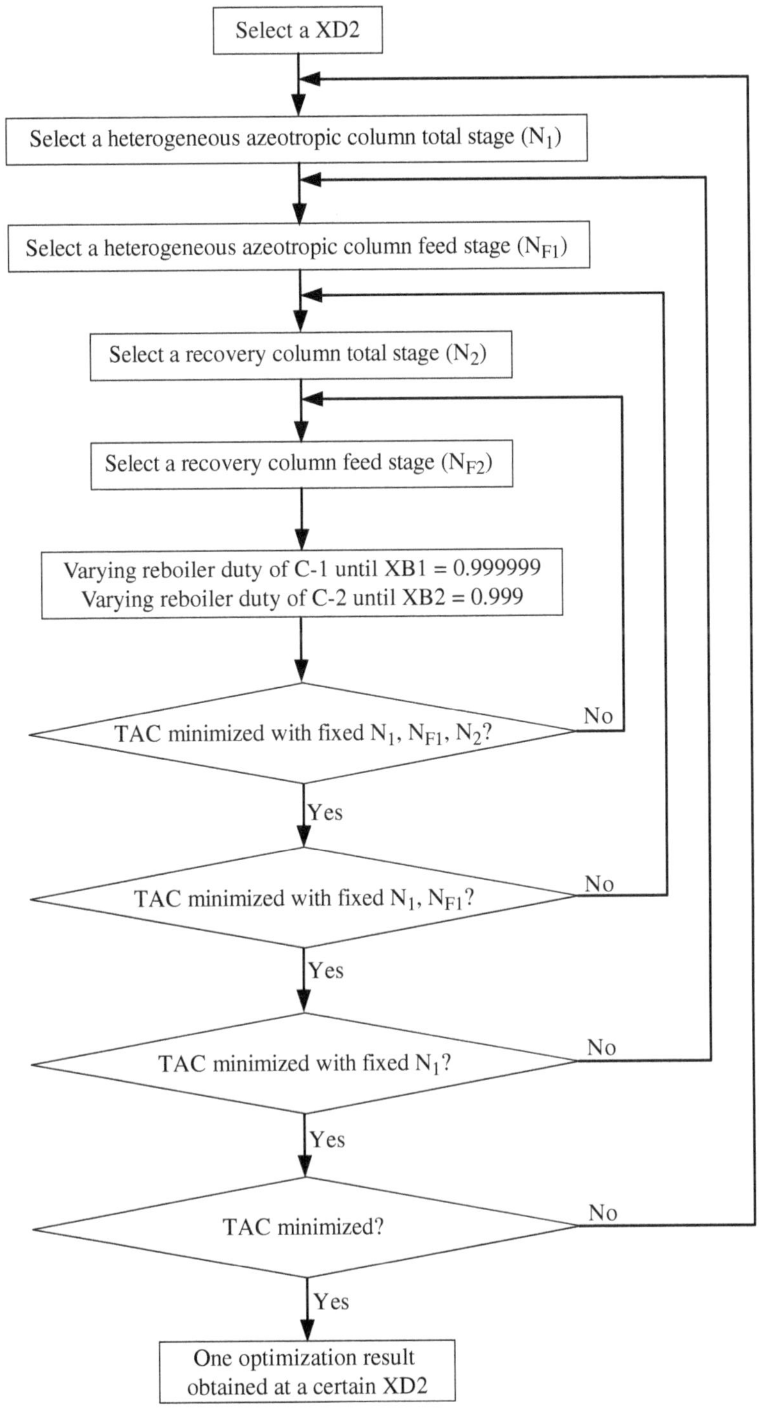


**Figure 8.15** Sequential iterative optimization procedure for this system.

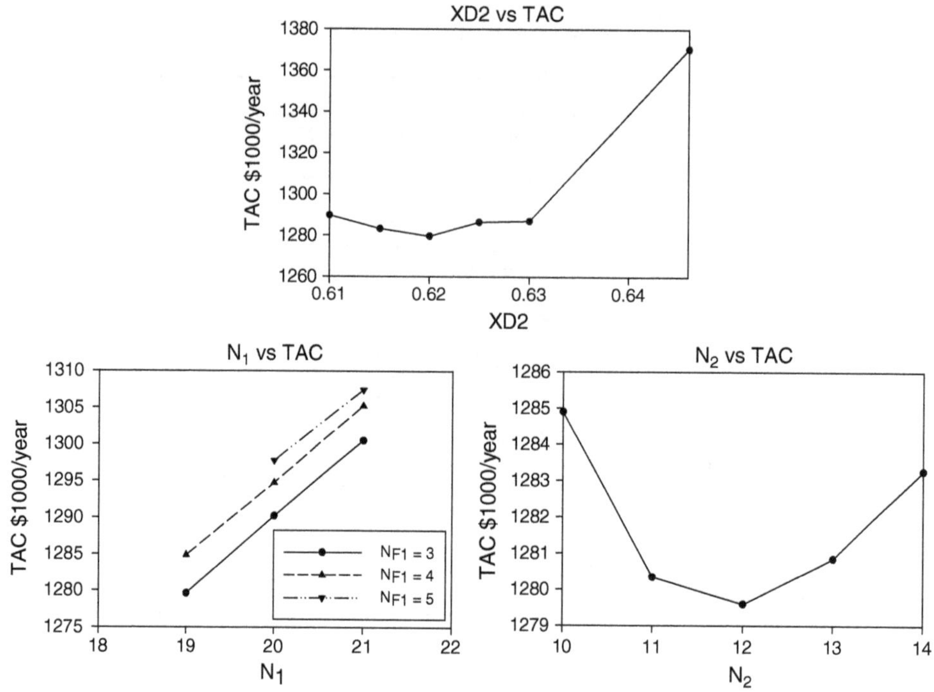


**Figure 8.16** TAC plots for alternative design 1.

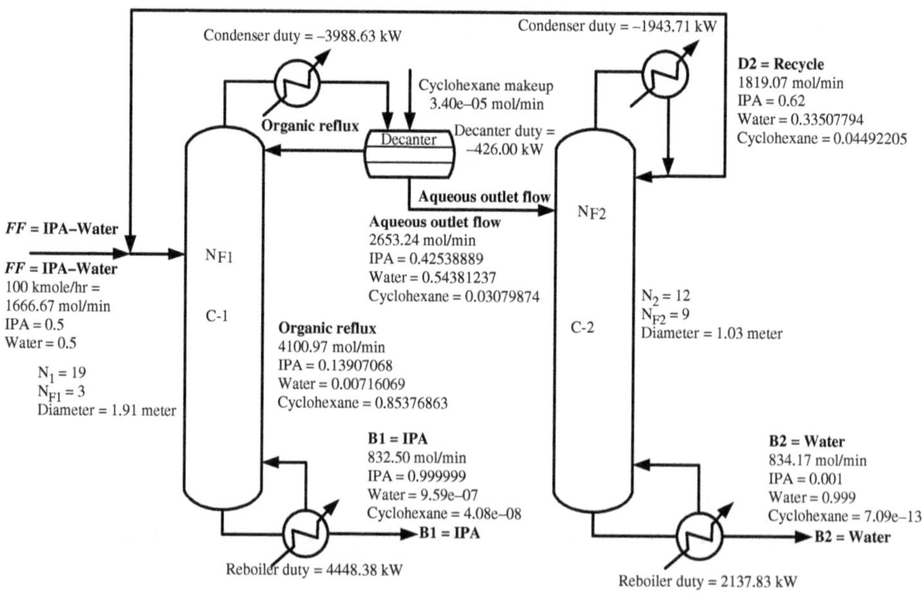


**Figure 8.17** Optimal design flowsheet for alternative design 1.

Fig. 8.14) is because the location of fresh feed is rich in water in the ternary diagram. Thus a preconcentrator should be helpful in moving the feed to the heterogeneous azeotropic column closer to the IPA–water azeotrope. In 1989 Ryan and Doherty[2] proposed an ingenious design flowsheet for their ethanol dehydration process with benzene as the entrainer. The design flowsheet combines the preconcentrator column and the recovery column into one column. Thus, the overall flowsheet still contains only two columns. In the following, we will apply this design to the IPA dehydration process.[3]

The process flowsheet for this design alternative with only two columns can be seen in Figure 8.18 with the material balance lines of the conceptual design shown in Figure 8.19. In this design flowsheet, fresh feed and the AO flow are combined and enter into the C2 column, which acts as the preconcentrator and also as the recovery column. The bottom of this column is designed to have a water purity of 99.9 mol%, and the top will feed into the heterogeneous azeotropic column (C1). The top composition of the C2 column (XD2), similar to previous design alternatives, is a design variable that will be determined to minimize the TAC of the overall system. The bottom IPA product purity of the heterogeneous azeotropic column is again set at a very high purity of 99.9999 mol% for semiconductor industrial usage.

The optimization procedure to be followed is exactly the same as the one for design alternative 1 with the total of five design variables (XD2, total number of stages of C1 and C2, and feed tray locations of C1 and C2). The procedure for the iterative sequential optimization search was outlined previously in Section 8.2.1. Figure 8.20 shows that the optimum value of XD2 is 0.620. With this XD2, the optimized total number of stages for C1 is 18 (including reboiler), with feed tray location at Stage 2 (counting from the top),

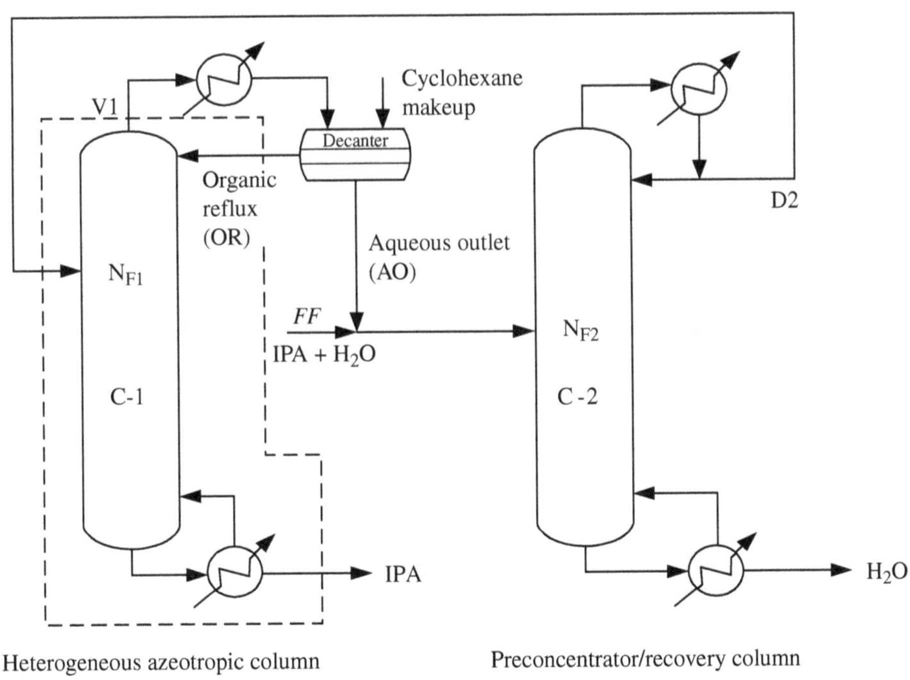


**Figure 8.18** Proposed design with two-column system.

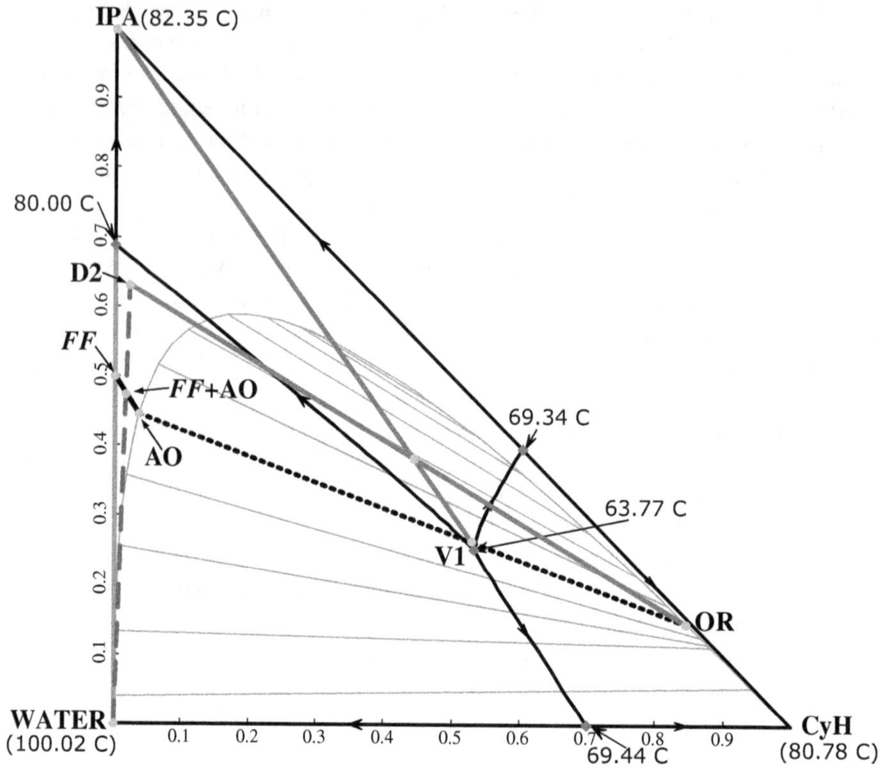


**Figure 8.19** Material balance lines for the proposed design.

and the optimized total number of stages for C2 is 10 (including reboiler and condenser) with feed tray location at Stage 6. The optimal design and operating conditions for this two-column system can be seen in Figure 8.21. Notice that the ratio of AO flow (1992.54 mol/min) to the bottom IPA product flow (832.50 mol/min) is 2.39, which is lower than the 3.19 required in design alternative 1.

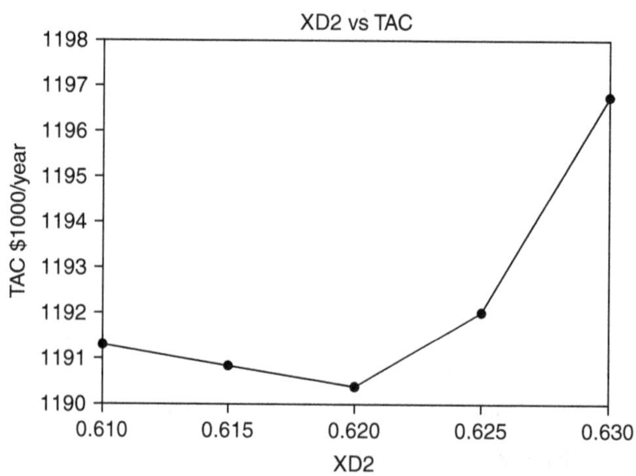


**Figure 8.20** Summary of TAC plot for the proposed design.

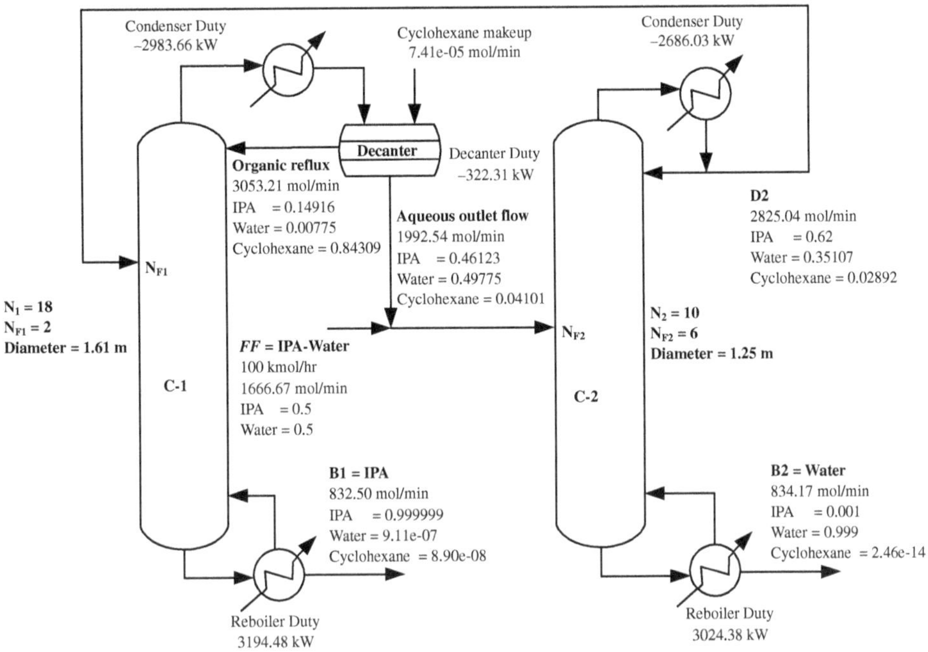


**Figure 8.21** Optimal design flowsheet for the proposed design.

### 8.2.3 Optimal Design Flowsheet for a Three-Column System

The optimal design flowsheet of the original three-column sequence, as shown in Figure 8.11, is also obtained and compared with the other alternative flowsheets. This design flowsheet can also reduce the circulation rate between the heterogeneous azeotropic column and the recovery column. The decision for the placement of point D1 is a tradeoff between the TAC of the preconcentrator column and lower flowrate from the heterogeneous azeotropic column to the recovery column. Placing D1 closer to the IPA–water azeotrope will increase the TAC of the preconcentrator column. However the benefit is that the top vapor flowrate of the heterogeneous azeotropic column can be lower, thus reducing the OR flowrate back to the same column, and also reducing the AO flowrate to the recovery column. The placement of point D1 is an additional design variable in this design flowsheet.

Since there is no recycle flow back to the first preconcentrator column, this stand-alone column can be optimized first with two product specifications and two design variables of the total number of stages and feed tray location. One of the product specifications is the selection of the XD1 composition by varying the reflux ratio. The other product specification is the bottoms water composition at 0.999 by varying the reboiler duty.

This stand-alone column can be optimized for each selection of the XD1 composition. For each XD1, we can go through the same procedure as outlined previously and obtain the optimal design and operating conditions for this overall three-column system. The optimized results can be seen in Figure 8.22. Notice that the ratio of AO flow (2056.01 mol/min) to the bottom IPA product flow (832.50 mol/min) has also been reduced from 3.19 to 2.47.

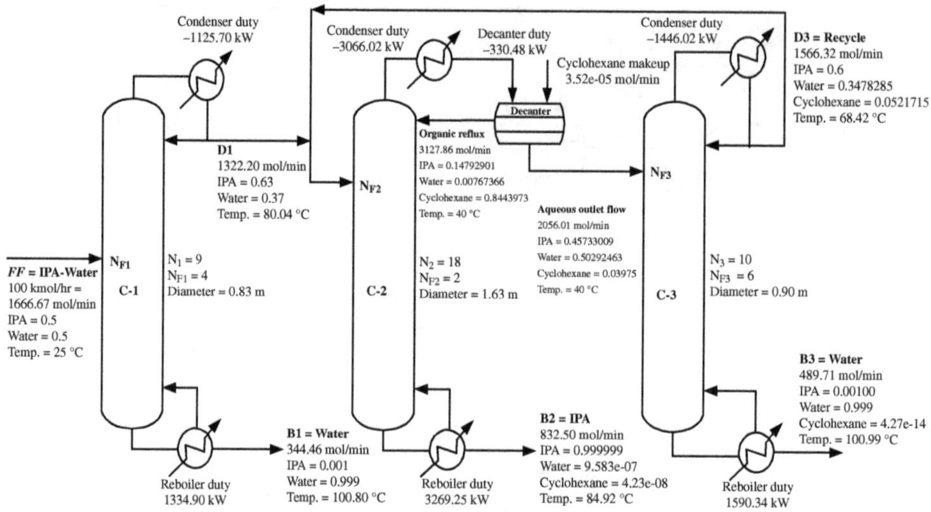


**Figure 8.22** Optimal design flowsheet for three-column system.

### 8.2.4 Comparison of Three Design Alternatives

The comparison of all the costs calculated for the three design alternatives are summarized in Table 8.3. Notice that the TAC for the proposed design is the lowest with 7.0% savings compared to alternative design 1 and 5.1% savings compared to the three-column system. The operating cost for the proposed design is slightly more than that of the three-column system; however, it is less than the cost of the other two-column system of alternative design 1 (2.1% less). It is decided that this proposed design, although with slightly more operating cost than the three-column system, would be the best design flowsheet for the IPA dehydration process because the TAC is the smallest and also there are fewer

**TABLE 8.3 Cost Comparison of the Three Design Alternatives.**

| Costs ($10^6$ \$/y) | Alternative Design 1 | Proposed Design | Three-Column System |
|---|---|---|---|
| C1 column annualized equipment cost | 0.5762 | 0.4640 | 0.1368 |
| C1 column steam cost | 0.2952 | 0.2119 | 0.1109 |
| C1 column cooling water cost | 0.01472 | 0.01102 | 0.003754 |
| C2 column annualized equipment cost | 0.2088 | 0.2424 | 0.4717 |
| C2 column steam cost | 0.1780 | 0.2518 | 0.2169 |
| C2 column cooling water cost | 0.006481 | 0.008956 | 0.01133 |
| C3 column annualized equipment cost | — | — | 0.1653 |
| C3 column steam cost | — | — | 0.1324 |
| C3 column cooling water cost | — | — | 0.004821 |
| Make-up entrainer cost | ~0 | ~0 | ~0 |
| TAC | 1.280 | 1.190 | 1.254 |

pieces of process equipment in the overall design flowsheet. In Chapter 10, this optimal two-column design flowsheet will be compared to the results of the same separation task using extractive distillation instead of azeotropic distillation.

The control of the overall process is also very important in determining which design should be used for this IPA dehydration process. If the design with the lowest TAC were not able to properly maintain the product purity in the face of typical fresh feed disturbances, this process design would not be recommended. We will conduct the control study in the next section.

A useful comment about this combined-column design is that it can be used with other types of RCM system. For example, it can be applied in the RCM shown in Figure 8.6 (pyridine–water system). Remember, in the design flowsheet of Figure 8.7, to make the design right, we need to place the top vapor very close to the toluene–water azeotrope. However, to reach this location, the column profile needs to go through a very narrow funnel (as can be seen in Fig. 8.8). Thus, the capital cost of the heterogeneous azeotropic column will be high, or it may even be infeasible to reach this location.

To improve the design, an alternative two-column system is proposed (see Fig. 8.23) with the material balance lines of the conceptual design shown in Figure 8.24. Note that the top vapor of the heterogeneous azeotropic column does not need to be strictly placed at the toluene–water azeotrope. Column C2 serves as a preconcentrator column and also as a recovery column to obtain a pure water product. The comparison of this proposed design versus the one in Figure 8.7 can be found in a recent paper by Wu and Chien.[4] Simulation results showed that the design in Figure 8.7 cannot place the top vapor of the heterogeneous azeotropic distillation column right at the toluene–water azeotrope even with a very large number of total stages. The pyridine loss from the AO stream is much greater compared to the loss of pyridine from the proposed flowsheet in Figure 8.23.

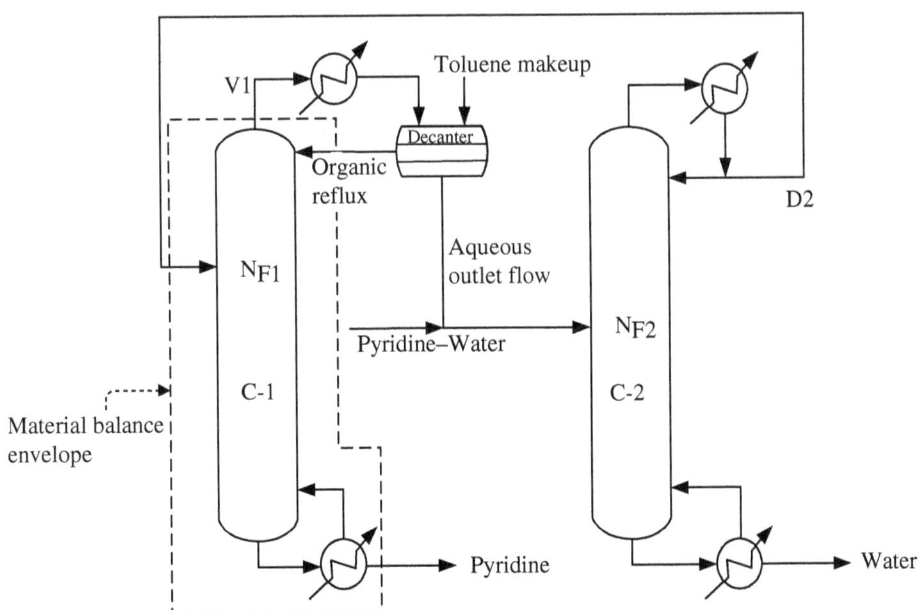


**Figure 8.23** Proposed two-column design for pyridine–water system with toluene as the entrainer.

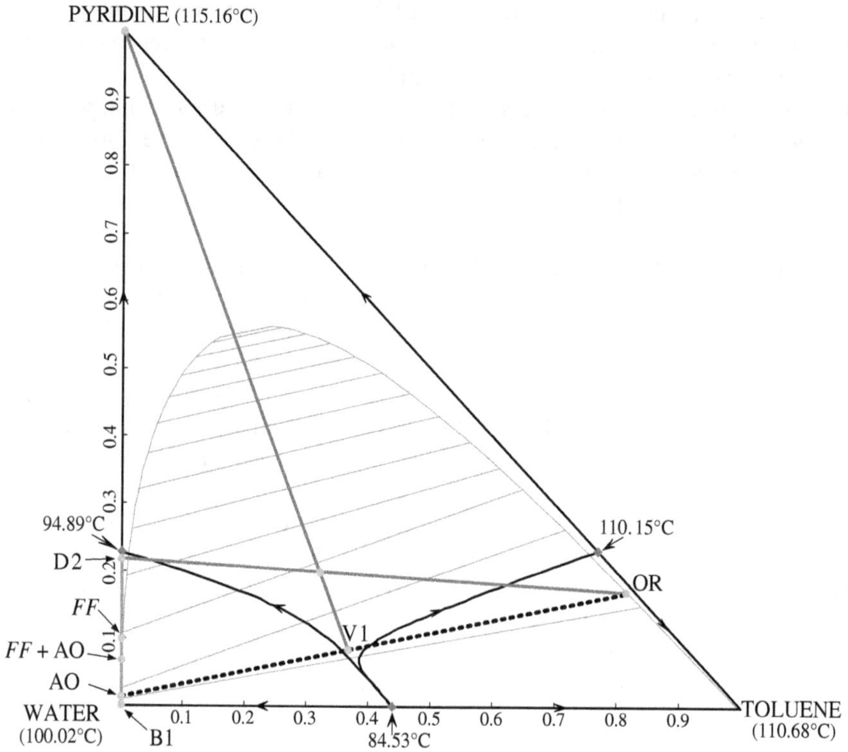


**Figure 8.24** Material balance lines for pyridine–water separation using proposed design.

## 8.3 OVERALL CONTROL STRATEGY DEVELOPMENT

In the following, we will investigate the proper overall control strategy for the proposed design with the lowest TAC. Only a tray temperature control loop(s) will be used in the overall control strategy for wider industrial applications.

The Aspen Plus steady-state simulation in the last section is exported to the dynamic simulation of Aspen Dynamics. The tray sizing option in Aspen Plus is utilized to calculate the column diameter of both columns with the assumed tray spacing of 0.6096 m. The diameter of the C1 and the C2 columns are calculated to be 1.61 and 1.25 m, respectively, and the weir heights of both columns are assumed to be 0.0508 m.

Other equipment sizing followed the recommendations in Chapter 4. The volumes of the column base for the C1 column and also for the C2 column, and the reflux drum of the C2 column, are all sized to give 10 min holdup with 50% liquid level. The decanter is sized to be bigger to allow for two liquid phases to separate. The holdup time of 20 min for the decanter is used in the dynamic simulation.

Pressure-driven simulation in Aspen Dynamics is used with the top stage pressures of the C1 column set at 1.1 atm to allow for some pressure drop in the condenser and decanter. The top pressure of the C2 column is set to be at atmospheric pressure. The pressure drops inside both columns are automatically calculated in Aspen Dynamics to account for liquid hydraulics and vapor traffic. The bottom pressure for the C1 and the C2 columns are calculated to be 1.200 and 1.056 atm, respectively.

### 8.3.1 Inventory Control Loops

The inventory control loops are designed as follows. For the heterogeneous azeotropic column with the decanter, the aqueous phase level is controlled by manipulating the AO flow, the column bottom level is controlled by manipulating the bottom IPA product flow, and the column top pressure is controlled by manipulating a control valve in the overhead vapor line.

One important loop pairing from past experience (see Chien et al.[5]) is that the organic phase level should be controlled by the entrainer makeup flow, not by an internal recycling flow of OR, so that the snowballing effect can be avoided. For the combined preconcentrator/recovery column, the reflux drum level is controlled by the distillate flow (feed to the heterogeneous azeotropic column); the column bottom level is controlled by the bottom water product flow; and the column pressure is controlled by the condenser duty.

In all the closed-loop simulation runs, P-only controllers are used in all level loops for both columns. The reason for using a P-only controller is to provide maximum flow smoothing and also because maintaining a liquid level at setpoint value is often not necessary. A controller gain of $K_C = 2$ as suggested in Chapter 4 is used in most of the level loops. For the organic phase level loop to manipulate the entrainer makeup flow, a gain of $K_C = 10$ is used to simulate very fast control behavior favorable in this control loop for quickly sending more entrainer or quickly reducing makeup entrainer into this system. For the top pressure control loops of both columns, tight PI controller tuning parameters of $K_C = 20$ and $\tau_I = 12$ min are used.

The remaining manipulated variables for the heterogeneous azeotropic column are the OR flow and the reboiler duty, and the remaining manipulated variables for the preconcentrator/recovery column are the reflux flow and the reboiler duty. We will investigate the simplest overall control strategy first with only one tray temperature control loop in each column.

As discussed in Chien et al.,[5] it is not possible to properly reject feed composition disturbances with only one tray temperature control loop in the heterogeneous azeotropic column. The reason for the difficulty in operating and controlling of this very sensitive heterogeneous azeotropic column was explained in detail in the previous studies (see Chien et al.).[5–7] The temperature profile (shown in left-hand plot of Fig. 8.25) of this heterogeneous azeotropic distillation column was very difficult to maintain under feed composition disturbances. In Chien et al.[5–7] dual temperature control loops were necessary to hold the bottom product at high-purity because fresh feed is directly entered into this column. With the proposed flowsheet as in Figure 8.21, the fresh feed composition disturbance will be dampened by the C2 column first. Thus, we want to investigate if a single-temperature loop is sufficient for this proposed flowsheet.

### 8.3.2 Tray Temperature Control Point(s)

Open-loop sensitivity analysis is used to determine the tray temperature control point for both the heterogeneous azeotropic column and the preconcentrator/recovery column.

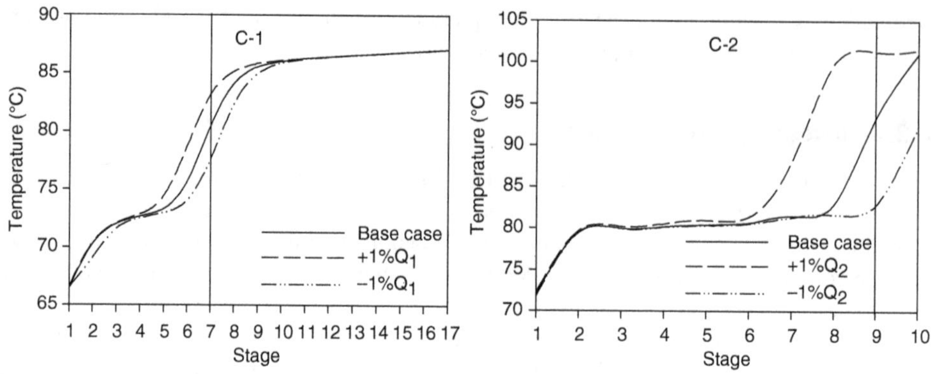


**Figure 8.25** Open-loop sensitivity plots.

Figure 8.25 shows the open-loop sensitivity plots for both columns. The one on the left-hand side is for the heterogeneous azeotropic column with $\pm 1\%$ changes in its reboiler duty, and the one on the right-hand side is for the preconcentrator/recovery column with $\pm 1\%$ changes in its reboiler duty. Tray 7 of the heterogeneous azeotropic column and tray 9 of the preconcentrator/recovery column are selected as the temperature control points because of high sensitivity and nearly linear behavior.

Proportional–integral controllers are used in these two temperature control loops. The tuning constants are determined using the IMC–PID tuning rules of Chien and Fruehauf.[8] From the open-loop step responses, the controlled temperature in C1 column is assumed to behave like an integrating plus deadtime model, and the controlled temperature in the C2 column is assumed to behave like a first-order plus deadtime model. The closed-loop time constants of these two loops are set to be twice their model apparent deadtimes. The resulting PI tuning constants for the tray 7 temperature loop of the heterogeneous azeotropic column are $K_C = 3.58$ and $\tau_I = 6.25$ min, and the tuning constants for the tray 9 temperature loop of the preconcentrator/recovery column are $K_C = 0.40$ and $\tau_I = 12.8$ min.

The other two manipulated variables, not used in the two tray temperature control loops, are the OR flow for the heterogeneous azeotropic column and the reflux flow for the preconcentrator/recovery column. Ratio control schemes are used so that these two manipulated variables can be adjusted in the face of disturbances. The OR flow is set to maintain a constant ratio to the feed flow of the heterogeneous azeotropic column, and the reflux ratio of the preconcentrator/recovery column is also maintained. The overall proposed control strategy is shown in Figure 8.26.

### 8.3.3 Simulation Results

Two types of disturbances are used to test the proposed control strategy. The first is $\pm 20\%$ changes in the fresh feed flowrate. With these changes, the OR flowrate will be adjusted to hold the R/F ratio (OR/feed to heterogeneous azeotropic column). The reflux flow of the preconcentrator/recovery column will also be adjusted to hold the reflux ratio at a constant value.

Figure 8.27 shows the dynamic responses for some of the important variables in the process. The top two graphs show that the two tray temperatures are smoothly controlled back to

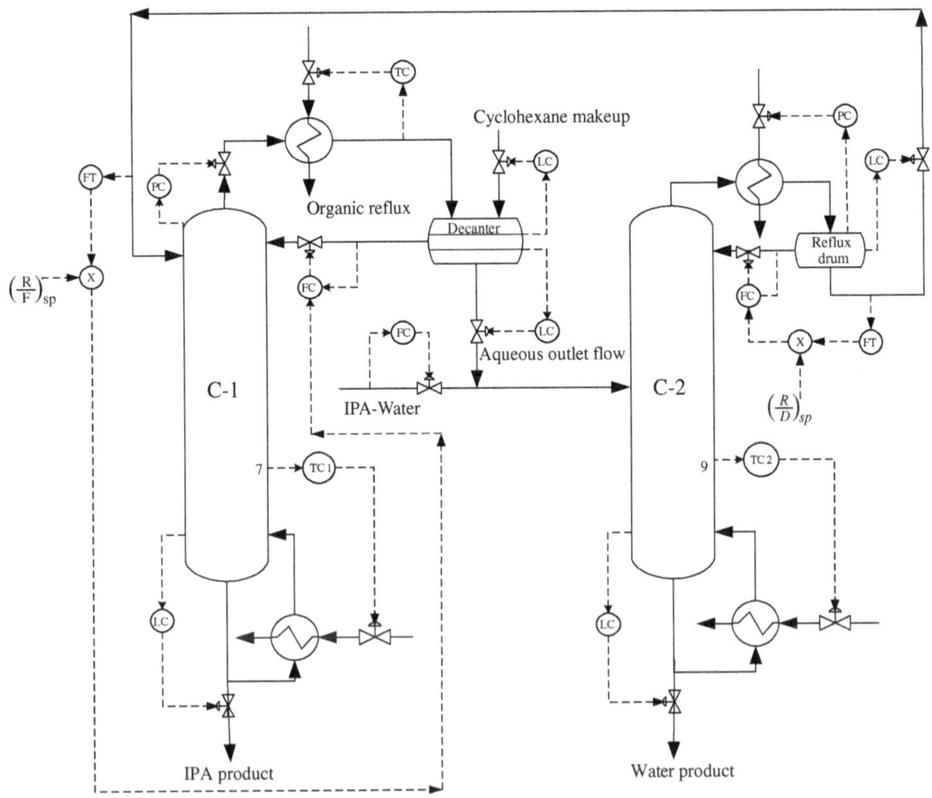


**Figure 8.26** Proposed overall control strategy.

their setpoint values. The middle two graphs show the feed composition and feed flowrate (D2) entering the heterogeneous azeotropic column. It is noticed that there are only slight changes in the feed composition to this "more sensitive" column, and the feed rate changes (assumed measurable) are also quite smooth. There is no snowballing effect happening in this overall system even though no recycling flowrate is fixed in the proposed control strategy. The lower two graphs show that the two product compositions are still very close to their purity specifications. The IPA product stream is still in the ultrapure region (>99.99986 mol%) and the water product stream returns back to 99.9 mol%. The transient deviation could be greatly reduced by using a steam-to-feed ratio, which would be reset by the temperature controller.

With the $\pm 20\%$ changes in the fresh feed flowrate, the IPA product flowrate and water product flowrate also increase/decrease correspondingly to their new values. For example, with $+20\%$ in the fresh feed flowrate, the IPA product flowrate changes from 832.50 to 998.62 mol/min (also a $+20\%$ increase) and the water product flowrate changes from 834.17 to 1000.83 mol/min (also a $+20\%$ increase).

The second type of disturbance tested is a change in the fresh feed composition, which is often considered as a tougher disturbance to reject because they are unmeasured. Large variations of $\pm 20\%$ changes in the fresh feed IPA composition are tested with the closed-loop results shown in Figure 8.28. Again the two tray temperatures are smoothly controlled back to their setpoint values. More importantly, the feed composition of the

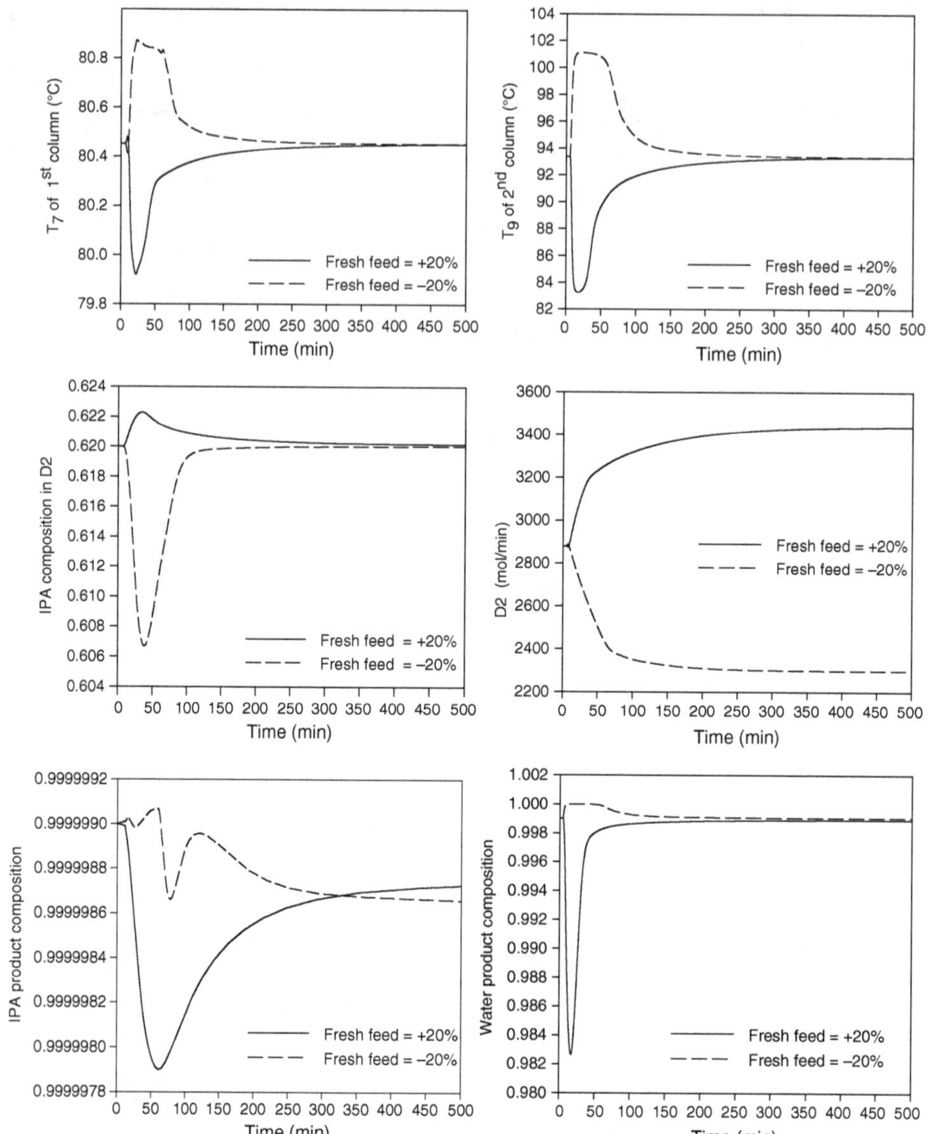


**Figure 8.27** Closed-loop responses with $\pm 20\%$ feed flow rate changes.

more sensitive heterogeneous azeotropic column does not vary much (only changes from 0.620 to 0.625 or from 0.620 to 0.615). The final IPA product composition is still kept in the ultrapure region (>99.9997 mol%) and the water product composition is returned back to 99.9 mol%.

This desirable result is mainly due to the combined preconcentrator/recovery column dampening the large disturbance from fresh feed, thus making the feed variations to the more sensitive C1 column small and easier to handle. The three-column system should also have this desirable feature. However, it requires more process, instrumentation, and control equipment, and the TAC is also higher than the proposed design.

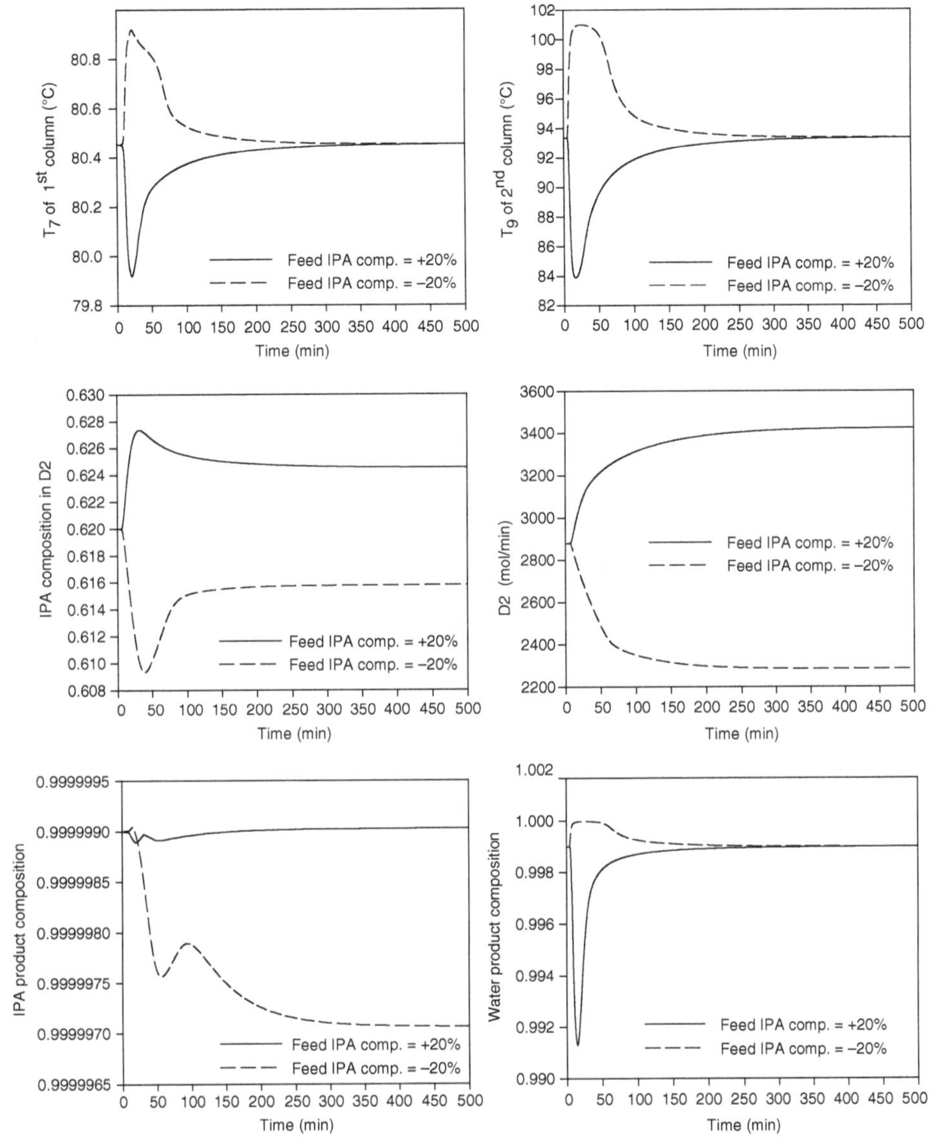


**Figure 8.28** Closed-loop responses with $\pm 20\%$ feed IPA composition changes.

## 8.4 CONCLUSION

The feasible designs of three different heterogeneous azeotropic column systems have been illustrated in this chapter with real industrial applications. With the aid of liquid–liquid separation, the products of a column sequence can be located in different distillation regions.

The design and control of a complete heterogeneous azeotropic distillation column system for IPA dehydration has also been studied in this chapter. The TAC and operating cost of three design alternatives are compared. The first design alternative having fresh feed directly entering into the heterogeneous azeotropic column gives the highest TAC and operating cost. The operating cost of the proposed design with two columns is comparable to that of a three-column system with the proposed two-column design giving a 5% savings in TAC.

The proposed design combining a preconcentrator column and a recovery column into a single column is very robust in the face of wide variations in the fresh feed flowrate and feed composition. Only one tray temperature control loop in the heterogeneous azeotropic column and one tray temperature in the combined preconcentrator/recovery column are needed in the proposed overall control strategy. The more complicated inverse double-loop control strategy discussed in previous papers (Chien et al.[5–7]) is not needed since the combined preconcentrator/recovery column "buffers" the disturbances from directly entering into the heterogeneous azeotropic column. The benefit of dampening the fresh feed disturbances by a preconcentrator column as in a three-column system is retained without extra process, instrumentation, and control equipments.

## REFERENCES


1. Wang C. J., D. S. H. Wong, I. L. Chien, R. F. Shih, W. T Liu, and C. S. Tsai, Critical reflux, parametric sensitivity, and hysteresis in azeotropic distillation of isopropyl alcohol + water + cyclohexane. *Ind. Engng. Chem. Res.*, **37**, 2835–2843 (1998).
2. Ryan P. J. and M. F. Doherty, Design/optimization of ternary heterogeneous azeotropic distillation sequences, *AIChE J.*, **35**, 1592–1601 (1989).
3. Arifin S. and I. L. Chien, Combined preconcentrator/recovery column design for isopropyl alcohol dehydration process, *Ind. Engng. Chem. Res.*, **46**, 2535–2543 (2007).
4. Wu Y. C. and I. L. Chien, Design and control of a heterogeneous azeotropic Column system for the separation of pyridine and water, *Ind. Engng. Chem. Res.*, **48**, 10564–10576 (2009).
5. Chien I. L., K. L. Zeng, and H. Y. Chao, Design and control of a complete heterogeneous azeotropic distillation column system, *Ind. Engng. Chem. Res.*, **43**, 2160–2174 (2004).
6. Chien I. L., C. J. Wang, and D. S. H. Wong, Dynamics and control of a heterogeneous azeotropic distillation column: conventional control approach, *Ind. Engng. Chem. Res.*, **38**, 468–478 (1999).
7. Chien I. L., C. J. Wang, D. S. H. Wong, C. H. Lee, S. H. Cheng, R. F. Shih, W. T. Liu, and C. S. Tsai, Experimental investigation of conventional control strategies for a heterogeneous azeotropic distillation column, *J. Process Control*, **10**, 333–340 (2000).
8. Chien, I. L. and P. S. Fruehauf. Consider IMC tuning to improve controller performance, *Chem. Eng. Prog.*, **86**, 33–41 (10 Oct., 1990).

# CHAPTER 9

# ACETIC ACID–WATER (ISOBUTYL ACETATE AS THE ENTRAINER)

Acetic acid (HAc) dehydration is an important operation in the production of aromatic acid, such as terephthalic acid or in the manufacture of cellulose acetate. Although acetic acid and water do not form an azeotrope, using simple distillation to separate these two components requires many equilibrium stages and thus is impractical. The reason is because the system has a tangent pinch on the pure water end (see Fig. 9.1) where the $x$ and $y$ curves are very close together. Therefore, it is more customary to use an entrainer via a heterogeneous azeotropic distillation column system for this separation if high-purity water must be produced.

In this chapter, a suitable entrainer for this acetic acid dehydration system will be selected from several candidate acetates. Steady-state tray-by-tray column simulation will be used to determine the best entrainer with minimum total annual cost (TAC). Optimum process design and operating conditions will be determined to keep a high-purity bottom acetic acid composition and also to keep a small acetic acid loss in top aqueous draw. The overall control strategy of this column system will also be proposed to hold both bottom and top product specifications in spite of feed rate and feed composition disturbances. In the control study, conventional control strategy using only tray temperature measurements will be considered so that the result of this study can easily be used directly in industry.

In the first half of this chapter, only three components (acetic acid, water, and the entrainer) in the heterogeneous distillation column will be assumed to be present.[1] However, in the production of aromatic acid, acetic acid is used as a solvent in the terephthalic acid or isophthalic acid processes. A terephthalic acid, or isophthalic acid slurry in acetic acid solvent, is produced by oxidizing p-xylene (PX) or m-xylene (MX), removing water by evaporation of a stream of mixture with water and acetic acid, and returning acetic acid to the oxidation step (Hindmarsh et al.[2] and Lee et al.[3]). To recover acetic acid as a cyclic solvent stream, an acetic acid dehydrating column is necessary. However, a tiny amount of one reactant (PX or MX) may also enter into this acetic acid dehydration column through the feed stream. The

*Design and Control of Distillation Systems for Separating Azeotropes.* By William L. Luyben and I-Lung Chien

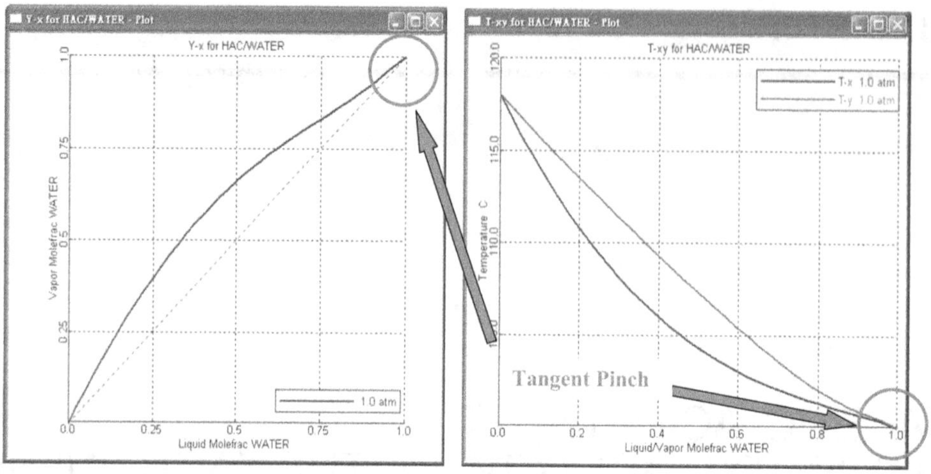


**Figure 9.1** *yx* and *Txy* plots of an acetic acid–water system.

existence of this tiny amount of impurity has a remarkable effect on the design and control of this dehydration column. The results of the study on an industrial column with feed impurity will be presented in the later part of this chapter.[4]

## 9.1 COMPARISON OF THREE CANDIDATE ENTRAINERS

Three candidate acetates will be studied in detailed process simulations to demonstrate the factors needed to be considered in determining the suitable entrainer for this system. The three candidate acetates to be considered are: ethyl acetate (Tanaka and Yamada[5] and Siirola[6]), isobutyl acetate (iBuAc) (Costantini et al.[7] and Parten and Ure[8]), and *n*-butyl acetate (Othmer[9,10]). The important experimental physical properties of these three acetates at atmospheric pressure are listed in Table 9.1. The azeotropic data are from Horsley[11] and Gmehling,[12] the VLE data are from Gmehling and Onken,[13] with the VLE data for acetic acid–isobutyl acetate system from Christensen and Olson.[14] The binary and ternary LLE data are from Sørensen and Arlt.[15]

The NRTL activity coefficient model (Renon and Prausnitz[16]) was used for the VLLE for the ternary system. The Hayden–O'Connell[17] second virial coefficient model, with association parameters, was used to account for the dimerization of acetic acid in the vapor phase. The Aspen Plus[18] built-in association parameters were employed to compute fugacity coefficients. The extended Antoine equation is used to calculate the vapor pressure of each component in the system. The Aspen Plus[18] built-in parameters were again used in the simulation.

The set of NRTL parameters used provides a good match between predicted and experimental data: binary and ternary VLE and LLE. The sets of NRTL parameters for the ternary systems of acetic acid–water–ethyl acetate, acetic acid–water–isobutyl acetate, and acetic acid–water–*n*-butyl acetate are listed in Tables 9.2 to 9.4. The NRTL model parameters are obtained by fitting all binary and ternary VLE and LLE data while trying to get a good prediction of the azeotropic temperature and composition. The pure component parameters used in the Hayden–O'Connell[17] model are shown in Table 9.5 while the association parameters

**TABLE 9.1 Experimental Properties of Three Candidate Acetates.**

| Acetate | Normal Boiling Point (°C) | Azeotropic Component | Azeotropic Temp. (°C) | Azeotropic Composition (Water, mol%) | Aqueous Phase (Acetate, mol%) | Organic Phase (Acetate, mol%) |
|---|---|---|---|---|---|---|
| Ethyl acetate | 77.15 | Water | 70.38–71.05 | 29.50–31.51% | 1.40% (@40°C) | 83.5% (@40°C) |
| Isobutyl acetate | 117.2 | Water | 87.40–87.90 | 56.02–64.40% | 0.127% (@40°C) | 90.47% (@40°C) |
| *n*-Butyl acetate | 126.2 | Water | 90.20–91.05 | 70.10–72.30% | 0.3638% (@25°C) | 86.31% (@25°C) |

**TABLE 9.2 NRTL Parameters for Acetic Acid–Water–Ethyl Acetate System.**

| Comp. $i$<br>Comp. $j$ | Acetic Acid<br>Water | Acetic Acid<br>Ethyl Acetate | Ethyl Acetate<br>Water |
|---|---|---|---|
| $a_{ij}$ | 0 | 0 | 0 |
| $a_{ji}$ | 0 | 0 | 0 |
| $b_{ij}$ (K) | −211.310 | −322.424 | 416.124 |
| $b_{ji}$ (K) | 652.995 | 576.234 | 1024.50 |
| $\alpha_{ij}$ | 0.3 | 0.3 | 0.3067 |

**TABLE 9.3 NRTL Parameters for Acetic Acid–Water–Isobutyl Acetate System.**

| Comp. $i$<br>Comp. $j$ | Acetic Acid<br>Water | Acetic Acid<br>Isobutyl Acetate | Isobutyl Acetate<br>Water |
|---|---|---|---|
| $a_{ij}$ | 0 | 0 | 0 |
| $a_{ji}$ | 0 | 0 | 0 |
| $b_{ij}$ (K) | −211.310 | 90.268 | 489.609 |
| $b_{ji}$ (K) | 652.995 | 194.416 | 1809.079 |
| $\alpha_{ij}$ | 0.3 | 0.3 | 0.2505 |

**TABLE 9.4 NRTL Parameters for Acetic Acid–Water–$n$-Butyl Acetate System.**

| Comp. $i$<br>Comp. $j$ | Acetic Acid<br>Water | Acetic Acid<br>$n$-Butyl Acetate | $n$-Butyl Acetate<br>Water |
|---|---|---|---|
| $a_{ij}$ | 0 | 0 | 0 |
| $a_{ji}$ | 0 | 0 | 0 |
| $b_{ij}$ (K) | −211.310 | −68.61 | 354.31 |
| $b_{ji}$ (K) | 652.995 | 397.85 | 2578.35 |
| $\alpha_{ij}$ | 0.3 | 0.3 | 0.219 |

**TABLE 9.5 Hayden–O'Connell Pure Component Parameters.**

| Component | $T_c$ (°C) | $P_c$ (KPa) | $V_c$ ($m^3$/kg mol) | Dipole Moment (debye) | Radius of Gyration (m) |
|---|---|---|---|---|---|
| Acetic acid | 318.80 | 5786 | 0.1797 | 1.7388 | 2.61E−10 |
| Water | 373.95 | 22064 | 0.05595 | 1.8497 | 6.15E−11 |
| Ethyl acetate | 250.15 | 3880 | 0.2860 | 1.7807 | 3.47E−10 |
| Isobutyl acetate | 287.65 | 3010 | 0.389 | 1.8707 | 4.28E−10 |
| $n$-Butyl acetate | 302.25 | 3090 | 0.389 | 1.8407 | 4.28E−10 |

**TABLE 9.6 Hayden–O'Connell Association Parameters for Acetic Acid–Water–Ethyl Acetate System.**

| | Acetic Acid | Water | Ethyl Acetate |
|---|---|---|---|
| Acetic acid | 4.50 | 2.50 | 2.00 |
| Water | 2.50 | 1.70 | 1.30 |
| Ethyl acetate | 2.00 | 1.30 | 0.53 |

**TABLE 9.7 Hayden–O'Connell Association Parameters for Acetic Acid–Water–Isobutyl Acetate System.**

| | Acetic Acid | Water | Isobutyl Acetate |
|---|---|---|---|
| Acetic acid | 4.50 | 2.50 | 0.00 |
| Water | 2.50 | 1.70 | 0.00 |
| Isobutyl acetate | 0.00 | 0.00 | 0.00 |

for the three ternary systems are shown in Tables 9.6 to 9.8. For completeness of the physical property model, the parameters in the extended Antoine equation for all the components are listed in Table 9.9.

All three candidate entrainers form minimum-boiling azeotropes with water. A heterogeneous azeotropic distillation column can be designed to obtain high-purity acetic acid product (b.p. of 118°C) at the column bottom while obtaining a minimum boiling entrainer–water azeotrope at the top of the column. With this column design, by adding entrainer into the system, the difficult tangent pinch of the pure water side can be avoided at the top of the column. Since this entrainer–water azeotrope is heterogeneous, the top column vapor stream forms two liquid phases in the decanter after condensation. The organic phase will be refluxed back to the heterogeneous azeotropic column to provide enough entrainer inside of the column. The aqueous phase containing mostly water will be assumed to be drawn out from the system for further treatment or discharged. Some of the aqueous phase can be refluxed back to the heterogeneous azeotropic column to fulfill the column specifications of small acetic acid loss through the aqueous outlet (AO) stream. The conceptual design of this heterogeneous azeotropic distillation column system is illustrated in Figure 9.2.

**TABLE 9.8 Hayden–O'Connell Association Parameters for Acetic Acid–Water–*n*-Butyl Acetate System.**

| | Acetic Acid | Water | *n*-Butyl Acetate |
|---|---|---|---|
| Acetic acid | 4.50 | 2.50 | 2.00 |
| Water | 2.50 | 1.70 | 1.30 |
| *n*-Butyl acetate | 2.00 | 1.30 | 0.53 |

**TABLE 9.9 Coefficients of Antoine Equation for the Studied System.**

| Components | Acetic Acid | Water | Ethyl Acetate | Isobutyl Acetate | *n*-Butyl Acetate |
|---|---|---|---|---|---|
| Temperature Units | °C | °C | °C | °C | °C |
| Property Units | atm | atm | atm | atm | atm |
| $C_1$ | 41.7439 | 62.123 | 55.298 | 60.784 | 111.294 |
| $C_2$ | −6304.5 | −7258.2 | −6227.6 | −6944.3 | −9253.2 |
| $C_3$ | 0 | 0 | 0 | 0 | 0 |
| $C_4$ | 0 | 0 | 0 | 0 | 0 |
| $C_5$ | −4.2985 | −7.3037 | −6.41 | −7.298 | −14.99 |
| $C_6$ | 8.8865E−18 | 4.1659E−6 | 1.7914E−17 | 3.7892E−6 | 1.047E−5 |
| $C_7$ | 6 | 2 | 6 | 2 | 2 |

Aspen Plus extended Antoine vapor pressure model is:

$$\ln P_i^* = C_{1i} + \frac{C_{2i}}{T + C_{3i}} + C_{4i}T + C_{5i}\ln T + C_{6i}T^{C_{7i}}, \quad i = \text{component}$$

The residue curve maps (RCMs) with the binodal curve of the LLE of the three entrainer systems studied in this chapter are shown in Figures 9.3 to 9.5. These figures show that the prediction for entrainer solubility in water matches with the experimental data given in Table 9.1 quite closely. The azeotropic temperature and composition for the three entrainers are all within the uncertainty range from various sources of the experimental data.

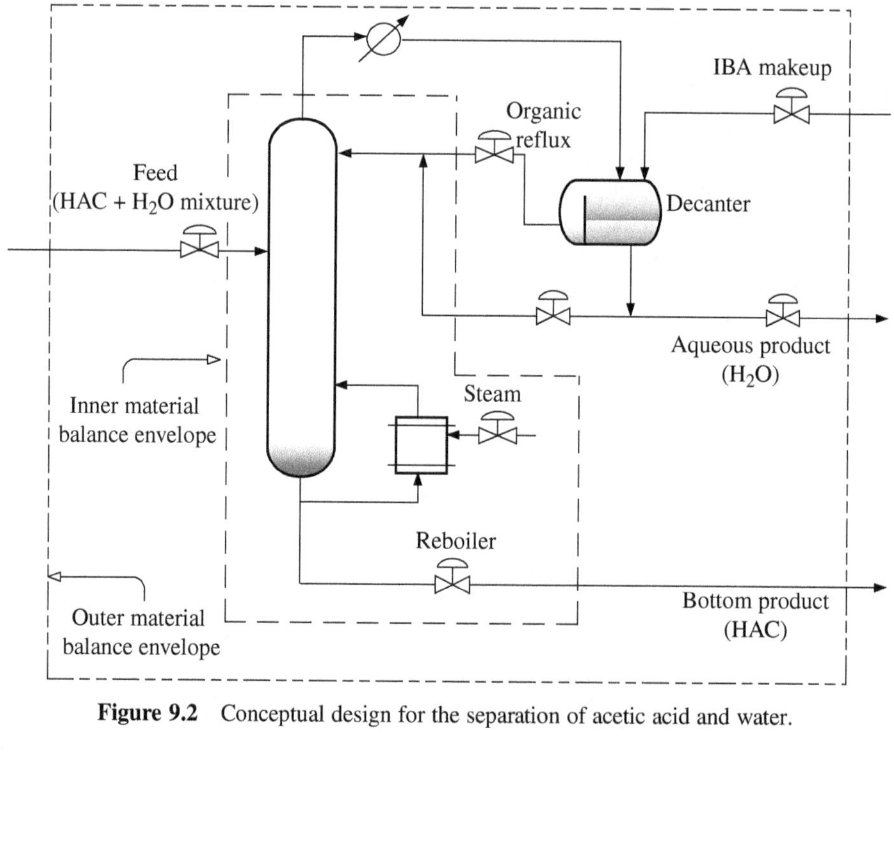


**Figure 9.2** Conceptual design for the separation of acetic acid and water.

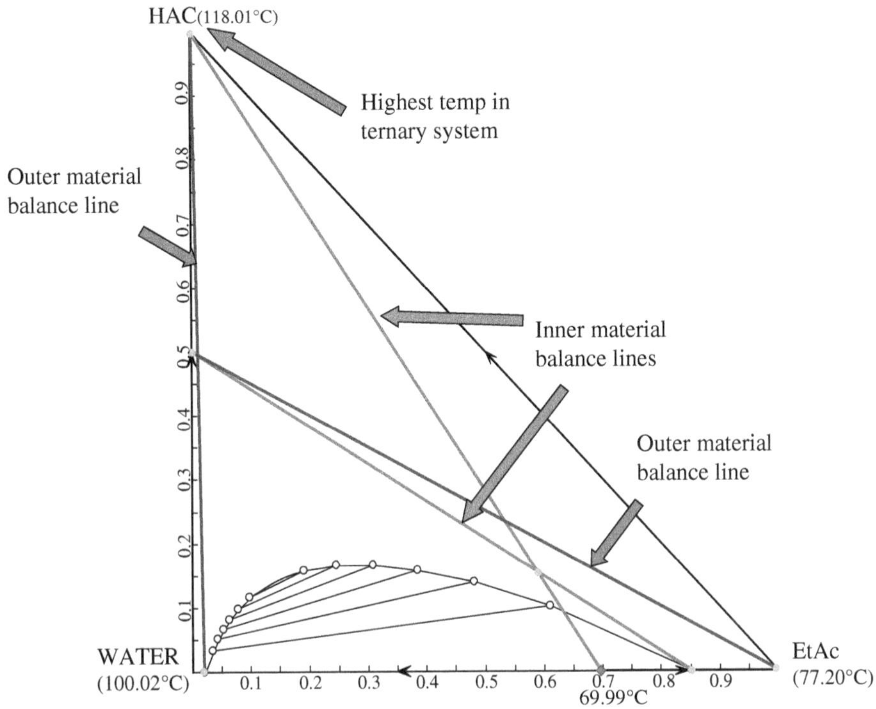


**Figure 9.3** RCM, LLE envelope, and material balance lines of an acetic acid–water system using ethyl acetate as the entrainer.

The RCMs for the ethyl acetate and the isobutyl acetate systems are similar in nature with the two-component azeotrope as the lowest temperature in the system and acetic acid as the highest temperature in the system. The RCM for the *n*-butyl acetate system is different than the other two systems. In the n-butyl acetate system, the highest temperature in the system is *n*-butyl acetate (b.p. 126°C), not acetic acid (b.p. 118°C). Slippage of the entrainer into the bottom product stream is a situation that must be avoided for this system. The other two systems do not need to worry about this situation because acetic acid is the highest temperature in the system and should come out of the column in the bottom stream.

Rigorous process simulation is performed to find the optimum design and operating conditions of these three entrainer systems. The feed composition of 50 mol% acetic acid and 50 mol% water is considered for the Aspen Plus[18] simulation. The feed rate is assumed to be 500 kg/h and it is a saturated liquid phase. The column is assumed to operate at atmospheric pressure. The decanter temperature is 40°C. In the Aspen Plus simulation, the column bottom product is kept at 99.9 mol% acetic acid purity by varying the reboiler heat duty, and the column top aqueous product is kept at 0.1 mol% loss of acetic acid by varying the entrainer makeup flowrate. If the high purity specifications cannot be met, a portion of the aqueous phase can be refluxed back to the column to fulfill the column specifications. This extra third degree of freedom (aqueous reflux flowrate) is fixed at a value that will meet both top and bottom product specifications while also minimizing reboiler heat duty of the column system.

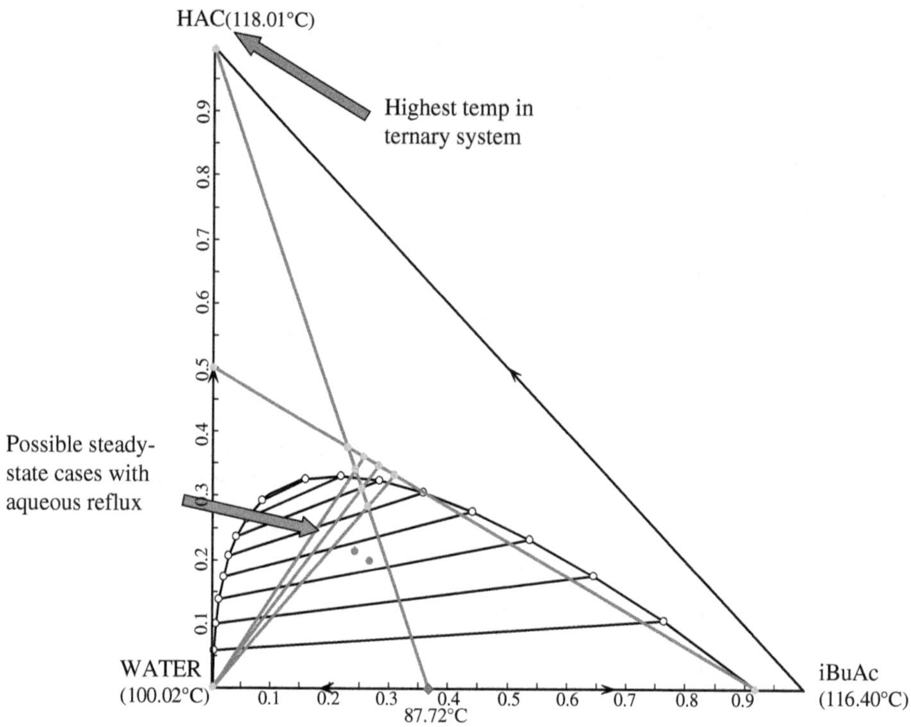


**Figure 9.4** RCM, LLE envelope, and material balance lines of acetic acid–water system using isobutyl acetate as the entrainer.

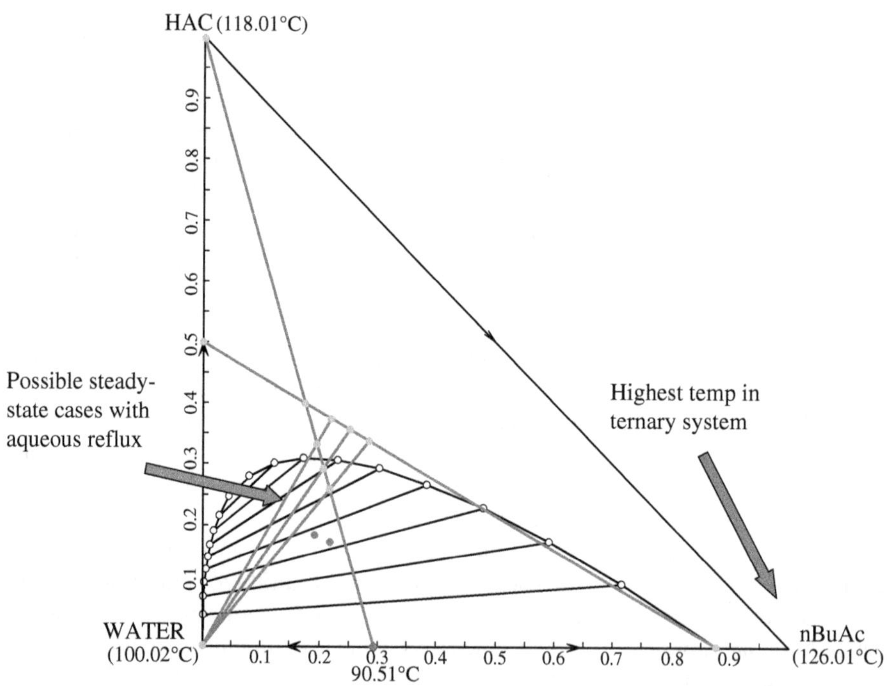


**Figure 9.5** RCM, LLE envelope, and material balance lines of acetic acid–water system using *n*-butyl acetate as the entrainer.

The total number of trays is a compromise between the total equipment cost and the total utility cost. The optimum total number of trays and the optimum feed-tray location are determined to minimize TAC. The calculation procedure of Douglas[19] is followed with the annual capital charge factor of 1/3. The utility costs include the steam and cooling water for the operation of reboiler and condenser. The total operating costs include the utility costs and the entrainer makeup cost. The Aspen Plus[18] simulation results for the three entrainers are summarized in Tables 9.10 to 9.12.

Several observations can be made by comparing these three tables. First, for the system of acetic acid–ethyl acetate–water, no aqueous reflux is necessary to meet the product specifications, while the other two systems need an aqueous reflux stream for the separation with higher aqueous reflux flowrate for the n-butyl acetate system. Second, the organic reflux (OR) flowrate and also the reboiler heat duty for the ethyl acetate system are much larger in comparison with the other two systems.

This high OR flowrate in the ethyl acetate system can actually be predicted by the inner molar balance envelope in Figure 9.2 with the RCM plot of the ethyl acetate system in Figure 9.3. Assuming ideal conditions, the column top-vapor composition should be at the ethyl acetate–water azeotrope, and the column bottom composition should be very close to the pure acetic acid corner in Figure 9.3. Since the feed composition is at 50 mol% acetic acid and 50 mol% water and the other inlet stream to the column for the inner molar balance envelope in Figure 9.2 is the OR (recall that no aqueous reflux is necessary for this system), the intersection of the two inlet and outlet molar balance lines can be used to estimate the OR flowrate. Since the intersection point is closer to the OR composition point, the OR flowrate is quite high. If the feed is much richer in acetic acid, the OR flowrate will be lower than the current case.

Another observation by comparing these three tables is that the makeup flowrate is the highest for the ethyl acetate system and is the lowest for the isobutyl acetate system. This can be explained by the outer molar balance envelope in Figure 9.2 with the knowledge of the aqueous phase composition in Figure 9.3. Assuming ideal conditions for the ethyl acetate system, the two outlet streams for the outer molar balance envelope are at the points of aqueous phase composition and pure acetic acid in Figure 9.3. The two inlet streams are at the points of feed composition and pure ethyl acetate (entrainer makeup) point. How close the intersection point of the two molar balance lines to the feed composition point can be used to determine the entrainer makeup flowrate since the feed flowrate is known. If this intersection point is very close to the feed composition point, the entrainer makeup flowrate will be small. From this explanation, it is not difficult to conclude that the ethyl acetate system will have the highest entrainer makeup flowrate and the isobutyl acetate system will have the lowest entrainer makeup flowrate.

The comparison of the minimum attainable TAC for these three systems, as well as the acetic acid dehydration system without any entrainer, is shown in Table 9.13. From the table, one can observe that the no-entrainer system has the highest TAC and the isobutyl acetate system is most favorable for this feed composition and product specification requirements. The TAC for the isobutyl acetate system is only about 55% of the no-entrainer system, which shows that a large saving can be made by using the isobutyl acetate system. Notice that this finding is in general agreement with the industrial applications. (See patents by Costantini et al.[7] and Parten and Ure[8]). The above two patents also found isobutyl acetate as a favorable entrainer for the separation of acetic acid and water. Another earlier patent by Othmer[20] found *n*-propyl acetate to be useful as an entrainer for this system. The patent by Mitsui Petrochemical Industries[21] found *n*-butyl acetate to be

**TABLE 9.10 Stream Information for the System of Acetic Acid–Water–Ethyl Acetate.**

| | Feed | Bottom Product | Top Product | Aqueous Reflux | Organic Reflux | Makeup Stream | Reboiler Energy |
|---|---|---|---|---|---|---|---|
| Flowrate (mol/min) | 213.48 | 106.74 | 108.98 | 0 | 573.95 | 2.24 | – |
| Acetic acid mole fraction | 0.5 | 0.999 | $1.00 \times 10^{-3}$ | – | $2.87 \times 10^{-3}$ | 0 | – |
| Water mole fraction | 0.5 | $9.90 \times 10^{-4}$ | 0.9785 | – | 0.17586 | 0 | – |
| Ethyl acetate mole fraction | 0.0 | $1.0 \times 10^{-5}$ | $2.05 \times 10^{-2}$ | – | 0.82127 | 1 | – |
| Heat duty (KW) | – | – | – | – | – | – | 401.17 |

**TABLE 9.11 Stream Information for the System of Acetic Acid–Water–Isobutyl Acetate.**

| | Feed | Bottom Product | Top Product | Aqueous Reflux | Organic Reflux | Makeup Stream | Reboiler Energy |
|---|---|---|---|---|---|---|---|
| Flowrate (mol/min) | 213.48 | 106.74 | 106.90 | 33.36 | 92.71 | 0.16 | – |
| Acetic acid mole fraction | 0.5 | 0.999 | $1.00 \times 10^{-3}$ | $1.00 \times 10^{-3}$ | $1.90 \times 10^{-3}$ | 0 | – |
| Water mole fraction | 0.5 | $5.90 \times 10^{-4}$ | 0.9979 | 0.9979 | $7.98 \times 10^{-2}$ | 0 | – |
| Isobutyl acetate mole fraction | 0.0 | $4.10 \times 10^{-4}$ | $1.10 \times 10^{-3}$ | $1.10 \times 10^{-3}$ | 0.9183 | 1 | – |
| Heat duty (KW) | – | – | – | – | – | – | 167.01 |

**TABLE 9.12 Stream Information for the System of Acetic Acid–Water–$n$-Butyl Acetate.**

| | Feed | Bottom Product | Top Product | Aqueous Reflux | Organic Reflux | Makeup Stream | Reboiler Energy |
|---|---|---|---|---|---|---|---|
| Flowrate (mol/min) | 213.48 | 106.74 | 107.44 | 98.78 | 102.32 | 0.70 | – |
| Acetic acid mole fraction | 0.5 | 0.999 | $1.00 \times 10^{-3}$ | $1.00 \times 10^{-3}$ | $1.50 \times 10^{-3}$ | 0 | – |
| Water mole fraction | 0.5 | $7.48 \times 10^{-4}$ | 0.9928 | 0.9928 | 0.1448 | 0 | – |
| $n$-Butyl acetate mole fraction | 0.0 | $2.52 \times 10^{-4}$ | $6.20 \times 10^{-3}$ | $6.20 \times 10^{-3}$ | 0.8537 | 1 | – |
| Heat duty (KW) | – | – | – | – | – | – | 259.68 |

**TABLE 9.13 Comparison of TAC for the Acetic Acid Dehydration Systems.**

| Entrainer | Optimal Total Stages | Optimal Feed Stage | Annualized Capital Cost | Utility Cost | Entrainer Cost | TAC ($) |
|---|---|---|---|---|---|---|
| Ethyl acetate | 16 | 2 | $6.84 \times 10^4$ | $4.20 \times 10^4$ | $5.40 \times 10^4$ | $1.64 \times 10^5$ |
| Isobutyl acetate | 30 | 9 | $6.81 \times 10^4$ | $1.80 \times 10^4$ | $1.70 \times 10^4$ | $1.03 \times 10^5$ |
| *n*-Butyl acetate | 31 | 11 | $8.44 \times 10^4$ | $2.78 \times 10^4$ | $6.08 \times 10^4$ | $1.73 \times 10^5$ |
| No entrainer | 50 | 37 | $1.42 \times 10^5$ | $4.37 \times 10^4$ | 0 | $1.86 \times 10^5$ |

favorable for this system. Notice that in all the above patents, the designed feed compositions and the specified column bottom and top purities are all different from this chapter, thus no direct comparison of results can be made.

The vapor and liquid profiles inside the column for the optimized isobutyl acetate system can be seen in Figure 9.6. Notice that the column behaves as it was designed. The first five stages, counting from the top of the column, have two liquid phases. The combined liquid compositions for these five stages are plotted in this figure. Another thing worth a mention in the figure is that the column composition profiles bypass the corner of pure water that is the region to be avoided because of the tangent pinch.

From this study, some important factors in determining the suitable entrainer for the acetic acid dehydration system are summarized below. The information needed for this qualitative comparison can be illustrated by the RCMs with the binodal curve of the LLE as

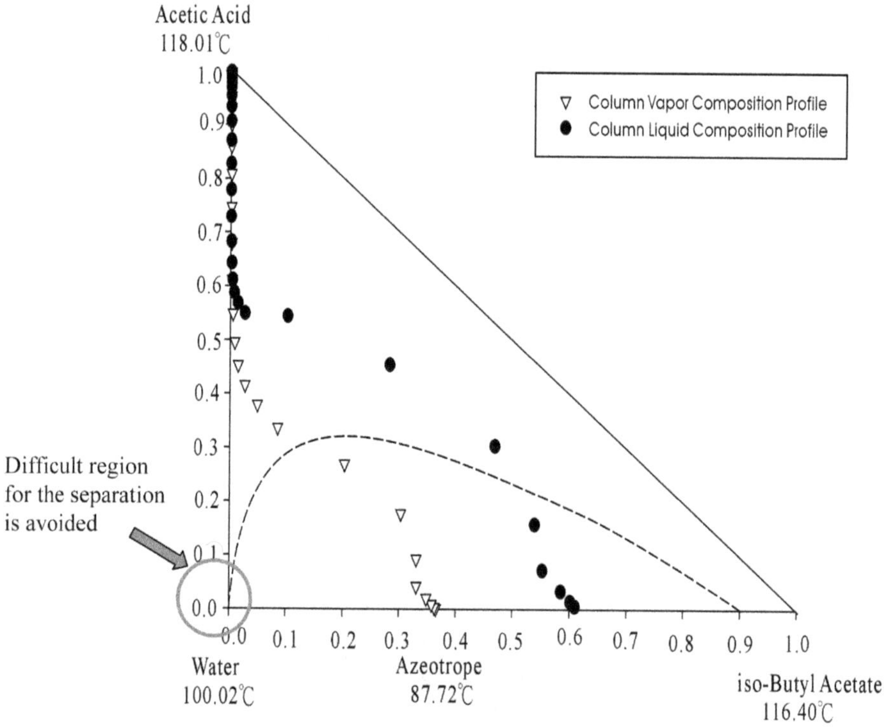


**Figure 9.6** Vapor and liquid profiles for the optimum system acetic acid–isobutyl acetate–water.

shown in Figures 9.3 to 9.5. The suitability of the entrainer is actually a combination of the following factors.

### 9.1.1 Azeotropic Composition and Organic Phase Composition

It is better to have the azeotropic composition containing more water in this mixture. This means that this entrainer is more capable of carrying water to the top of the column; thus less entrainer is needed inside the column. It is better to have the distance between the azeotropic composition point and organic phase composition point large. This means that besides the azeotropic composition containing more water, the organic phase composition contains more entrainer. The locations of these two points in Figures 9.3 to 9.5 dictate the required OR flowrate into the heterogeneous column (as explained above when estimating the OR flowrate for the ethyl acetate system). From Figures 9.3 to 9.5, the ethyl acetate is the worst entrainer if only considering this factor.

### 9.1.2 Azeotropic Temperature

The azeotropic temperature determines the temperature difference between the top and the bottom of the column. A large delta T of the azeotropic temperature to the pure acetic acid temperature implies easy separability. Fewer column stages will be needed for specific product purity specifications. In this regard, ethyl acetate is the best entrainer. This interpretation is confirmed by Table 9.13 because the ethyl acetate system requires the least total number of stages for the same separation.

### 9.1.3 Aqueous Phase Composition and Entrainer Pricing

The aqueous phase should contain as little entrainer as feasible. The reason is because the aqueous phase stream will be drawn out of the system, and any entrainer loss should be compensated by the makeup stream in Figure 9.2. This will correspond to a stream cost of the system as seen in Table 9.13. The makeup flowrate can actually be estimated using the outer molar balance envelope in Figure 9.2, assuming ideal conditions, as explained previously. In this regard, an isobutyl acetate system results in the least makeup flowrate while an ethyl acetate system requires the most makeup flowrate. This is confirmed by Tables 9.10 to 9.12.

The annual cost of this stream is not only related to its flowrate but also related to the entrainer pricing. Ethyl acetate is the cheapest and isobutyl acetate is the most expensive entrainer. With the knowledge of the entrainer pricing and the calculation of the makeup flowrates for the three systems, the entrainer cost can be estimated as seen in Table 9.13, even without any rigorous simulation.

Since the system with isobutyl acetate as the entrainer results in the most economical process design, we will study the dynamics and control strategy of this system in detail in the next section. Before doing that, let's first explore the necessity of the aqueous reflux stream under various feed compositions for the system using isobutyl acetate as the entrainer. Figure 9.7 shows the collection of many simulation results under various feed composition conditions. In all the simulation runs, the total number of stages for the column is fixed at the number shown in Table 9.13 (30 stages including reboiler but not the condenser). The column bottom product is kept at a high purity of 99.9 mol% acetic acid by varying the reboiler heat duty, and the column top aqueous product is kept at 0.1 mol% loss of acetic acid by

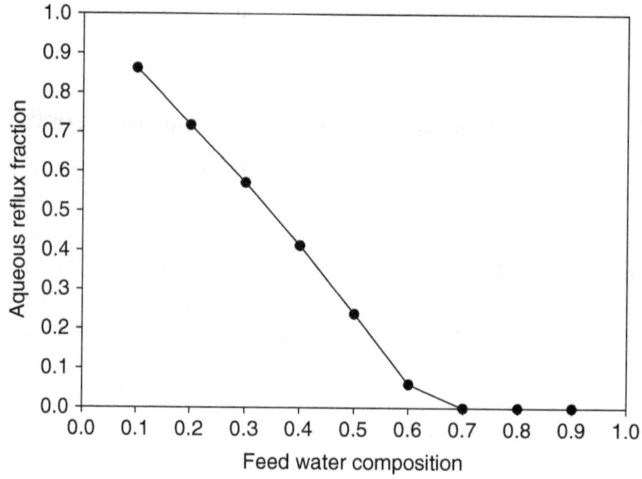


**Figure 9.7** Minimum aqueous reflux fraction under various feed compositions.

varying the entrainer makeup flowrate. The aqueous reflux flowrate is fixed at a value that will meet both top and bottom product specifications while also minimizing reboiler heat duty of this column system. From the figure, one can observed that, for feed water composition above 70 mol%, the aqueous reflux stream is not needed. For the feed composition studied here, the aqueous reflux stream is necessary in order to properly hold the bottom and top product specifications.

Table 9.14 shows the values of the main operating variables in keeping the top and bottom product purity at their specifications under various feed composition conditions. In this table, the operating conditions are not unique for feed water contents from 10% to 60%. For these feed composition cases, there are three degrees of freedom (extra one is the aqueous reflux) for the system with only two product purity specifications. The ones included in the table are the operating conditions that minimize reboiler heat duty by varying aqueous reflux flowrate.

One important thing that needs to be pointed out from the table is that in order to hold product specifications, the aqueous reflux flowrate needs to be adjusted in a very wide

**TABLE 9.14 Desired Operating Condition Under Various Feed Compositions.**

| Feed Water Composition (mol%) | Aqueous Reflux (mol/min) | Organic Reflux (mol/min) | Reboiler Duty (KW) | Entrainer Makeup (mol/min) | Aqueous Draw (mol/min) | Bottom Product (mol/min) |
|---|---|---|---|---|---|---|
| 10 | 133 | 102 | 189 | 0.07 | 21 | 192 |
| 20 | 109 | 100 | 186 | 0.08 | 43 | 171 |
| 30 | 85 | 99 | 182 | 0.11 | 64 | 150 |
| 40 | 60 | 96 | 175 | 0.13 | 86 | 128 |
| 50 | 33 | 93 | 167 | 0.16 | 107 | 107 |
| 60 | 8 | 90 | 159 | 0.16 | 128 | 85 |
| 70 | 0 | 99 | 175 | 0.16 | 150 | 64 |
| 80 | 0 | 113 | 200 | 0.19 | 171 | 43 |
| 90 | 0 | 127 | 226 | 0.21 | 193 | 21 |

range. This means that this manipulated variable (aqueous reflux) should not be fixed in the control strategy when trying to reject an unmeasured feed composition disturbance. This manipulated variable is preferable to be used in the inferred composition loop to hold product specifications.

Another observation is that the reboiler duty goes up dramatically for the cases with no aqueous reflux (70%, 80%, and 90% feed water compositions). This implies that it may be better to add a preconcentrator column before the heterogeneous azeotropic column to increase the acetic acid content in the feed to the heterogeneous azeotropic column if the fresh feed water composition is too high. Another possible advantage of adding a preconcentrator column may be to have an extra degree of freedom (aqueous reflux) to be manipulated in the control strategy. The detailed investigation of adding a preconcentrator column to the design and control of the acetic acid dehydration system for diluted fresh feed stream was studied in Chien and Kuo.[22]

## 9.2 CONTROL STRATEGY DEVELOPMENT

The heterogeneous azeotropic column system using isobutyl acetate as the entrainer will be studied in detail in this section. The overall control strategy of this system will be developed in order to maintain bottom and top product specifications in spite of feed flowrate and feed composition changes. In the control strategy development, we will assume no online composition measurement is available. The composition control loops will be inferred by some tray temperature control strategy. This type of control strategy can easily be implemented in industry for wider applications.

The Aspen Plus steady-state simulation in the last section is exported to the dynamic simulation of Aspen Dynamics. The tray sizing option in Aspen Plus is utilized to calculate the column diameter to be 0.3259 m and the tray spacing is 0.6096 m. Other equipment sizing recommended by Luyben[23] is used here. The volume of the reboiler is sized to give 10 min holdup with 50% liquid level. The decanter is sized to be bigger to allow for two liquid phases to separate. The holdup time of 20 min is used in the dynamic simulation. Pressure-driven simulation in Aspen Dynamics is used with the top pressure of the azeotropic column controlled at 1.1 atm to allow for some pressure drop in the condenser and decanter to give the decanter at atmospheric pressure. The pressure drop inside the column is automatically calculated in Aspen Dynamics. Since the tray pressures in the columns are different than the constant atmospheric pressure assumption used in steady-state simulation, the established base-case condition in Aspen Dynamics will be slightly different than Table 9.11. The final base-case steady-state condition used for control study can be seen in Table 9.15.

There are two inventory control strategies that can be used for this system. The first inventory control strategy (Inventory Strategy 1) uses entrainer makeup flow to control the organic phase level in the decanter. This inventory control strategy was successfully used in Chien et al.[24] and Arifin and Chien[25] when controlling an IPA dehydration column. The second inventory control strategy (Inventory Strategy 2) uses OR flow to control the organic phase level in the decanter. This second inventory control strategy is more intuitively sound because the OR flowrate is much larger than the entrainer makeup flowrate, so the organic phase level control should be more effective. The other inventory control loops are (1) using top aqueous product flow to control the aqueous phase level in the decanter and (2) using bottom product flow to control the column bottom level. The column top

**TABLE 9.15 Base-Case Condition of the Optimal Flowsheet for Dynamic Tests.**

| | |
|---|---|
| Total number of stages for the azeotropic column | 30 (including reboiler but not including condenser) |
| Feed stage | 9 (counting from the top tray) |
| Fresh feed flowrate | 213.48 mol/min |
| Fresh feed composition | 50 mol% acetic acid and 50 mol% water |
| Column reboiler duty | 162.27 KW |
| Organic reflux flowrate | 92.21 mol/min |
| Aqueous reflux flowrate | 33.40 mol/min |
| Entrainer makeup flowrate | 0.165 mol/min |
| Bottom flowrate | 106.76 mol/min |
| Bottom composition | 99.89 mol% acetic acid<br>0.0665 mol% water<br>0.0458 mol% isobutyl acetate |
| Top aqueous outlet flowrate | 106.89 mol/min |
| Top aqueous outlet composition | 0.0997 mol% acetic acid<br>99.79 mol% water<br>0.11 mol% isobutyl acetate |

pressure is controlled at 1.1 atm by manipulating the top vapor flow, and the decanter temperature is controlled at 40°C by manipulating the condenser duty.

After deciding the inventory control strategy, there are three variables left that can be used in some composition control strategy. The three candidate variables for Inventory Strategy 1 are OR flow, aqueous reflux flow, and the reboiler duty. The three candidate variables for Inventory Strategy 2 are entrainer makeup, aqueous reflux flow, and the reboiler duty. The control objective is to hold the bottom and the top aqueous product specifications at base-case condition under ±10% feed flow and ±10% feed water composition changes.

### 9.2.1 Dual-Temperature Control Strategy

Since product specifications at both bottom and top ends are specified, we will consider dual-point temperature control structures first. Sensitivity analysis with small perturbations of the three manipulated variables will be performed next in order to determine the two temperature control points. Figure 9.8 shows the sensitivity analysis of the three manipulated variable changes using Inventory Strategy 1, and Figure 9.9 shows the sensitivity analysis of the three manipulated variable changes using Inventory Strategy 2. The numbering of the stages in this column is counting from top to bottom with Stage 1 as the top stage and Stage 30 as the reboiler. When perturbing one manipulated variable, the other two manipulated variables are fixed at base-case conditions. The final steady-state conditions of delta T in Figures 9.8 and 9.9 are obtained by running dynamic simulation with the perturbations mentioned in the figures and then waiting until the dynamic simulation reaches a final steady state, followed by subtracting the new temperature values from the base-case values.

For OR changes shown in the lower graph in Figure 9.8, a process gain sign reversal is observed between Stages 12 and 13. Dynamically, a large inverse response was observed for column stages between Stages 13 to the column bottom. Similarly for entrainer makeup changes shown in the lower graph in Figure 9.9, a process gain sign reversal is

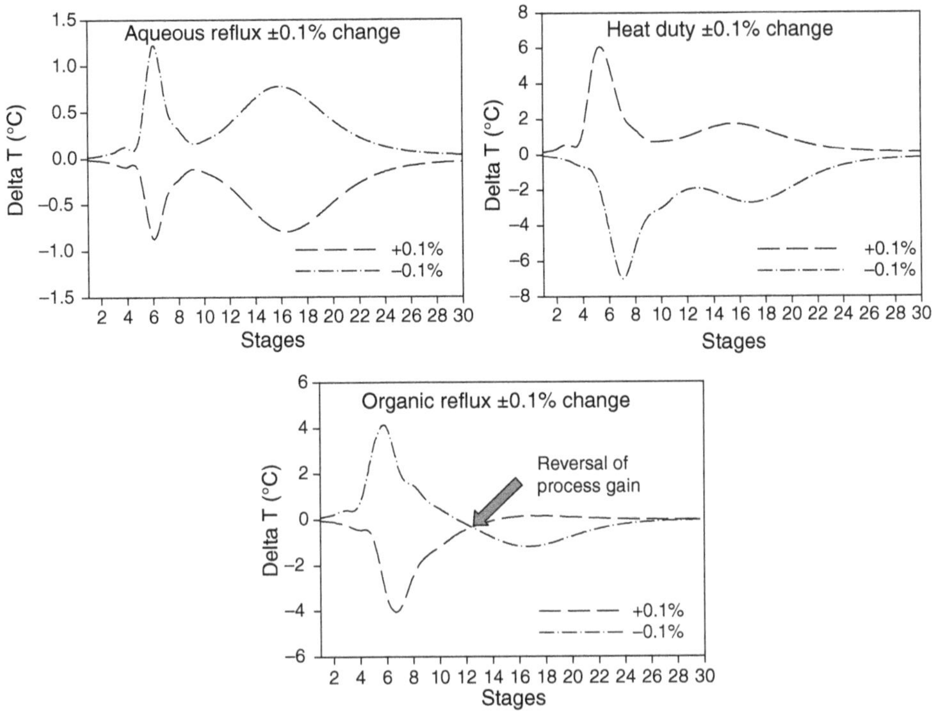


**Figure 9.8** Sensitivity analysis of the three manipulated variables under Inventory Strategy 1.

also observed between Stages 12 and 13. This also indicates that dynamically a large inverse response was also observed for column stages between Stages 13 to the column bottom.

There are three candidate manipulated variables for each inventory control strategy. If we want to control two temperatures, there will be three alternative temperature control structures. From the open-loop data in Figures 9.8 and 9.9, a steady-state gain matrix for each alternative overall control strategy can be obtained by averaging the positive and negative manipulated variable changes. Each elements of the steady-state gain matrix is made dimensionless by using the spans of the temperature sensors and the manipulated variables. Singular-value decomposition (SVD) as described by Moore[26] can be made on the steady-state gain matrices as follows:

$$K = U\Sigma V^T \tag{9.1}$$

$K$ is a 30 × 2 steady-state gain matrix for each control strategy. $U = [\mathrm{U}_1|\mathrm{U}_2]$ is an 30 × 2 orthonormal matrix, the columns of which are called the left singular vectors. $V^T$ is a 2 × 2 orthonormal matrix, the columns of which are called the right singular vectors. $\Sigma$ is a 2 × 2 diagonal matrix of scalars called the singular values ($\sigma_1$ and $\sigma_2$) that are arranged in descending order. The condition number is the ratio of the maximum singular value to the minimum singular value ($\sigma_1/\sigma_2$). To trade off between sensor sensitivity and loop interaction, a function was defined as the difference between the absolute values of the elements of the **U** vectors as:

$$Z_i = |\mathrm{U}_{1i}| - |\mathrm{U}_{2i}| \tag{9.2}$$

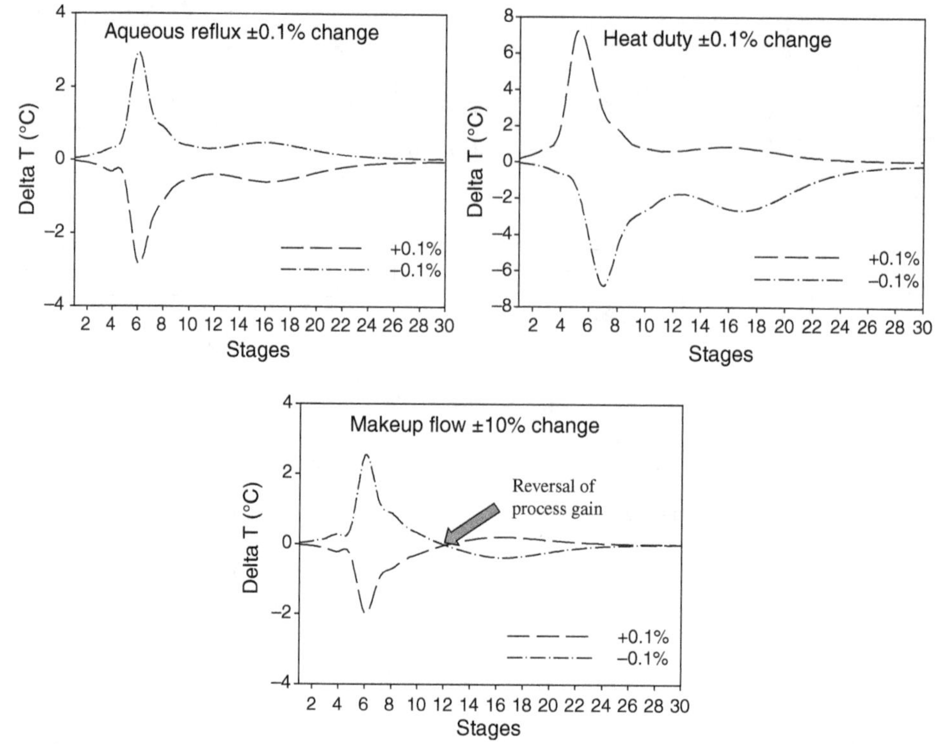


**Figure 9.9** Sensitivity analysis of the three manipulated variables under Inventory Strategy 2.

The maximum and the minimum of this function as suggested by Moore[26] are selected as the two tray locations for the temperature control points. For an example, Figure 9.10 shows the $\boldsymbol{Z}_i$ for the Inventory Strategy 1 with two manipulated variables of aqueous reflux and reboiler duty. From this figure, temperatures at Stage 6 and Stage 16 are selected as the two controlled

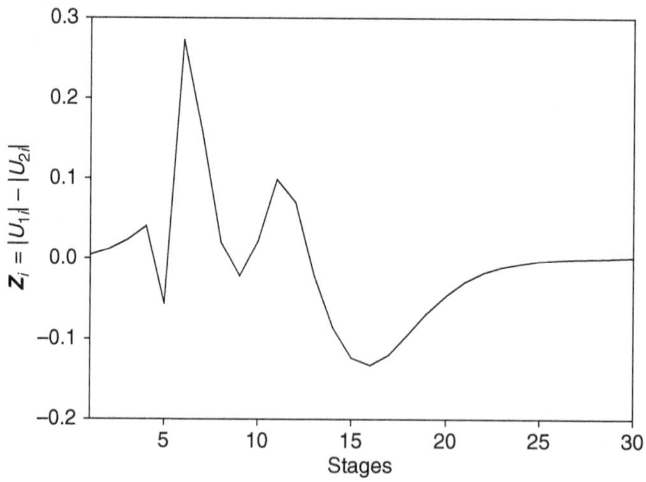


**Figure 9.10** Plot of $\boldsymbol{Z}_i$ for control structure CS1QA.

variables for the above two manipulated variables. To compare among the alternative control structures, the condition number (CN) and the relative gain array (RGA) are also calculated for each control structure. The candidate control structures are listed below.

CS1QA: Using Inventory Strategy 1 with reboiler duty and aqueous reflux as two manipulated variables for dual-point temperature control.

CS1QO: Using Inventory Strategy 1 with reboiler duty and organic reflux as two manipulated variables for dual-point temperature control.

CS1AO: Using Inventory Strategy 1 with aqueous reflux and organic reflux as two manipulated variables for dual-point temperature control.

CS2QA: Using Inventory Strategy 2 with reboiler duty and aqueous reflux as two manipulated variables for dual-point temperature control.

CS2QE: Using Inventory Strategy 2 with reboiler duty and entrainer makeup as two manipulated variables for dual-point temperature control.

CS2AE: Using Inventory Strategy 2 with aqueous reflux and entrainer makeup as two manipulated variables for dual-point temperature control.

Table 9.16 summarizes the results of SVD and RGA analysis for the above six control structures. From the results of this table, several guidelines, as suggested in Moore,[26] are followed to screen out the undesirable control structures from this steady-state analysis. The guidelines are: (1) select the structure with the largest "minimum" singular value, (2) select the structure with the smallest condition number, and (3) select the structure with RGA ($\lambda_{11}$) values close to unity. From these guidelines, CS2QE and CS2AE are dropped from further comparison. For the remaining four control structures (CS1QA, CS1QO, CS1AO, and CS2QA), further closed-loop dynamic evaluation will be made to determine which control structure is the best.

For the manipulated variables not used for temperature control purpose, it is preferable to design some kind of ratio scheme in order to move these manipulated variables according to the measured disturbance changes. For example, with control structure CS1QA, a constant ratio of OR flowrate to feed flowrate is maintained throughout the closed-loop simulation

**TABLE 9.16 SVD and RGA Analysis for each Control Structure.**

| Control Structure | Controller Pairing | Singular Values | CN | RGA ($\lambda_{11}$) |
|---|---|---|---|---|
| CS1QA (Inventory Strategy 1) | $T_{16} \leftrightarrow$ aqueous reflux | $\sigma_1 = 102.9$ | 10.05 | 2.64 |
| | $T_6 \leftrightarrow$ reboiler duty | $\sigma_2 = 10.24$ | | |
| CS1QO (Inventory Strategy 1) | $T_7 \leftrightarrow$ organic reflux | $\sigma_1 = 112.2$ | 3.14 | 0.673 |
| | $T_{17} \leftrightarrow$ reboiler duty | $\sigma_2 = 35.76$ | | |
| CS1AO (Inventory Strategy 1) | $T_{16} \leftrightarrow$ aqueous reflux | $\sigma_1 = 60.72$ | 3.27 | 0.874 |
| | $T_7 \leftrightarrow$ organic reflux | $\sigma_2 = 18.57$ | | |
| CS2QA (Inventory Strategy 2) | $T_6 \leftrightarrow$ aqueous reflux | $\sigma_1 = 108.0$ | 7.89 | 2.24 |
| | $T_{15} \leftrightarrow$ reboiler duty | $\sigma_2 = 13.68$ | | |
| CS2QE (Inventory Strategy 2) | $T_7 \leftrightarrow$ entrainer makeup | $\sigma_1 = 101.6$ | 529.2 | 0.594 |
| | $T_{16} \leftrightarrow$ reboiler duty | $\sigma_2 = 0.192$ | | |
| CS2AE (Inventory Strategy 2) | $T_{16} \leftrightarrow$ aqueous reflux | $\sigma_1 = 38.93$ | 282.1 | 0.595 |
| | $T_6 \leftrightarrow$ entrainer makeup | $\sigma_2 = 0.138$ | | |

run. In order to provide effective dynamic compensation for this feed flowrate disturbance, a first-order lag with adjustable time constant is also included in the ratio scheme.

Table 9.16 only shows the steady-state characteristics of each control structure. However, good steady-state characteristics are not a sufficient condition for good dynamic control system performance. Thus, Aspen Dynamics will be used to evaluate the control system performance for the alternative control structures. Since the unmeasured feed composition changes are the more severe closed-loop test in comparison with the feed rate changes, these load changes will be made in the closed-loop dynamic simulations for comparison.

All level loops are assumed to be controlled by a P-only controller in order to smooth out their manipulated variables in the system. A controller gain of 2.0, as suggested in Luyben,[23] is used in all the level loops. The PID tuning constants for all the stage temperature control loops are tuned using the same multiloop tuning guideline (see Chien et al.[27]), so fair comparison can be made of the closed-loop dynamic responses among the four candidate control structures. Besides achieving a stable dynamic response, the evaluation of which control structure is most effective will be based on the final steady-state product purities in the face of load disturbances.

The closed-loop dynamic responses of the four control structures (CS1QA, CS1QO, CS1AO, and CS2QA) with ±10% changes in the feed water composition are shown in Figures 9.11 through 9.14 respectively. The disturbances are introduced at time = 0.5 h. With Inventory Strategy 1 (organic phase level to manipulate the entrainer makeup flow), the maximum makeup flow is assumed to be larger than twice of the steady-state flowrate in order to have better control of the organic phase level when this level is dropping.

Since the control loop pairing of CS1QA is unconventional (reboiler duty to control temperature at a stage closer to the top of the column), it is very difficult to find proper tuning constants for the two temperature control loops. The tuning guideline in Chien, et al.[27] has to be further detuned in order to make the system stable. Notice that for the −10% change in feed water shown in the bottom graphs in Figure 9.11, the entrainer makeup flow valve has to be fully open at time = 2.4 h and stays fully open until time = 10.2 h. The controlled temperatures have not come to steady state at the final simulation time of 20 h.

For CS1QO in Figure 9.12, the dynamic response of –10% feed water change is quite satisfactory. Two controlled temperature points are returned back to setpoints well before time = 20 h. However the dynamic response is unacceptable for +10% feed water change. Although the two controlled temperatures reach their setpoints by the final simulation time, the dynamic response of organic phase shown in the bottom graphs in Figure 9.12 is very bad. The manipulated variable switches from fully close to fully open for the entire simulation run. Similar unacceptable dynamic responses are observed in Figure 9.13 for CS1AO.

In comparison with the other three control structures, it is quite obvious from Figure 9.14 that CS2QA is the best control structure in terms of the dynamic response. All controlled and manipulated variables reach a new steady state within 8 h. The organic phase level is also maintained much better than in the other three control structures.

Although not shown here, dynamic closed-loop tests for CS2QE and CS2AE were also performed for ±10% changes in the feed water composition. The closed-loop performance was also not satisfactory, particularly for the temperature loop using an entrainer makeup flow as the manipulated variable. This ineffectiveness of the entrainer makeup flow on the controlled tray temperature can actually be seen in a previous sensitivity plot (shown in the bottom graph in Figure 9.9). Large ±10% changes in the entrainer makeup only

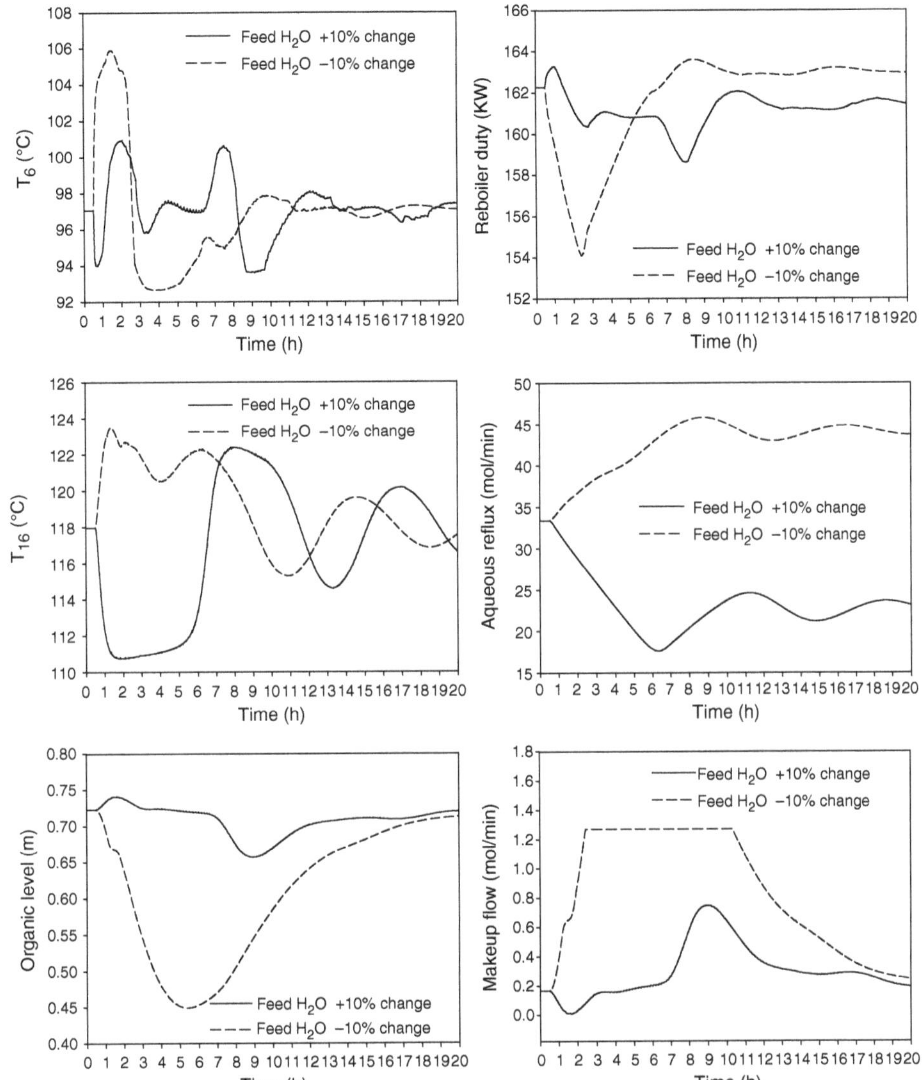


**Figure 9.11** CS1QA with $\pm10\%$ changes in the feed water composition.

give tray temperature perturbations that are comparable to $\pm0.1\%$ changes in the aqueous reflux flow.

The large variations of the entrainer makeup flowrate as well as the controlled tray temperature have detrimental effects on the product composition specifications. One thing worth mentioning is that, although the closed-loop dynamic response of CS2AE is much worse than CS2QA, its performance is somewhat better than CS1QA, CS1QO, and CS1AO. This is another proof that good steady-state characteristics are not a sufficient condition for good dynamic control system performance of nonlinear systems. The condition number of CS2AE (282.1 in Table 9.16) is much larger than CS1QA, CS1QO, and CS1AO but it gives better dynamic control performance.

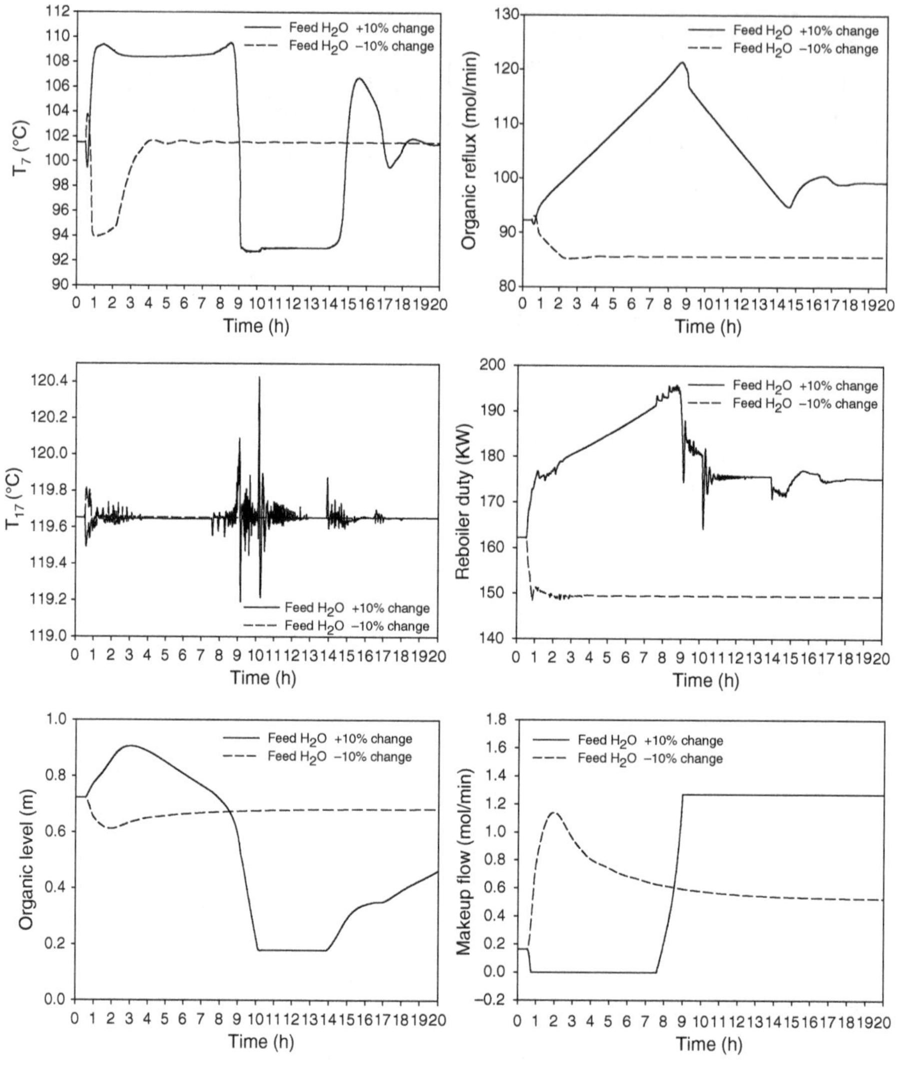


**Figure 9.12** CS1QO with ±10% changes in the feed water composition.

The final objective of the alternative control structures is to maintain the bottom and top products specifications in spite of the load disturbances. Figures 9.15 and 9.16 compare the dynamic responses of these four control structures with +10% and –10% feed water composition changes, respectively. It is obvious from these two figures that CS2QA is the best control structure to reject feed water composition disturbances. Both bottom and top product compositions are quickly returned back to specifications much faster than when using the other three alternative control structures.

Control structure CS1AO dynamically departs the furthest for both bottom and top product compositions from their specifications for the +10% feed water composition change. Control structure CS1QA dynamically departs the furthest for the –10% feed water composition change. In terms of the final steady-state value, control structure CS1QO gives the largest departure in bottom product purity (see Fig. 9.16). The CS1QO control structure keeps the aqueous reflux flowrate fixed during the dynamic runs. The ability to adjust the aqueous

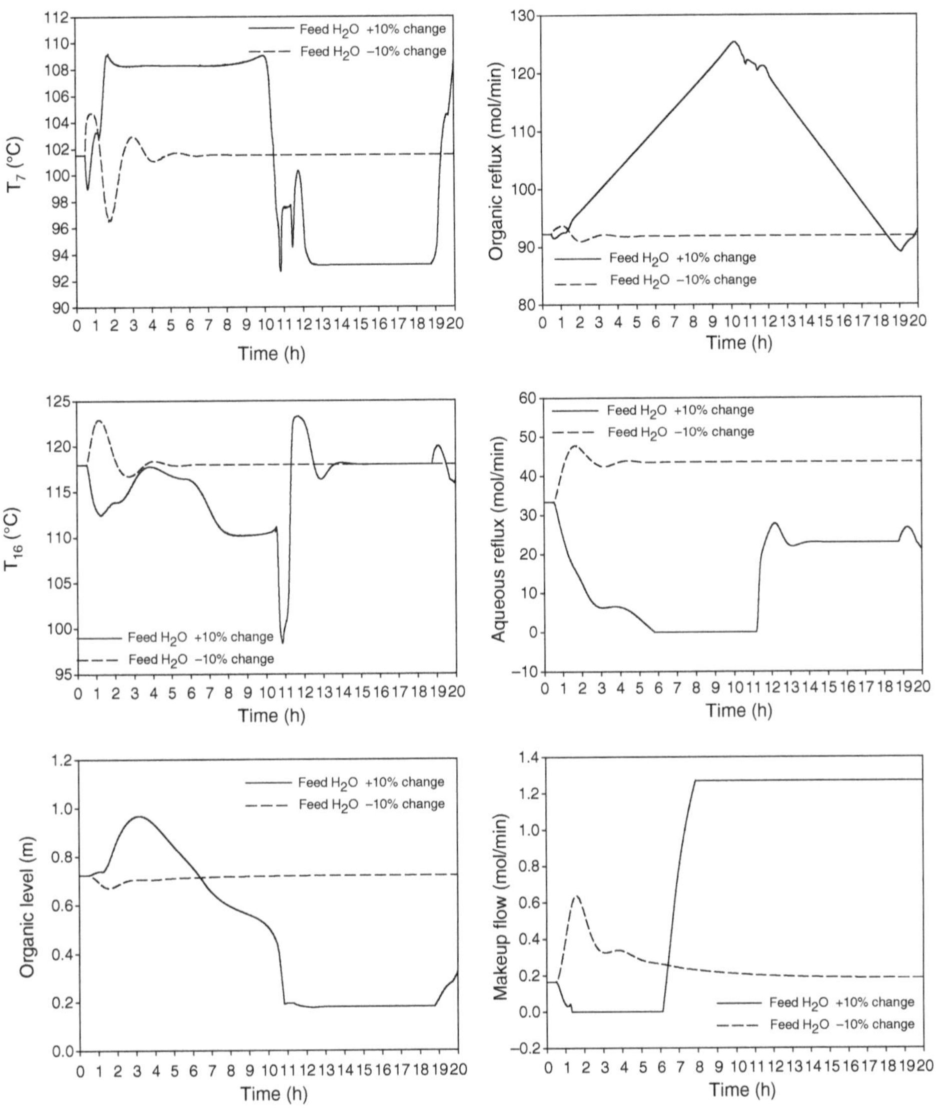


**Figure 9.13** CS1AO with $\pm 10\%$ changes in the feed water composition.

flowrate upward or downward to cope with the feed water composition changes is lost. Attempts have been made to introduce a ratio scheme to maintain constant aqueous reflux ratio instead of ratio-to-feed flow. The dynamic results are even worse than fixing the aqueous reflux flowrate during load disturbances.

Another important closed-loop test for this control system is the feed flowrate changes. These changes are necessary in order to adjust the production rate upward or downward. Figure 9.17 shows the dynamic responses for the proposed control structure CS2QA for $\pm 10\%$ feed rate changes. Notice again that the closed-loop dynamic response is very satisfactory. Although not shown here, both product specifications are maintained in spite of the production rate changes.

Some concluding remarks can be made in light of the above dynamic results. First, the good steady-state characteristics of CS1QA, CS1QO, and CS1AO (see Table 9.16) are

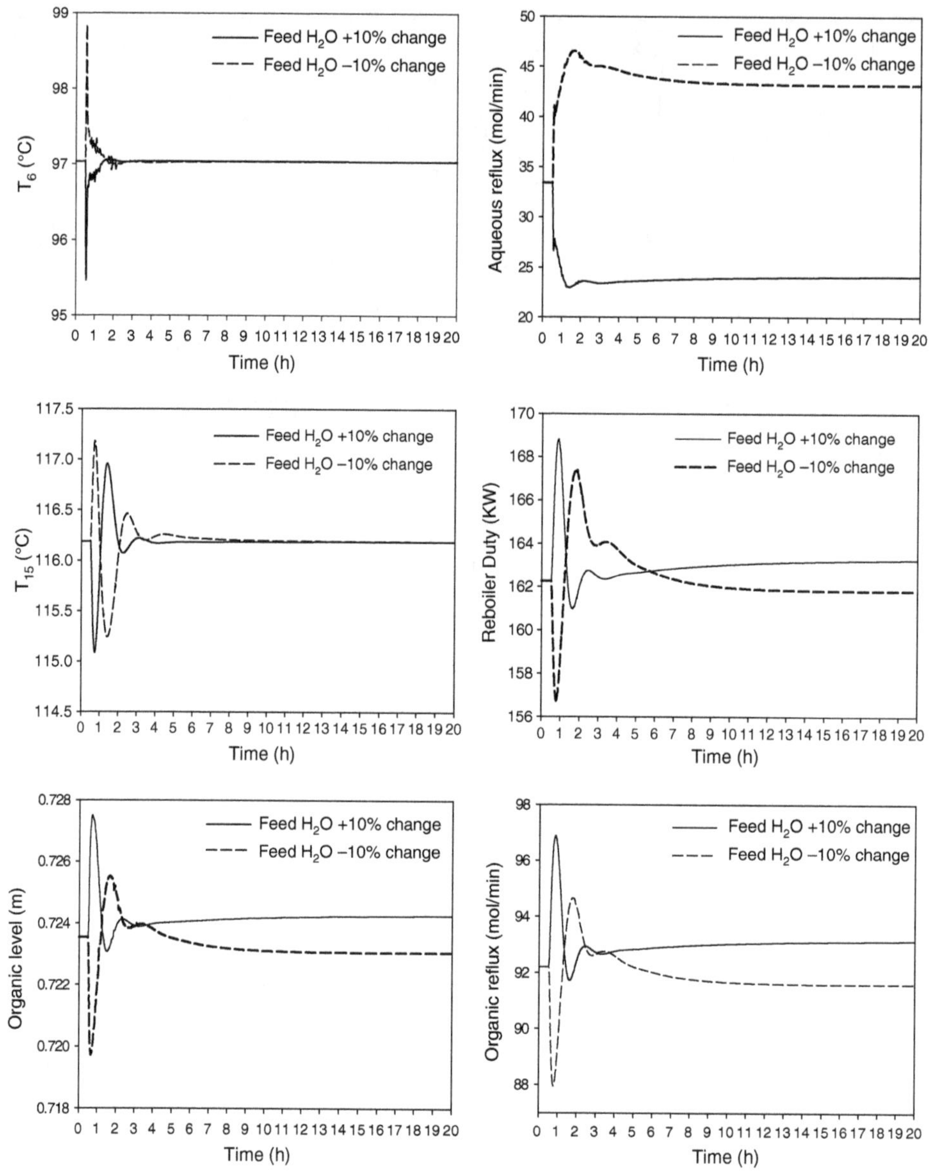


**Figure 9.14** CS2QA with $\pm 10\%$ changes in the feed water composition.

not a sufficient condition for good dynamic control system performance of this nonlinear system. These three control structures gave much less acceptable closed-loop performance than that achieved with CS2QA. Second, dynamic simulations show that it is better to avoid using entrainer makeup flow as the manipulated variable for either of the organic phase level loop (CS1QA, CS1QO, and CS1AO) or the controlled tray temperature loop (CS2QE and CS2AE). In this system, there are three "free" manipulated variables that can be chosen for the dual temperature loop. If the organic phase level loop is included, there are a total of four manipulated variables that can be chosen from. Thus, it is possible to

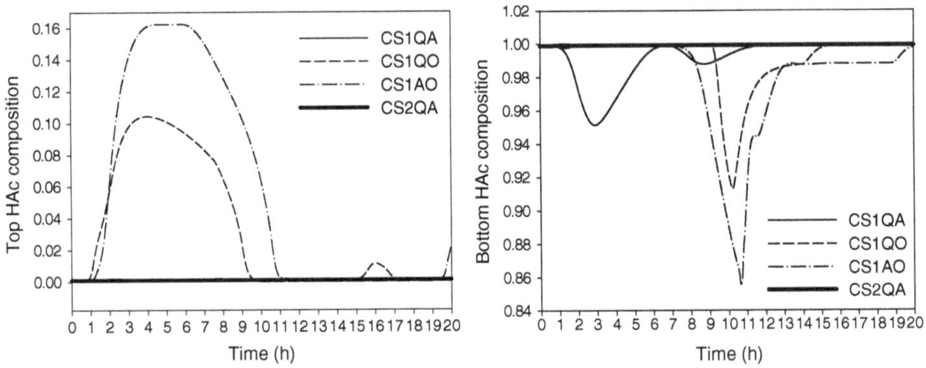


**Figure 9.15** Comparison under +10% feed water composition change.

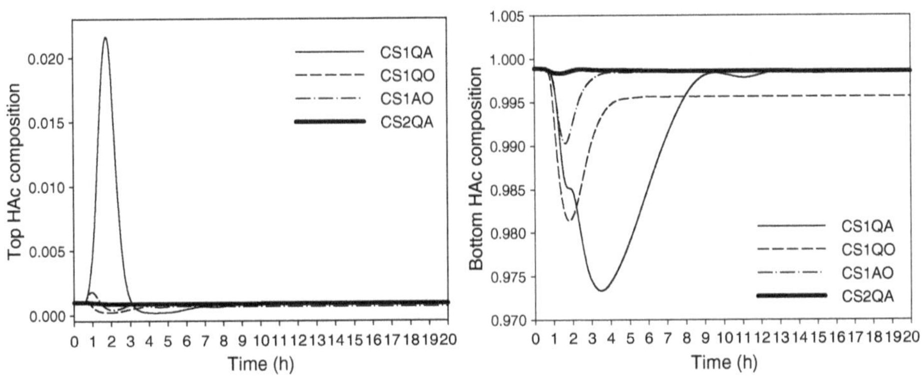


**Figure 9.16** Comparison under −10% feed water composition change.

select a control strategy not using an entrainer makeup flow as a manipulated variable. This entrainer makeup flowrate is fixed at the base-case value and ratioed to fresh feed rate changes.

Fixing this entrainer makeup flow at base-case value under various feed composition changes can still meet product purity specifications. This can be demonstrated by some steady-state simulation runs as given in Table 9.14. In that table, the desirable (optimum) entrainer makeup flowrate is at 0.16 mol/min for the 50% feed water composition. If feed water composition is changed to 40%, the desirable (optimum) entrainer makeup flowrate is 0.13 mol/min. This does not mean that the entrainer makeup flowrate has to be at this value to meet the two product purity specifications. As mentioned previously, there are multiple steady-state conditions that can meet both product purity specifications because there are three degrees of freedom. In fact, a steady-state run at 40% feed water composition can be made with the two product purities held constant by varying reboiler duty and aqueous reflux, with the entrainer makeup flowrate fixed at 0.16 mol/min. The resulting steady-state condition shows a slightly higher reboiler duty (176 KW versus 175 KW in Table 9.14) but still holding two product purity specifications. Notice that the impurity of the bottom product shifted to more isobutyl acetate and less water, but the total impurity

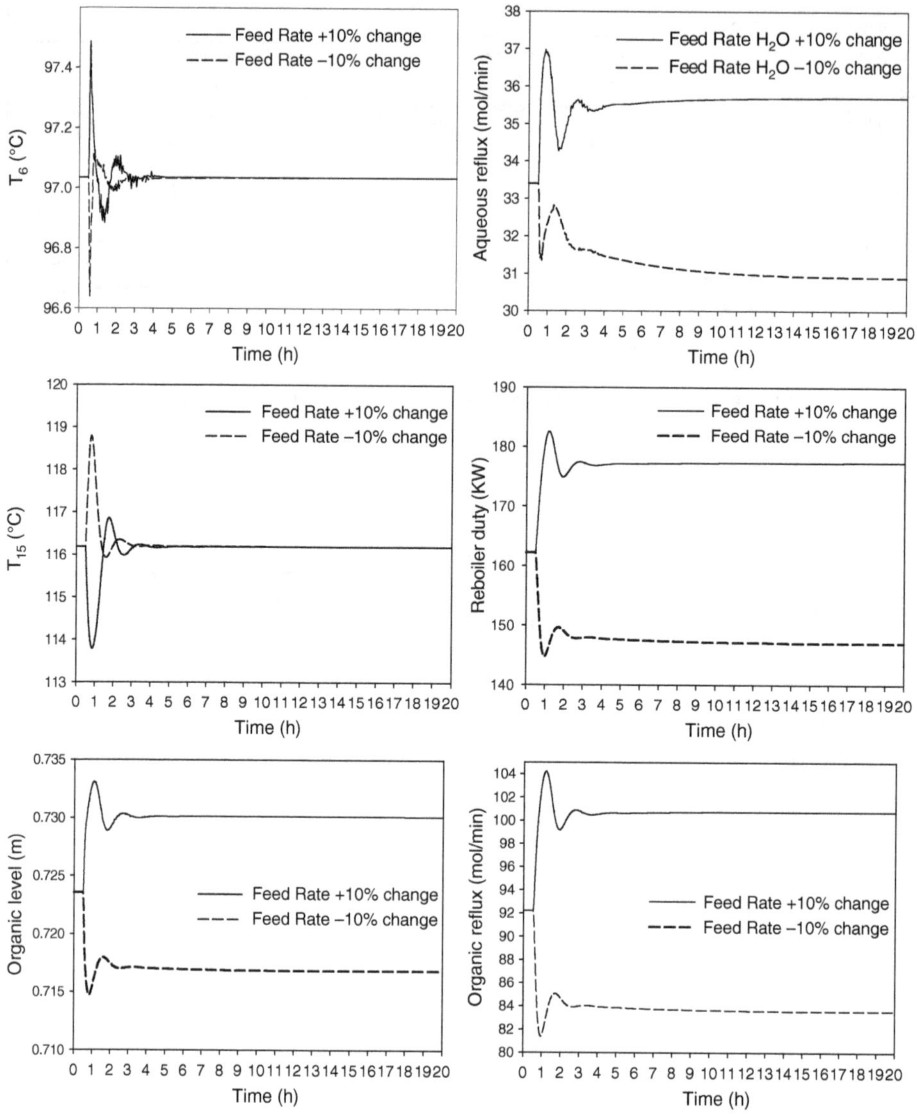


**Figure 9.17** CS2QA with ±10% changes in the feed flowrate.

(iBuAc + $H_2O$) is still at 0.1 mol% as desired. This demonstrates that the control structure with a fixed entrainer makeup flowrate is workable.

There is a dynamic advantage in having the extra degree of freedom of aqueous reflux. It can be used to improve the closed-loop dynamic response of the system. Since Figure 9.7 shows that aqueous reflux is only needed for water feed composition <70 mol%, the use of a preconcentrator column before the azeotropic column may improve dynamic behavior.

From the dynamic responses of CS2QA (Fig. 9.14), another important observation can be made. With disturbances like ±10% changes in the feed water composition, the aqueous reflux flowrate is adjusted upward or downward in order to maintain about the same overall water composition into the column. However, the reboiler duty actually returns back almost to its original steady-state value after some dynamic transients. This inspires the thinking that a simple single-temperature control strategy may work. Attempts to use a single-temperature control strategy are explored in the next section.

### 9.2.2 Single-Temperature Control Strategy

The idea of the simpler single-temperature loop-control strategy is to use an aqueous reflux flowrate to hold some tray temperature inside the column and to ratio the other two manipulated variables (reboiler duty and entrainer makeup) to the feed flowrate. From the earlier sensitivity analysis in Figure 9.9*a*, the most sensitive control point inside the column is Stage 6, which is close to the top of the column. Since the main acetic acid product is drawn out from the bottom of the column, an alternative control point is to select the second most sensitive control point, which is closer to the bottom of the column. This alternative control point will be Stage 16. Thus two single-loop-control structures will be evaluated using closed-loop dynamic tests. They are:

CS3: Using Inventory Strategy 2 in the previous subsection, with temperature at Stage 6 controlled by manipulating aqueous reflux flow, while the other two manipulated variables, reboiler duty and entrainer makeup, are ratioed to the feed flowrate.

CS4: Using Inventory Strategy 2 in the previous subsection, with temperature at Stage 16 controlled by manipulating aqueous reflux flow, while the other two manipulated variables, reboiler duty and entrainer makeup, are ratioed to the feed flowrate.

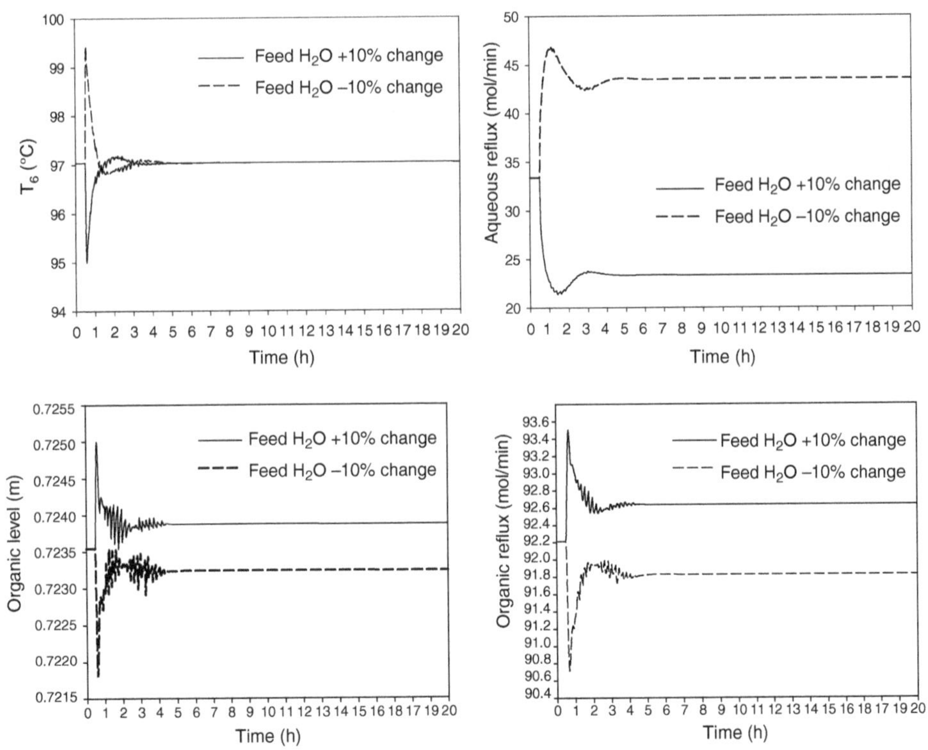


**Figure 9.18** CS3 with $\pm 10\%$ changes in the feed water composition.

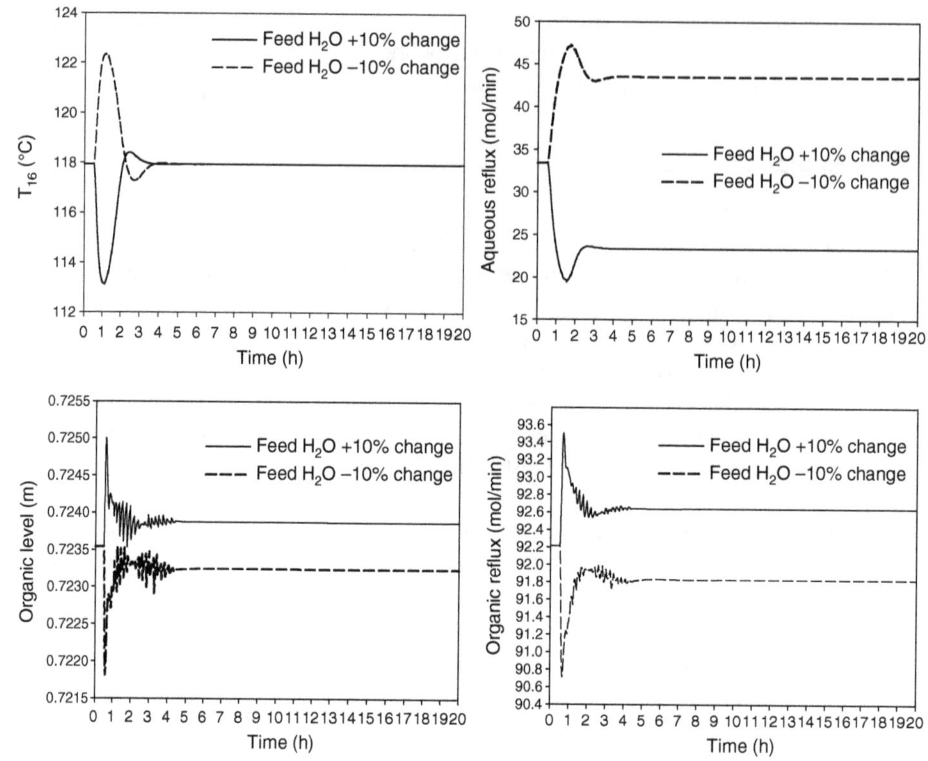


**Figure 9.19** CS4 with $\pm10\%$ changes in the feed water composition.

Figures 9.18 and 9.19 show the closed-loop dynamic responses for CS3 and CS4 for $\pm10\%$ feed water composition changes. Notice that the dynamic responses are all quite satisfactory with all variables settled out at new steady-state values even faster than CS2QA (comparing to Fig. 9.14). The dynamic responses of the important bottom and top product compositions when using CS3 or CS4 are compared with results using CS2QA in Figures 9.20 and 9.21 for $+10\%$ and $-10\%$ changes in the feed water composition. Notice first that the scales of Figures 9.20 and 9.21 are much smaller than those in previous

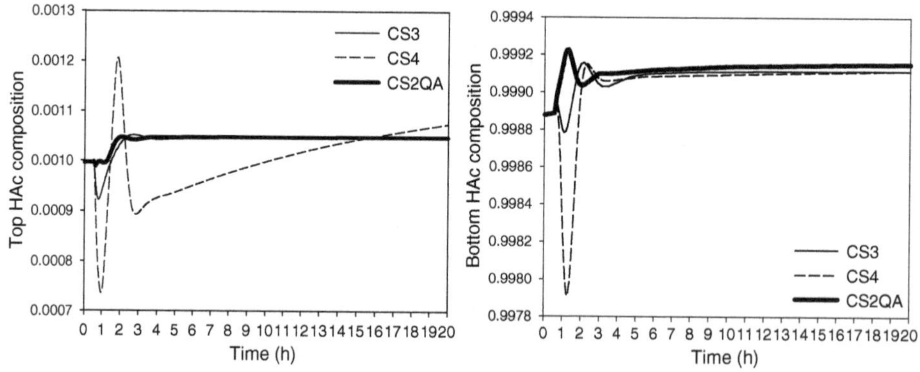


**Figure 9.20** Comparison under $+10\%$ feed water composition change.

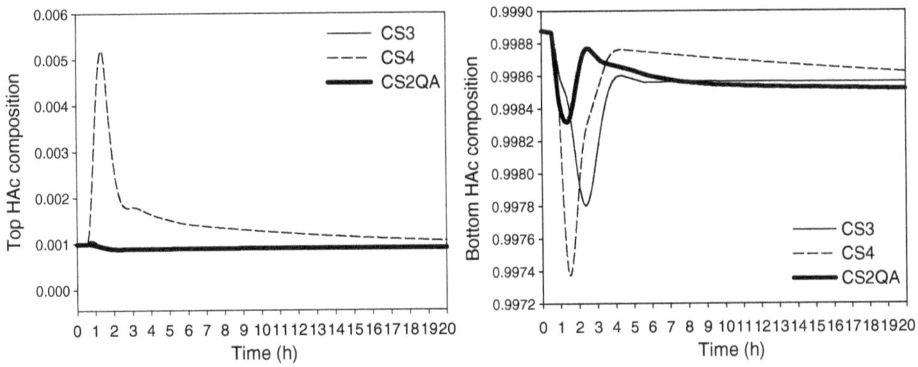


**Figure 9.21** Comparison under −10% feed water composition change.

Figures 9.15 and 9.16 indicating these control structures perform much better than CS1QA, CS1QO, and CS1AO.

Compared to the more complex two temperature control structure CS2QA, the single-temperature control structure CS3 performs very satisfactorily. Both product compositions are maintained at tight specifications even though the temperature control point is far away from the column bottom. However, control structure CS4 is not able to keep the acetic acid loss out of the column top close to the desired level.

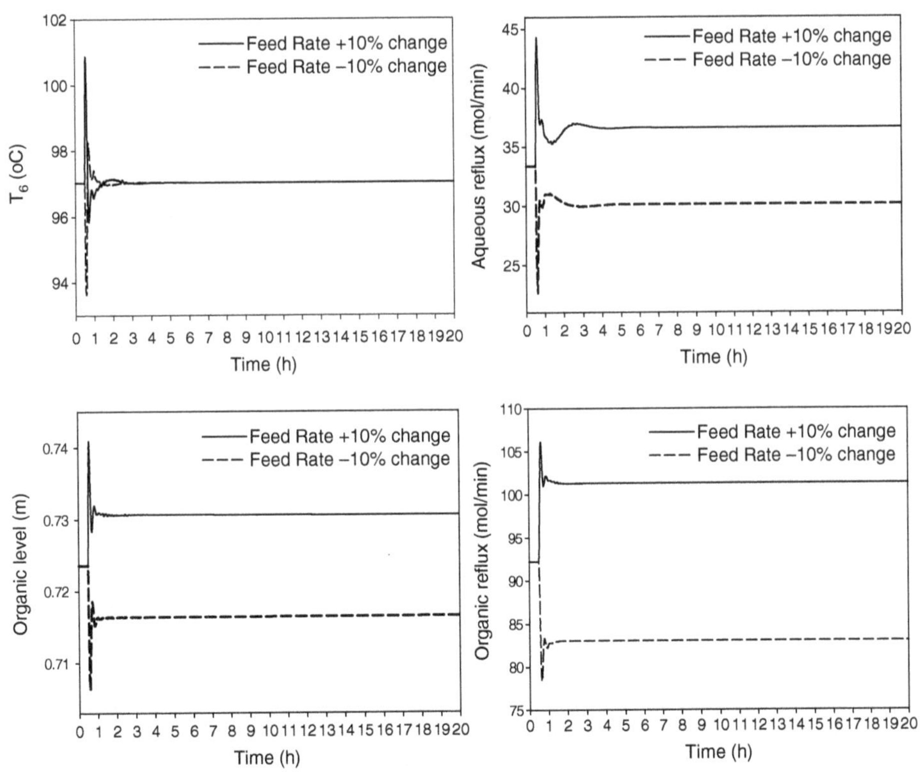


**Figure 9.22** CS3 with ±10% changes in the feed flowrate.

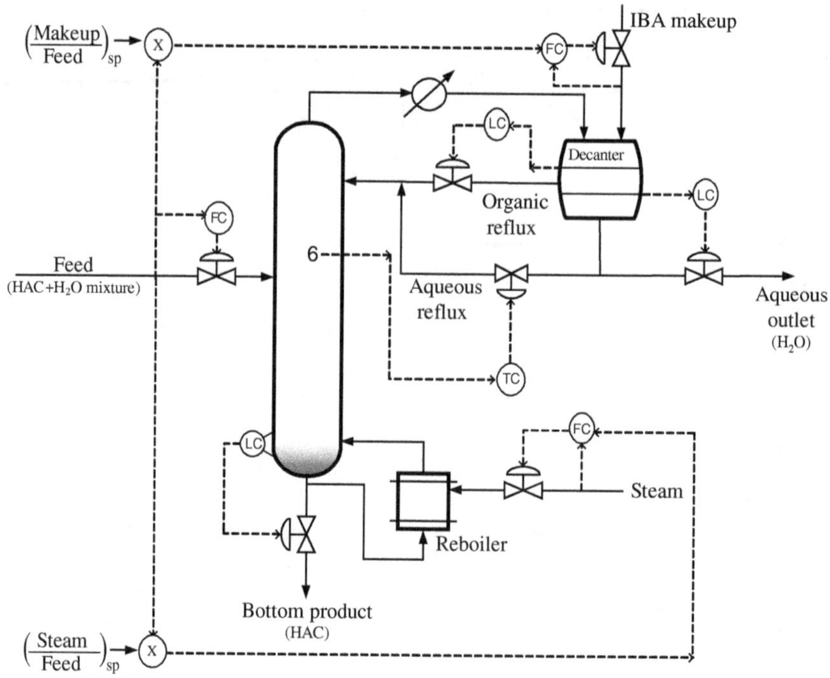


**Figure 9.23** Schematic diagram of the proposed control structure CS3.

The proposed control structure CS3 also performs very well for $\pm10\%$ feed rate disturbances. The dynamic responses for the temperature loop and the organic level loop are shown in Figure 9.22. All controlled and manipulated variables reach new steady-state values after a short dynamic transient. Although not shown here, both product compositions are also maintained very close to their specifications.

The final proposed simple single-temperature loop-control structure of CS3 is shown in Figure 9.23. The two product purities are assumed not to be measured online, but if they can be measured infrequently off-line, small trimmings of the controlled temperature setpoint or small changes of the reboiler heat duty can be made even more precisely to hold the product purities at their specifications during sustained feed disturbances.

## 9.3 INDUSTRIAL COLUMN WITH PRELIMINARY DYNAMIC SIMULATIONS

In this section, design and operation of an industrial column for acetic acid dehydration via heterogeneous azeotropic distillation is investigated. The entrainer used for this industrial column to aid the acetic acid and water separation is also isobutyl acetate. This entrainer is circulating inside the column through OR stream from a decanter. Multiple column feed streams from various parts of the upstream process are fed into this column. The feed components besides acetic acid and water also include small amount of methyl acetate and m-xylene. These components are intermediate boilers and tend to accumulate inside the column. They cannot leave the column system through either the top decanter AO stream or

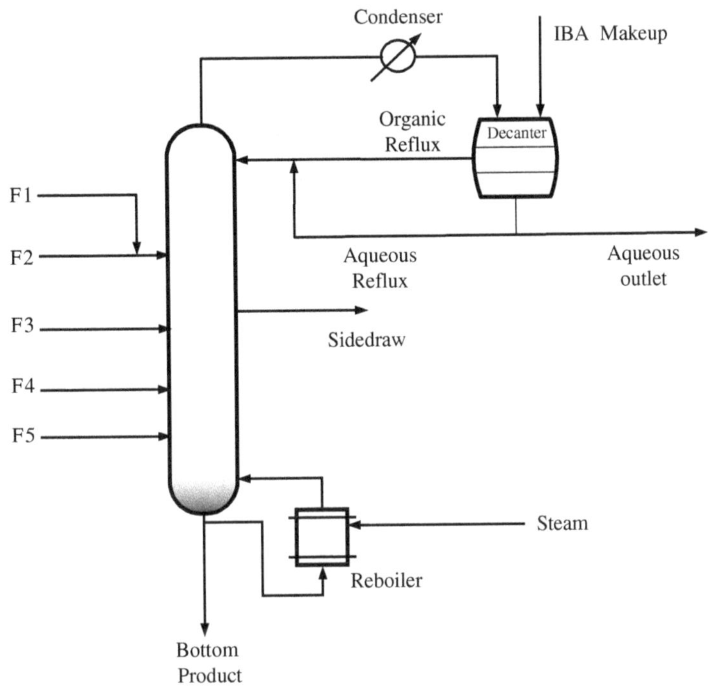


**Figure 9.24** Acetic acid dehydration column of an industrial unit.

the column bottom stream. Therefore, the proper design and operation of this system needs to be investigated to prevent the accumulation.

### 9.3.1 Column Simulation

The layout of the studied industrial column can be seen in Figure 9.24. There are five feed streams into a heterogeneous azeotropic distillation column. These five feed streams are all coming from the upstream process. The feed flowrates and their compositions are listed in Table 9.17. From this table, one can observe that there are four components in the feed streams including acetic acid, water, small amounts of methyl acetate, and m-xylene. The top vapor of the heterogeneous distillation column, after condensing to a temperature of 42.5°C, enters a decanter to separate into two liquid phases. The organic phase containing mostly the entrainer isobutyl acetate is recycled back to the column. The aqueous phase

**TABLE 9.17 Feed Stream Information of an Industrial Column.**

| Stream Number | F1 | F2 | F3 | F4 | F5 |
|---|---|---|---|---|---|
| Flowrate (kg mol/h) | 23.11 | 11.40 | 78.61 | 63.09 | 41.05 |
| Acetic acid composition (mol%) | 7.93 | 0.59 | 43.56 | 80.68 | 51.32 |
| Water composition (mol%) | 92.07 | 99.41 | 55.51 | 18.08 | 47.94 |
| Methyl acetate composition (mol%) | $5.84 \times 10^{-4}$ | 0.0 | 0.64 | 1.24 | 0.74 |
| Isobutyl acetate composition (mol%) | 0.0 | 0.0 | 0.0 | 0.0 | 0.0 |
| m-Xylene composition (mol%) | 0.0 | 0.0 | 0.29 | 0.0 | 0.0 |

partly is recycle back to the column acting as reflux stream and partly is withdrawn from the system as the AO stream. A small amount of isobutyl acetate makeup enters the decanter to balance the small loss of isobutyl acetate through the AO stream and the column bottoms stream. A possible side stream is also included in Figure 9.24 to remove m-xylene from the system. The proper design of the side stream location and its flowrate will be investigated in the next section. In this section, the accumulation of this m-xylene component within the column will be demonstrated via rigorous dynamic simulation of this industrial column.

Aspen Plus and Aspen Dynamics will be used for the rigorous simulation. Several preliminary runs were made to fit the plant data from the column outlet streams. In the simulation, the side stream is left out first so that accumulation of the m-xylene component can be demonstrated. A steady-state flowsheet is obtained first with the same feed condition as in Table 9.17, but without the small m-xylene component in the column through feed stream F3. The two specifications of this flowsheet are set to be a bottoms acetic acid purity of 99.5 wt% and an AO stream impurity of 0.1 wt% acetic acid. The isobutyl acetate makeup flowrate is set to be the same as in the plant (0.123 kg mol/h or 14.3 kg/h).

In the simulation, the NRTL activity coefficient model was used for the VLLE of this system. The Hayden–O'Connell[17] second virial coefficient model with association parameters was used to account for the dimerization of acetic acid in the vapor phase. The set of NRTL parameters for acetic acid, water, and isobutyl acetate is given in Table 9.3. Other NRTL parameters are obtained mostly from the Aspen Plus data bank. Four pairs

**TABLE 9.18 Base-Case Condition of the No Feed Impurity Case.**

| | |
|---|---|
| Total number of stages for the azeotropic column | 21 (including reboiler) |
| Feed stages | F1 at fifth stage (counting from the top stage)<br>F2 at fifth stage<br>F3 at tenth stage<br>F4 at thirteenth stage<br>F5 at fourteenth stage |
| Feed flowrate and feed composition | Same as in Table 9.17 |
| Entrainer makeup flowrate (kg mol/h) | 0.1231 (14.3 kg/h) |
| Reboiler duty (KW) | 2855 |
| Organic reflux flowrate (kg mol/h) | 128.9 |
| Aqueous reflux flowrate (kg mol/h) | 11.79 |
| Bottom flowrate (kg mol/h) | 109.9 |
| Bottom composition | HAc: 98.39 mol% (99.5 wt%)<br>$H_2O$: 1.61 mol%<br>MA: $2.27 \times 10^{-5}$ mol%<br>IBuAc: $6.49 \times 10^{-3}$ mol%<br>MX: 0.0 mol% |
| Aqueous outlet flowrate (kg mol/h) | 107.3 |
| Top composition | HAc: $3.16 \times 10^{-2}$ mol% (0.1 wt%)<br>$H_2O$: 98.38 mol%<br>MA: 1.48 mol%<br>IBuAc: $1.08 \times 10^{-1}$ mol%<br>MX: 0.0 mol% |

of NRTL parameters are estimated using UNIFAC (Dortmund modified[28]) group contribution method because no experimental data can be found in the open literature or from the Aspen Plus data bank. These pairs are isobutyl acetate–methyl acetate, isobutyl acetate–m-xylene, acetic acid–m-xylene, and methyl acetate–m-xylene. All other physical properties, such as vapor pressure, enthalpies, and densities of liquid and vapor, are obtained from the Aspen Plus data bank. The steady-state simulation results of each stream are listed in Table 9.18.

### 9.3.2 Control Strategy Used in the Dynamic Simulation

In order to dynamically show the influence of adding a small flowrate of the m-xylene component into feed stream F3 to the overall column performance, the Aspen Plus simulation was converted to a dynamic simulation using Aspen Dynamics. The tray sizing option in Aspen Plus is utilized to calculate the column diameter to be 1.58 m and the tray spacing is 0.6096 m. Other equipment sizing recommended by Luyben[23] is used. The volume of the reboiler is sized to give 10 min holdup with 50% liquid level. The decanter is sized to be bigger to allow for two liquid phases to separate. A holdup time of 20 min is used in the dynamic simulation. A pressure-driven simulation in Aspen Dynamics is used with the top pressure of the azeotropic column controlled at 1.1 atm to allow for some pressure drop in the condenser and decanter so that the decanter can be operated at atmospheric pressure. The pressure drop inside the column is automatically calculated in Aspen Dynamics.

The overall control strategy is adapted from the previous section when there were only three main components in the system. No online composition analyzer is needed in this recommended control strategy. This overall control strategy is very simple, requiring only one tray temperature control loop to hold the product specifications by manipulating the aqueous reflux flowrate. The inventory control loops are using OR flow to control the organic phase level in the decanter, using aqueous product flow to control the aqueous phase level in the decanter, and using bottom acetic acid product flow to control the column bottom

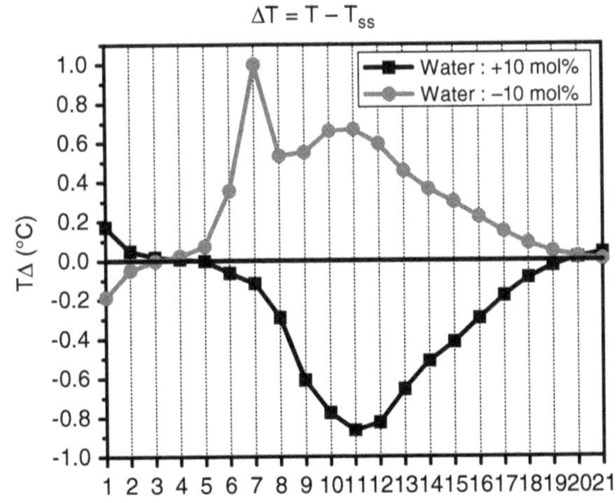


**Figure 9.25** Closed-loop sensitivity plot for the system without m-xylene impurity.

level. The column top pressure is controlled at 1.1 atm by manipulating the top vapor flow, and the decanter temperature is controlled at 42.5°C by manipulating the condenser duty.

The determination of the temperature control point inside the column is from a combination of closed-loop sensitivity analysis and verification by open-loop sensitivity analysis. For the closed-loop sensitivity analysis, closed-loop simulation runs are made in which both the top and bottom product purities are held at their specifications for different feed compositions. This is to mimic ideal (although not practical) situation with two online composition measurements and two composition control loops setting aqueous reflux flow and reboiler duty. The tray temperature with the least amount of variation for changes in feed composition is selected as the temperature control point. The specific feed composition changes are feed F3 water molar composition $\pm 10\%$ changes while keeping the total molar flowrate of F3 the same by adjusting the acetic acid molar composition.

Figure 9.25 shows the deviation of temperature under this ideal control situation. Temperature at the fourth stage is selected as the control point, since it shows the least deviation from base-case condition. Therefore, by holding the temperature at this control point, both bottom and top product purities should be maintained near their specifications. The other two "free" manipulated variables (reboiler heat duty and isobutyl acetate makeup flow) are fixed and only adjusted according to the total molar feed flowrate using a ratio scheme. Notice that the selected temperature control point should be verified by open-loop sensitivity analysis similar to those shown in Figure 9.9, so that enough open-loop sensitivity between the manipulated variable to the controller variable is confirmed.

The overall control strategy adapted in the previous section is shown in Figure 9.26. This simple overall control strategy was demonstrated to perform very well in spite of feed flowrate and feed composition changes. A dynamic simulation will be shown next to illustrate the

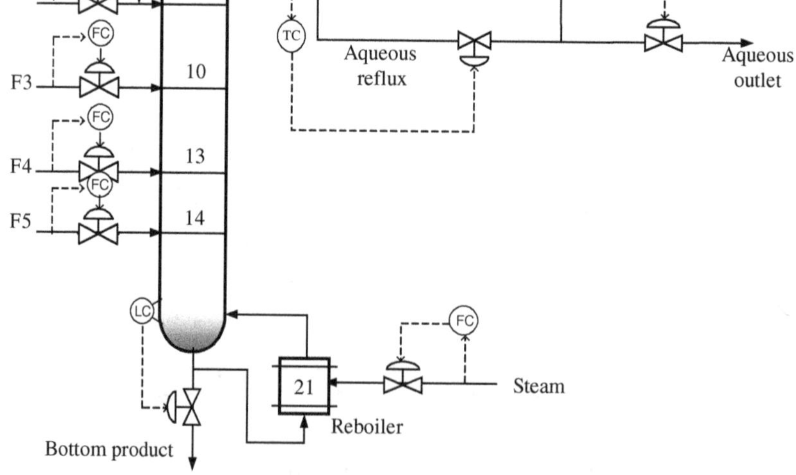


**Figure 9.26** Recommended overall control strategy for the system without m-xylene impurity.

accumulation of the m-xylene component inside the column with the overall control strategy in operation. In the dynamic simulation, all three level loops are assumed to be controlled by a P-only controller in order to smooth out their manipulated variables in the system. A controller gain of 2.0, as suggested in Luyben,[23] is used in the aqueous level and bottom level loops. Higher controller gain of 10.0 is used in the organic level loop because the faster dynamic response of adjusting OR flow into the system is desirable. The PID tuning constants for the stage temperature control loops are tuned using the IMC–PID tuning guideline by Chien and Fruehauf[29] with the initial dynamic open-loop response modeled as an integrating system plus deadtime. The same tuning guideline will be used in all the closed-loop simulation runs for fair comparison.

### 9.3.3 Dynamic Simulation Results

Before showing the run with impurity accumulation inside the column, the normal feed composition load disturbance changes are shown. Changes in the feed F3 water molar composition of ±10% are made at time = 5 h. When feed F3 water molar composition is changed, the acetic acid molar composition is adjusted so that the total molar feed rate of F3 is maintained constant.

Figure 9.27 shows the closed-loop dynamic simulation with these feed composition load disturbances. From the figure, the fourth stage temperature is controlled back to setpoint during load changes with smooth manipulation of the aqueous reflux flowrate. More importantly, the two product purities (in the bottom half of Fig. 9.27) are maintained quite close

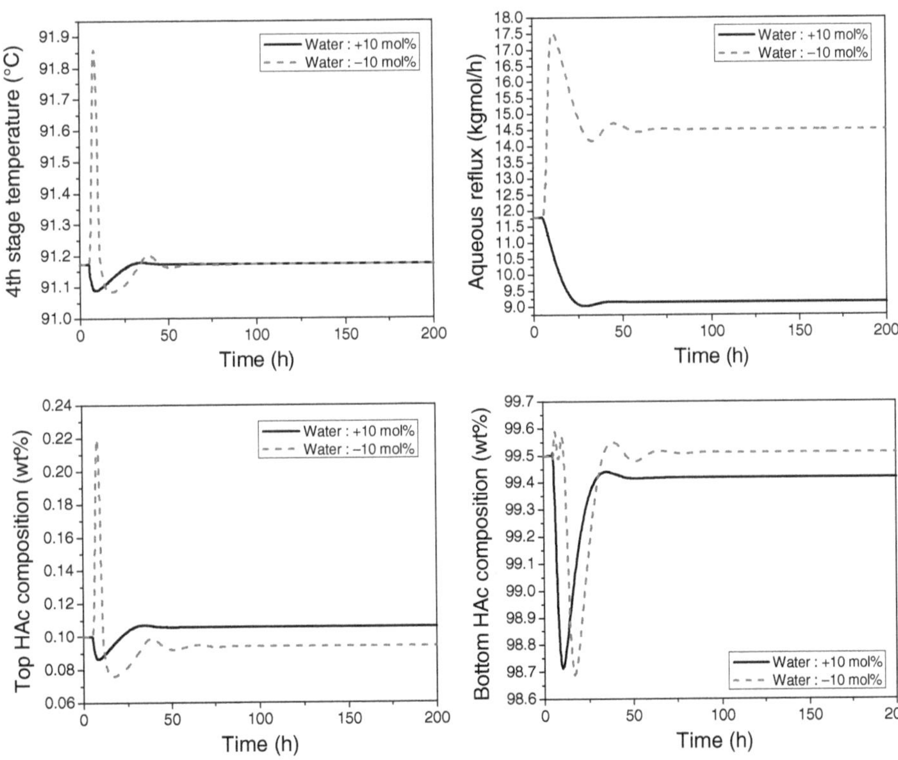


**Figure 9.27** Closed-loop test with ±10% feed F3 water composition change.

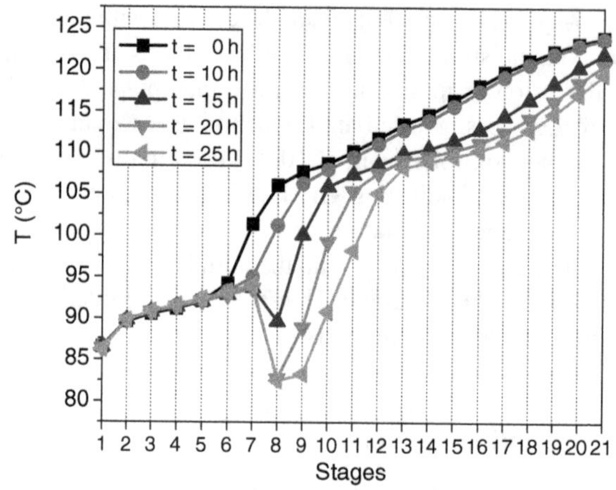


**Figure 9.28** Dynamic response of the temperature profile with impurity added at time = 5 h.

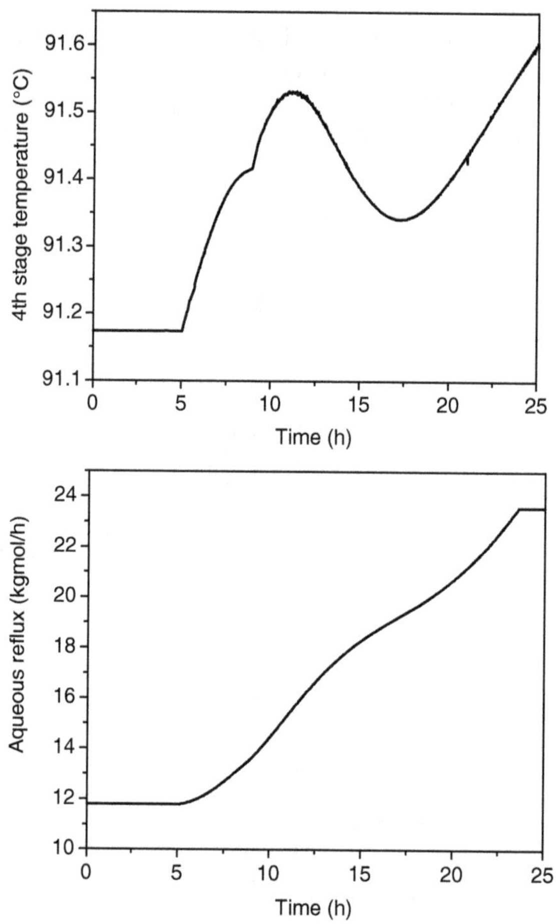


**Figure 9.29** Closed-loop response of the controlled temperature loop.

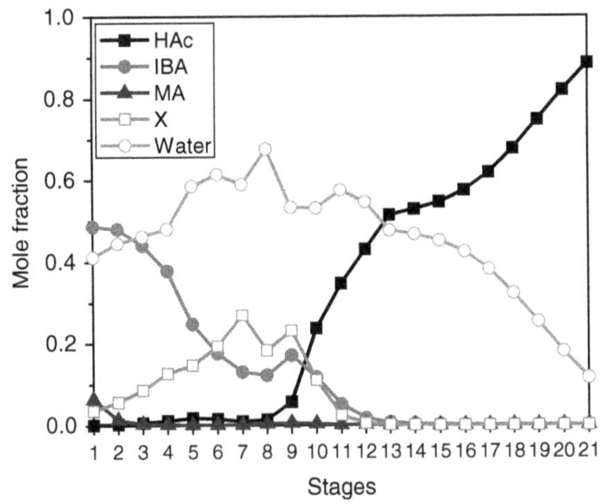


**Figure 9.30** Column composition profile at time = 25 h.

to their high-purity specifications of 99.5 wt% acetic acid in the column bottoms and 0.1 wt% acetic acid in the AO stream.

With this proper control strategy in operation, the next dynamic simulation run will put a small amount of m-xylene (24.29 kg/h, 0.23 kg mol/h) component into feed stream F3 at time = 5 h. Figure 9.28 shows the shifting in the column temperature profile at 10 h, 15 h, 20 h, and 25 h of closed-loop operation. The controlled and manipulated variables of the fourth stage temperature control loop are shown in Figure 9.29. Notice that the aqueous reflux flowrate continuously goes up until the control valve is fully opened at time = 23 h. The fourth stage temperature controller is not able to maintain its setpoint value.

The column composition profile of all components at time = 25 h is shown in Figure 9.30. From the figure, the greatest accumulation of the m-xylene component occurs at the seventh stage. The dynamic accumulation of the m-xylene component at seventh stage is shown in Figure 9.31.

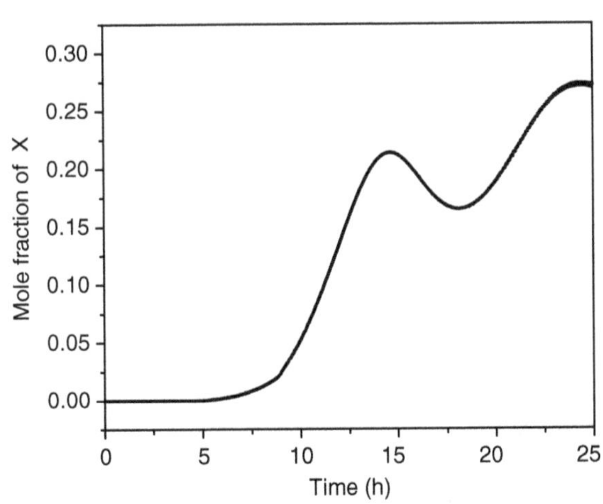


**Figure 9.31** Dynamic accumulation of impurity m-xylene at the seventh stage.

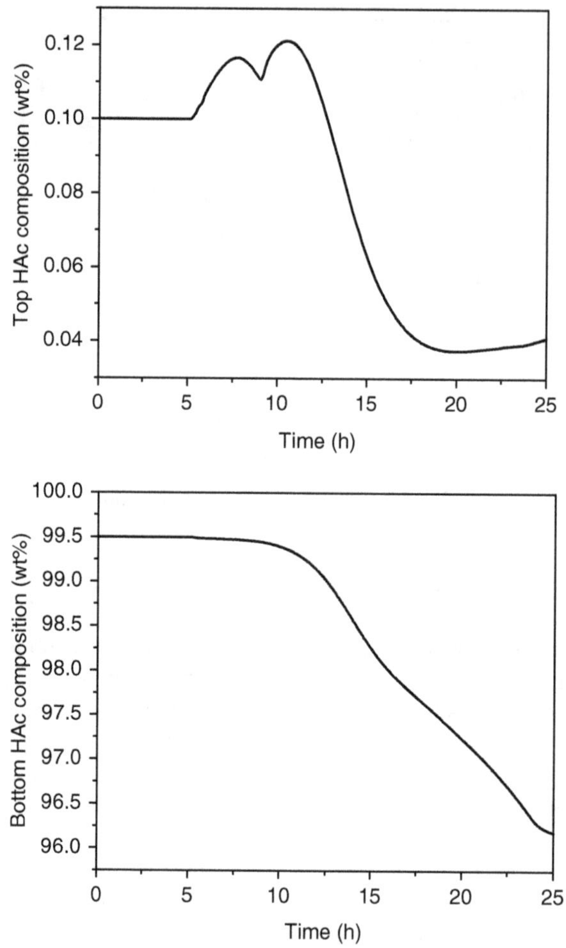


**Figure 9.32** Dynamic response of top and bottom product compositions.

The top and bottom product specifications cannot be met with this small addition of this component into the column from feed F3. The dynamic responses of the top and bottom acetic acid compositions can be seen in Figure 9.32. Notice that the top acetic acid composition is purer than the specification of 0.1 wt% acetic acid loss. However, the important bottoms acetic acid composition continuously drops from the specification of 99.5 wt%. At time = 25 h, the bottom product acetic acid composition had dropped to an unacceptable value of 96.3 wt%. Because the m-xylene component will be accumulated inside the column, a side stream is necessary to take out this feed impurity.

## 9.4 INDUSTRIAL COLUMN WITH CONTINUOUS SIDE-STREAM DRAW OFF

The previous section demonstrated that the operation of the acetic acid dehydration column encountered problems when the feed stream contains small amounts of m-xylene or p-xylene impurity. In this section, a side stream will be included in this dehydration column in order to avoid the operational problem.

### 9.4.1 Optimal Side-Stream Location and Flowrate

The optimal side-stream location and its flowrate will be investigated in this section. Aspen Plus steady-state simulation will be used in this study. The simulated flowsheet is similar to Figure 9.26 but with two differences. The first one is that the feed stream F3 will have 24.29 kg/h of component m-xylene, as was used in the dynamic simulation previously shown. The second difference is that a side stream will be added to the flowsheet of Figure 9.26. The two product specifications are still the same as in the last section. They are bottoms acetic acid purity of 99.5 wt% and AO stream impurity of 0.1 wt% acetic acid. The isobutyl acetate makeup flowrate is again set to be the same as in the plant (14.3 kg/h). Exhaustive simulation runs are made to satisfy the above two product specifications (by manipulating the reboiler duty and aqueous reflux) while varying the side-stream location and flowrate. The TAC is used as the objective function to be minimized. The calculation procedure of Douglas[19] is followed with the annual capital charge factor of 1/3 used. The utility costs include the steam and cooling water for the operation of reboiler and condenser. The total operating costs include the utility costs, the isobutyl acetate makeup stream cost, and also an additional term for the cost of side-draw acetic acid loss.

Figure 9.33 shows the summary of many simulation runs with varying side-stream locations and its flowrate. The two product specifications cannot be met when the side-draw flowrate is below 30 kg/h because most of the impurity m-xylene has to come out in the side stream. The minimum TAC is achieved for a side-stream location of Stage 10, counting from the top stage, and a flowrate of 50 kg/h. Notice that the optimum side-stream location is right at the F3 feed location where the impurity m-xylene enters the column. When the side-stream location is fixed at Stage 10, smaller purge flows result in more reboiler duty and aqueous reflux to hold the product specifications. Consequently, the energy cost will be higher. Furthermore, with more vapor rate inside the column, the column diameter needs to be larger, causing the capital cost to be higher also. However, because the purge

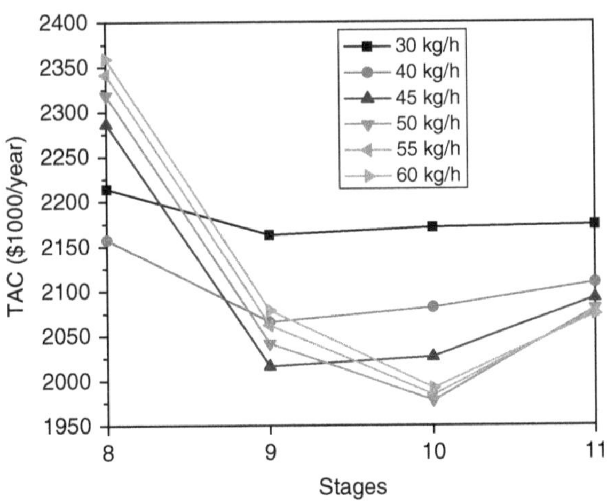


**Figure 9.33** TAC plot to determine the side stream location and its flowrate.

rate is smaller, the acetic acid loss through the side-draw is lower. This optimum side-stream purge rate of 50 kg/h is a trade-off among all of the above-mentioned costs.

### 9.4.2 Closed-Loop Dynamic Simulations with Load Disturbances

In this section, two load disturbances are introduced to test the proposed column configuration. The overall control strategy, shown in Figure 9.26, is also implemented here. The only difference is a side stream at Stage 10 with the flowrate controlled at 50 kg/h. The two load disturbances considered are feed F3 water molar composition ±10% changes and feed F3 m-xylene component flowrate ±20% changes.

Closed-loop sensitivity analysis is again used to choose the proper tray temperature as the control point. Runs are made holding the two product purities constant for the two above-mentioned load disturbances. Similar to Figure 9.25, Figure 9.34 shows the deviation of each tray temperature under the ideal control situation when holding two product purities.

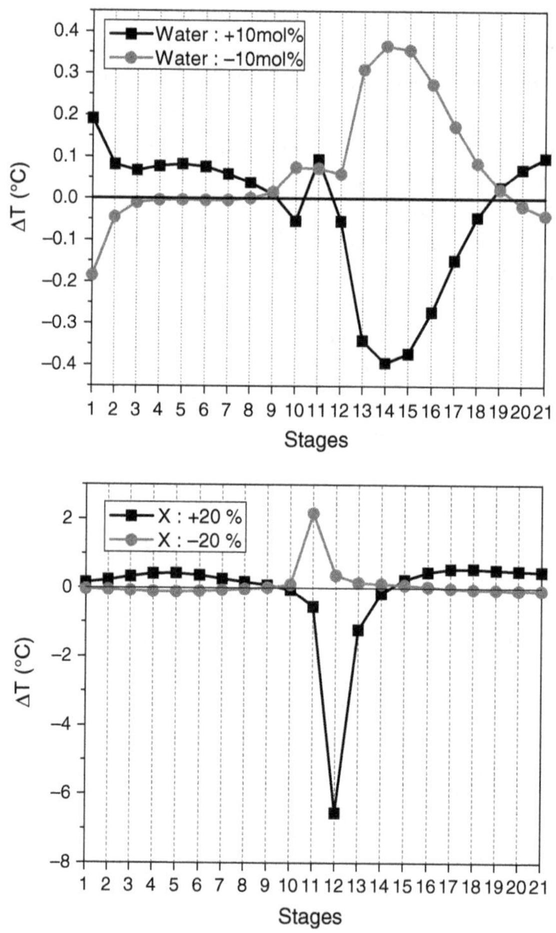


**Figure 9.34** Closed-loop sensitivity plot for the system with continuous side stream drawn off.

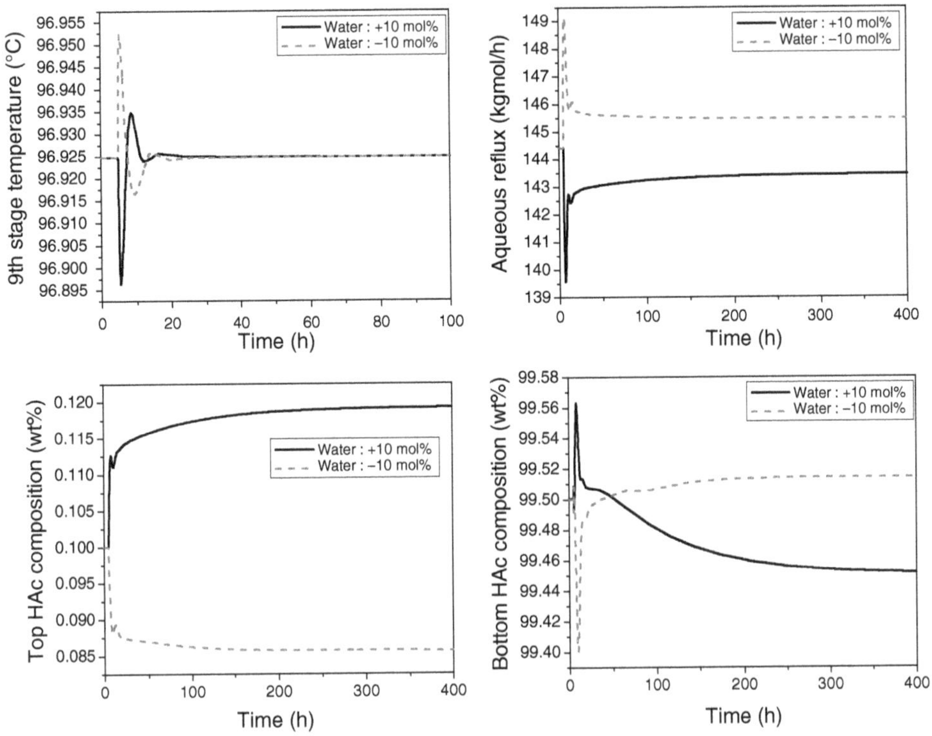


**Figure 9.35** Side-draw system with ±10% feed F3 water composition change.

The temperature at Stage 9 is selected as the control point since it shows the least deviation from base-case conditions.

Figure 9.35 shows the closed-loop simulation runs with ±10% changes in the feed F3 water molar composition. Similar to the result in Figure 9.27, this overall control strategy is able to maintain bottoms product purity very close to its 99.5 wt% acetic acid specification. The acetic acid loss in the top aqueous product is a little more for the +10% water composition load change. The acetic acid composition in the top aqueous product goes up to 0.118 wt% for this load change.

Figure 9.36 shows the closed-loop simulation results with ±20% changes in the m-xylene flowrate in feed F3. The closed-loop response is not as good as in Figure 9.35 with the load disturbance when feed water composition changes. This small impurity is coming from an upstream extraction process. Variations of ±20% in the feed F3 impurity (m-xylene) component flowrate, although unlikely, are still possible. With these small F3 impurity changes, the product purities of both top and bottom products are further away from their specifications in comparison with the feed water disturbances. For example, with small +20% changes in feed F3 impurity component flowrate (from 24.29 kg/h to 29.15 kg/h), the main bottom purity goes down to 99.33 wt% acetic acid and top aqueous purity goes up to 0.134 wt% acetic acid.

Since there are two other potential manipulated variables (reboiler heat duty and side-draw flowrate) for product quality control, a more complicated two-point or even

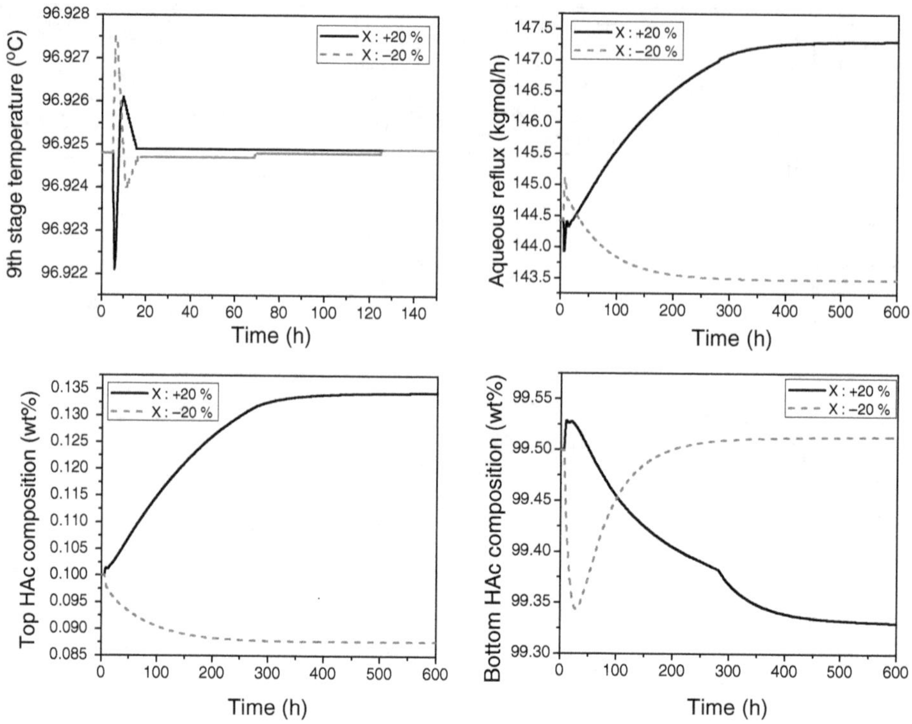


**Figure 9.36** Side-draw system with $\pm 20\%$ changes in the impurity flowrate (side stream flowrate is set at 50 kg/h).

three-point temperature-control strategy can be considered. However, in this study, a more practical and more simple control strategy, which can easily be used in industry, is considered. Thus, in the next section, a more conservative base case with a higher side-draw flowrate is selected to demonstrate its capability to better handle variations in feed impurity.

### 9.4.3 Higher Side-Stream Flowrate

Since, in the last section, the base-case operating condition that optimizes TAC is not able to handle properly the impurity disturbance changes in the $\pm 20\%$ range, the side-stream flowrate is set to be higher than the optimum case of 50 kg/h in order to handle these disturbances better in the feed F3. A side-stream flowrate of 60 kg/h is used here for illustration purposes. Notice that when the side-stream flowrate increases from 50 kg/h to 60 kg/h, the TAC only increases by 0.66% (see Fig. 9.33).

Figure 9.37 shows the closed-loop simulation with $\pm 20\%$ changes in the feed F3 impurity (m-xylene) component flowrate at this new base-case condition. Notice that for the $+20\%$ case, the bottom product purity returns to the product specification of 99.5 wt% acetic acid and the top aqueous product purity is also much better than the one shown in Figure 9.36 (0.113 wt% in this figure versus 0.134 wt% in Fig. 9.36). With the alternative selection of the base-case operating condition only slightly different to the optimum TAC case, the capability of handling the feed impurity variations is improved.

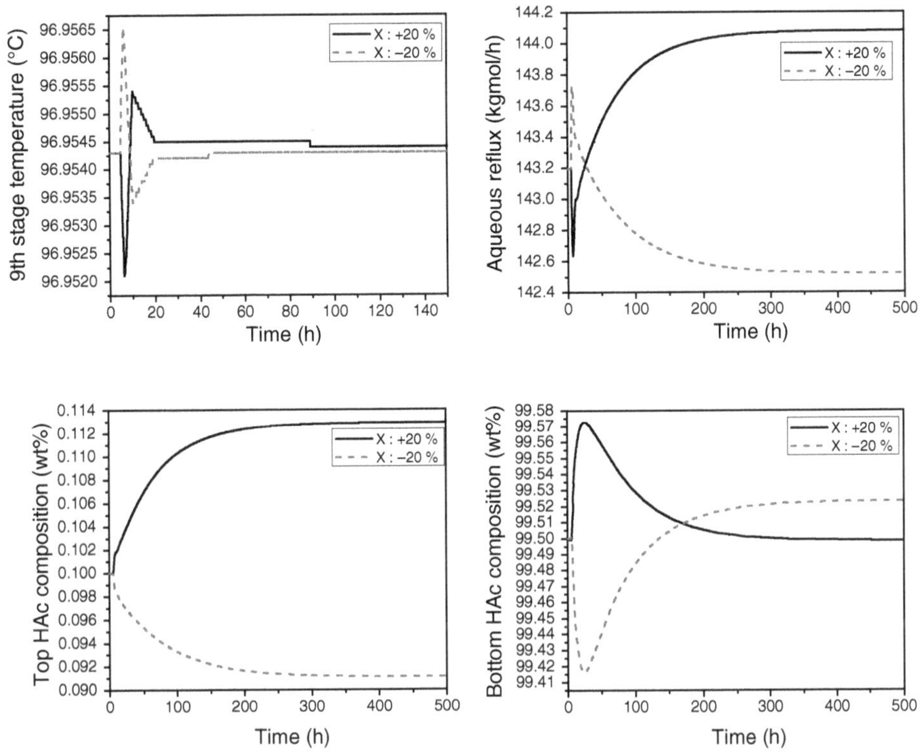


**Figure 9.37** Side-draw system with ±20% changes in the impurity flowrate (side stream flowrate is set at 60 kg/h).

### 9.4.4 Comparison with the Base Case

The base-case condition in Section 9.3 with no side-draw and no feed impurity is compared with the ones in this section. Figure 9.38 displays the column composition profiles for the base case in Section 9.3 and the optimum TAC case in this section with a side-draw flowrate of 50 kg/h.

With the same specifications of 99.5 wt% acetic acid in the bottom product and 0.1 wt% acetic acid in the top aqueous product, the acetic acid and water composition profiles of the two base cases are similar. The major difference is the isobutyl acetate composition profiles inside the column. The isobutyl acetate composition inside the top parts of the column is lower than the case with no side-draw and no feed impurity. Also, even with a very small amount of impurity in the feed, the impurity composition inside the column is quite high.

Table 9.19 compares the operating conditions of the three industrial base cases studied in this chapter. Notice that, with an addition of a tiny amount of impurity in the feed, the operating conditions are drastically changed. The reboiler heat duty is more than doubled for the two base cases with side-draw in comparison with the one without side-draw and feed impurity. The OR flowrate and the aqueous reflux flowrate are also increased considerably. Due to the difference of the vapor traffic inside the column, the column diameters will be quite different too.

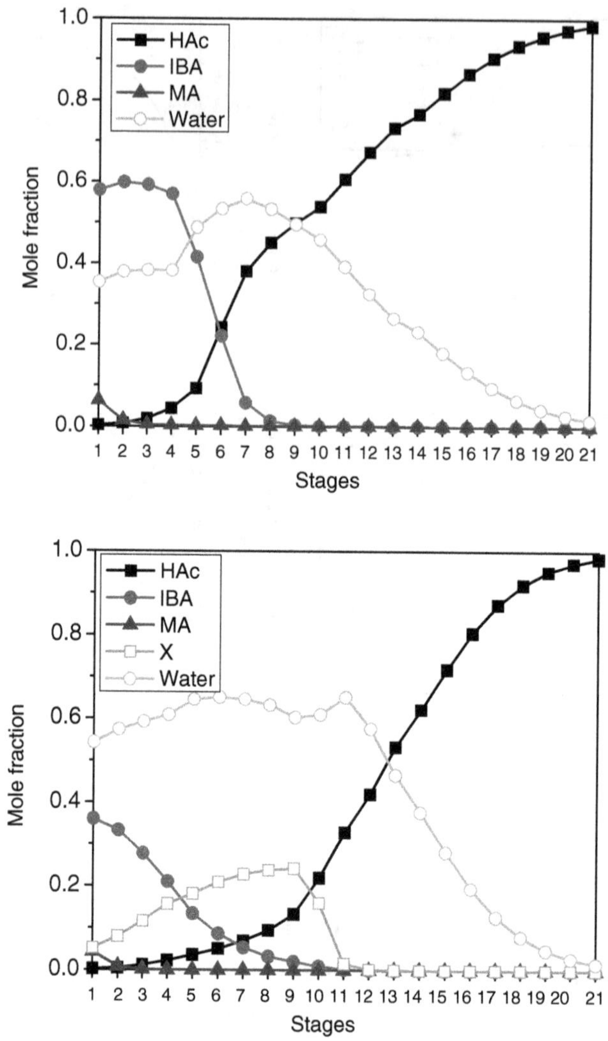


**Figure 9.38** Compare of column composition profiles (with or without impurity).

From the above information, one can calculate the TAC and operating cost saving if, somehow, this impurity can be designed to not come into this column from the upstream extraction process. With no feed impurity, the TAC and operating cost saving are calculated to be 28% and 24%, respectively, of the optimum TAC case with feed impurity and side stream. This shows that a large incentive exists for the development of some innovative method in the upstream process to get rid off this impurity. Possible techniques include permeation, adsorption, and reaction. Since the alternative method of not letting this impurity enter this column is beyond the scope of this study, we will not discuss this method any further. If this impurity is inevitably to be present in the feed stream, the side stream is then needed with a higher reboiler heat duty and higher organic and aqueous reflux flowrates.

**TABLE 9.19 Comparison of Base-Case Operating Conditions in Sections 9.3 and 9.4.**

| | Base Case in Section 9.3 | Base Case in Section 9.4 (Side-Draw with 50 kg/h) | Base Case in Section 9.4 (Side-Draw with 60 kg/h) |
|---|---|---|---|
| Column diameter (m) | 1.58 | 2.31 | 2.31 |
| Entrainer makeup flowrate (kg mol/h) | 0.1231 | 0.1231 | 0.1231 |
| Reboiler duty (KW) | 2855 | 5914 | 5885 |
| Organic reflux flowrate (kg mol/h) | 128.9 | 248.5 | 247.8 |
| Aqueous reflux flowrate (kg mol/h) | 11.79 | 144.4 | 143.2 |
| Side-draw flowrate (kg mol/h) | 0.0 | 1.09 (50 kg/h) | 1.41 (60 kg/h) |
| Side-draw composition | – | HAc: 20.08 mol%<br>$H_2O$: 58.11 mol%<br>MA: 0.06 mol%<br>IBuAc: 1.13 mol%<br>MX: 20.62 mol% | HAc: 22.04 mol%<br>$H_2O$: 60.96 mol%<br>MA: 0.05 mol%<br>IBuAc: 0.98 mol%<br>MX: 15.97 mol% |
| Bottom flowrate (kg mol/h) | 109.9 | 109.6 | 109.5 |
| Bottom composition | HAc: 98.39 mol% (99.5 wt%)<br>$H_2O$: 1.61 mol%<br>MA: $2.27 \times 10^{-5}$ mol%<br>IBuAc: $6.49 \times 10^{-3}$ mol%<br>MX: 0.0 mol% | HAc: 98.39 mol% (99.5 wt%)<br>$H_2O$: 1.61 mol%<br>MA: $3.38 \times 10^{-6}$ mol%<br>IBuAc: $5.24 \times 10^{-3}$ mol%<br>MX: $2.81 \times 10^{-3}$ mol% | HAc: 98.39 mol% (99.5 wt%)<br>$H_2O$: 1.61 mol%<br>MA: $3.38 \times 10^{-6}$ mol%<br>IBuAc: $3.98 \times 10^{-3}$ mol%<br>MX: $1.85 \times 10^{-3}$ mol% |
| Aqueous outlet flowrate (kg mol/h) | 107.3 | 106.7 | 106.4 |
| Top composition | HAc: $3.16 \times 10^{-2}$ mol% (0.1 wt%)<br>$H_2O$: 98.38 mol%<br>MA: 1.48 mol%<br>IBuAc: $1.08 \times 10^{-1}$ mol%<br>MX: 0.0 mol% | HAc: $3.16 \times 10^{-2}$ mol% (0.1 wt%)<br>$H_2O$: 98.38 mol%<br>MA: 1.49 mol%<br>IBuAc: $9.84 \times 10^{-2}$ mol%<br>MX: $5.05 \times 10^{-4}$ mol% | HAc: $3.16 \times 10^{-2}$ mol% (0.1 wt%)<br>$H_2O$: 98.38 mol%<br>MA: 1.49 mol%<br>IBuAc: $9.85 \times 10^{-2}$ mol%<br>MX: $4.97 \times 10^{-4}$ mol% |

## 9.5 SIDE DRAW WITH A PRACTICAL AUTOMATIC PURGING STRATEGY

Since the side-draw flowrates of 50 or 60 kg/h are small for a large industrial column, the piping of the side stream may be too small to be practical. Very small pipes may lead to plugging problems. For comparison purposes, the other two outlet streams (bottom and AO) have flowrates of 6510 kg/h and 2022 kg/h, respectively. A more practical operating method for this very small side stream is to use an automatic purging strategy with larger piping.

A side-stream system capable of permitting a maximum flowrate of 200 kg/h when the valve is fully opened is used as an example in this section. The side stream will be scheduled to fully open periodically to purge out the troublesome feed impurity. In order to select a tray temperature as an indication for such an action, the following simulation run is made to start from the base case with continuous side-draw flowrate of 50 kg/h in previous Section 9.4.2. In Figure 9.39, the side stream valve is fully closed at time = 5 h with the Stage 9 temperature control loop in automatic control mode. From the figure, one observes that the temperature at the Stage 12 is the stage with the largest deviation from its steady-state value. Therefore, the temperature of this stage is selected to trigger the opening mode of the side-stream valve. A very simple purging strategy is used here for demonstration purposes of the feasibility of such an operation. The opening of the side-stream valve is activated when there is a 4°C drop in the Stage 12 temperature from its original steady-state value. This valve is returned to a fully closed mode when the temperature at Stage 12 is back to its original steady-state value.

Figure 9.40 shows the dynamic run with such automatic purging strategy. Starting from the base case in Section 9.4.2, at time = 5 h the side-stream valve is fully closed and the automatic purging strategy is activated. From the figure, one observes that the side-stream valve is switching back and forth from fully opened to fully closed modes to periodically purge out the impurity. The Stage 9 temperature control loop is still in automatic control

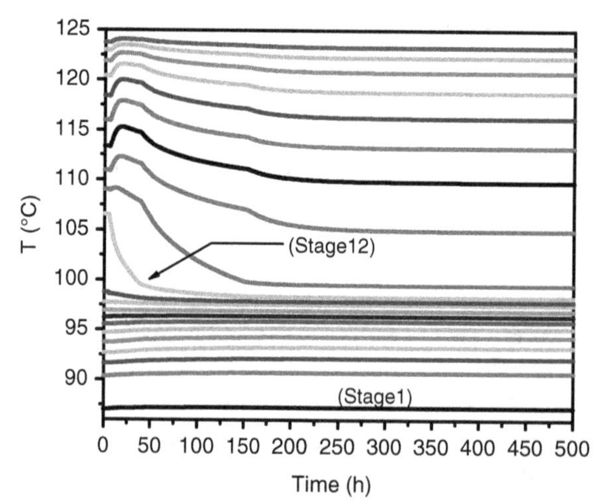


**Figure 9.39** Temperature at each stage when side stream valve is closed at time = 5 h.

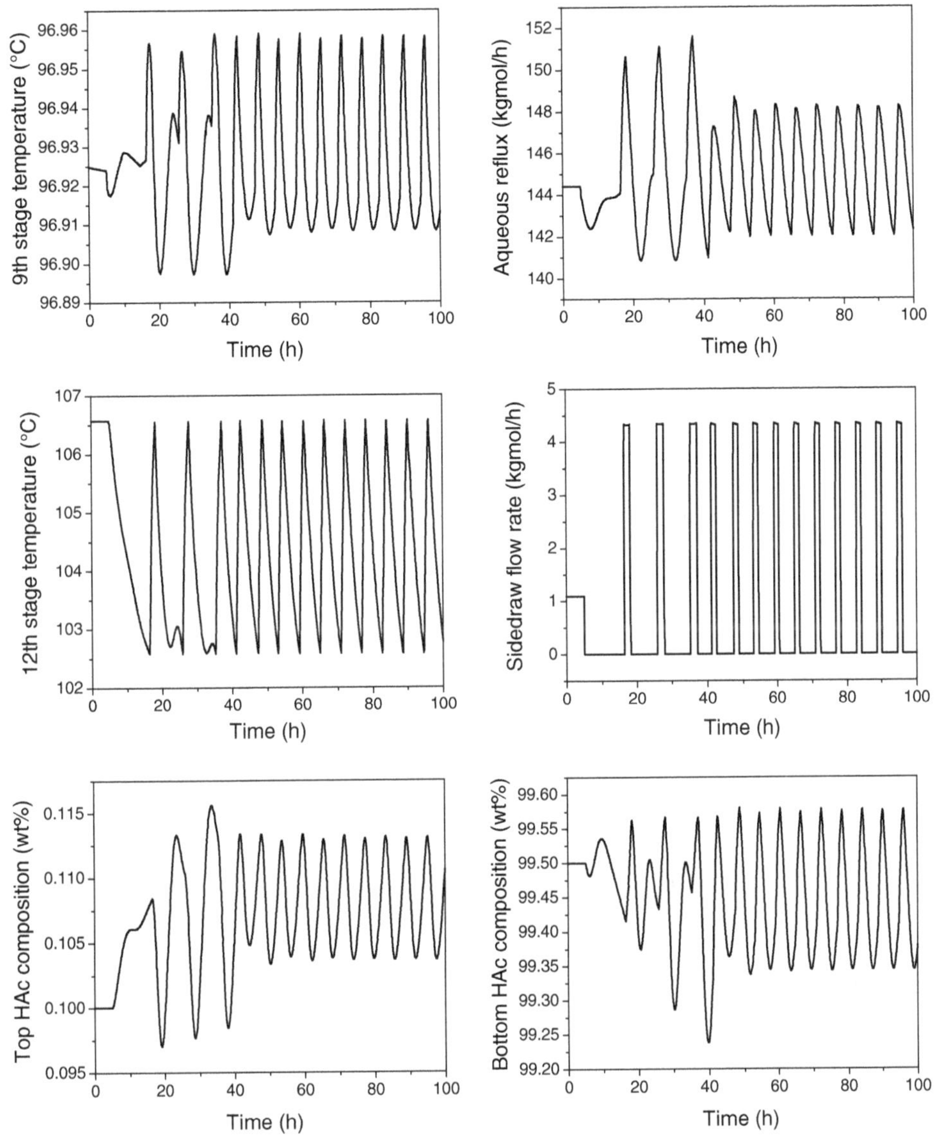


**Figure 9.40** Automatic purging at side-stream location with base-case impurity flowrate.

mode to handle other feed disturbances. The top and bottom product compositions are cycling around their product specifications. The "averaged" bottom acetic acid composition is at 99.45 wt% and the "averaged" top acetic acid composition is at 0.1076 wt%. Both are very close to their high-purity specifications.

The following two figures show the dynamic results with changes in the feed impurity component flowrate at time = 5 h. Figures 9.41 and 9.42 show the +20% and −20% changes in this feed impurity flowrate, respectively. Notice that, from Figure 9.41, because the feed impurity component flowrate is higher, the opening of the side-stream valve is

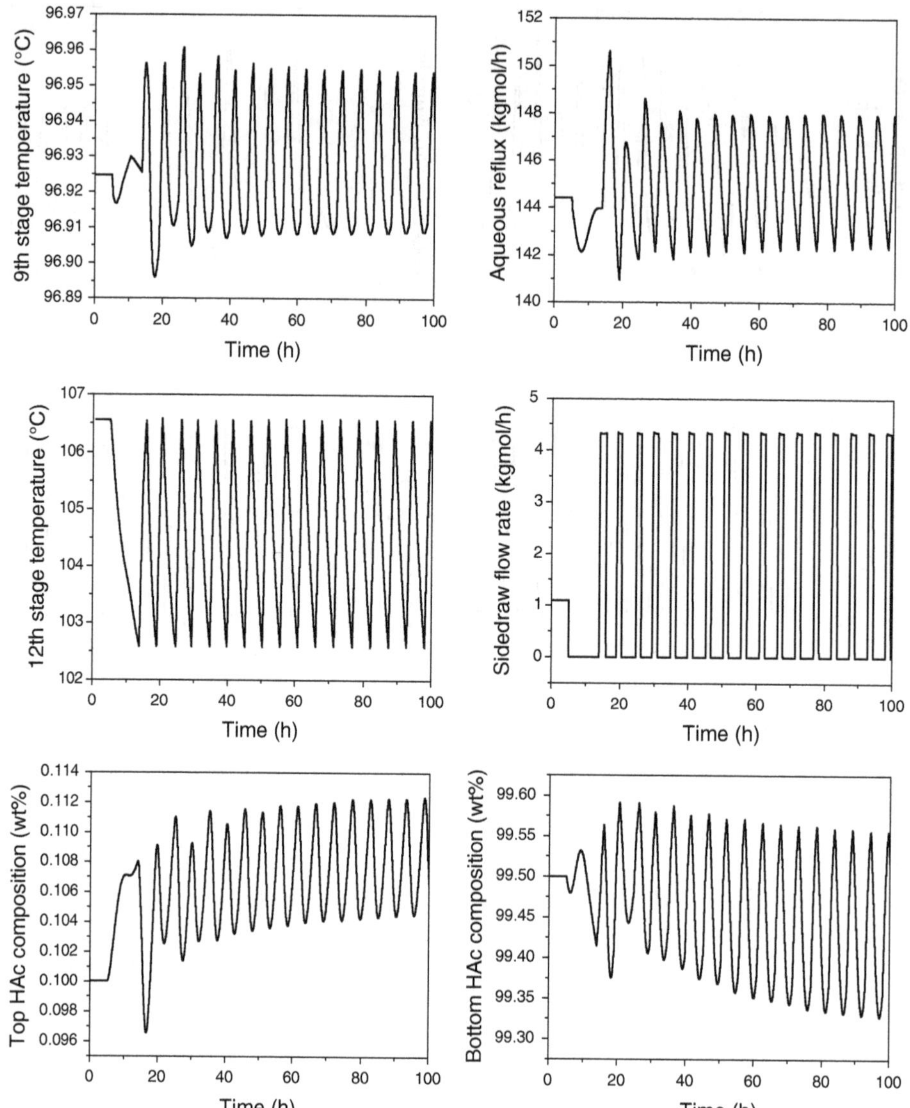


**Figure 9.41** Automatic purging at side-stream location with +20% change of the impurity flowrate.

automatically more frequent with this purging strategy. Similarly, in Figure 9.42, because the feed impurity component flowrate is lower, the opening of the side-stream valve is less frequent. The "averaged" bottom and top acetic acid compositions are 99.43 wt% and 0.1075 wt%, respectively, for +20% feed impurity component flowrate. The "averaged" bottom and top acetic acid compositions are 99.43 wt% and 0.1088 wt%, respectively, for −20% feed impurity component flowrate. All three simulation runs result in quite acceptable bottom and top product purities with this simple purging strategy.

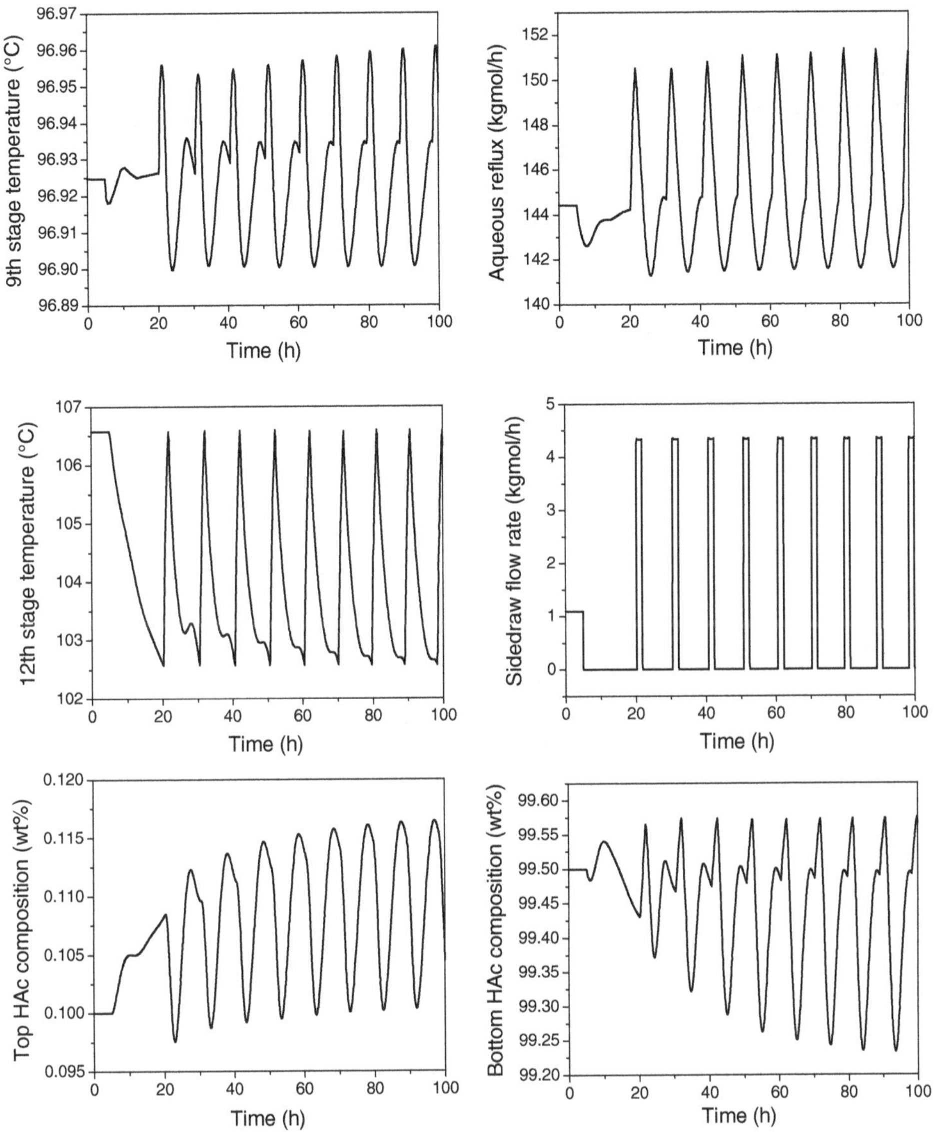


**Figure 9.42** Automatic purging at side-stream location with −20% change of the impurity flowrate.

## 9.6 CONCLUSION

Three candidate entrainers (ethyl acetate, isobutyl acetate, and *n*-butyl acetate) are considered for acetic acid dehydration via heterogeneous azeotropic distillation. The factors that need to be considered in selecting the proper entrainer are illustrated for this example system. Optimum column designs and operating conditions are obtained for these three candidate systems using rigorous process simulation. The TAC is used as the objective function in determining the optimum column designs and operating conditions for these three candidate systems. Isobutyl acetate was found to be the best entrainer with resulting TAC only at about 55% of the system with no entrainer.

An optimum overall control strategy is also proposed for this column system to hold both bottom and top product specifications in spite of $\pm10\%$ feed rate and $\pm10\%$ feed water composition load disturbances. Several alternative control structures are compared using dynamic simulation. The proposed overall control strategy is very simple, requiring only one tray temperature control loop in the column. This simple overall control strategy can easily be implemented in industry.

In the latter part of this chapter, design and operation of an industrial column for acetic acid dehydration with feed impurity are investigated. This feed impurity component during normal operation will not leave the column system through either top decanter AO stream or column bottoms stream. The accumulation of this component inside the column system is illustrated via rigorous dynamic simulation. A side stream is proposed to solve the accumulation problem of this component. Appropriate side-stream location and flowrate are determined by TAC analysis.

This small feed impurity greatly influences the design and operation of the column. With the same purity specifications on both top and bottom products, the column system with feed impurity has a much larger TAC than the design without feed impurity. Flow controlling the side stream is a simple way to purge the impurity from the system. In order to handle wider ranges of the feed impurity variations, the flowrate of the side stream should be set to be a little higher than the optimum flowrate.

Since the flowrate of the side stream is quite small, an alternative way to periodically purge out this impurity is also purposed. This automatic purging strategy does not rely on any online composition measurement but only uses a temperature measurement. Closed-loop dynamic simulation demonstrates that the proposed purging strategy can successfully keep the column under normal operation with both product purities at their specifications despite feed impurity variations.

## REFERENCES

1. Chien I. L., K. L. Zeng, H. Y. Chao, and J. H. Liu, Design and control of acetic acid dehydration system via heterogeneous azeotropic distillation, *Chem. Eng. Sci.*, **59**, 4547–4567 (2004).
2. Hindmarsh E., J. A. Turner, and A. M. Ure, Process for the production of terephthalic acid, U.S. Patent 5,563,293 (1996).
3. Lee F. M., W. Lamshing, and R. W. Wytcherley, Method and apparatus for preparing purified terephthalic acid and isophthalic acid from mixed xylenes, U.S. Patent 6,054,610 (2000).
4. Chien I. L., H. P. Huang, T. K. Gau, and C. H. Wang, Influence of feed impurity on the design and operation of an industrial acetic acid dehydration Column, *Ind. Engng Chem. Res.*, **44**, 3510–3521 (2005).
5. Tanake S. and J. Yamada, Graphical calculation method for minimum reflux ratio in azeotropic distillation, *J. Chem. Eng. Japan*, **5**, 20–26 (1972).
6. Siirola J. J., An industrial perspective on process synthesis. In *AIChE Symposium Series*, Eds. Biegler L. T. and M. F. Doherty, CACHE, Austin, TX, 1995; No. 304, pp. 222–233.
7. Costantini G., M. Serafini, and P. Paoli, Process for the recovery of the solvent and of the by-produced methylacetate in the synthesis of terephthalic acid, U.S. Patent 4,250,330 (1981).
8. Parten W. D. and A. M. Ure, Dehydration of acetic acid by azeotropic distillation in the production of an aromatic acid, U.S. Patent 5,980,696 (1999).
9. Othmer D. F., Azeotropic distillation for dehydrating acetic acid, *Chem. Metall. Eng.*, **40**, 91–95 (1941).

10. Othmer D. F., Azeotropic and extractive distillation. In *Kirk–Othmer Encyclopedia of Chemical Technology*, Eds. Kirk R. E. and D. F. Othmer, Wiley, New York, 1978, pp. 352–377.
11. Horsley L. H., *Azeotropic Data III*, Advances in Chemistry Series No. 116, American Chemical Society, Washington, DC, 1973.
12. Gmehling J. *Azeotropic Data*, Second Edition, Wiley-VCH, Weinheim, 2004.
13. Gmehling J. and U. Onken, Vapor–liquid equilibrium data collection. In *DECHEMA Chemistry Data Series*, Eds. Behrens D. and R. Eckermann, DECHEMA, Frankfurt, Germany, 1977.
14. Christensen S. P. and J. D. Olson, Phase equilibria and multiple azeotropy of the acetic acid–isobutyl acetate system, *Fluid Phase Equilibria*, **79**, 187–199 (1992).
15. Sørensen J. M. and W. Arlt, Liquid–liquid equilibrium data collection binary systems. In *DECHEMA Chemistry Data Series*, Eds. Behrens D. and R. Eckermann, DECHEMA, Frankfurt, Germany, 1979.
16. Renon H. and J. M. Prausnitz, Local compositions in thermodynamics excess functions for liquid mixtures, *AIChE J*, **14**, 135–144 (1968).
17. Hayden J. G. and J. P. O'Connell, A generalized method for predicting second virial coefficients, *Ind. Eng. Chem. Process Des. Dev.*, **14**, 209–216 (1975).
18. Aspen Plus, Version 2006.5. Aspen Technology, Inc., Cambridge, MA, 2007.
19. Douglas J. M., *Conceptual Design of Chemical Processes*, McGraw-Hill, New York, NY, 1988.
20. Othmer D. F., Process for dehydration of acetic acid and other lower fatty acids. U.S. Patent 2,050,234 (1936).
21. Mitsui Petrochemical Industries, Ltd. Process for azeotropic distillation. U.K. Patent 1,576,787 (1980).
22. Chien I. L. and C. L. Kuo, Investigating the need of a pre-concentrator column for acetic acid dehydration system via heterogeneous azeotropic distillation, *Chem. Eng. Sci.*, **61**, 569–585 (2006).
23. Luyben W. L., *Plantwide Dynamic Simulators in Chemical Processing and Control*, Marcel Dekker, New York, NY, 2002.
24. Chien I. L., K. L. Zeng, and H. Y. Chao, Design and control of a complete heterogeneous azeotropic distillation column system, *Ind. Engng Chem. Res.*, **43**, 2160–2174 (2004).
25. Arifin S. and I. L. Chien, Combined preconcentrator/recovery column design for isopropyl alcohol dehydration process, *Ind. Engng Chem. Res.*, **46**, 2535–2543 (2007).
26. Moore C. F. Selection of controlled and manipulated variables. In *Practical Distillation Control*, Ed. Luyben W. L., Van Nostrand Reinhold, New York, 1992.
27. Chien I. L., H. P. Huang, and J. C. Yang, A simple multiloop tuning method for pid controllers with no proportional kick, *Ind. Engng Chem. Res.*, **38**, 1456–1468 (1999).
28. Weidlich U. and J. Gmehling, A modified UNIFAC model. 1. prediction of VLE, $h^E$, and $\gamma^\infty$, *Ind. Engng Chem. Res.*, **26**, 1372–1381 (1987).
29. Chien I. L. and P. S. Fruehauf, Consider IMC tuning to improve controller performance, *Chem. Eng. Prog.*, **86**, 33–41 (Oct., 1990).

# PART 4

# SEPARATIONS USING HEAVY ENTRAINER (EXTRACTIVE DISTILLATION)

According to the definition from Doherty and Malone, "extractive distillation" is a method of separating minimum-boiling binary azeotropes by use of an entrainer that is the heaviest species in the mixture, does not form any azeotropes with the original components, and is completely miscible with them in all proportions. Adding this heavy entrainer into the system causes the relative volatility of the original two components to be enhanced so that one original component can go overhead and the other component will go with the heavy entrainer to the column bottoms. Three industrial examples will be studied in the following three chapters to illustrate the design and control of such systems. Chapter 10 will study the isopropanol–water separation using DMSO as the heavy entrainer. The optimal design of this separation using extractive distillation will be compared with the same separation using heterogeneous azeotropic distillation. Chapter 11 will study the acetone–methanol separation using various heavy entrainers. The effect of different entrainers on the controllability of the distillation system will be studied. Chapter 12 will study a system with different RCM types. The system is the separation of a mixture of acetone and chloroform having a maximum-boiling instead of a minimum-boiling azeotrope. Separation of this mixture via pressure-swing and extractive distillation will be compared.

*Design and Control of Distillation Systems for Separating Azeotropes.* By William L. Luyben and I-Lung Chien

# CHAPTER 10

# ISOPROPANOL–WATER (DIMETHYL SULFOXIDE AS THE ENTRAINER)

In Chapter 8 we used a light entrainer (cyclohexane) via heterogeneous azeotropic distillation to separate the mixture of isopropanol–water. An alternative way for the separation of this mixture is to use extractive distillation by adding a heavy-boiling entrainer. The purpose of the entrainer in this extractive distillation column is to alter the relative volatility between IPA and water, making IPA go to the top of the column and water go to the bottom of the column. The upper section of the column (above the entrainer feed location) is called the rectifying section, and its purpose is to separate IPA and the entrainer. The middle section of the column (between the entrainer feed stage and the fresh feed stage) is called the extractive section. The purpose of this section is to suppress water from going up the column. The bottom section of the column (below the fresh feed location) is called the stripping section, and its purpose is to keep IPA from going down the column. The bottoms product of the column is the mixture of water and the entrainer, and it is fed to another downstream entrainer recovery column to separate these two components, so the entrainer can be recycled back to the extractive column. The conceptual design of the overall process, via extractive distillation, is shown in Figure 10.1.

In this chapter, design and control of the IPA dehydration process via extractive distillation will be studied.[1] Since, in this distillation system, entrainer selection is an important step before working on the optimal design of the column sequence, we will start by comparing two alternative entrainers for this separation system in the following section.

*Design and Control of Distillation Systems for Separating Azeotropes*. By William L. Luyben and I-Lung Chien

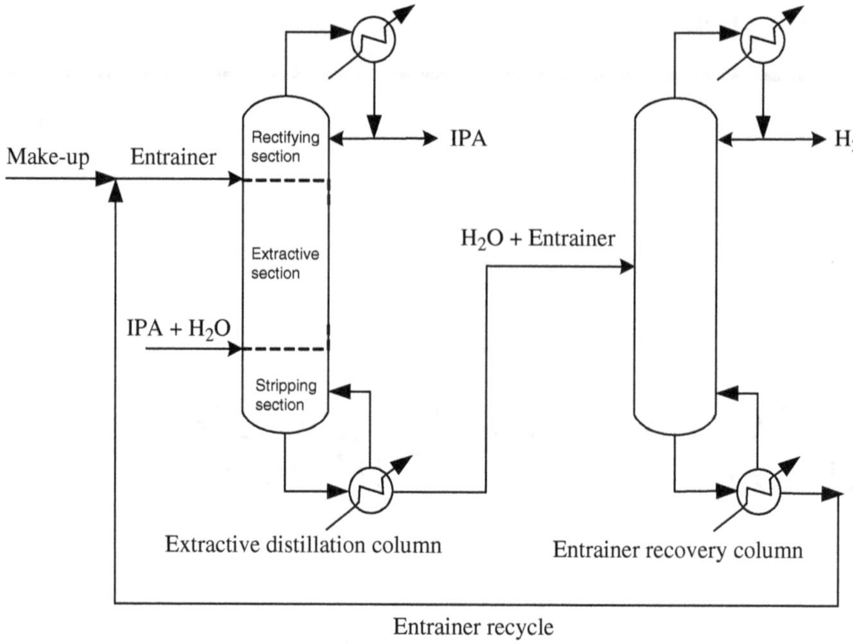


**Figure 10.1** Conceptual design flowsheet of IPA dehydration process via extractive distillation column.

## 10.1 COMPARING ENTRAINERS: DIMETHYL SULFOXIDE VERSUS ETHYLENE GLYCOL

We use IPA dehydration as an example to demonstrate what factors are important to make a good entrainer. Two candidate entrainers, DMSO and ethylene glycol (EG), are considered for this system.

### 10.1.1 Isovolatility Curves

In Chapter 2, we demonstrated that it is very easy to generate an isovolatility curve for a ternary system in Aspen. Figure 10.2 displays the isovolatility curve for the IPA–water system at 1 atm using DMSO as an entrainer. Remember that the volatility between IPA and water is unity in the binary system when we are at the azeotrope (shown on the ordinate in Fig. 10.2 with a temperature of 80°C). As DMSO is added, the isovolatility line moves towards the hypotenuse of the triangle, where the concentration of water goes to zero. The location where this isovolatility curve intercepts the IPA–DMSO edge of the triangle is important in determining if DMSO is an effective entrainer. The closer the intersection point is to the IPA corner, the less entrainer is required, which means lower operating and capital costs. As shown in Figure 10.2, this location is at $x_E = 0.15$ (15 mol% DMSO).

Another observation from Figure 10.2 is to determine the feasible distillate product of the extractive distillation column. Because the isovolatility curve in Figure 10.2 intercepts the IPA–DMSO edge of the triangle, it is concluded that IPA is the distillate product. The reasoning is that by adding DMSO into the system, the relative volatility will become greater than unity and result in IPA going up the column. Note that the relative volatility between IPA and water is greater than unity in the region below the isovolatility curve.

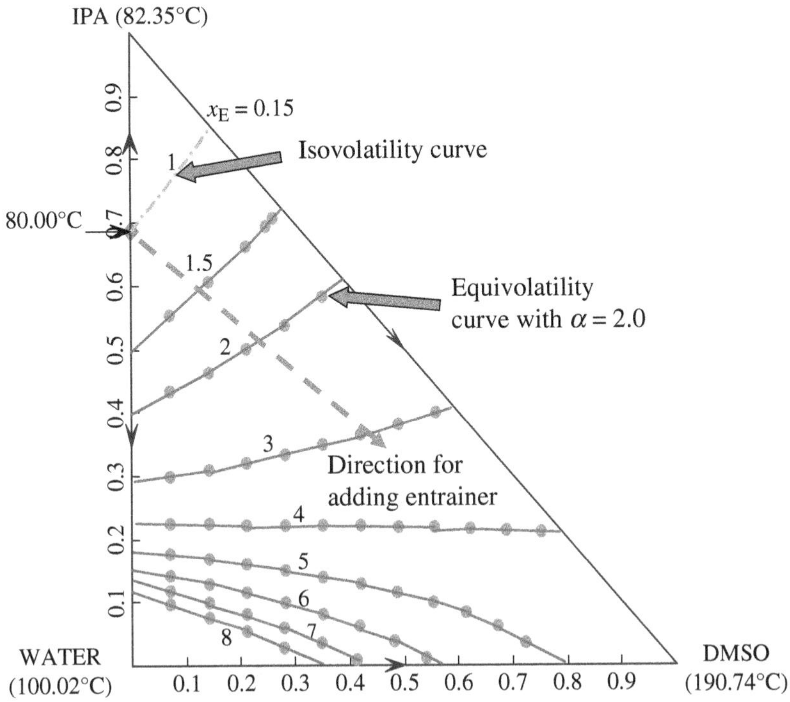


**Figure 10.2** Iso- and equivolatility plots at 1 atm for the IPA dehydration process using DMSO as an entrainer.

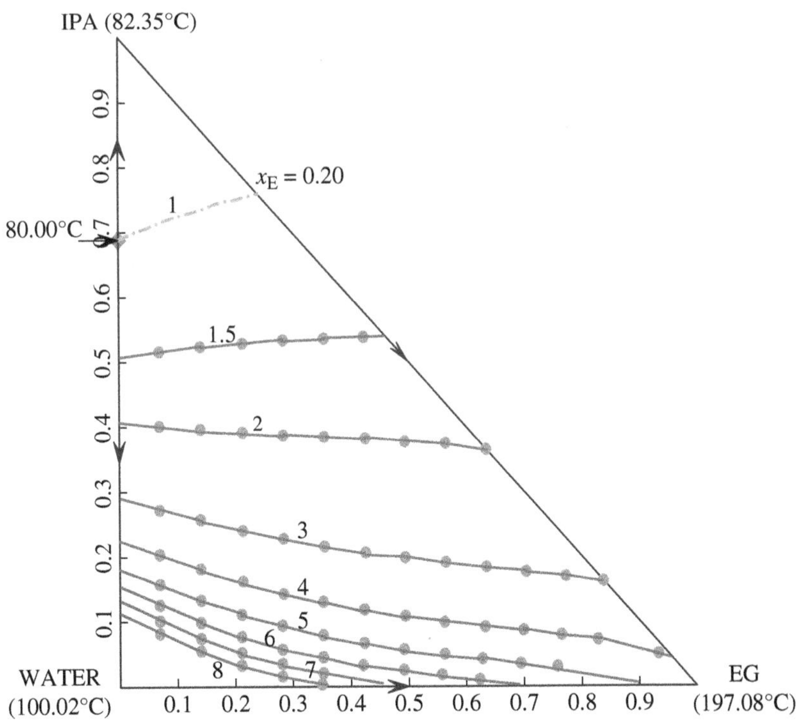


**Figure 10.3** Iso- and equivolatility plots at 1 atm for the IPA dehydration process using EG as an entrainer.

**TABLE 10.1 NRTL Parameters for IPA–Water–DMSO or EG System.**

| Comp. $i$<br>Comp. $j$ | IPA<br>Water | IPA<br>DMSO | Water<br>DMSO | IPA<br>EG | Water<br>EG |
|---|---|---|---|---|---|
| $a_{ij}$ | 0 | 0 | −1.2449 | −8.6577 | 0.3479 |
| $a_{ji}$ | 0 | 0 | 1.7524 | 11.683 | −0.0567 |
| $b_{ij}$ (K) | 185.40 | 115.28 | 586.80 | 3473.7 | 34.823 |
| $b_{ji}$ (K) | 777.30 | −25.012 | −1130.2 | −4342.5 | −147.14 |
| $\alpha_{ij}$ | 0.5 | 0.3 | 0.3 | 0.3 | 0.3 |

Aspen Plus NRTL:

$$\ln \gamma_i = \frac{\sum_j x_j \tau_{ji} G_{ji}}{\sum_k x_k G_{ki}} + \sum_j \frac{x_j G_{ij}}{\sum_k x_k G_{kj}} \left[ \tau_{ij} - \frac{\sum_m x_m \tau_{mj} G_{mj}}{\sum_k x_k G_{kj}} \right]$$

where $G_{ij} = \exp(-\alpha_{ij} \tau_{ij})$

$$\tau_{ij} = a_{ij} + \frac{b_{ij}}{T}$$

$\alpha_{ij} = c_{ij}$, $\tau_{ii} = 0$, $G_{ii} = 1$

For comparison, Figure 10.3 displays the isovolatility curve at 1 atm using EG as entrainer. The location has been moved to $x_E = 0.20$ (20 mol% EG) showing that EG is a less effective entrainer than DMSO. For completeness of the information for generating the above plots, the NRTL parameters and the coefficients of the Antoine equation for calculating the vapor pressure are listed in Tables 10.1 and 10.2, respectively.

### 10.1.2 Equivolatility Curves

The main purpose of adding the entrainer into the system is to alter the relative volatility between IPA and water. Thus we would like to check this capability for using DMSO or

**TABLE 10.2 Coefficients of Antoine Equation for IPA–Water–DMSO or EG System.**

| Components<br>Temperature Units<br>Property Units | IPA<br>°C<br>atm | Water<br>°C<br>atm | DMSO<br>°C<br>atm | EG<br>°C<br>atm |
|---|---|---|---|---|
| $C_1$ | 64.904 | 62.123 | 44.747 | 72.564 |
| $C_2$ | −7607 | −7258.2 | −7620.6 | −10,411 |
| $C_3$ | 0 | 0 | 0 | 0 |
| $C_4$ | 0 | 0 | 0 | 0 |
| $C_5$ | −7.4086 | −7.3037 | −4.6279 | −8.1976 |
| $C_6$ | 4.3986E−18 | 4.1653E−6 | 4.3819E−7 | 1.6536E−18 |
| $C_7$ | 6 | 2 | 2 | 6 |

Aspen Plus extended Antoine vapor pressure model is:

$$\ln P_i^* = C_{1i} + \frac{C_{2i}}{T + C_{3i}} + C_{4i}T + C_{5i} \ln T + C_{6i} T^{C_{7i}}, \quad i = \text{component}$$

EG. The use of equivolatility curves to screen entrainers has been proposed in Laroche et al.[2] The way to generate equivolatility curves in Aspen Plus is to use the *Flash2* module in the unit operation library. With any arbitrary feed composition in the ternary diagram, under adiabatic flash operating at 1 atm and with negligible vapor flow, the vapor and liquid composition in equilibrium can be calculated by Aspen Plus. Extracting x and y information about the IPA and water, the relative volatility between these two components can be calculated with any amounts of addition of the entrainer component. Collecting the feed compositions that give the same relative volatility, a curve can be drawn with equal relative volatility. Figure 10.2 displays various equivolatility curves when using DMSO as the entrainer. For comparison, Figure 10.3 also displays various equivolatility curves when using EG as the entrainer. From these two figures, it is clear that DMSO is a more effective entrainer to enhance the relative volatility between IPA and water. Gmehling and Möllmann[3] used a separation factor calculated at infinite dilution that also showed that DMSO is a more effective entrainer than EG.

One final comment about the iso- and equivolatility curves is that we should not solely rely on the location where the isovolatility curve intercepts the IPA-entrainer edge of the triangle to determine which one is the more effective entrainer. We will show a

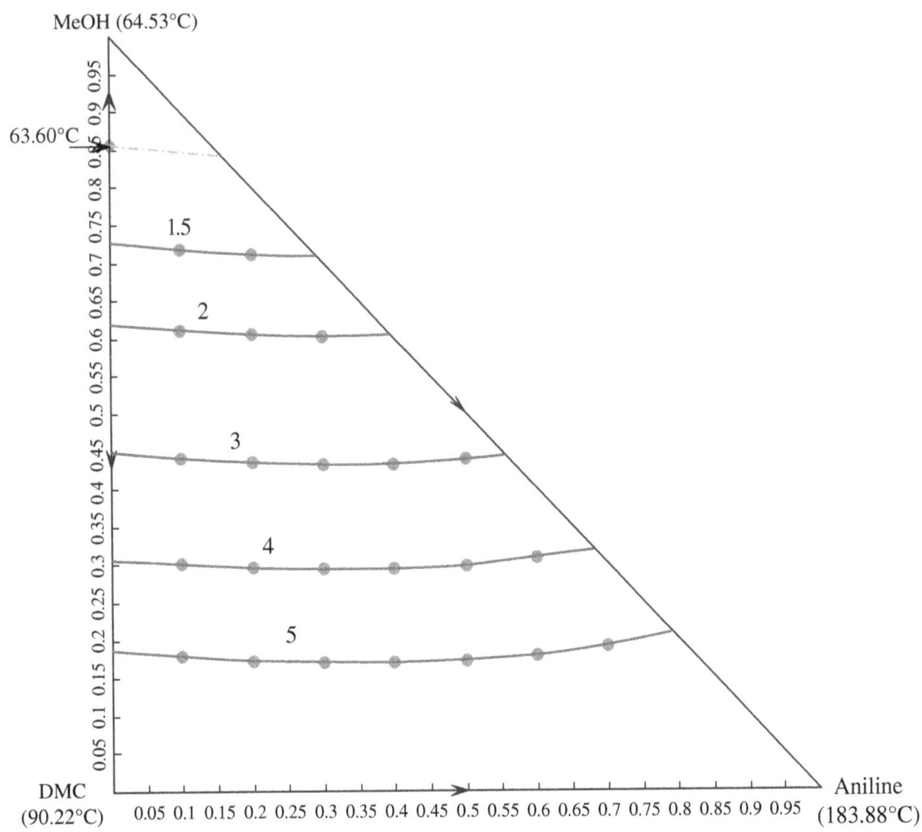


**Figure 10.4** Iso- and equivolatility plots at 1 atm for MeOH–DMC separation using aniline as an entrainer.

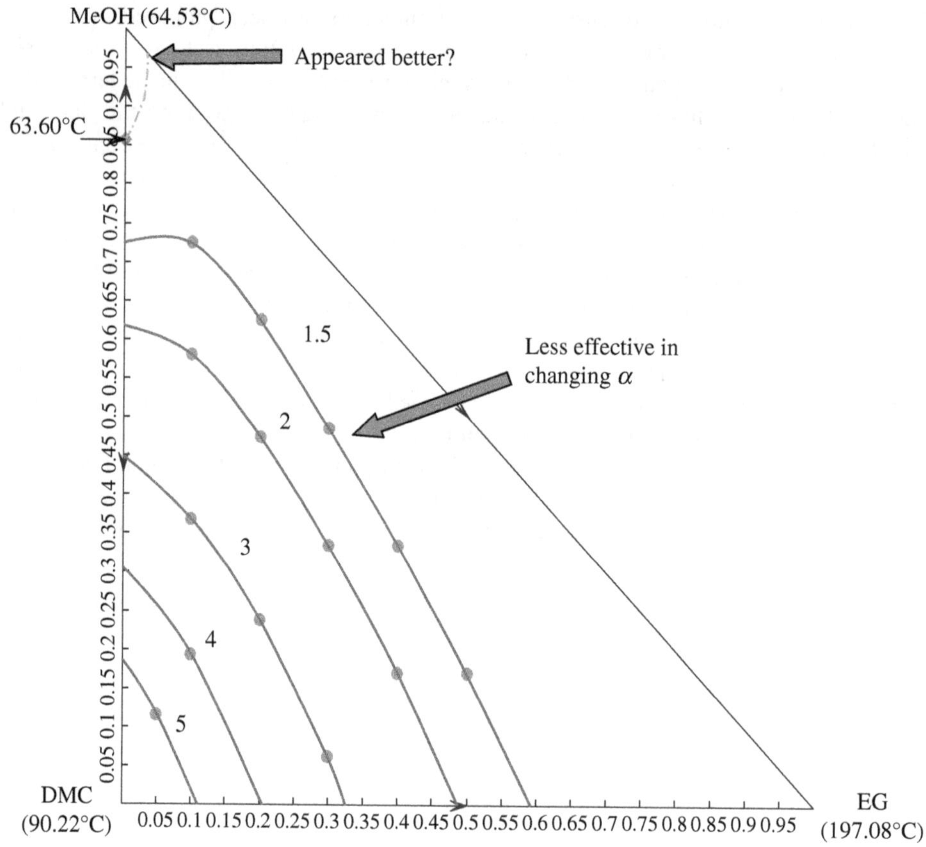


**Figure 10.5** Iso- and equivolatility plots at 1 atm for MeOH–DMC separation using EG as an entrainer.

counterexample where $x_E$ has a smaller value, but the maps of the equivolatility curves are not as favorable.

The example is to separate a mixture containing dimethyl carbonate (DMC) and methanol via extractive distillation. Figures 10.4 and 10.5 show the iso- and equivolatility curves by using either aniline or EG as the entrainer. It is demonstrate that although the $x_E$ for using EG as the entrainer is much smaller than that for using aniline, the enhancement of relative volatility between DMC and methanol is rather poor. Thus EG is not a good entrainer for this separation. For completeness of the information for generating the above plots, the UNIQUAC parameters used in this system for calculating the liquid activity coefficient and the coefficients of the Antoine equation for calculating the vapor pressure are listed in Tables 10.3 and 10.4, respectively.

### 10.1.3 Binary VLE Diagrams

Other factors that affect the total annual cost (TAC) of the overall design flowsheet are the *yx* and *Txy* plots for the IPA–entrainer pair and the *yx* or *Txy* plots for the water–entrainer pair. The *yx* and *Txy* plots for the IPA–entrainer pair can be used to determine the ease of

**TABLE 10.3 UNIQUAC Parameters for DMC–MeOH–Aniline or EG System.**

| Comp. $i$<br>Comp. $j$ | MeOH<br>DMC | MeOH<br>Aniline | DMC<br>Aniline | MeOH<br>EG | DMC<br>EG |
|---|---|---|---|---|---|
| $a_{ij}$ | 0 | 0 | 0 | −32.5987 | 0 |
| $a_{ji}$ | 0 | 0 | 0 | 2.2712 | 0 |
| $b_{ij}$ (K) | −24.323 | 88.415 | 101.44 | 10,000 | −244.31 |
| $b_{ji}$ (K) | −268.98 | −291.40 | −32.733 | −599.11 | −139.43 |

Aspen Plus UNIQUAC:

$$\ln \gamma_i = \ln \frac{\Phi_i}{x_i} + \frac{z}{2} q_i \ln \frac{\theta_i}{\Phi_i} - q_i' \ln t_i' - q_i' \sum_j \frac{\theta_j' \tau_{ij}}{t_j'} + l_i + q_i' - \frac{\Phi_i}{x_i} \sum_j x_j l_j$$

where $\theta_i = \frac{q_i x_i}{q_T}$; $q_T = \sum_k q_k x_k$; $\theta_i' = \frac{q_i' x_i}{q_T'}$; $q_T' = \sum_k q_k' x_k$; $\Phi_i = \frac{r_i x_i}{r_T}$; $r_T = \sum_k r_k x_k$

$$l_i = \frac{z}{2}(r_i - q_i) + 1 - r_i;\ t_i' = \sum_k \theta_k' \tau_{ki};\ \tau_{ij} = \exp\left(a_{ij} + \frac{b_{ij}}{T}\right);\ \text{and } z = 10$$

**TABLE 10.4 Coefficients of Antoine Equation for DMC–MeOH–Aniline or EG System.**

| Components<br>Temperature Units<br>Property Units | MeOH<br>°C<br>atm | DMC<br>°C<br>atm | Aniline<br>°C<br>atm | EG<br>°C<br>atm |
|---|---|---|---|---|
| $C_1$ | 71.192 | 46.507 | 54.761 | 72.564 |
| $C_2$ | −6904.5 | −5991.3 | −8207.1 | −10411 |
| $C_3$ | 0 | 0 | 0 | 0 |
| $C_4$ | 0 | 0 | 0 | 0 |
| $C_5$ | −8.8622 | −5.0971 | −6.0132 | −8.1976 |
| $C_6$ | 7.4664E−6 | 1.3402E−17 | 2.8414E−18 | 1.6536E−18 |
| $C_7$ | 2 | 6 | 6 | 6 |

Aspen Plus extended Antoine vapor pressure model is:

$$\ln P_i^* = C_{1i} + \frac{C_{2i}}{T + C_{3i}} + C_{4i} T + C_{5i} \ln T + C_{6i} T^{C_{7i}}, \quad i = \text{component}$$

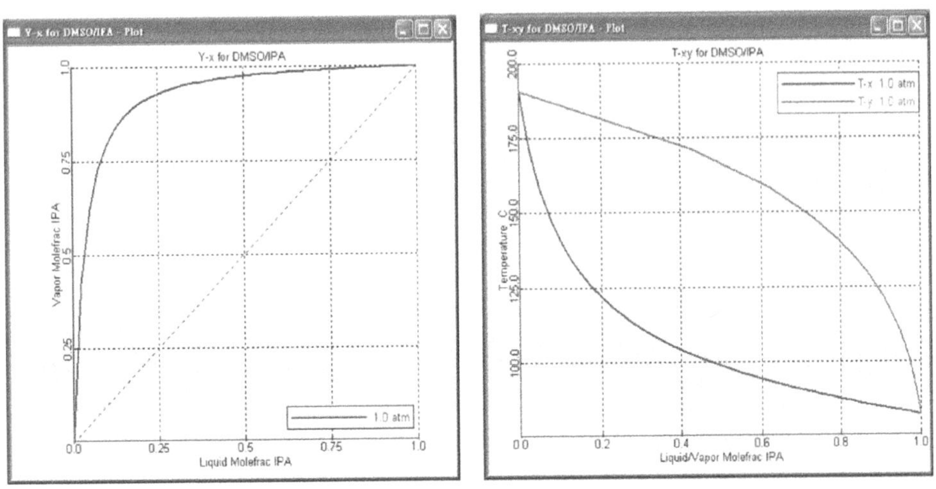


**Figure 10.6** *yx* and *Txy* plots of the IPA–DMSO pair at 1 atm.

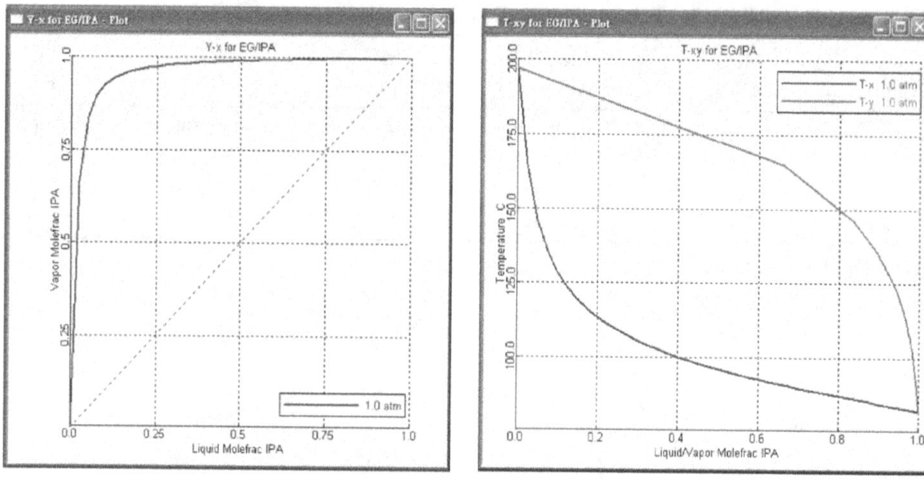


**Figure 10.7** *yx* and *Txy* plots of the IPA–EG pair at 1 atm.

separation in the rectifying section of the extractive distillation column. The *yx* and *Txy* plots for the water–entrainer pair can be used to determine if the separation in the entrainer recovery column is easy or not.

Figures 10.6 and 10.7 show the *yx* and *Txy* plots for the IPA–entrainer pair for using DMSO or EG as the entrainer, respectively. From these two figures, it is concluded that both DMSO and EG exhibit no problem for the separation of IPA and entrainer. Figures 10.8 and 10.9 show the *yx* and *Txy* plot for the water–entrainer pair for using DMSO or EG as the entrainer, respectively. From these two figures, it is also concluded that both DMSO and EG are good entrainers when we focus on the separation in the entrainer recovery column.

From the above analysis, by looking into the iso- and equivolatility curves and the binary VLE diagrams for the IPA–entrainer and the water–entrainer pairs, it is concluded that

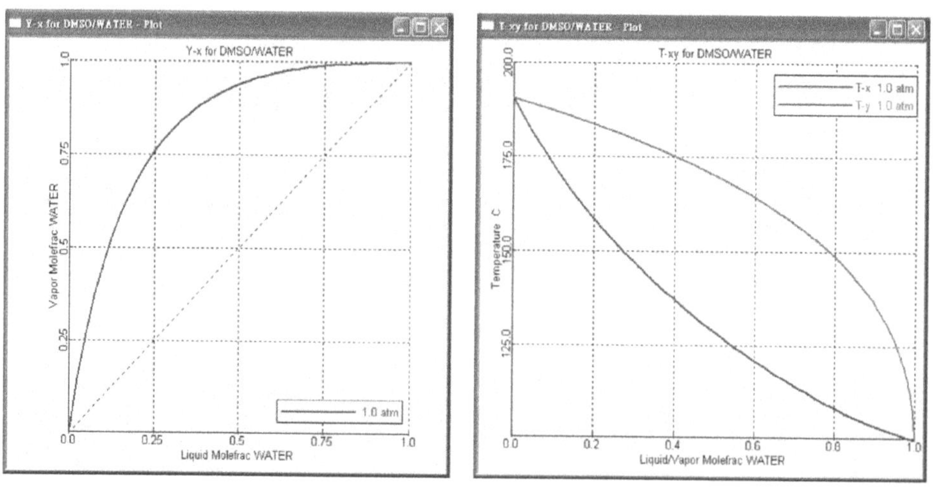


**Figure 10.8** *yx* and *Txy* plots of the water–DMSO pair at 1 atm.

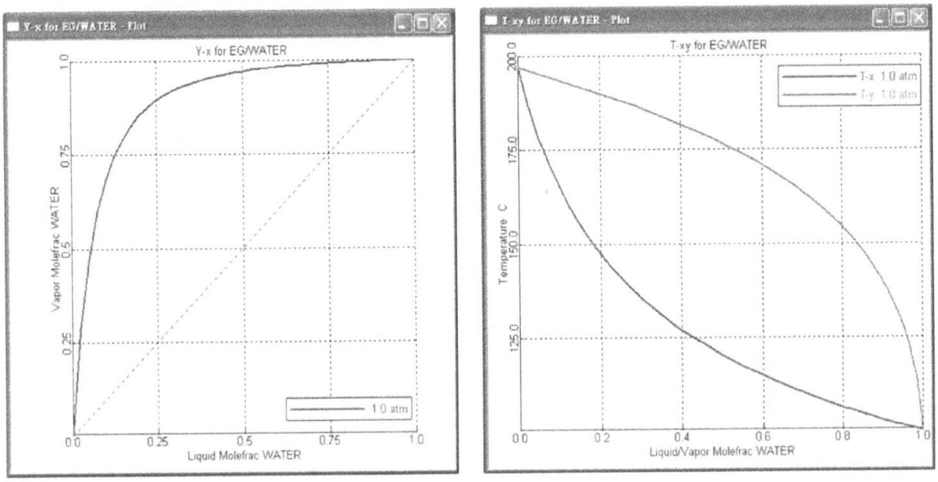


**Figure 10.9** *yx* and *Txy* plots of the water–EG pair at 1 atm.

DMSO is a better entrainer for this system. In the following, the economic optimal design flowsheet of this IPA dehydration process when using DMSO as the entrainer will be established to minimize the TAC of the overall process.

It is worth mentioning that, apart from the important factors above relating to the phase equilibrium behavior, other factors such as thermally stable, nontoxic, low price, and other physical properties should also be considered in the entrainer selection. A good paper by Gmehling and Möllmann[3] used four examples to demonstrate the entrainer selection procedure for extractive and azeotropic distillation.

## 10.2 STEADY-STATE DESIGN AND ECONOMICAL ANALYSIS

### 10.2.1 Design Flowsheet via Extractive Distillation

The proposed design flowsheet for IPA–water separation with DMSO as the entrainer can be seen in Figure 10.10. The IPA–water mixture and the entrainer are fed into the extractive distillation column at different feed locations. The presence of DMSO alters the relative volatility between IPA and water causing IPA to move toward the top part and water to move toward the bottom part of the column. In the rectifying section, there is essentially no water; thus the simple separation between IPA and DMSO is performed with high-purity IPA going into the distillate and DMSO returning to the extractive section for entrainer usage. In the stripping section, IPA is the lightest component and is driven up toward the extractive section of the column, which results in only negligible IPA in the column bottoms stream. This column bottoms stream is fed into the entrainer recovery column to produce almost pure water in the distillate and almost pure DMSO in the column bottoms. DMSO, as a heavy entrainer, is recycled back to the extractive distillation column. To balance the tiny entrainer losses in both the D1 and D2 streams, a small makeup stream of entrainer should be added.

The product specifications are set to be exactly the same as in Chien et al.[4] and in Arifin and Chien.[5] They are 99.9999 mol% of IPA in D1 and 99.9 mol% of water in D2. The reason for the ultra-high purity in IPA is for semiconductor industry usage. The feed composition is typical of a waste IPA stream in the semiconductor industry and contains equal molar

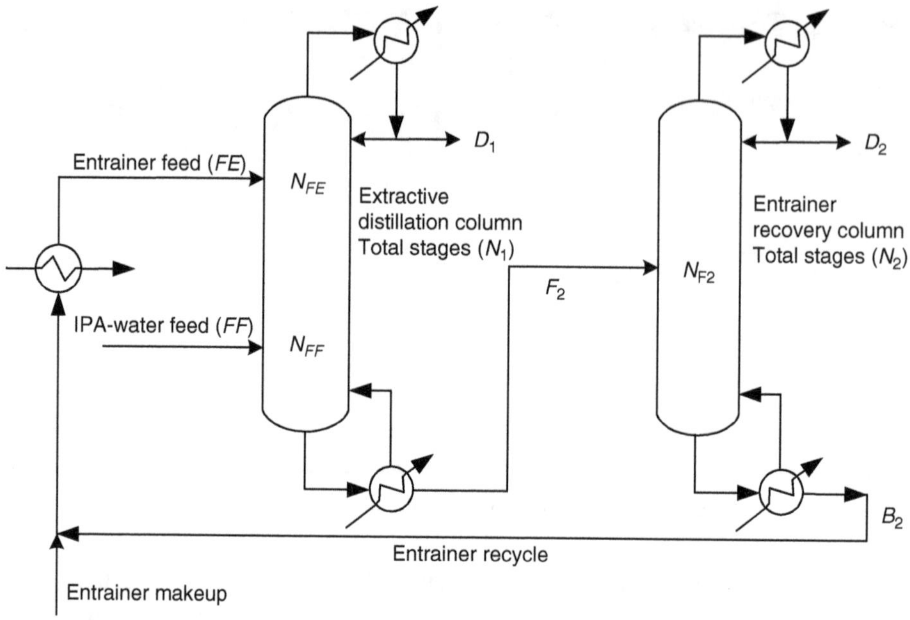


**Figure 10.10** Proposed design flowsheet of the IPA dehydration process.

amounts of IPA and water. With this feed composition not far from the azeotropic composition, a flowsheet arrangement with a preconcentrator column was not considered for the sake of having a simpler process as well as simpler instrumentation and control equipment. For a more dilute feed stream, a preconcentrator column may need to be considered to withdraw most of the water component before going into this two-column system. The feed flowrate is assumed to be 100 k mol/h (=1666.67 mol/min), also exactly the same as in Arifin and Chien.[5] Reflux drums of both columns are operated at atmospheric pressure for ease of operation. Column pressure drops are automatically calculated in the Aspen simulation.

The design variables to be determined in the flowsheet include the solvent-to-feed ratio ($FE/FF$), total stages of the extractive distillation column ($N_1$), entrainer and fresh feed tray locations ($N_{FE}$ and $N_{FF}$), total stages of the entrainer recovery column ($N_2$), and feed tray location of entrainer recovery column ($N_{F2}$). As can be seen in Knight and Doherty[6] and also summarized in Chapter 5 of Doherty and Malone,[7] the entrainer feed temperature can also been considered as another design variable, so a cooler is included in Figure 10.10.

There are many design variables to be determined, so to simplify the optimization the more important and complex extractive distillation column is optimized first. We considered three different cases for the entrainer feed temperature.

*Case 1*: Operating the entrainer feed temperature 5–15°C below the top temperature of the extractive distillation column (as suggested in Knight and Doherty[6]). We use 72°C in the following simulation.

*Case 2*: Operating the entrainer feed temperature the same as temperature in the B2 stream (196.6°C) with no cooler.

*Case 3*: Subcooling the entrainer feed temperature (from B2) to 40°C.

In each case, there are four design variables to be determined ($FE/FF$ feed ratio, $N_1$, $N_{FE}$, and $N_{FF}$). A sequential iterative optimization search is used to find the optimal design with feed ratio as the outer iterative loop, $N_1$ as the middle loop, and $N_{FE}$ and $N_{FF}$ as the inner iterative loop. The two design specifications for all the Aspen simulations are setting top composition at 99.9999 mol% IPA and setting the ratio of IPA to the sums of water and IPA in the bottom stream to be 0.001. The reason for this bottom specification is to set the IPA loss through the column bottoms. The above two design specifications can be met by varying the remaining two degrees of freedom in this column (e.g., reboiler duty and reflux flowrate).

The TAC is used as the objective function to be minimized, which includes annualized capital costs and operating costs. The capital costs include column shell, trays, reboiler, and condenser. A payback period of three years is assumed. The calculation formula for the above equipment can be found in Douglas.[8] The operating costs include the steam and cooling water for the operation of reboiler and condenser. The optimization procedure to minimize the TAC, by varying the four design variables, is summarized as follows.

1. Guess the $FE/FF$ feed ratio.
2. Guess the total stages of the extractive distillation column ($N_1$).
3. Guess the entrainer feed location ($N_{FE}$) and the fresh feed location ($N_{FF}$).
4. Change the reboiler duty and the reflux flowrate until the two design specifications can be met.
5. Go back to (3) and change the $N_{FE}$ and $N_{FF}$ until the TAC is minimized.
6. Go back to (2) and change $N_1$ until the TAC is minimized.
7. Go back to (1) and change the $FE/FF$ feed ratio until the TAC is minimized.

Figure 10.11 shows the summarized TAC plot at various feed ratios for Case 1. It is observed that the best feed ratio is at 1.025. The right-hand side plot of Figure 10.12 shows the results for varying $N_{FE}$ and $N_{FF}$ with $N_1$ fixed at 41 and the $FE/FF$ feed ratio fixed at 1.025. From this plot, the best $N_{FE}$ is Stage 7 and the best $N_{FF}$ is Stage 35 (stages are counting from top

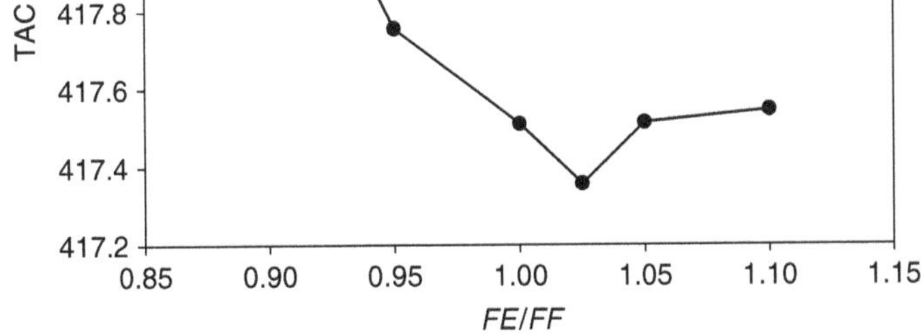


**Figure 10.11** Summarized TAC plots at various feed ratio.

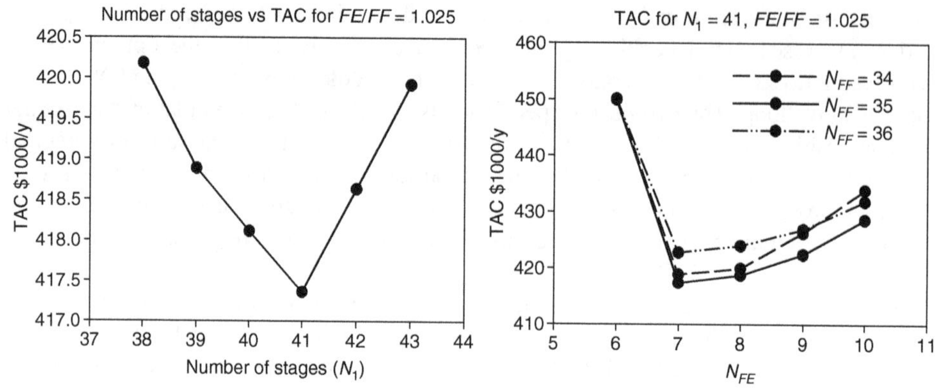


**Figure 10.12** TAC plots for the extractive distillation column.

to bottom with condenser as Stage 1 and reboiler as the last stage). The left-hand side plot of Figure 10.12 shows the collection of all the similar plots at various $N_1$ and fixing the feed ratio at 1.025. The column with $N_1 = 41$ is the best. For this Case 1, the final optimal design variables are $FE/FF$ feed ratio at 1.025, $N_1 = 41$, $N_{FE} = 7$, and $N_{FF} = 35$. Similar studies can be done for Cases 2 and 3 to find the optimal design variables for this extractive distillation column.

After the optimal design variables for the extractive distillation column are determined, the total TAC can be calculated with the entrainer recovery column and the recycle stream included. Additional costs in the TAC include the annualized capital cost for the entrainer recovery column, the costs associated with the cooler from *B*2 on the entrainer feed, the operating costs of the steam and cooling water to operate the entrainer recovery column, and the entrainer makeup cost. As an example, Figure 10.13 shows the results of the optimization runs for Case 1 with $N_2$ and $N_{F2}$ as the design variables. The *y* axis is the TAC of the complete flowsheet. From the figure, $N_2$ should be 24 and $N_{F2}$ should be at Stage 9.

Table 10.5 shows the optimal design variables and also the minimized TAC results for Cases 1, 2, and 3. The TAC of Case 1 is the lowest, which verifies the recommendation

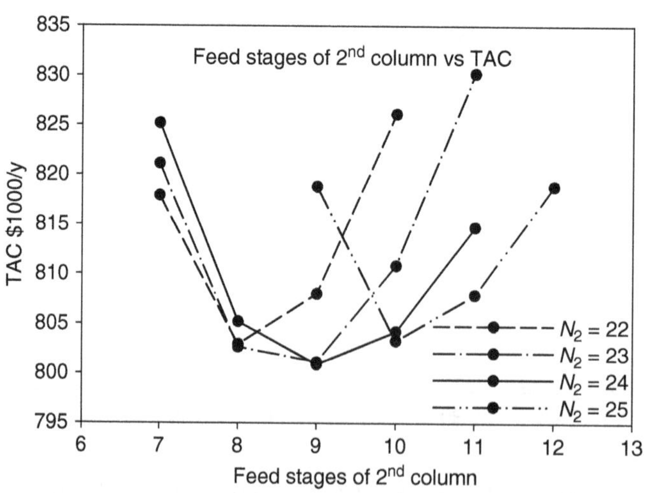


**Figure 10.13** TAC plots for the entrainer recovery column.

**TABLE 10.5 The Optimal Design Variables and the Breakdown of TAC (TAC $ × 1000).**

| | Case 1 | Case 2 | Case 3 |
|---|---|---|---|
| *First column* | | | |
| $FE/FF$ feed ratio | 1.025 | 0.95 | 1.00 |
| Total stages ($N_1$) | 41 | 40 | 41 |
| Entrainer feed stage ($N_{FE}$) | 7 | 6 | 8 |
| Fresh feed stage ($N_{FF}$) | 35 | 34 | 35 |
| Reboiler cost | 55.67 | 55.42 | 55.84 |
| Condenser cost | 32.51 | 45.84 | 29.72 |
| Column cost | 153.53 | 185.41 | 154.42 |
| Tray cost | 14.52 | 19.19 | 14.65 |
| Steam cost | 158.26 | 157.18 | 158.99 |
| Cooling water cost in condenser | 2.87 | 4.86 | 2.50 |
| | **417.36** | **467.90** | **416.12** |
| *Second column* | | | |
| Total stages ($N_2$) | 24 | 23 | 22 |
| Feed stage ($N_{F2}$) | 9 | 10 | 8 |
| Reboiler cost | 44.47 | 45.20 | 44.59 |
| Condenser cost | 25.07 | 25.47 | 25.19 |
| Column cost | 101.77 | 99.34 | 94.65 |
| Tray cost | 8.79 | 8.55 | 8.03 |
| Steam cost | 164.73 | 168.94 | 165.41 |
| Cooling water cost in condenser | 2.74 | 2.81 | 2.76 |
| | **347.57** | **350.31** | **340.63** |
| Entrainer cost | 0.29 | 0.27 | 0.70 |
| Cooler cost | 33.67 | – | 53.96 |
| Cooling water cost in cooler | 2.01 | – | 2.40 |
| TAC | **800.90** | **818.48** | **813.81** |

of Knight and Doherty.[6] Subcooling the entrainer makes the TAC of the extractive distillation column lower; however, an additional cooler cost is incurred. Therefore, this trade-off results in Case 1 being better than Case 3. Compared to the case with no cooling at all (Case 2), there is only a slight reduction (2.1%) of the TAC in the best case (Case 1).

The effect of the feed ratio is also surprisingly small if for each case the optimal design is carefully determined. For example, Figure 10.11 shows the effect of changing the feed ratio on the TAC of the extractive distillation column. In a feed ratio range from 0.9 to 1.1, the best case ($FE/FF = 1.025$) only reduces the TAC from the worse case ($FE/FF = 0.90$) by 0.27%. On the other hand, the determination of the entrainer feed location is relatively more important. With the feed ratio, $N_1$, and $N_{FF}$ at their optimal values, the reduction of the TAC from changing $N_{FE}$ from sixth stage to seventh stage can be as large as 7.3% (see right-hand plot of Fig. 10.12). This reveals that a sufficient number of stages are needed in the rectifying section to satisfy the high-purity specification.

In this study, heat-integration was not considered. Presumably feed-effluent heat exchangers can be placed for preheating of $B1$ into $F2$ by cooling of $B2$ or for preheating of fresh feed into the extractive distillation column by the cooling of $B2$. However, additional heat exchange equipment will be needed, so detailed calculations need to be made to see if it is worthwhile.

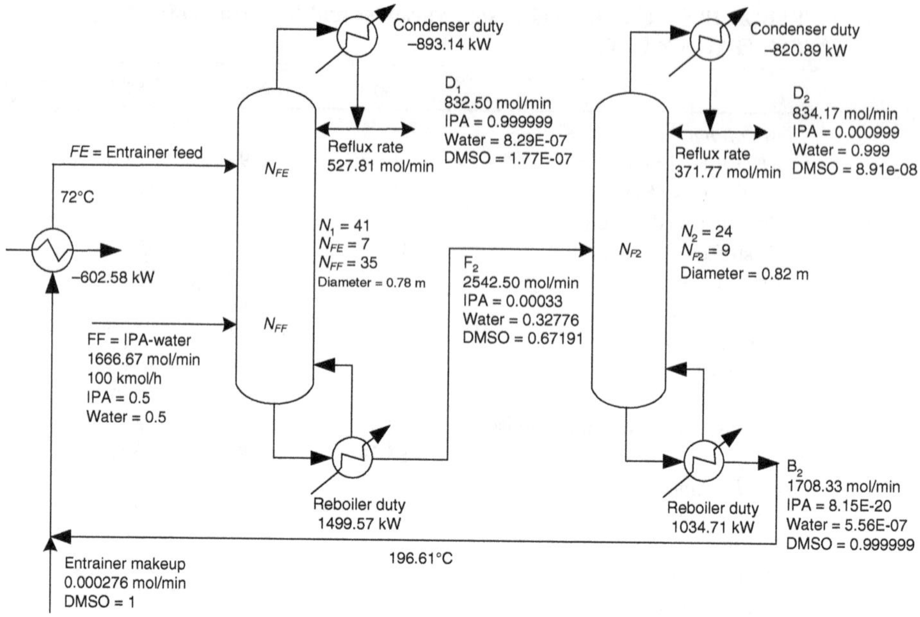


**Figure 10.14** Optimal design flowsheet via extractive distillation.

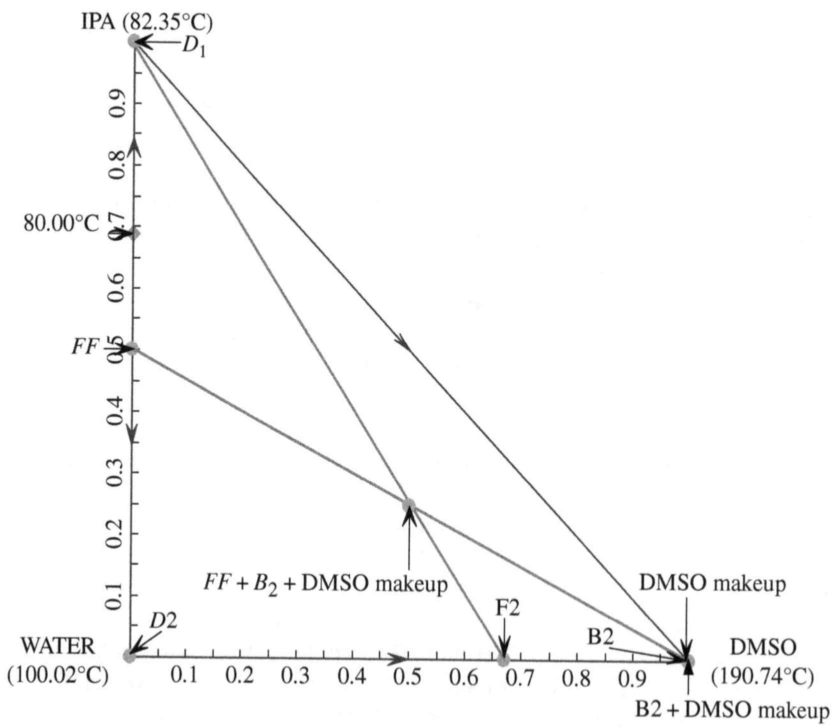


**Figure 10.15** Material balance lines for this IPA dehydration process.

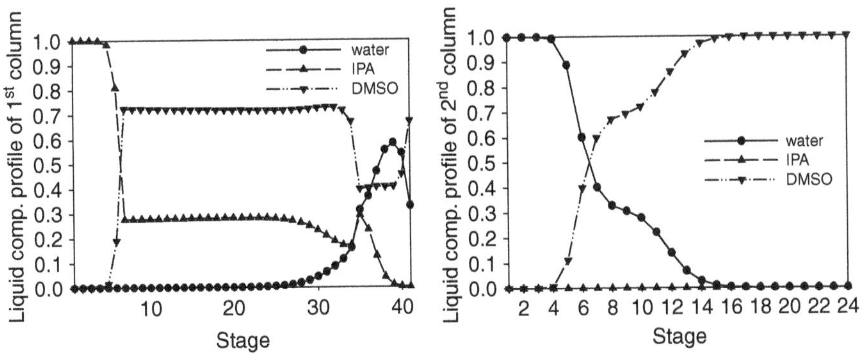


**Figure 10.16** Liquid composition profile for the two columns.

The final optimal flowsheet for this system is shown in Figure 10.14, and the material balance lines for this separation process are demonstrated in Figure 10.15. The liquid composition profiles of the two columns for the flowsheet in Figure 10.14 are shown in Figure 10.16. From the left-hand plot of Figure 10.16, it is noticed that the goal of the extractive section is fulfilled in preventing water from going into the rectifying section. Because the product specification is ultra-pure, quite a few stages are needed to serve this purpose. We will compare this flowsheet to the optimized design flowsheet via heterogeneous azeotropic distillation next.

### 10.2.2 Comparison to the Design Flowsheet Using Heterogeneous Azeotropic Distillation

In Chapter 8, and also in a paper by Arifin and Chien,[5] an overall design flowsheet using cyclohexane as the entrainer, via heterogeneous azeotropic distillation, was developed. Two columns were also used in their overall design flowsheet with one column serving as the combined preconcentrator/recovery column and another one as the heterogeneous azeotropic column with a decanter. The optimal design flowsheet of the two-column system in Chapter 8 is again shown in Figure 10.17.

Since the feed rate, feed composition, and the two product specifications are exactly the same, a direct comparison can be made. Comparing the optimal flowsheet developed in this chapter using extractive distillation to that of Figure 10.17 in which heterogeneous azeotropic distillation is used, it is found that the extractive distillation column is much taller ($N_T = 40$ if not including condenser) than the heterogeneous azeotropic column ($N_T = 18$). However, the extractive distillation column is much thinner (column diameter of 0.78 m) than the heterogeneous azeotropic column (column diameter of 1.61 m). The TAC of the overall process using extractive distillation is $8.009 × $10^5$/y (see Table 10.5). This can be directly compared to the TAC of the overall process using heterogeneous azeotropic distillation ($1.190 × $10^6$/y, see Table 8.3 in Chapter 8). This shows that the separation using extractive distillation reduces the TAC by as much as 32.7%. There is also a significant reduction in steam cost for the separation. The total steam cost for the overall process using extractive distillation from Table 10.5 is $3.230 × $10^5$/y while using heterogeneous azeotropic distillation from Table 8.3 of Chapter 8 is $4.637 × $10^5$/y. This represents a reduction of 30.3% of the cost of energy.

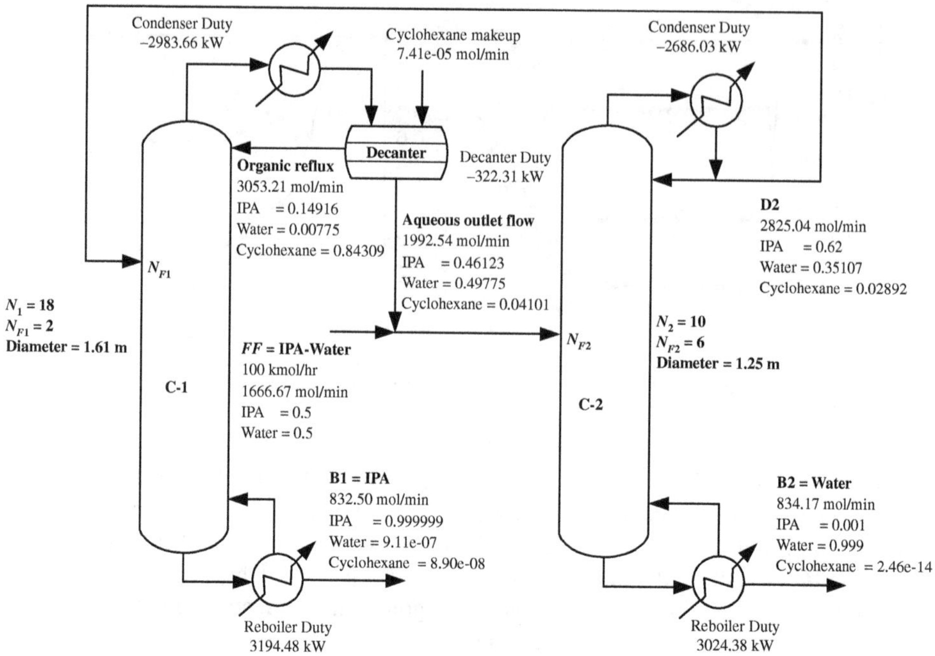


**Figure 10.17** Optimal design flowsheet via heterogeneous azeotropic distillation.

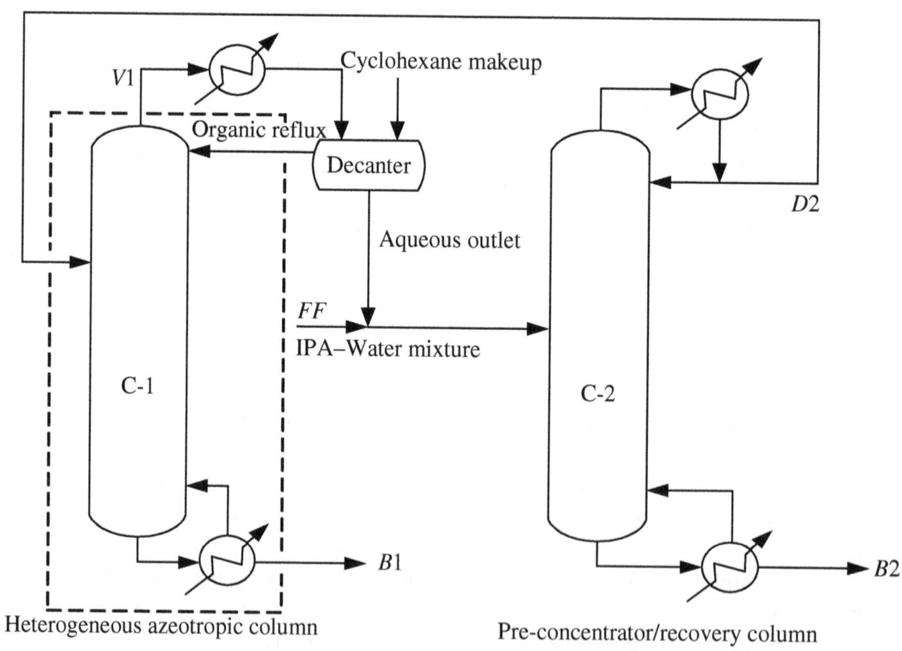


**Figure 10.18** Conceptual design flowsheet via heterogeneous azeotropic distillation.

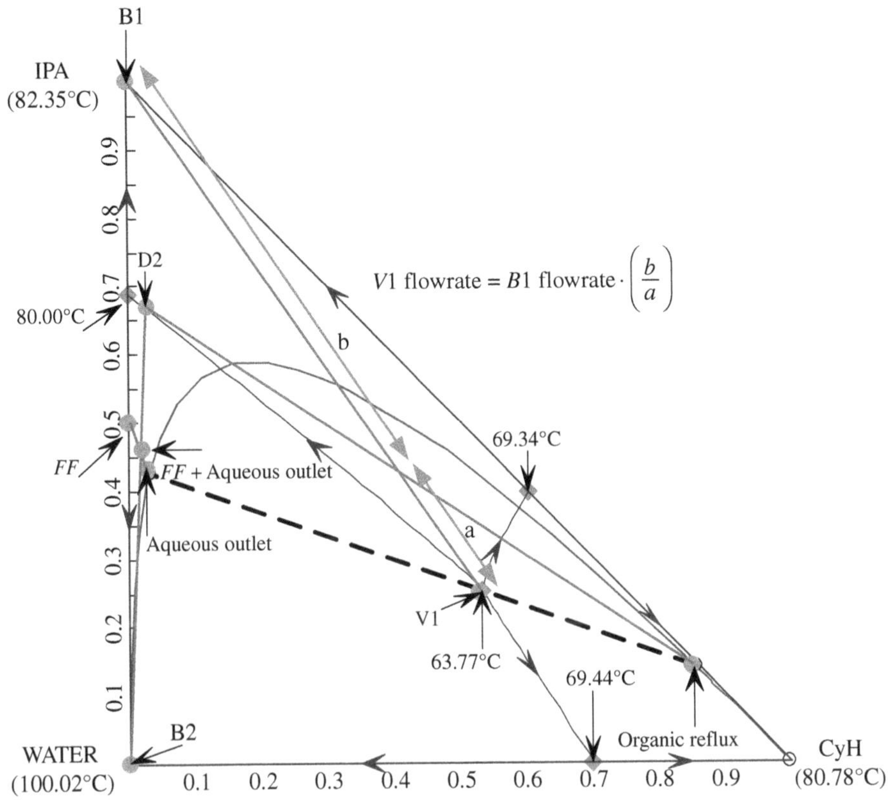


**Figure 10.19** Material balance lines via heterogeneous azeotropic distillation.

The main reason for the high TAC and energy requirement via heterogeneous azeotropic distillation for this chemical system can be explained by the particular RCM and material balance lines for the feasible separation. A conceptual diagram of the separation via heterogeneous azeotropic distillation can be seen in Figure 10.18 with the dashed-line showing the material balance around the heterogeneous azeoptropic distillation column. The material balance lines in Figure 10.19 represent the ideal case with the column bottoms reaching the composition of pure IPA, the column top vapor approaching the ternary azeotrope, and $D2$ is right at the distillation boundary. With this ideal case, we can easily estimate top vapor flow using the formula noted in Figure 10.19. In this calculation of the top vapor flowrate, $B1$ can be estimated to be half of the fresh feed flowrate, and "$b$" and "$a$" can easily be determined by a line between $D2$ and organic reflux (OR) and another line between $B1$ and $V1$. This means that the top vapor flowrate will be quite large, resulting in a high OR flowrate back into the column and also a high aqueous outlet (AO) flowrate into the $C2$ column. This, in turn, requires more reboiler duty for the two columns to generate a vapor rate going up the column and larger diameters of the two columns.

On the other hand, there is no distillation boundary in the RCM for the separation using extractive distillation. In fact, the reflux flowrate for the extractive distillation column is only 527.81 mol/min (comparing an OR flowrate of 3053.21 mol/min in the heterogeneous azeotropic distillation system). The two only possible drawbacks for the extractive distillation process are that the extractive distillation column needs to be very tall (this may cause some difficulty for the construction of this column) and the two columns need to be operated at higher temperatures (higher pressure steam will be needed for the two reboilers).

### 10.2.3 Ideal Entrainer Using Heterogeneous Azeotropic Distillation

The above analysis shows that the particular RCM and material balance lines for the feasible separation using heterogeneous azeotropic distillation with cyclohexane as entrainer has its deficiency. We would like to discuss further what will be a more ideal entrainer for the IPA dehydration process. Figure 10.20 shows the material balance lines for a heterogeneous azeotropic distillation when benzene is used as the entrainer. The ternary azeotrope is located lower in the ternary diagram, and the tie-line that passes through the ternary azeotrope has a different slope than in the cyclohexane entrainer system. With the ternary azeotrope and the OR located more favorably when using benzene as the entrainer, the estimated *V*1 flowrate is much less, which results in less OR flowrate back to the heterogeneous azeotropic column, and also much less AO flowrate into the combined preconcentrator/recovery column.

Although benzene is a more favorable entrainer for this separation or similar ethanol dehydration process, its use is prohibited because of the carcinogenic nature of this entrainer. Having insight into what constitutes a good entrainer will help to identify other potential candidates as entrainers.

The best ideal entrainer for the separation using heterogeneous azeotropic distillation is one that only forms one additional minimum-boiling heterogeneous azeotrope with water.

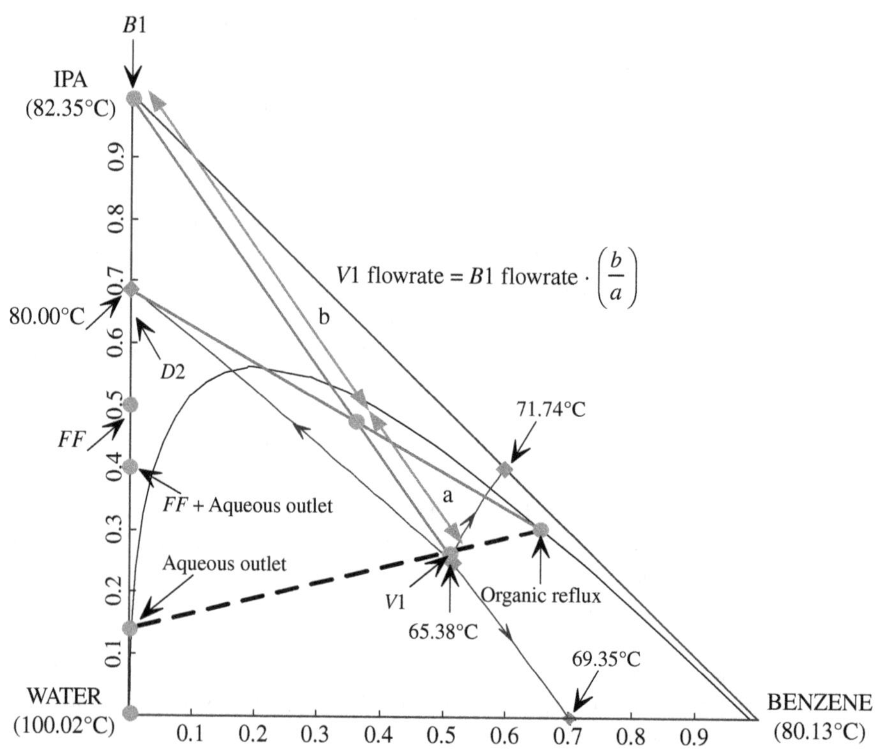


**Figure 10.20** Favorable location of the ternary azeotrope via heterogeneous azeotropic distillation using benzene as entrainer.

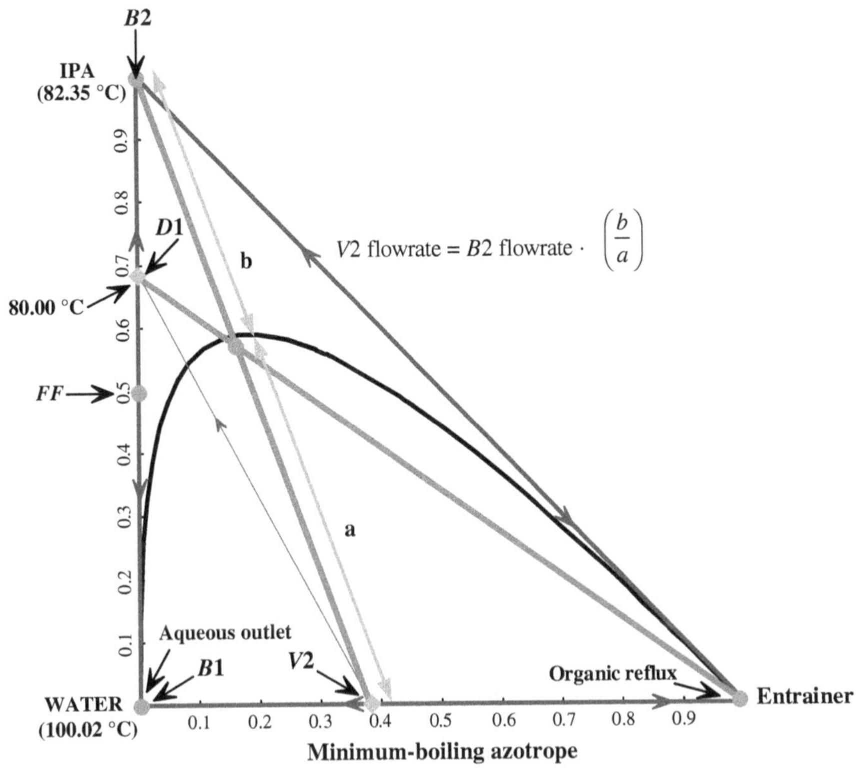


**Figure 10.21** Favorable RCM type for the heterogeneous azeotropic distillation.

Figure 10.21 shows the material balance lines of a feasible design flowsheet with a preconcentrator and a heterogeneous azeotropic distillation column (see Fig. 8.3). Analysis of Figure 10.21 demonstrates that the top vapor flowrate (*V*2) of this design flowsheet is further reduced, which makes this design flowsheet more competitive with extractive distillation. However, this ideal entrainer is yet to be found.

## 10.3 OVERALL CONTROL STRATEGY DEVELOPMENT

The steady-state economics of the overall extractive distillation process are much better than the economics of the heterogeneous azeotropic distillation process. In the following, we will investigate the control of the extractive distillation process using DMSO as shown in Figure 10.14. We would like to restrict ourselves to use tray temperature control loop(s) to indirectly hold the product purity to facilitate wider industrial application. Large feed disturbances ($\pm 20\%$ changes in the feed composition and also $\pm 20\%$ changes in the fresh feed flowrate) will be used to test the proposed overall control strategy.

Pressure-driven simulation in Aspen Dynamics is used in the control strategy development. Before converting the Aspen Plus simulation to Aspen Dynamics, sizing of all equipments is needed. The tray sizing tool in Aspen Plus is used to calculate the column diameters of both columns to be 0.78 m and 0.82 m for the first and the second column, respectively. Tray spacing and weir height of both columns are assumed to be 0.6096 m and 0.0508 m, respectively. A 10 min holdup time with 50% liquid level is used to calculate the volume of each column base and reflux drum. Top pressures of both columns are set at atmospheric pressure. The tray rating tool in Aspen Plus is used to automatically calculate the pressure drop in the columns.

### 10.3.1 Inventory Control Loops

Some simple regulatory control loops are determined first. The levels of the reflux drums for both columns are controlled by manipulating the distillate. Top pressures of both columns are controlled by the condenser duty. The bottom level of the extractive distillation column is controlled by manipulating the bottoms flow. The entrainer feed temperature is controlled at 72°C by manipulating the cooler duty. The initial control structure tested is one in which the reflux ratios in the two columns are controlled by manipulating the reflux. This was used in the overall control structure in Luyben.[9]

An important inventory control loop in this overall process is the bottom level of the entrainer recovery column. The control of this level was suggested by Grassi[10] and Luyben[9] to be held by the entrainer makeup flow. However, because this flow is very small, the bottom level essentially floats as changes in the entrainer flowrate occur. With this control pairing, the entrainer feed to the first column is flow-controlled in Grassi.[10] We adapted this control pairing for the overall control strategy in our study.

Luyben[9] also suggested ratioing the entrainer flow to the column to the feed flow. In addition, a control structure that always maintained an excess of solvent was recommended. A first-order lag, a high selector, and a low selector were used to always have an excess of entrainer. The entrainer "leads" the feed for an increase in feed and "lags" the feed for a decrease.

### 10.3.2 Tray Temperature Control Point(s)

The above control strategy leaves two reboiler duties that can be used in a tray temperature control loop in each column. Figures 10.22 and 10.23 display the results of open-loop

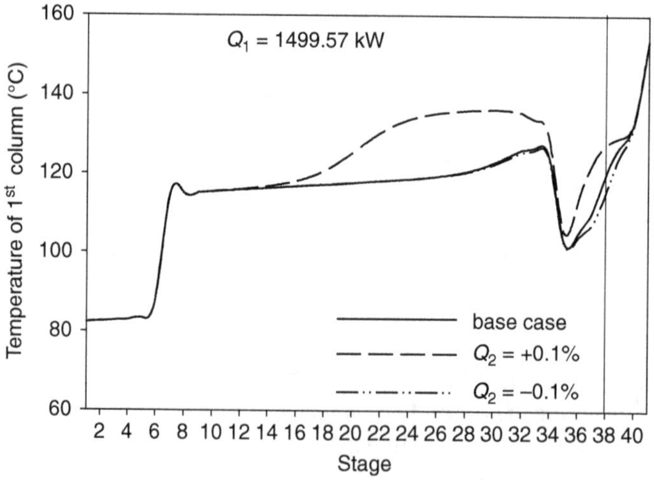


**Figure 10.22** Open-loop sensitivity plots for ±0.1% changes in $Q_1$.

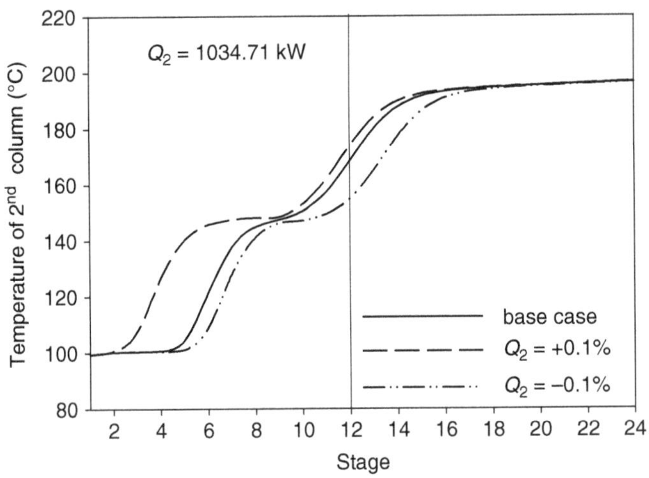


**Figure 10.23** Open-loop sensitivity plots for $\pm 0.1\%$ changes in $Q_2$.

sensitivity analysis with $\pm 0.1\%$ changes in reboiler duty of the extractive distillation column and the entrainer recovery column, respectively. The temperature control point is selected at a particular stage with high sensitivity and also with near linear behavior. For the extractive distillation column, temperature at Stage 38 (location shown with a vertical line in Fig. 10.22) is chosen as the control point. For the entrainer recovery column, temperature at Stage 12 (location also shown with a vertical line in Fig. 10.23) is chosen as the control point. There are two temperature breaks for the entrainer recovery column. However, the temperature control point in the rectifying section (Stage 6) is not chosen because of an inferior open-loop dynamic response owing to a larger apparent deadtime.

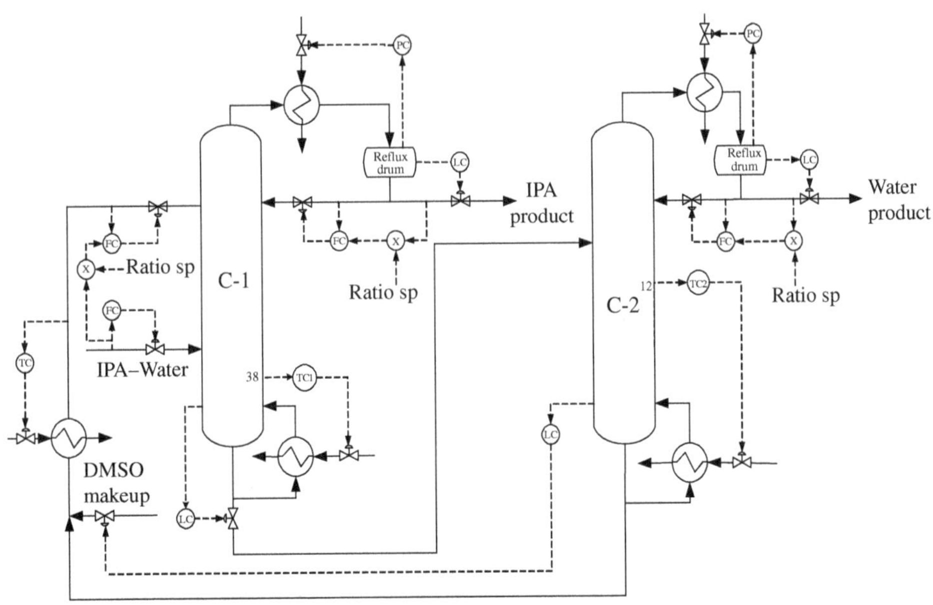


**Figure 10.24** Overall control strategy with reflux ratio fixed.

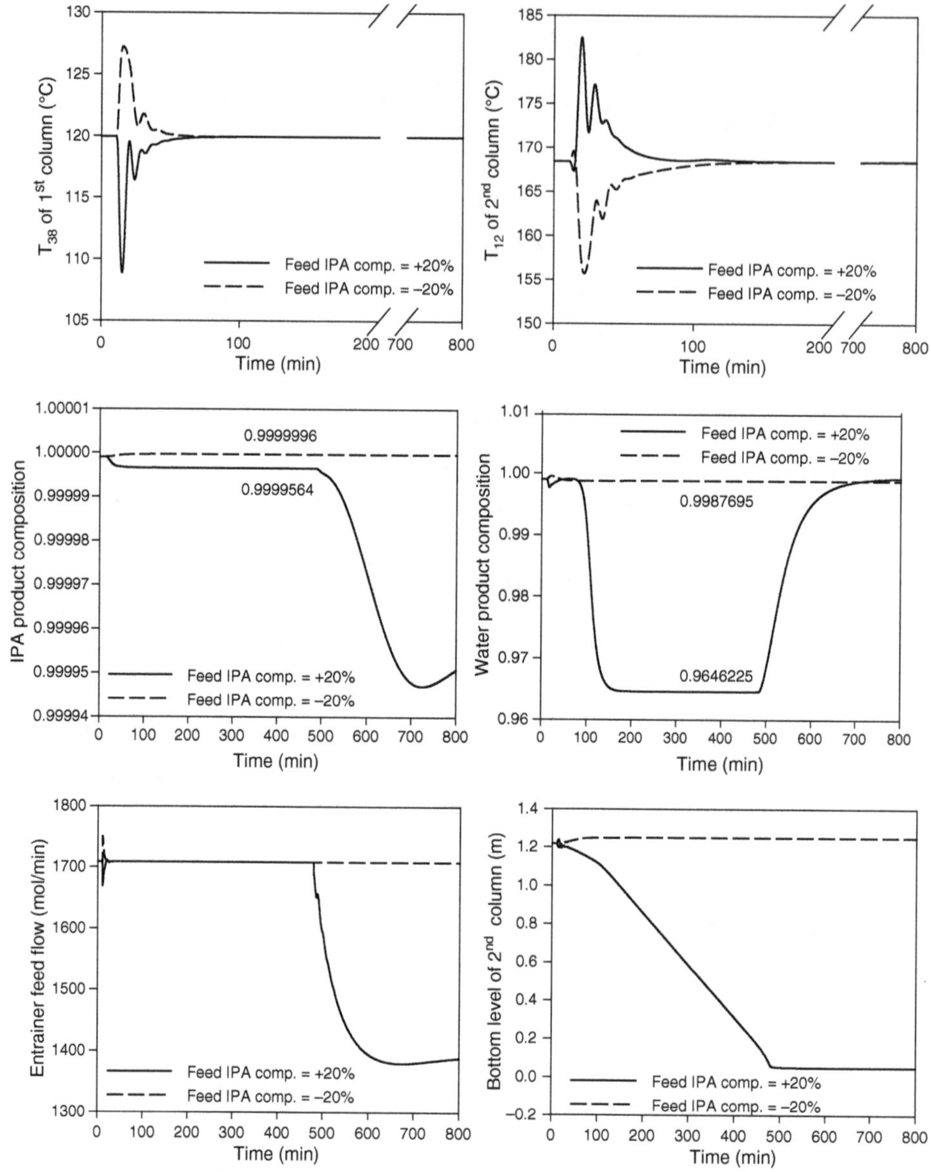


**Figure 10.25** ±20% changes in the feed IPA composition using above control strategy.

Another reason is that it is more important to keep water from dropping out the bottom and messing up the extractive column than it is to keep DMSO out of the water product. Thus, a temperature lower in the column is selected. The control strategy of the overall process is summarized in Figure 10.24.

P-only controllers with $K_C = 2$ are used for both reflux drum levels as suggested in Luyben[11] and $K_C = 10$ is used for both base bottom levels for faster dynamics of the internal flow of the overall process and also for faster increase or decrease of entrainer makeup into the system. The PI settings of the top pressure control loops for both columns are set at $K_C = 20$ and $\tau_I = 12$ min. For the two crucial tray temperature loops, open-loop tests were performed for determining the PI tuning constants following the IMC–PI tuning rule of Chien and Fruehauf[12] with the assumption of integrating plus deadtime model form for the initial dynamic response. The results of those calculations are $K_C = 1.54$ and $\tau_I = 7.5$ min for the tray temperature loop in the extractive distillation column and $K_C = 1.72$ and $\tau_I = 13.75$ min for the tray temperature loop in the entrainer recovery column.

### 10.3.3 Simulation Results

The unmeasured feed IPA composition disturbances will be used to test this control strategy with $\pm 20\%$ changes in the IPA in fresh feed composition at time = 10 min. Figure 10.25 shows the closed-loop results for this unmeasured disturbance. From the top graphs in Figure 10.25, it is found that the two tray temperature control loops perform well and bring the temperatures back to their setpoints. The middle graphs in Figure 10.25 shows that the IPA product composition is still maintained at high-purity for the first 500 min but becomes less pure after that time. The purity of the water product drops to unacceptable levels between 100 to 500 min for the +20% change in the feed IPA composition. The bottom graphs in Figure 10.25 show the problem with this overall control strategy. The bottom level of the entrainer recovery column drops continuously for the +20% change of the feed IPA composition until almost empty. At that time, the entrainer fresh feed valve is wide open and the entrainer feed to the extractive column cannot be maintained at its setpoint.

For the +20% change in the feed IPA composition, the water product flowrate should decrease. By holding the reflux ratio at a nominal constant value, the reflux flowrate will decrease, which results in less materials going down the column and a drop in the purity of the water product.

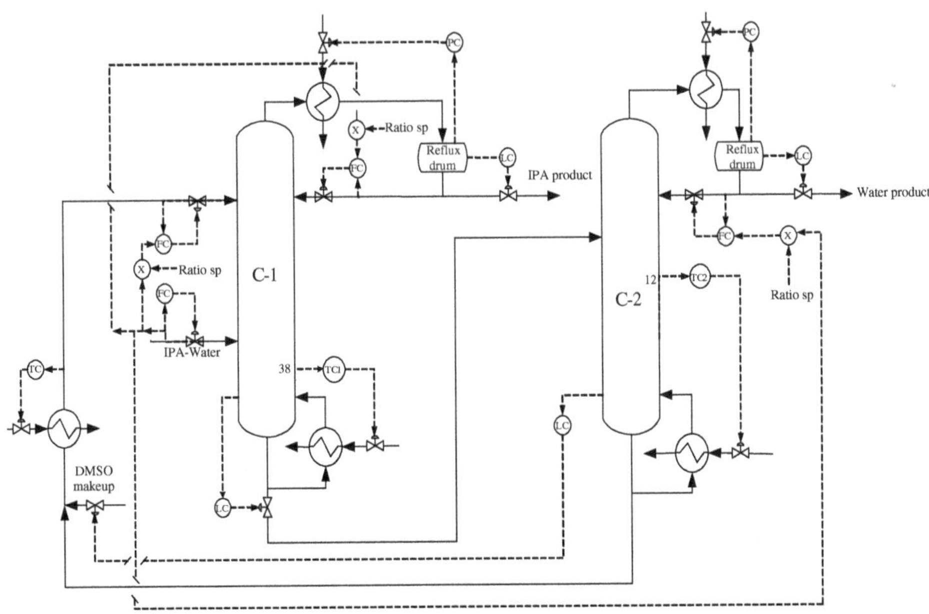


**Figure 10.26** Proposed overall control strategy.

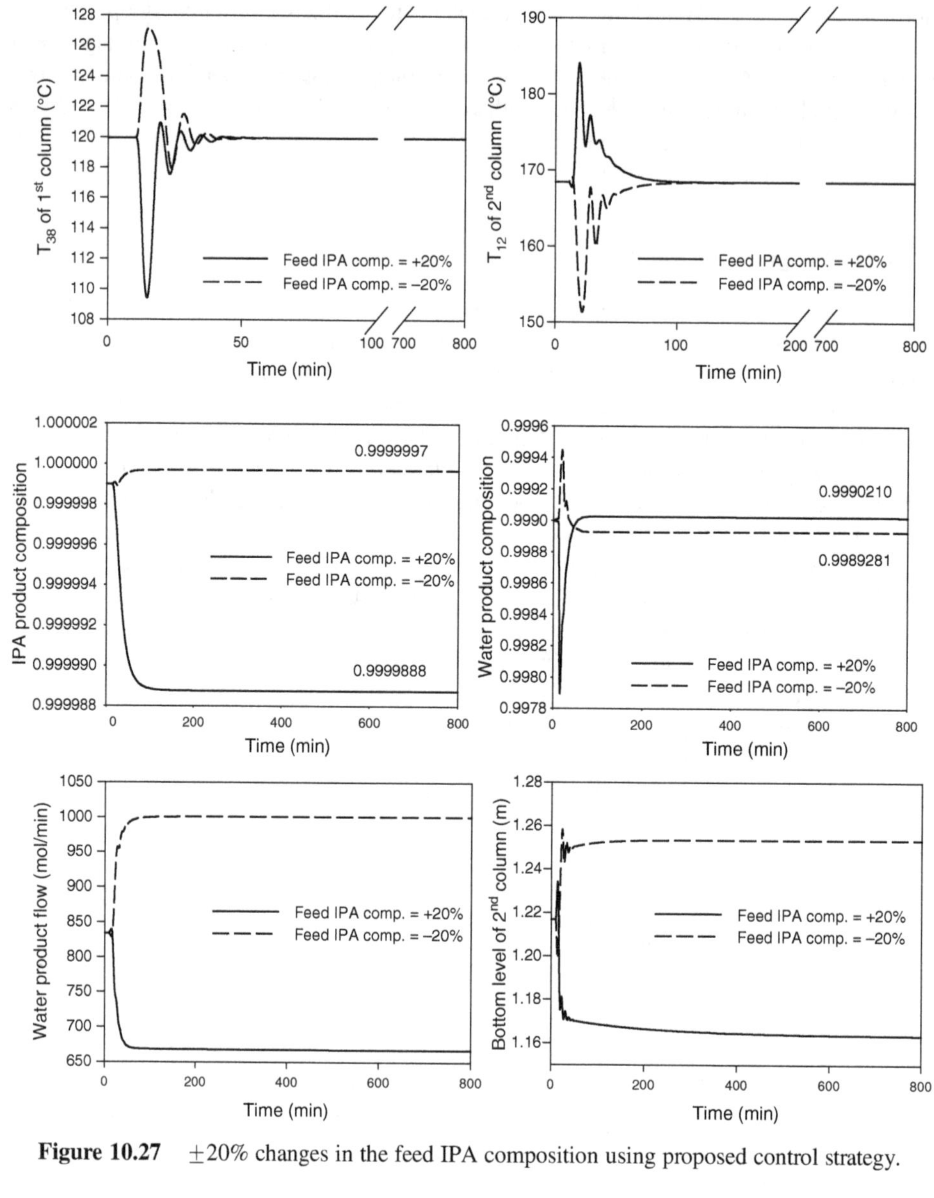


**Figure 10.27** ±20% changes in the feed IPA composition using proposed control strategy.

A modified overall control strategy is to ratio both reflux flowrates to the fresh feed flowrate. This modified overall control strategy was recommended in Grassi[10] and is shown in Figure 10.26. Notice that the temperature control points for both columns are the same as with the other reflux-ratio structure because there was little change in the open-loop sensitivity analysis from those shown in Figures 10.22 and 10.23.

Figure 10.27 shows the closed-loop responses when this overall control strategy is used with the same unmeasured feed composition disturbances as in Figure 10.25. As shown in the middle graphs in Figure 10.27, the two product compositions are maintained much closer to their specifications. From the bottom graphs in Figure 10.27, the bottom level of the

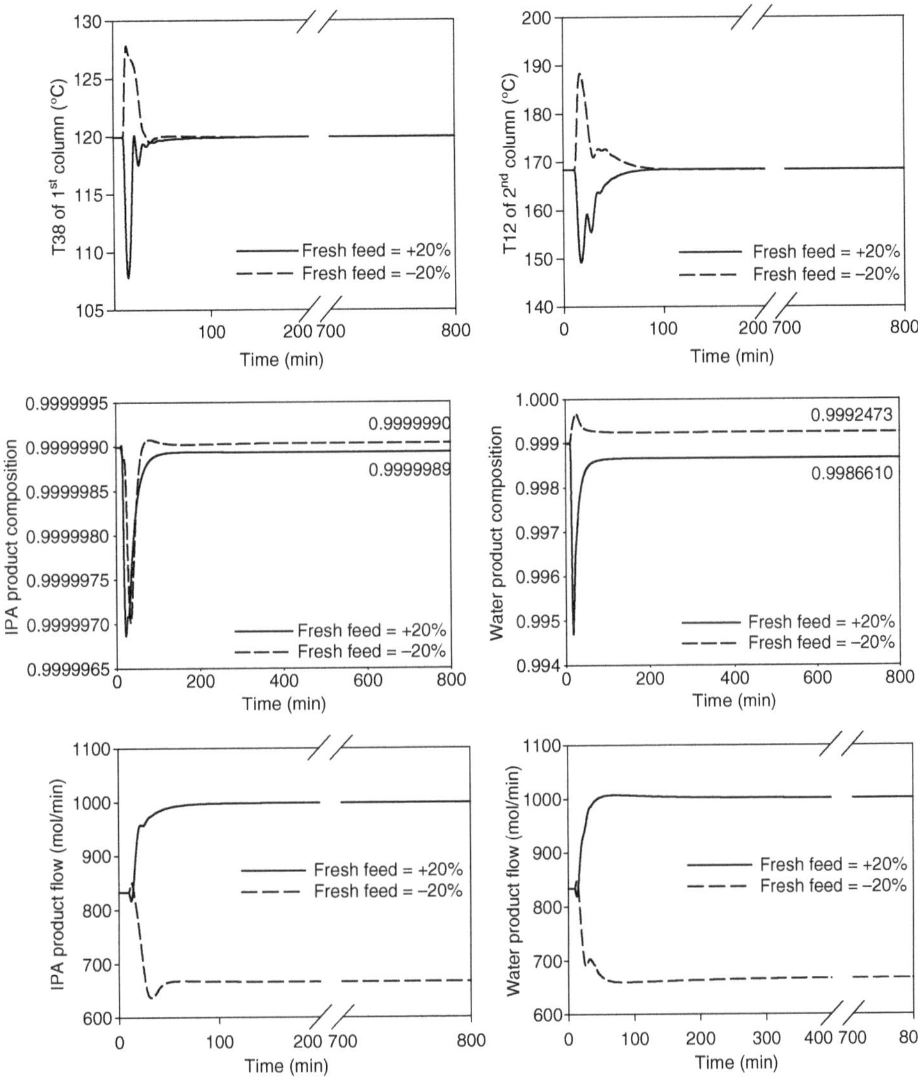


**Figure 10.28** ±20% changes in the fresh feed flow rate using proposed control strategy.

entrainer recovery column is maintained without the problem exhibited when the reflux ratio scheme was used. The bottom left graph in Figure 10.27 also shows that the water product flow increased or decreased accordingly to the feed IPA composition changes.

Another feed disturbance to test the overall control strategy is throughput changes. These changes are made by increasing or decreasing of the fresh feed flowrate. Figure 10.28 shows the closed-loop responses using the overall control strategy in Figure 10.26 with ±20% changes of the fresh feed flowrate. With this control strategy, the two reflux flowrates and the entrainer feed flowrate will be changed in accordance with the fresh feed flowrate. From the middle graphs in Figure 10.28 it is found that the overall control strategy performed nicely in maintaining high-purity of the two products. As shown in the bottom graphs in Figure 10.28, both IPA and water product flowrates increase or decrease as expected to new values.

## 10.4 CONCLUSION

In the first part of this chapter, the important factors of an effective entrainer for separation of an azeotrope using extractive distillation have been demonstrated using IPA dehydration as an example. The isovolatility curve can be used to determine the feasible distillate product in the extractive distillation column. The isovolatility and equivolatility curves can be used to determine which candidate entrainer is more effective in enhancing the relative volatility of the azeotropic mixture. The VLE information of the binary pairs between the candidate entrainer and either of the two original components should also be plotted to make sure there is no problem for the separation in the rectifying section of the extractive distillation column and in the entrainer recovery column.

In the latter part of this chapter, design and control of an IPA dehydration process using extractive distillation is thoroughly investigated. Using TAC as the objection function, the optimal design flowsheet of the process of a two-column system is developed. This optimal design flowsheet can be compared directly to an earlier study that used heterogeneous azeotropic distillation. The extractive distillation process is much more economical than the heterogeneous azeotropic distillation process. The TAC is reduced by as much as 32.7%, and the steam cost is cut by 30.3%. The two drawbacks for the extractive distillation process are a taller extractive distillation column and the higher base temperatures, which require higher pressure steam in the two reboilers.

As for the overall control strategy of the process, it is found that fixing of the two reflux ratio does not work for +20% changes of the feed IPA composition. The recommended overall control strategy is the one that ratios the reflux flowrates of the two columns to the fresh feed flowrate. The high-purity of the two products can be maintained despite very large feed disturbance variations.

## REFERENCES

1. Arifin S. and I. L. Chien, Design and control of an isopropyl alcohol dehydration process via extractive distillation using dimethyl sulfoxide as an entrainer, *Ind. Engng Chem. Res.*, **47**, 790–803 (2008).
2. Laroche L., N. Bekiaris, H. W. Andersen, and M. Morari, Homogeneous azeotropic distillation. Comparing entrainers, *Can. J. Chem. Eng.*, **69**, 1302–1319 (1991).
3. Gmehling J. and C. Möllmann, Synthesis of distillation process using thermodynamic models and the Dortmund Data Bank, *Ind. Engng Chem. Res.*, **37**, 3112–3123 (1998).
4. Chien I. L., K. L. Zeng, and H. Y. Chao, Design and control of a complete heterogeneous azeotropic distillation column system, *Ind. Engng. Chem. Res.*, **43**, 2160–2174 (2004).
5. Arifin S. and I. L. Chien, Combined preconcentrator/recovery column design for isopropyl alcohol dehydration process, *Ind. Engng Chem. Res.*, **46**, 2535–2543 (2007).
6. Knight J. R. and M. F. Doherty, Optimal design and synthesis of homogeneous azeotropic distillation sequences, *Ind. Engng Chem. Res.*, **28**, 564–572 (1989).
7. Doherty M. F. and M. F. Malone, *Conceptual Design of Distillation Systems*, McGraw-Hill, New York, NY, 2001.
8. Douglas J. M., *Conceptual Design of Chemical Processes*, McGraw-Hill, New York, NY, 1988.

9. Luyben W. L., Plantwide control of an isopropyl alcohol dehydration process, *AIChE J.*, **52**, 2290–2296 (2006).
10. Grassi V. G., Process design and control of extractive distillation. In *Practical Distillation Control*, Van Nostrand Reinhold Press, New York, 1992, pp. 370–404.
11. Luyben W. L., *Plantwide Dynamic Simulators in Chemical Processing and Control*, Marcel Dekker, New York, 2002.
12. Chien I. L. and P. S. Fruehauf, Consider IMC tuning to improve controller performance, *Chem. Eng. Prog.*, **86**, 33–41 (Oct., 1990).

# CHAPTER 11

# EXTRACTIVE DISTILLATION OF THE ACETONE–METHANOL SYSTEM

In Chapter 10 the use of extractive distillation for the separation of the homogeneous minimum-boiling azeotropic isopropanol–water system was explored. In this chapter we look at another system in which extractive distillation can be effectively used. The separation of the acetone–methanol azeotrope was discussed in Chapter 5 using pressure-swing distillation. Extractive distillation is an alternative for this separation. A comparison of the two designs is presented in this chapter. Both steady-state economics and dynamic control are compared. Heat integration can be used in either system, and heat-integrated designs are also considered. Purities of the two products are the same in all flowsheets (99.5 mol%).

In a final section of this chapter, we look at the use of other solvents beside water. The purpose is to see the effect of solvent selection on dynamic controllability.

## 11.1 ACETONE–METHANOL–WATER PHASE EQUILIBRIUM

Acetone and methanol have very similar normal boiling points (329.2 and 337.5 K) and form a homogeneous minimum-boiling azeotrope at 1 atm with a composition 77.6 mol% acetone at 328 K. Figure 11.1 gives the *Txy* diagram for the binary system.

Extractive distillation is feasible using water as the solvent. Figure 11.2 gives the ternary diagram for acetone–methanol–water.

*Design and Control of Distillation Systems for Separating Azeotropes.* By William L. Luyben and I-Lung Chien

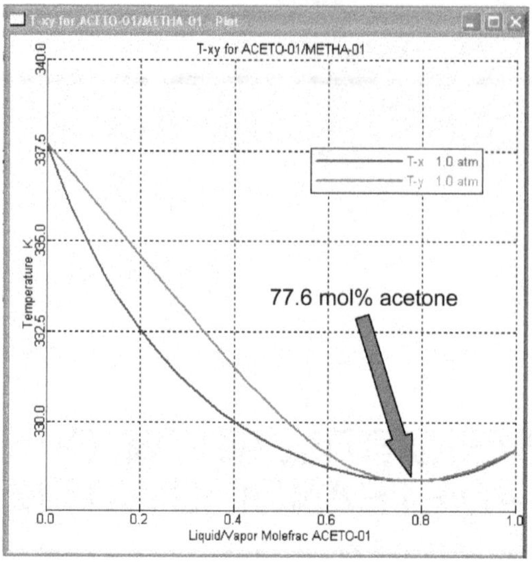


**Figure 11.1** *Txy* diagram for acetone–methanol at 1 atm.

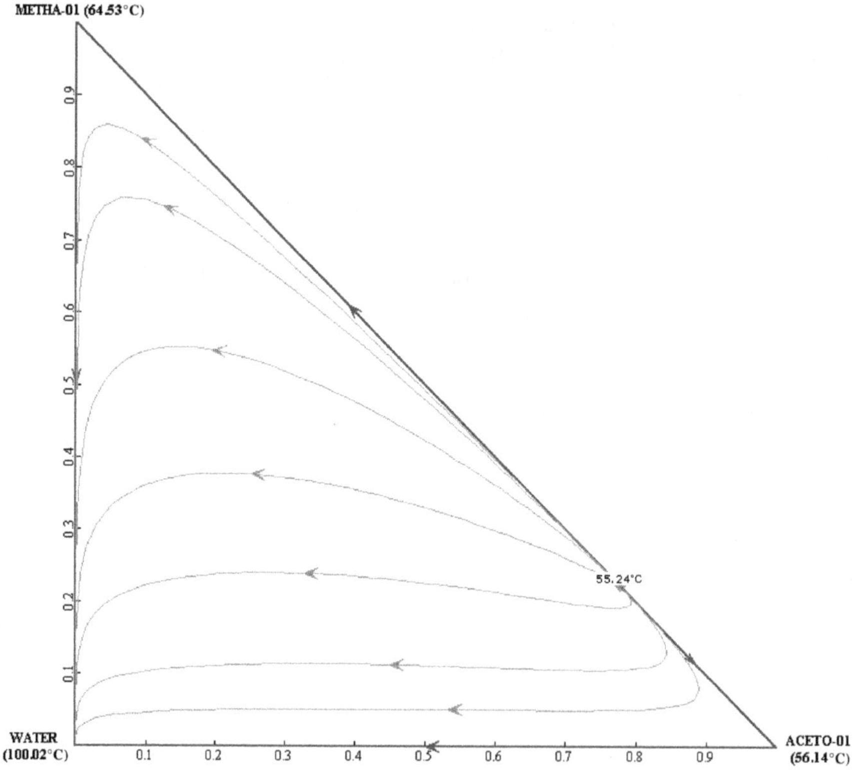


**Figure 11.2** Ternary diagram for acetone–methanol–water.

## 11.2 STEADY-STATE DESIGN

Knapp and Doherty[1] present the economic optimum design of an acetone–methanol separation using water as the extractive solvent. The design used in this chapter is based on their work. Kossack et al.[2] presented a systematic synthesis framework for extractive distillation systems and the acetone–methanol system was considered.

The Knapp–Doherty optimum design had a 57-stage extractive column and a 26-stage methanol column, both operating at 1 atm. They do not provide information about feed tray locations, so these were found empirically by finding what locations give minimum energy consumption. Note that the solvent is fed to the extractive column on a tray above the feed tray and below the top of the column so that high-purity acetone can be produced in the distillate.

Figure 11.3 shows the flowsheet used in this chapter with stream information, heat duties, equipment sizes, and operating conditions. Notice that a small make-up water stream is required to account for the small losses of water in both product streams. Notice also, that there is more water impurity in the acetone product leaving in the distillate from the extractive column than methanol (0.39 mol% water versus 0.11 mol% methanol), despite the fact that water has a higher boiling point than methanol. A cooler is also required to cool down the water recycle from the bottom of the methanol column from 378 to 320 K before it is fed into the extractive column.

This flowsheet differs from that given in the Knapp–Doherty[1] design in one important feature: the solvent to feed ratio. Knapp and Doherty use a ratio of unity. For a feed flowrate of 540 kmol/h, this corresponds to a solvent flowrate to the extractive column of 540 kmol/h. However, as Figure 11.3 shows, a solvent flowrate of 1100 kmol/h is used

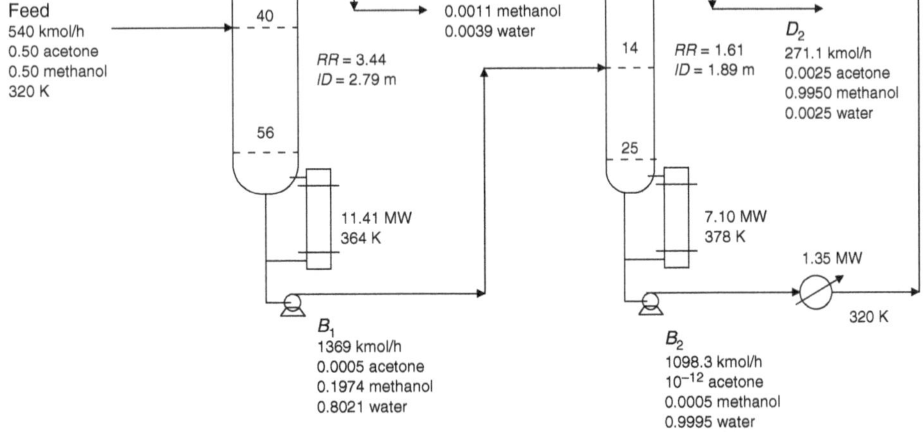


**Figure 11.3** Flowsheet of extractive distillation system; acetone–methanol.

in this paper. We are unable to get the desired purity of the acetone product at lower solvent flowrates. The acetone–methanol system using water as the solvent is also explored by Kossack et al.,[2] and they report (their Table 5) that the economic optimum design had a solvent-to-feed ratio of 1.886. This is much closer to our results.

The other difference between this design and the Knapp and Doherty design is the reflux ratios in the two columns. They report 2.76 and 1.06, while the present design values are 3.44 and 1.61. This difference is probably due to our higher solvent flowrate.

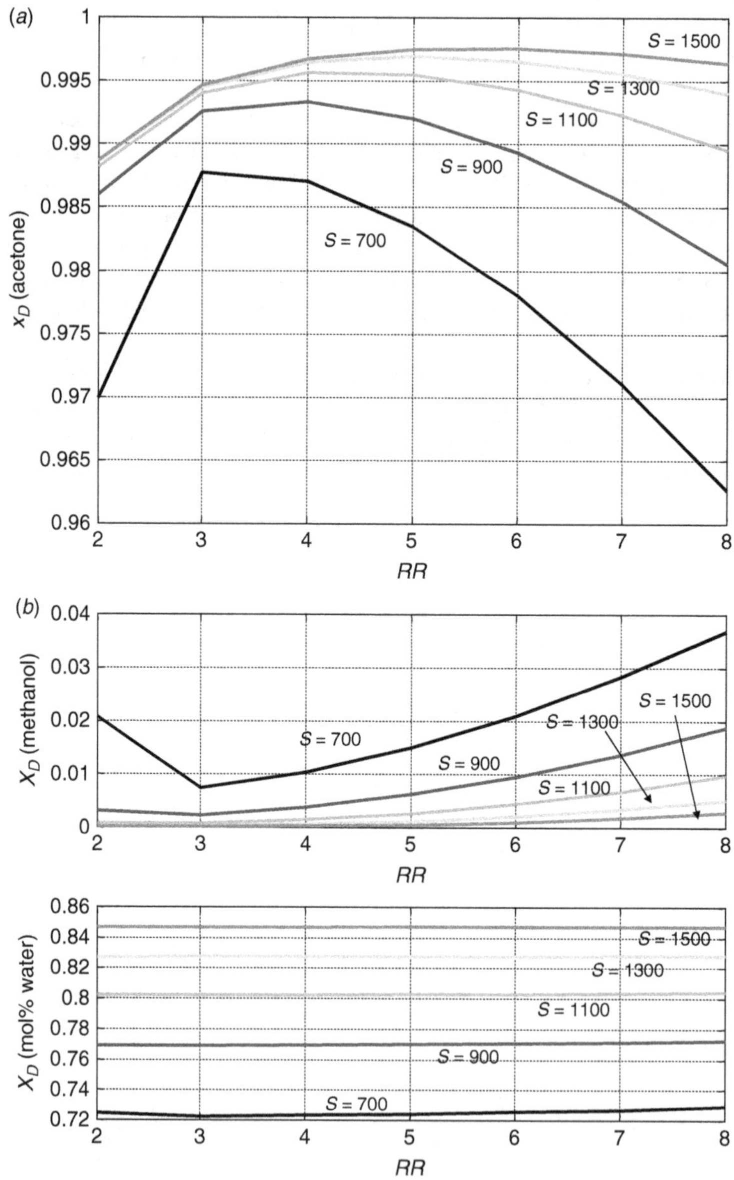


**Figure 11.4** (*a*) Effect of *RR* and solvent in extractive column on acetone purity. (*b*) Effect of *RR* and solvent in extractive column on impurities.

The extractive column has three design degrees of freedom once the total stages and feed locations are fixed: reflux ratio, solvent flowrate, and reboiler heat input. Figure 11.4 shows the effects of changing the solvent flowrate and the reflux ratio on the compositions of the distillate stream from the extractive column, which is the acetone product. The composition of acetone in the bottoms is held at 0.05 mol% acetone by manipulating the reboiler heat input for all the cases shown in Figure 11.4. Figure 11.4*a* shows that for a given solvent flowrate *S*, there is an optimum reflux ratio *RR* that gives a maximum acetone purity. The higher the solvent flowrate, the higher the acetone purity. To achieve the desired 99.5 mol% acetone purity, the solvent flowrate must be about 1100 kmol/h at a reflux ratio of about 4. Figure 11.4*b* shows how the impurities of methanol and water are affected by solvent and reflux ratio. Increasing the solvent flow decreases the methanol impurity but increases the water impurity.

These results illustrate that it can be difficult to achieve the desired acetone product purity in this extractive distillation because of the competing effects of solvent flowrate and reflux ratio. This problem becomes less severe as the volatility difference between the light key component and the solvent increases. The acetone–water system does not have a large volatility, which leads to difficulty in attaining the desired acetone purity. Other solvents with higher boiling points should permit higher acetone purities. However, the base temperature in the solvent recovery column would be higher, which would require a higher-cost energy source. The nonmonotonic relationship between acetone purity and reflux ratio can also lead to difficult operating problems, as pointed out by Knapp and Doherty.[1]

This inherent problem of an extractive distillation should be compared to the ease of attaining higher purity levels in a pressure-swing distillation system, as discussed in Chapter 5. As Figure 11.5 illustrates, very high product purities can easily be obtained with only modest increases in energy costs in the pressure-swing distillation system.

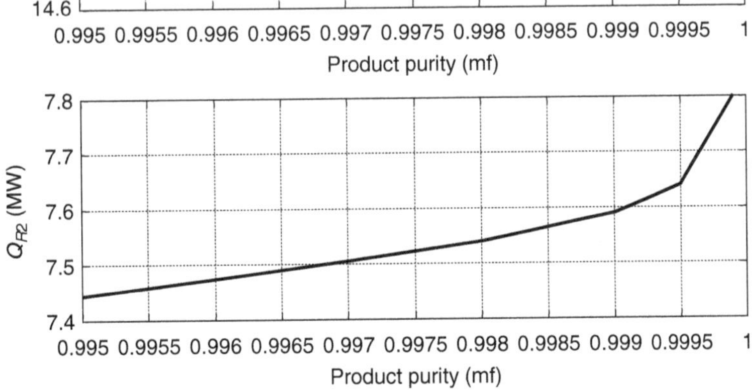


**Figure 11.5** Energy demand in pressure-swing distillation for high purities.

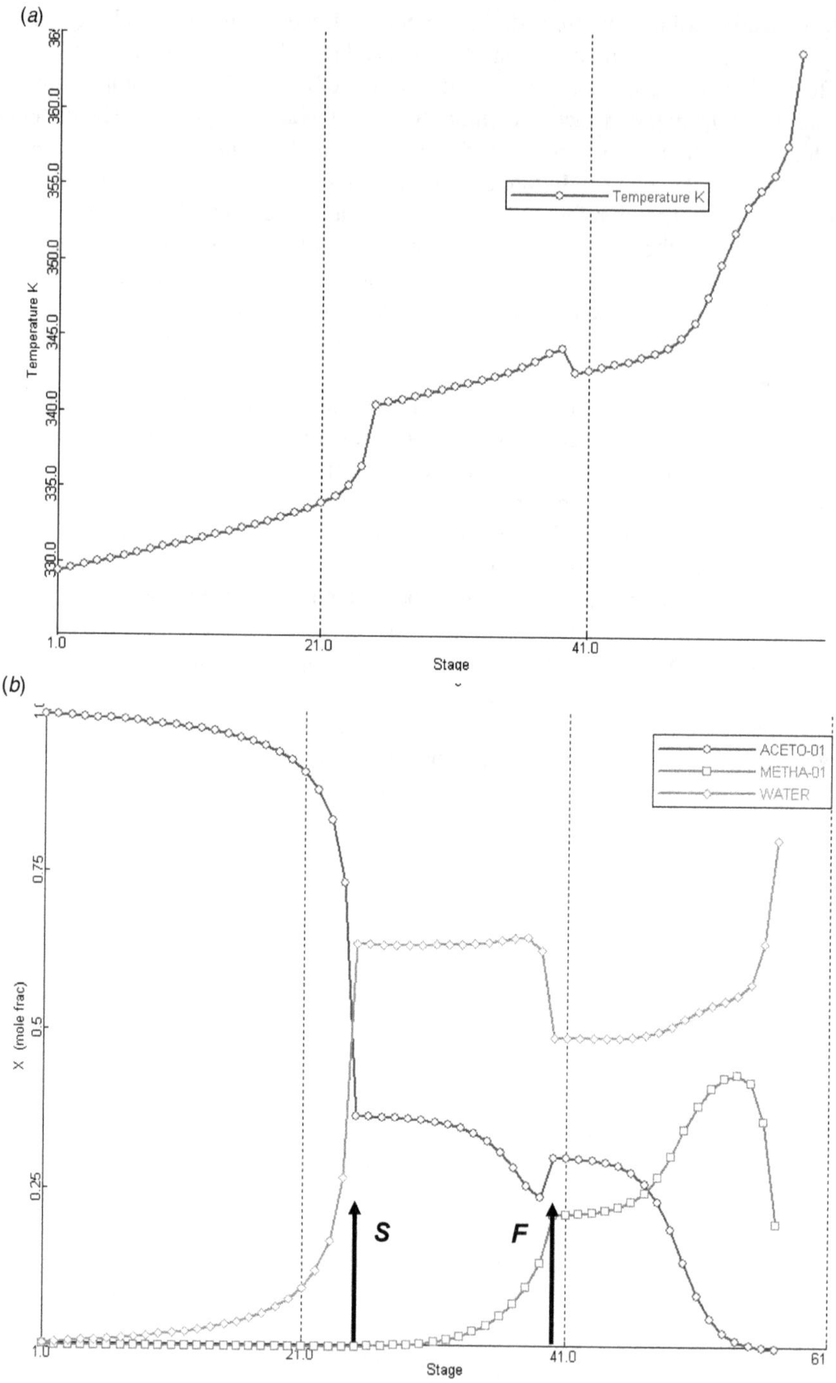


**Figure 11.6** (*a*) Extractive column temperature profile. (*b*) Extractive column composition profiles.

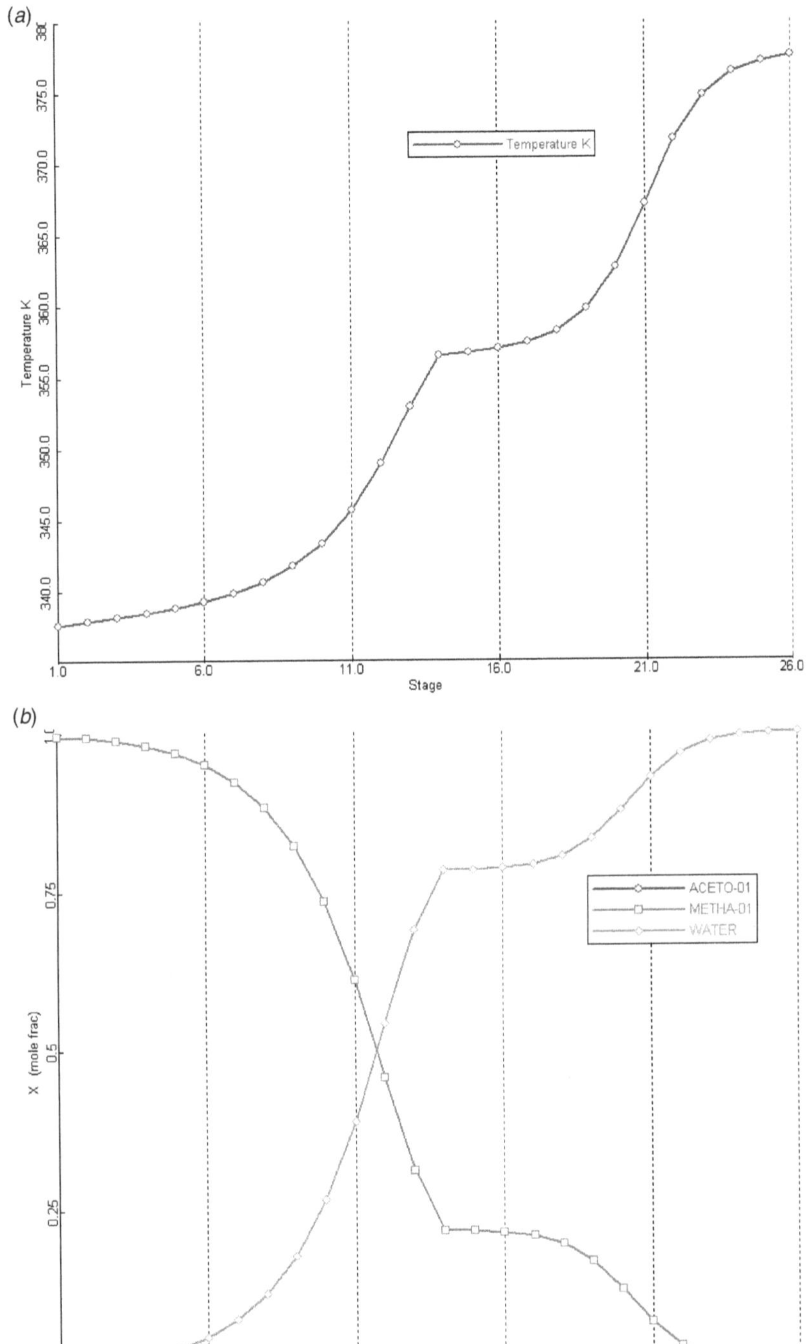


**Figure 11.7** (*a*) Methanol column temperature profile. (*b*) Methanol column composition profiles.

These results are generated with a fixed number of trays in each column. The reboiler heat duties are $Q_{R1}$ and $Q_{R2}$ in the low and high-pressure columns, respectively. This feature is a clear advantage of the pressure-swing process.

Temperature and composition profiles in the two columns are given in Figures 11.6 and 11.7 for the extractive distillation flowsheet. The extraction column composition profiles given in Figure 11.6*b* are quite unusual. The solvent is fed on Stage 25 and the feed is introduced on Stage 40.

The economics of the extractive distillation process are presented in the second column of Table 11.1. Notice that the column diameters are smaller, heat exchanger areas are smaller, and energy consumption is smaller in the extractive system compared with the pressure-swing system. The TAC is about 17% lower for the extractive system.

The TAC given by Knapp and Doherty for their extractive system is \$2,790,000/y, compared with the \$3,750,000/y calculated in this work. The difference is probably due to our much higher solvent circulation rate.

**TABLE 11.1 Design Results.**

| | | Pressure Swing | Extractive |
|---|---|---|---|
| C1 | *N*stage | 52 | 57 |
| | *ID* (m) | 3.30 | 2.79 |
| | $Q_R$ (MW) | 15.8 | 11.4 |
| | $Q_C$ (MW) | 15.1 | 9.87 |
| | $Q_{HX}$ (MW) | – | 1.35 |
| | $T_{reflux}$ (K) | 328 | 329 |
| | $A_C$ (m$^2$) | 996 | 609 |
| | $A_R$ (m$^2$) | 758 | 576 |
| | $A_{HX}$ (m$^2$) | – | 52.7 |
| | Shell (10$^6$ \$) | 1.13 | 1.03 |
| | HX (10$^6$ \$) | 1.19 | 1.02 |
| | Total capital (10$^6$ \$) | 2.32 | 2.06 |
| | Energy (\$10$^6$/y) | 2.22 | 1.69 |
| | TAC (\$10$^6$/y) | 3.00 | 2.38 |
| C2 | *N*stage | 62 | 26 |
| | *ID* (m) | 1.89 | 1.89 |
| | $Q_R$ (MW) | 7.42 | 7.10 |
| | $Q_C$ (MW) | 5.86 | 6.89 |
| | $T_{reflux}$ (K) | 407 | 338 |
| | $A_C$ (m$^2$) | 72.2 | 288 |
| | $A_R$ (m$^2$) | 383 | 350 |
| | Shell (10$^6$ \$) | 0.722 | 0.346 |
| | HX (10$^6$ \$) | 0.466 | 0.624 |
| | Total capital (10$^6$ \$) | 1.19 | 0.970 |
| | Energy (\$10$^6$/y) | 1.13 | 1.03 |
| | TAC (\$10$^6$/y) | 1.52 | 1.38 |
| Total system | Total capital (10$^6$ \$) | 3.51 | 3.03 |
| | Total energy (\$10$^6$/y) | 3.35 | 2.72 |
| | TAC (\$10$^6$/y) | 4.52 | 3.75 |

## 11.3 DYNAMICS AND CONTROL

The Aspen Plus file of this extractive distillation system is exported to Aspen Dynamics after dynamic parameters are specified (equipment sizes). Figure 11.8 shows the control structure developed for this system, which is based on the extractive distillation control structure proposed by Grassi.[3] Relay-feedback testing and Tyreus–Luyben tuning of the temperature loops give the controller parameters given in Table 11.2. The temperature controllers have 1 min deadtimes in the loops. Reflux ratios are held constant in each column (3.44 in the extractive column and 1.61 in the methanol column).

The temperature control tray location is selected by choosing the tray where there is a large temperature change from stage to stage (Stage 50 at 347.6 K in the extractive column and Stage 22 at 371.7 K in the methanol column). See Figures 11.6*a* and 11.7*a*.

The two essential features of this control structure are as follows (see the block arrows in Figure 11.8).

1. The solvent flow is ratioed to the feed flow (the "ratio" multiplier).
2. The base level in the methanol column (the *LC21* controller) is controlled by manipulating the make-up water flowrate.

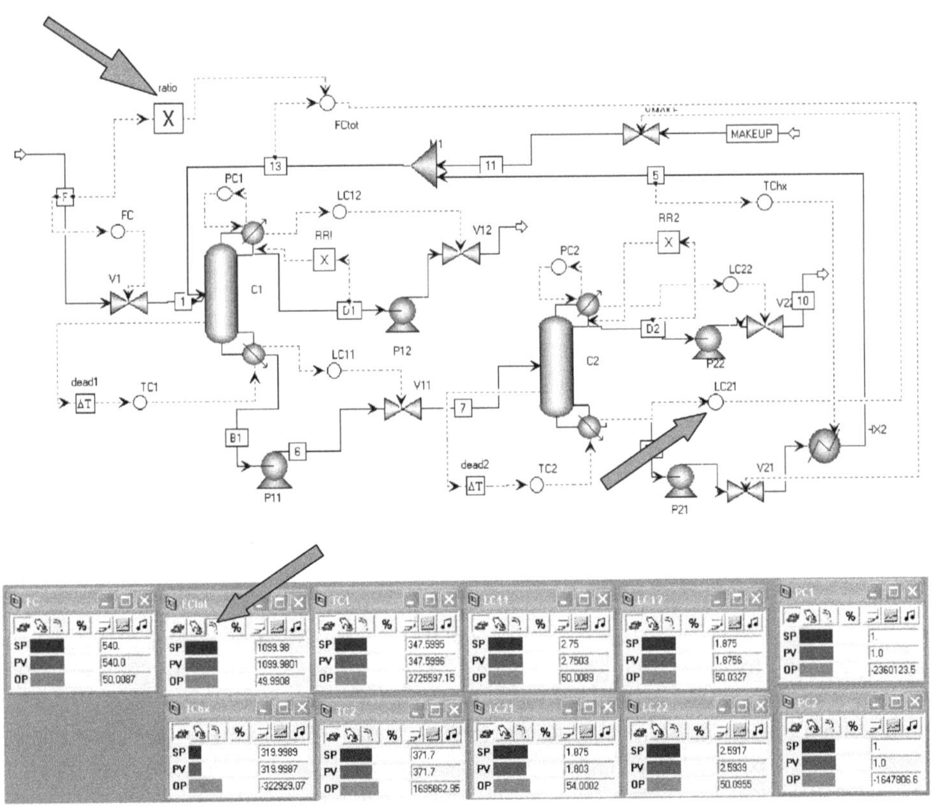


**Figure 11.8** Control structure for extractive distillation system.

**TABLE 11.2 Controller Tuning Parameters.**

| | Pressure Swing | Extractive | Pressure Swing with Heat Integration | Extractive with Heat Integration |
|---|---|---|---|---|
| | | *TC1* | | |
| Controlled variable | $T_{1,48} = 342.8$ K | $T_{1,50} = 347.6$ K | $T_{1,48} = 342.8$ K | $T_{1,50} = 347.6$ K |
| Manipulated variable | $Q_{R1}$ | $Q_{R1}$ | $Q_{AUX}$ | $Q_{AUX}$ |
| Transmitter range | 300–400 K | 300–400 K | 300–400 K | 300–400 K |
| Controller output range | $0–6.6 \times 10^6$ cal/s | $0–6 \times 10^6$ cal/s | 0–40 GJ/h[(1)] | –30–30 GJ/h[(2)] |
| Ultimate gain | 26.1 | 7.15 | 44 | 10.3 |
| Ultimate period | 2.4 min | 4.2 min | 3.6 | 4.8 |
| $K_C$ | 8.16 | 2.23 | 13.7 | 3.2 |
| $\tau_I$ | 5.28 min | 9.24 min | 7.9 | 10.6 |
| | | *TC2* | | |
| Controlled variable | $T_{2,53} = 413$ K | $T_{2,22} = 371.7$ K | $T_{2,53} = 140$°C | $T_{2,21} = 136.3$°C |
| Manipulated variable | $Q_{R2}$ | $Q_{R2}$ | $Q_{R2}$ | $Q_{R2}$ |
| Transmitter range | 350–450 K | 325–425 K | 350–450 K | 300–400 K |
| Controller output range | $0–3.6 \times 10^6$ cal/s | $0–3.6 \times 10^6$ cal/s | $0–3.6 \times 10^6$ cal/s | $0–6 \times 10^6$ cal/s |
| Ultimate gain | 13.1 | 4.23 | 24.9 | 1.20 |
| Ultimate period | 3.6 min | 4.8 min | 4.8 | 6.0 |
| $K_C$ | 4.11 | 1.32 | 7.8 | 0.375 |
| $\tau_I$ | 7.92 min | 10.6 min | 10.6 | 13.2 |

*Notes*: (1) Faceplate shows heat input in metric (GJ/h)
(2) Auxiliary condenser needed to handle increase in methanol in feed.

Although the water make-up stream is much smaller than the total solvent flow, the 5 min holdup time in the base of the methanol column is able to handle dynamic changes in the water solvent circulation rate. When the feed is increased, the ratio immediately increases the solvent fed to the extractive column. The level in the base of the methanol column immediately begins to drop. However, the increased flows to the extractive column eventually produce an increase in the feed to the methanol column, which brings the base level back up.

Controller faceplates are shown at the bottom of Figure 11.8. Notice that the controller *FCtot* is on "cascade" because it receives its remote setpoint from the multiplier "ratio" so that the solvent flowrate is ratioed to the feed flowrate.

Figure 11.9 demonstrates the effectiveness of this control structure. Stable regulatory control is achieved for several large disturbances. The responses of the system to large 20% increases and decreases in feed flowrate are shown in Figure 11.9*a*. Responses for feed composition disturbances from 50 to 60 mol% acetone and from 50 to 40 mol% acetone are shown in Figure 11.9*b*. Product purities ($x_{D1(A)}$ and $x_{D2(M)}$) are held quite close to their specifications. The system takes about 1.5 h to come to a new steady state.

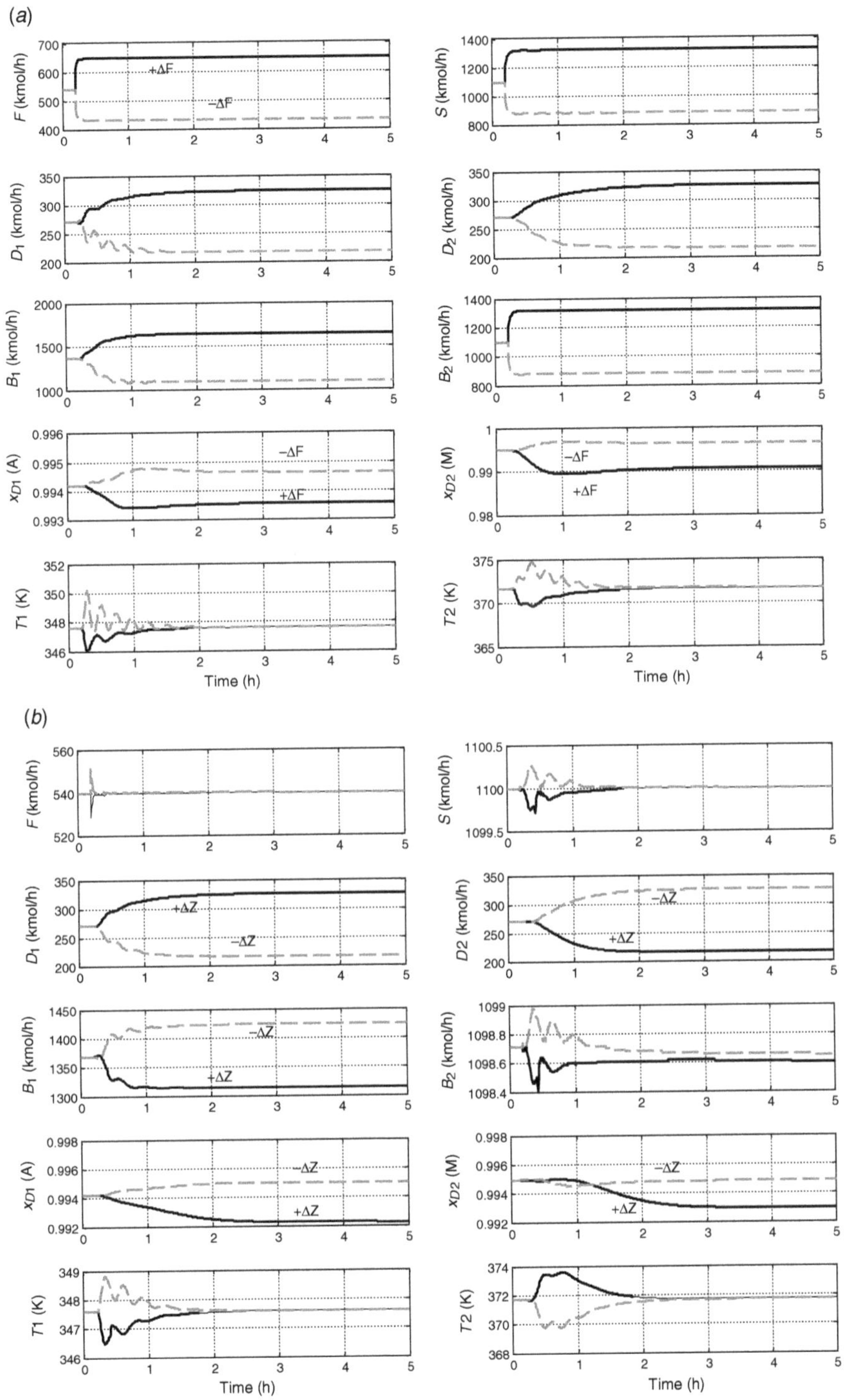


**Figure 11.9** (*a*) Extractive distillation system; 20% feed flow disturbances. (*b*) Extractive distillation system; feed composition disturbances.

(*a*)

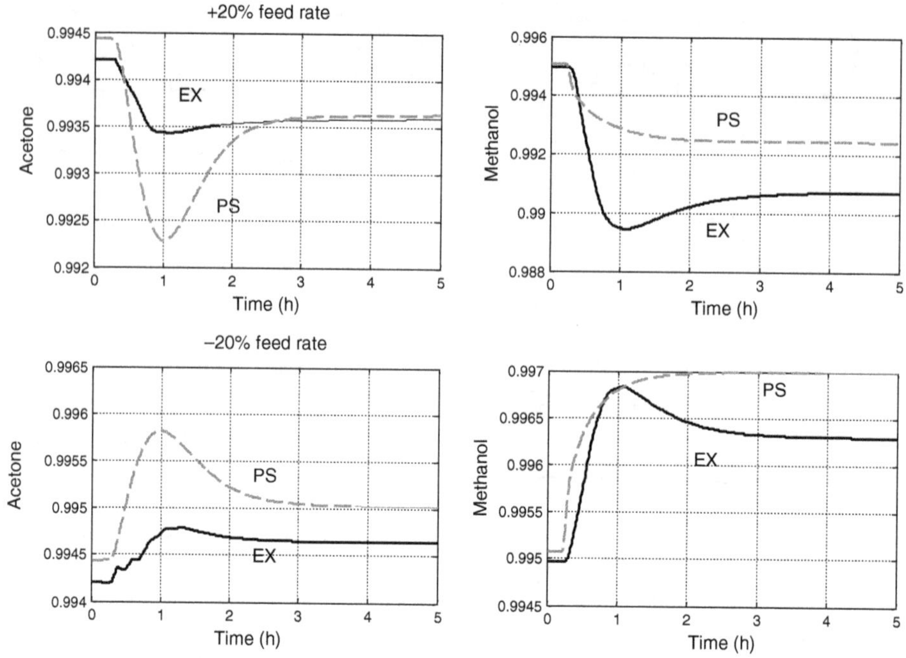


(*b*)

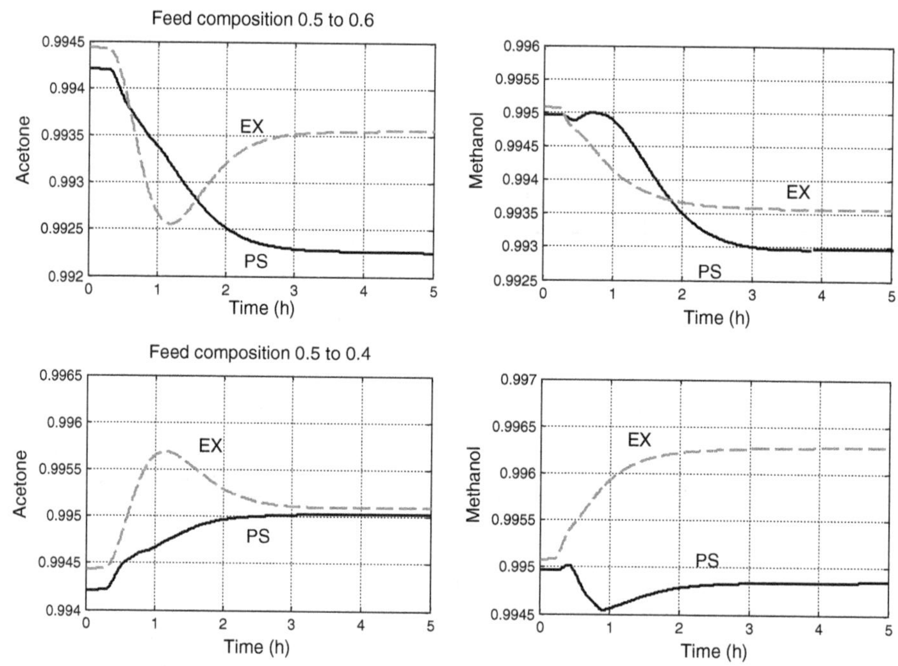


**Figure 11.10** (*a*) Comparison of extractive distillation and pressure-swing systems; feed flowrate disturbances. (*b*) Comparison of extractive distillation and pressure-swing systems; feed composition disturbances.

At this point it is interesting to give a direct comparison of the dynamic performances of the extractive flowsheet considered in this chapter with the pressure-swing flowsheet discussed in Chapter 5. Figure 11.10 provides this comparison. The solid lines are for the extractive distillation system and the dashed lines are for the pressure-swing distillation system. Overall, these responses are quite similar. The deviations in acetone purity are larger in the pressure-swing system (bottoms from the high-pressure column) than those seen in the extractive system (distillate from the extractive column). This difference could be explained by the response of a column to a change in feed flowrate. If the feed is liquid, the bottoms product is more disturbed than the distillate. Since the acetone product comes out the bottom of the high-pressure column, it is disturbed more than in the extractive distillation system in which the acetone product comes off the top of the extractive column.

If this explanation is valid, one would expect to see larger disturbances in the methanol product coming out the bottom of the low-pressure column in the pressure-swing system than those seen in the methanol product coming out the top of the methanol column in the extractive system. However, this is not observed in Figure 11.10*a*.

From a purely conceptual dynamic point of view, one would expect the extractive distillation process to exhibit less product quality variability because both products are removed as distillate streams from columns that have liquid feed streams. The results shown in Figure 11.10 do not give a clear demonstration of this principle.

## 11.4 HEAT-INTEGRATED SYSTEM

The extractive distillation system considered so far in this chapter has considered a flowsheet with independent reboilers and condensers on both columns, that is, no heat integration. In pressure-swing distillation, the large differences in the temperature between the high and low-pressure columns make heat integration a natural. In the extractive process, Knapp and Doherty[1] suggest raising the pressure in the methanol column so that its overhead vapor can be condensed in the reboiler of the extractive column. They point out that the alternative of raising the pressure in the extractive column is infeasible because of distillation boundaries that occur in the ternary diagram at higher pressures.

In this section we compare the steady-state design and the dynamic control of heat-integrated extractive and pressure-swing processes. The same numbers of trays used in the base-case designs are used in both systems. The systems have not been reoptimized for heat integration. Only partial heat integration is considered in which an auxiliary reboiler is used.

### 11.4.1 Design

Figure 11.11 gives the flowsheet for the revised extractive distillation process in which the pressure in the methanol column is increased from 1 to 5 atm. This raises the reflux-drum temperature to 385 K with the 99.5 mol% methanol distillate. The base of the extractive column is 364 K, so there is a reasonable 21 K differential temperature driving force to size the reboiler/condenser heat exchanger.

The higher pressure in the methanol column requires a higher reflux ratio and more heat input than the low-pressure operation because the higher pressure makes the separation more difficult. The reflux ratio increases from 1.61 to 2.91, and the reboiler heat input increases from 7.1 to 11.4 MW.

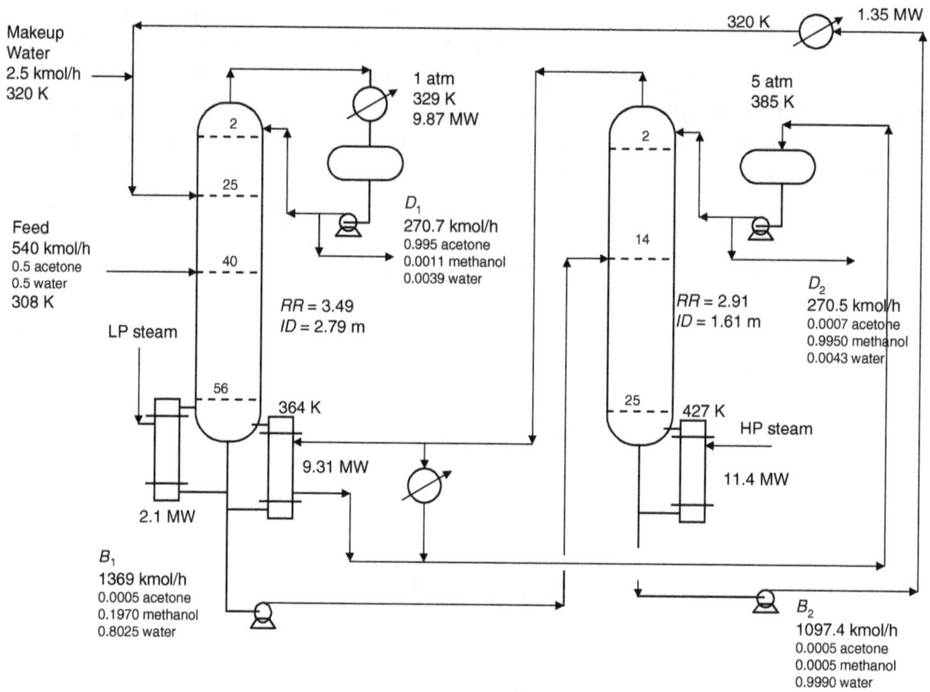


**Figure 11.11** Flowsheet of extractive distillation with heat integration.

The condenser duty in the methanol column is now 9.31 MW, which can be used to produce vapor in the base of the extractive column. The total energy needed in the extractive column is 11.4 MW, so an auxiliary steam-driven reboiler is required with a duty of 2.1 MW. The extractive column itself is unchanged except it has two reboilers.

The reboiler/condenser is sized by using an overall heat transfer coefficient of 0.00306 GJ $h^{-1}$ $m^{-2}$ $K^{-1}$. The heat transfer rate is 33.53 GJ/h (9.31 MW) and the temperature difference is 21.2 K, giving an area of 514 $m^2$.

The second column in Table 11.3 gives sizing and cost information. Notice that the diameter of the methanol column (1.61 m), which is operating at 5 atm, is smaller than the diameter in the nonheat-integrated case at 1 atm (1.89 m), despite having a larger reboiler heat input (11.3 versus 7.59 MW). The larger vapor density at higher pressure results in a smaller diameter column.

The TAC of the heat-integrated extractive system is \$2,980,000/y. This should be compared with the base-case TAC of \$3,750,000/y. Total capital investment is essentially the same in the base and heat-integrated designs (\$3,000,000), while energy cost is reduced from \$2,720,000 to \$1,980,000.

### 11.4.2 Control

Implementing heat integration in Aspen Dynamics requires the used of *Flowsheet Equations*, as discussed in detail in Chapter 6. There are two conditions that must be satisfied at each point in time during the dynamic simulation.

**TABLE 11.3 Design Results for Heat-Integrated Systems.**

| | | Pressure Swing with Heat Integration | Extractive with Heat Integration |
|---|---|---|---|
| C1 | *Nstage* | 52 | 57 |
| | *ID* (m) | 3.30 | 2.79 |
| | $Q_{CR}$ (MW) | 5.86 | 9.31 |
| | $Q_{AUX}$ (MW) | 8.93 | 2.10 |
| | $Q_C$ (MW) | 15.1 | 9.87 |
| | $Q_{HX}$ (MW) | – | 1.35 |
| | $T_{reflux}$ (K) | 328 | 329 |
| | $A_{RC}$ (m$^2$) | 110 | 514 |
| | $A_C$ (m$^2$) | 982 | 609 |
| | $A_{HX}$ (m$^2$) | – | 52.7 |
| | $A_{AUX}$ (m$^2$) | 450 | 106 |
| | Shell (10$^6$ \$) | 1.13 | 1.02 |
| | Condenser (10$^6$ \$) | 0.643 | 0.567 |
| | Condenser/Reboiler (10$^6$ \$) | 0.155 | 0.421 |
| | Aux. Reboiler (10$^6$ \$) | 0.388 | 0.151 |
| | Cooler HX (10$^6$ \$) | – | 0.096 |
| | Aux. condenser (10$^6$ \$) | – | ? |
| | Total capital (10$^6$ \$) | 1.58 | 2.26 |
| | Energy (\$10$^6$/y) | 1.32 | 0.311 |
| C2 | *Nstage* | 62 | 26 |
| | *ID* (m) | 1.89 | 1.61 |
| | $Q_R$ (MW) | 7.59 | 11.3 |
| | $A_C$ (m$^2$) | – | – |
| | $A_R$ (m$^2$) | 383 | 570 |
| | Shell (10$^6$ \$) | 0.722 | 0.292 |
| | Condenser (10$^6$ \$) | – | – |
| | Reboiler (10$^6$ \$) | 0.349 | 0.451 |
| | Total capital (10$^6$ \$) | 1.07 | 0.743 |
| | Energy (\$10$^6$/y) | 1.13 | 1.67 |
| Total system | Total capital (10$^6$ \$) | 2.65 | 3.00 |
| | Total energy (\$10$^6$/y) | 2.45 | 1.98 |
| | TAC (\$10$^6$/y) | 3.33 | 2.98 |

1. The heat transfer in the reboiler/condenser $Q_{RC}$ must be equal to the product of the fixed area $A_{RC}$, the fixed heat-transfer coefficient $U$, and the temperature difference between the condenser temperature in the methanol column and the base temperature in the extractive column. This heat transfer rate is the heat removed in the condenser of the methanol column.
2. The total heat transfer in the base of the extractive column is the sum of the heat removed in the condenser of the methanol column $Q_{RC}$ and the heat added in the auxiliary reboiler $Q_{AUX}$ in the extractive column.

These conditions correspond to the first and second laws of thermodynamics. Figure 11.12 shows the flowsheet equations for the extractive heat-integrated system. The two conditions listed above are specified in the top two equations in the *Text Editor* window.

```
Text Editor - Editing Flowsheet
CONSTRAINTS
  // Flowsheet variables and equations...

Blocks("C2").condenser(1).Q = -514.1*0.00306*(Blocks("C2").Stage(1).T - Blocks("C1").TReb);
Blocks("C1").QReb= Blocks("TC1").OP - Blocks("C2").condenser(1).Q;

Blocks("dead2").Input_=Blocks("C2"). Stage(21).T -  (Blocks("C2").Stage(21).P - 5.136)*5.47/0.864;
END
```

**Figure 11.12** Aspen Dynamics flowsheet equations for heat-integrated extractive process and pressure-compensated temperature.

The third equation is used to provide a "pressure-compensated" temperature measurement in the methanol column. This is needed because, in the heat-integrated system, the pressure in the methanol column is not controlled. It floats with operating conditions. If more heat transfer is required in the reboiler/condenser, a larger temperature difference is required, and this is achieved by the pressure in the methanol column increasing, which raises the bubblepoint temperature in the reflux drum.

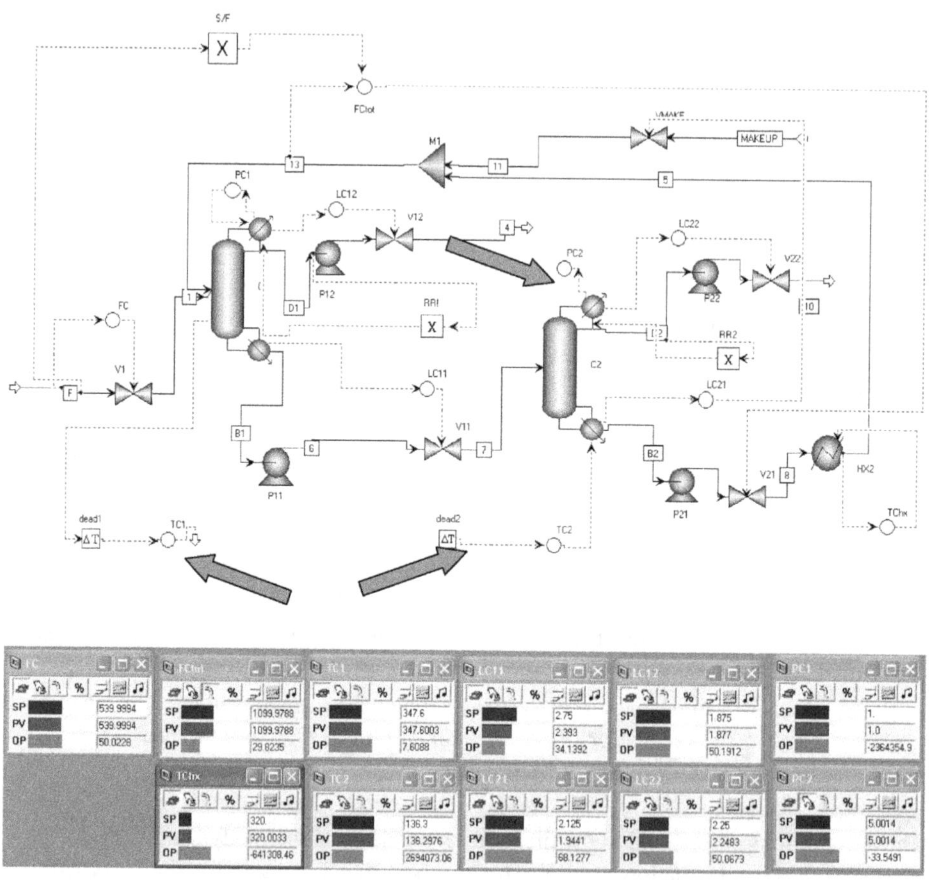

**Figure 11.13** Control structure for extractive distillation with heat integration.

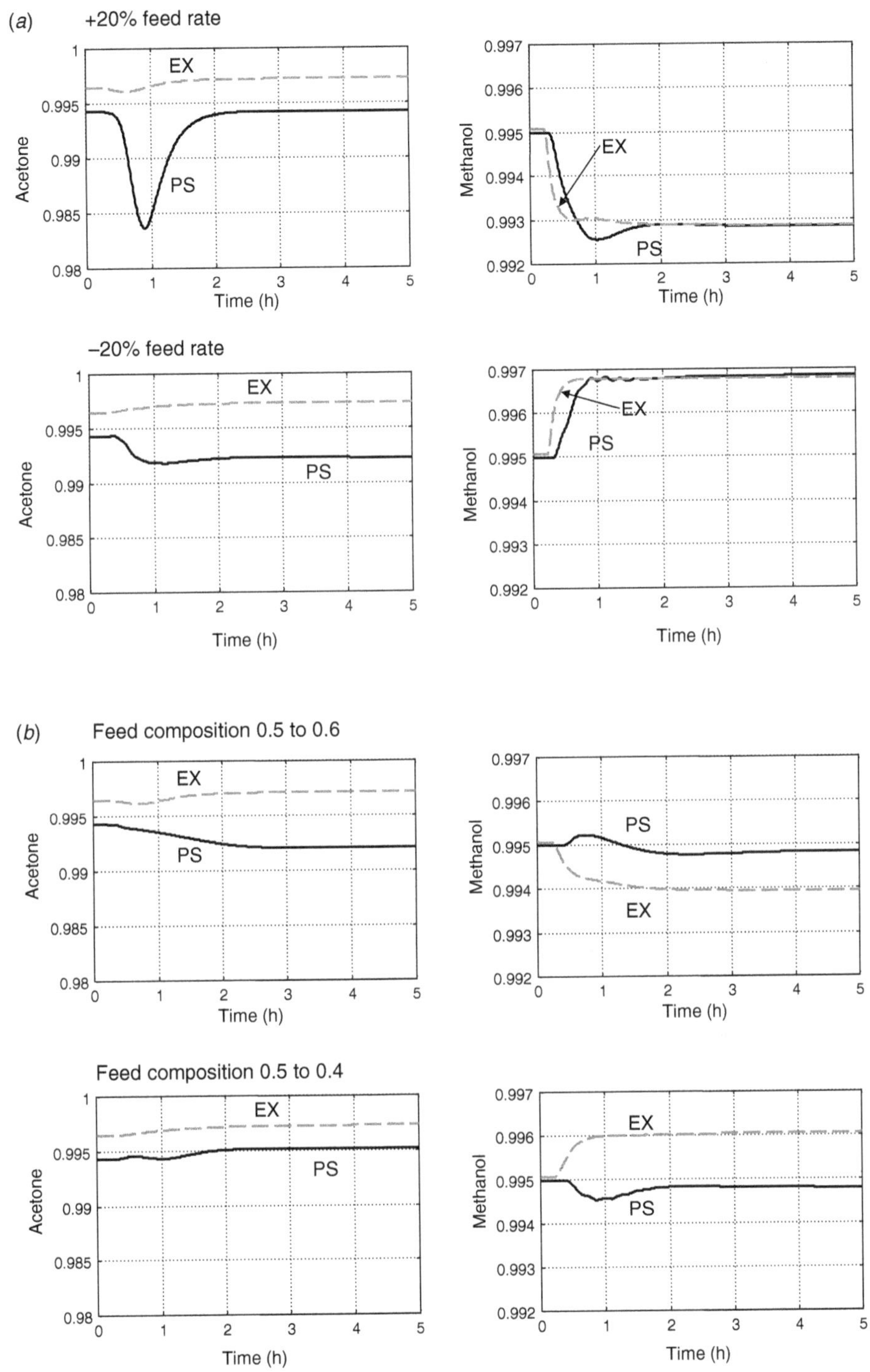


**Figure 11.14** (*a*) Comparison of extractive and pressure-swing distillation with heat integration: feed rate disturbances. (*b*) Comparison of extractive and pressure-swing distillation with heat integration: feed composition disturbances.

Figure 11.13 shows the control structure for the heat-integrated extractive system. The three block arrows point to the changes that have been made from the base-case control structure.

1. The output signal from the TC1 controller in the extractive column is the auxiliary heat input $Q_{AUX}$, which is 7.61 GJ/h (note Aspen uses metric units).
2. The PC2 pressure controller in the methanol column has no input signal and is on manual.
3. The deadtime element "dead2" has no input shown in Figure 11.13, but its input signal is the pressure-compensated temperature signal as defined in the flowsheet equation given in Figure 11.12.

The temperature and pressure at Stage 21 in the methanol column are measured, and a "pressure-compensated" temperature "$T_{PC}$" is fed into the deadtime element as its input signal. The design pressure on Stage 21 is 5.136 atm with a composition of 45 mol% methanol and a temperature of 396.19 K. Generating a $Txy$ diagram at 6 atm gives a temperature of 401.66 K at the same composition. The following equation is used to calculate the signal fed to the deadtime element and subsequently into the temperature controller as the PV signal.

$$T_{PC} = T_{21} - (P_{21} - 5.136)\left(\frac{401.66 - 396.19}{6 - 5.136}\right)$$

The two temperature controllers are tuned using the methods discussed previously in Section 11.3. Tuning parameters are given in Table 11.2. Remember that the temperature controller in the extractive column is manipulating the energy input $Q_{AUX}$ from the auxiliary reboiler (2.10 MW or 7.6 GJ/h as given in the TC1 faceplate in Fig. 11.13).

The effectiveness of this control structure on the heat-integrated extractive distillation system is compared with that of the partially heat-integrated pressure-swing distillation system discussed in Chapter 6 in Figure 11.14. The controllability is quite similar.

## 11.5 EFFECT OF SOLVENT ON CONTROLLABILITY

The solvent used in the discussions presented so far in this chapter was water. In this section we look at the possibility of using other solvents. Solvent selection is critical to the successful design and operation of extractive distillation systems. Different solvents affect the phase equilibrium in different ways.

An informative paper by Kossack et al.[2] discusses in detail the steady-state economic design issues and presents several examples that use a variety of solvents for the separation of acetone and methanol. Some solvents drive the acetone overhead in the extractive column, while others drive the methanol overhead in the extractive column. The authors discuss the effect of solvent selection on issues such as selectivity, capacity, and boiling point. In this section we extend this work to compare the dynamic performance of the acetone–methanol system with different solvents.

The numerical cases are based on those presented by Kossack et al.[2] Three solvents are explored: water, DMSO, and chlorobenzene. The first and second solvents drive the acetone overhead in the extractive column. The last solvent drives the methanol overhead in the extractive column.

Control structures are developed for each system. Dynamic simulation results show that all systems are controllable, but product quality variability is poorest when the methanol is driven overhead in the extractive column. One explanation for this difference is that methanol is higher boiling than acetone and preferentially wants to go out the bottom.

### 11.5.1 Systems Studied

The acetone–methanol binary homogeneous minimum-boiling azeotropic system is considered with some of the solvents studied by Kossack et al.[2] Three solvents are explored that have different normal boiling points (373 K for water, 464 K for DMSO, and 405 K for chlorobenzene). These solvents have different effects on the azeotropic mixture. The first and second solvents drive the acetone overhead in the extractive column. The chlorobenzene solvent drives the methanol overhead in the extractive column. The normal boiling points of acetone and methanol are 329 and 338 K, respectively, so acetone is the lighter component and would preferentially go overhead. The composition of the acetone–methanol azeotrope is 77.6 mol% acetone at atmospheric pressure as shown in Figure 11.1.

Figure 11.15*a* gives *Txy* diagrams for acetone–water and methanol–water. The first comes into play in the top of the extractive column. The second applies in the solvent

(*a*)

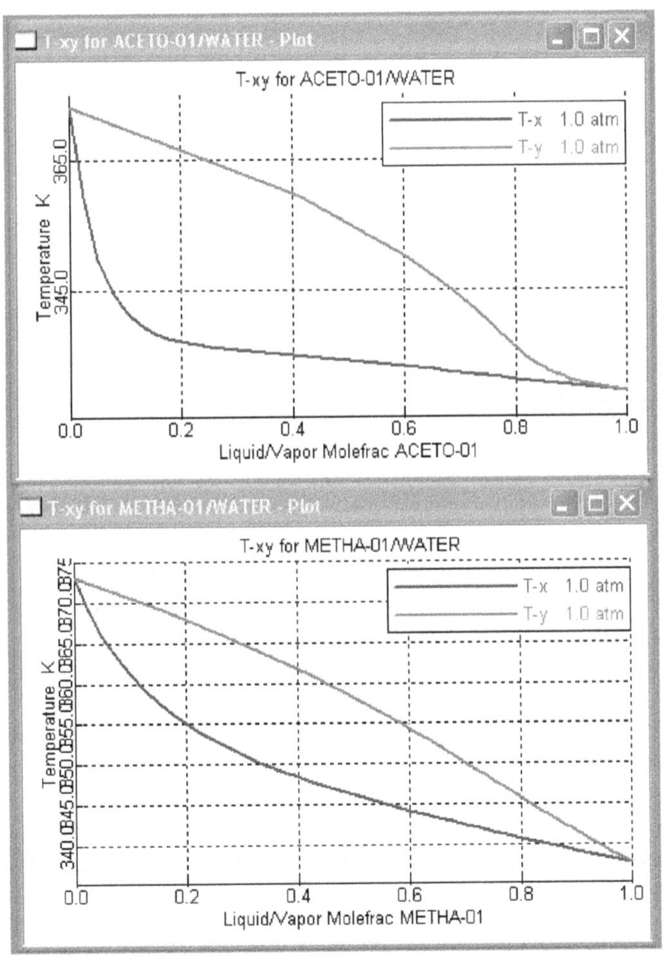


**Figure 11.15** (*a*) *Txy* diagrams for acetone and methanol with water.

(b)

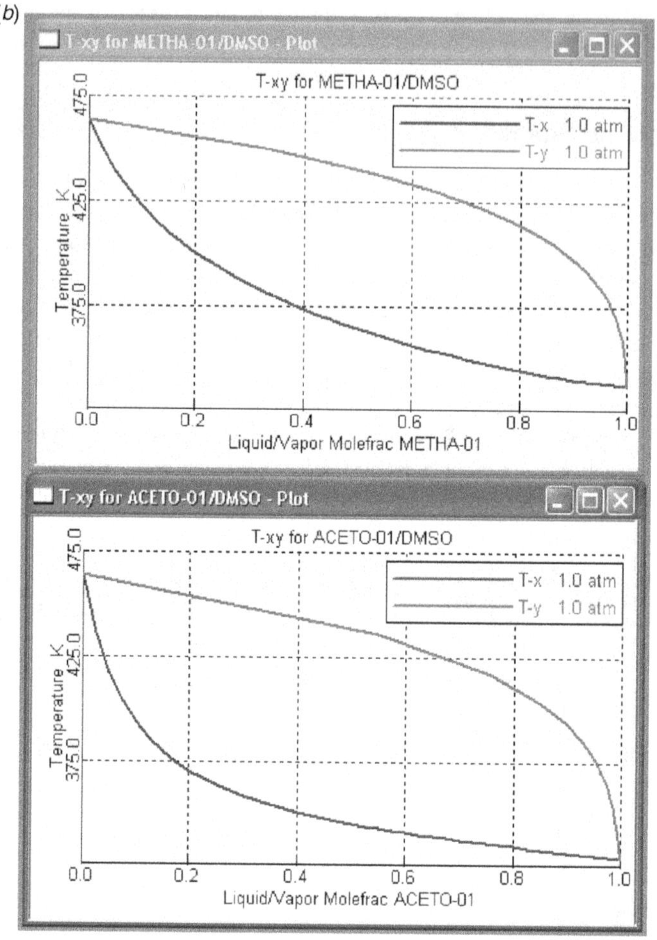


**Figure 11.15** (*b*) *Txy* diagrams for acetone and methanol with DMSO.

recovery column. The pinch at the high acetone end of the curves means the separation to get high purity acetone may be somewhat difficult.

Figure 11.15*b* gives *Txy* diagrams for acetone–DMSO and methanol–DMSO. Both of these separations should be easy. Figure 11.15*c* gives *Txy* diagrams for acetone–chlorobenzene and methanol–chlorobenzene. These separations are not as easy as in the DMSO system, particularly because of the pinch in the methanol curves. These binary vapor–liquid relationships impact the energy and capital costs of the systems.

The UNIQUAC physical property package in Aspen Plus is used for the design of all systems. Aspen notation is used for numbering stages (the condenser is Stage 1). All designs are based on the case given by Knapp and Doherty[1]: feed is 540 kmol/h of 50/50 mol% acetone–methanol. All columns operate at atmospheric pressure, which is high enough to permit the use of cooling water in all condensers. The Aspen *Design Spec/Vary* feature is used to drive overhead and bottoms compositions to their desired values by manipulating distillate flowrate and reflux ratio. Solvent flowrate is fixed at a level high enough to achieve the specified product purities.

(c)

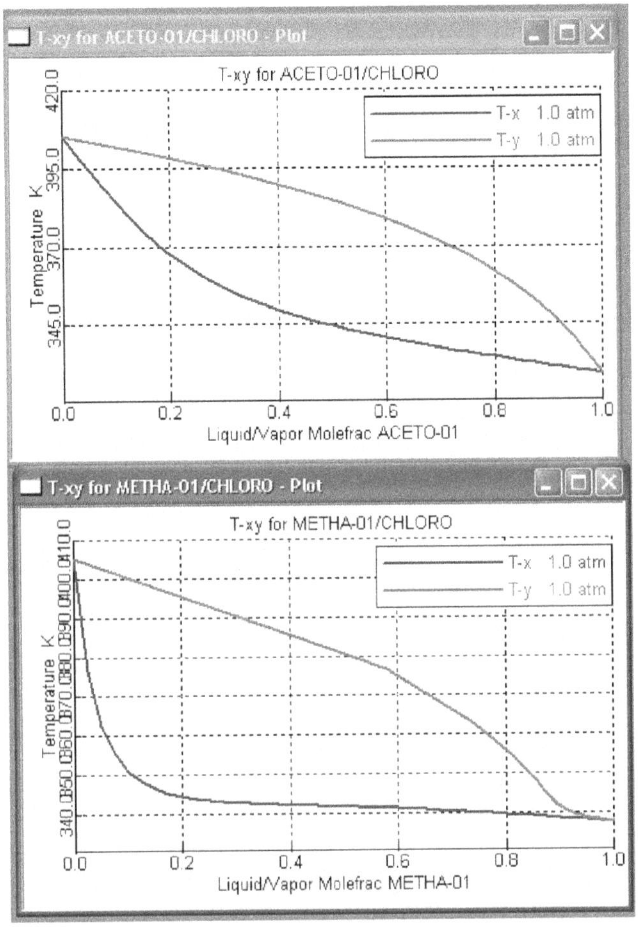


**Figure 11.15** (*c*) *Txy* diagrams for acetone and methanol with chlorobenzene.

### 11.5.2 Water Solvent

The use of water as the solvent is discussed in Section 11.1. The flowsheet is shown in Figure 11.3. Notice that there is more water impurity in the acetone product leaving in the distillate from the extractive column than methanol (0.39 mol% water versus 0.11 mol% methanol), despite the fact that water is a higher boiler than methanol. A cooler is also required to cool down the water recycle from the bottom of the methanol column. With water as the solvent, the base of the solvent recovery column runs at 378 K, so low-pressure steam can be used. Higher-boiling solvents will have higher base temperatures, require higher-pressure, and more expensive sources of energy.

The economics of the extractive distillation process with the three solvents are presented in Table 11.4. The sizing and economic factors used are those presented by Luyben.[4] The TAC using water solvent is \$3,750,000/y. The TAC given by Knapp and Doherty[1] for their extractive system is \$2,790,000/y. The difference is probably owing to our higher solvent circulation rate.

**TABLE 11.4 Design Results with Different Solvents.**

| | | Water | DMSO | Chlorobenzene |
|---|---|---|---|---|
| C1 | *Nstage* | 57 | 37 | 45 |
| | *ID* (m) | 2.79 | 2.00 | 3.31 |
| | $Q_R$ (MW) | 11.4 | 7.93 | 15.97 |
| | $Q_C$ (MW) | 9.87 | 4.99 | 6.73 |
| | $Q_{HX}$ (MW) | 1.35 | 5.12 | 8.29 |
| | $T_{reflux}$ (K) | 329 | 329 | 338 |
| | $A_C$ (m$^2$) | 609 | 308 | 282 |
| | $A_R$ (m$^2$) | 576 | 401 | 806 |
| | $A_{HX}$ (m$^2$) | 52.7 | 104 | 233 |
| | Shell ($10^6$ \$) | 1.03 | 0.498 | 1.004 |
| | HX ($10^6$ \$) | 1.02 | 0.810 | 1.103 |
| | Total capital ($10^6$ \$) | 2.06 | 1.308 | 2.107 |
| | Energy (\$$10^6$/y) | 1.69 | 1.177 | 2.367 |
| | TAC (\$$10^6$/y) | 2.38 | 1.613 | 3.06 |
| C2 | *Nstage* | 26 | 17 | 18 |
| | *ID* (m) | 1.89 | 2.12 | 3.54 |
| | $Q_R$ (MW) | 7.10 | 6.10 | 9.50 |
| | $Q_C$ (MW) | 6.89 | 3.95 | 10.11 |
| | $T_{reflux}$ (K) | 338 | 338 | 329 |
| | $A_C$ (m$^2$) | 288 | 165 | 623 |
| | $A_R$ (m$^2$) | 350 | 308 | 480 |
| | Shell ($10^6$ \$) | 0.346 | 0.268 | 0.488 |
| | HX ($10^6$ \$) | 0.624 | 0.504 | 0.882 |
| | Total capital ($10^6$ \$) | 0.970 | 0.7726 | 1.370 |
| | Energy (\$$10^6$/y) | 1.03 | 0.9041 | 1.408 |
| | TAC (\$$10^6$/y) | 1.38 | 1.162 | 1.865 |
| Total system | Total capital ($10^6$ \$) | 3.03 | 2.08 | 3.48 |
| | Total energy (\$$10^6$/y) | 2.72 | 2.08 | 3.78 |
| | TAC (\$$10^6$/y) | 3.75 | 2.77 | 4.93 |

### 11.5.3 DMSO Solvent

DMSO has a much higher boiling point than either acetone or methanol, and it is a very effective solvent. So it is quite easy to attain very high product purities. The optimum design presented by Kossack et al.[2] provides the number of stages in each column and the feed stage locations. Figure 11.16 shows the flowsheet developed based on these columns. The acetone product and the methanol product are specified to be 99.95 mol% pure, which is higher than the purities used by Kossack et al.[2] Higher purities are selected because the solvent is so effective that high purities are easily obtained with little increase in energy or capital costs. Figure 11.17 shows that this high acetone purity can be achieved if the solvent flowrate is greater than 500 kmol/h. These results are generated holding a bottoms composition in the extractive column of 0.01 mol% acetone so that the desired high purity of the methanol leaving in the distillate of the solvent recovery column can be attained. Any acetone that enters the second column must go overhead and nothing can be done in the second column to affect distillate purity in terms of acetone.

The solvent rate used in the flowsheet shown in Figure 11.16 is 750 kmol/h, which gives a S/F ratio of 1.39. The second column in Table 11.5 shows that the Kossack et al.[2] design

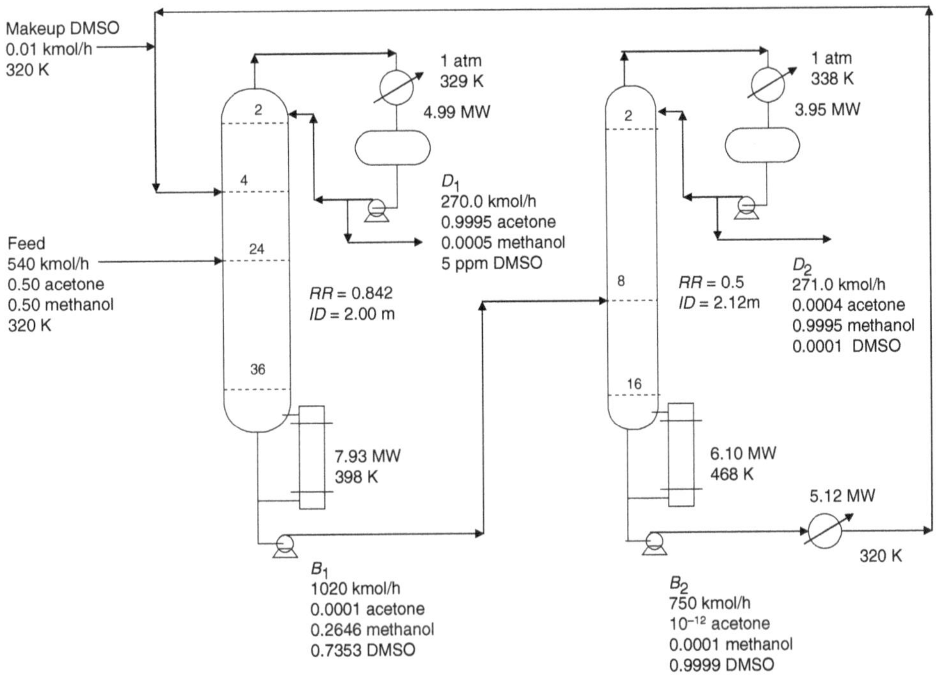


**Figure 11.16** Acetone–methanol extractive distillation with DMSO solvent.

uses a S/F of 0.992. Their system is designed for much less pure products (98.9 mol% methanol and 99.5 mol% acetone). Thus their system has somewhat lower costs, as shown in Table 11.5.

Figure 11.18 gives temperature and composition profiles for both columns. The shape of the temperature profile in the extractive column is quite unusual. It is not obvious what stage to use for temperature control. We return to this issue in the development of a control structure for this system later in this chapter.

Notice that the base temperature in the methanol column is much higher in the DMSO system (468 K) compared to that in the water system (378 K). The temperature profile shown in Figure 11.18*c* shows a very large temperature change from the top to the bottoms of the methanol column (130 K differential). This feature will impact the control structure, as discussed later, requiring the use of an average temperature control structure. The acetone–DMSO separation is quite easy, so a very small reflux ratio is required to achieve high purities. The reflux ratio in the methanol column is heuristically fixed at 0.5 in this system as a reasonable minimum operational value.

### 11.5.4 Chlorobenzene Solvent

Chlorobenzene has a different effect on the acetone–methanol system than water or DMSO. In both of the previous two solvent systems, acetone is driven overhead in the extractive column. Methanol is captured by the solvent and later separated in the solvent-recovery methanol column. In the chlorobenzene system, *methanol* is driven overhead in the extractive column. Acetone is captured by the solvent and later separated in the solvent-recovery acetone column. Since acetone is lighter than methanol, the solvent must fight against the

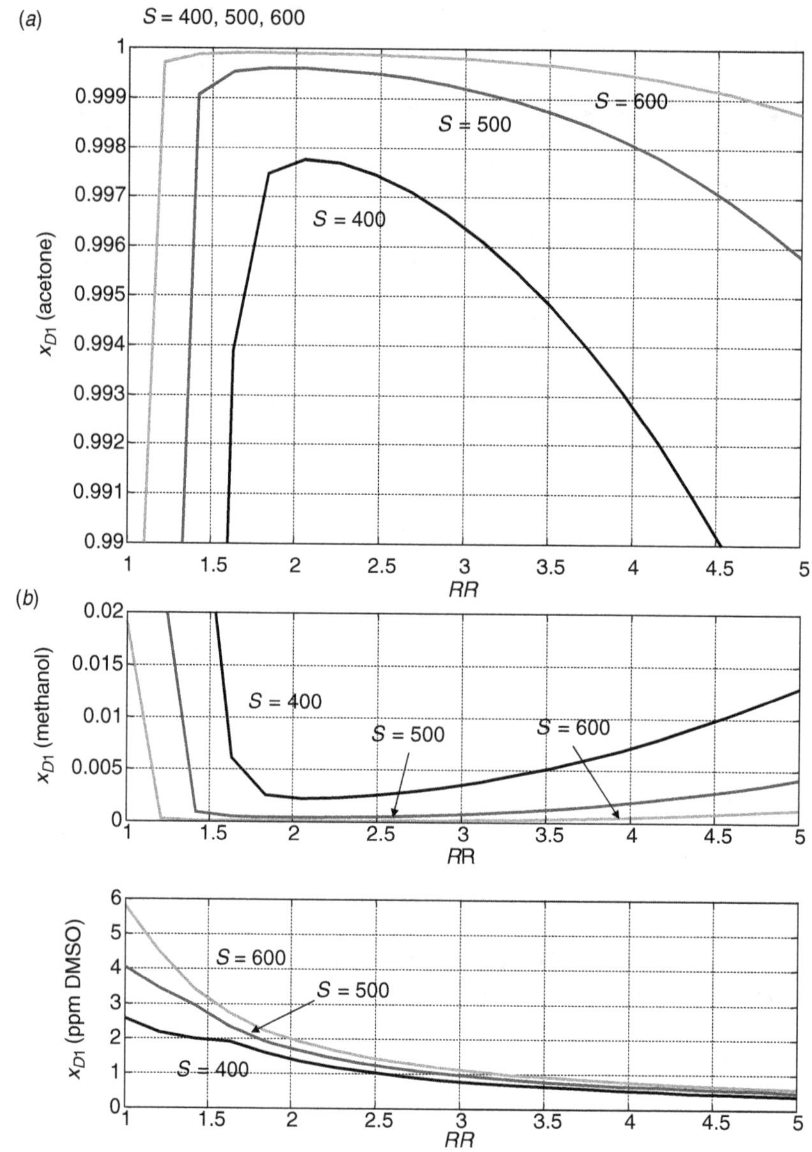


**Figure 11.17** (*a*) Effect of *RR* and solvent on acetone product purity with DMSO solvent. (*b*) Effect of *RR* and *S* on impurities in acetone product with DMSO solvent.

natural tendency of acetone to be more volatile. We will see the impact of this on the dynamic controllability later in this chapter.

The optimum design presented by Kossack et al.[2] provides the number of stages in each column and feed locations. Figure 11.19 shows the flowsheet developed based on these columns. The acetone product and the methanol product are specified to be 99.5 mol% pure. Figure 11.20 shows that this methanol purity can be achieved in the extractive column distillate if the solvent flowrate is greater than 1800 kmol/h. These results are generated holding a bottoms composition in the extractive column of 0.01 mol% methanol.

TABLE 11.5 **Comparison of Design Results.**

| | Water | DMSO | Chlorobenzene |
|---|---|---|---|
| | | *S/F* | |
| Kossack et al.[2] | 1.886 | 0.992 | 2.382 |
| Luyben[4] | 2.04 | 1.39 | 3.52 |
| | | *Total* $Q_R/F$ *($10^7$ J/kmol)* | |
| Kossack et al.[2] | 11.11 | 7.57 | 8.37 |
| Luyben[4] | 12.3 | 9.35 | 16.9 |
| | | *TAC ($10^5$ €/kmol)* | |
| Kossack et al.[2] | 4.46 | 3.42 | 3.256 |
| Luyben[4] | 5.28 | 3.90 | 6.95 |

*Notes*: (1) In the DMSO system, purities 0.9995 in this work; 0.995 in Kossack, et al.[2]
(2) Methanol purities 0.995 in this work; 0.982 in water, 0.989 in DMSO and 0.986 in chlorobenzene in Kossack, et al.[2]

The solvent rate used in the flowsheet shown in Figure 11.19 is 1900 kmol/h, which gives a S/F ratio of 3.52. The third column in Table 11.5 shows that the Kossack et al.[2] design uses a S/F ratio of 2.382. Their system is designed for a less pure methanol product (98.6 mol%) because they were not able to achieve 99.5 mol%. We did not encounter this limitation in our simulations.

There are major differences in the energy consumption and economics between the Kossack et al.[2] design and the present results. Their energy consumption is much lower than that found in this study, which makes their TAC much lower, as shown in Table 11.5. The reasons for this large difference are unclear, particularly since the comparisons are reasonably close in the other systems. The high solvent flowrate in the chlorobenzene system would be expected to result in higher energy and capital costs, but the Kossack et al.[2] design results actually show a lower TAC in the chlorobenzene system compared to the DMSO.

Notice that the reflux ratio in the acetone column ($RR = 3.52$ in Fig. 11.19) is much higher in this chlorobenzene system than that required in the methanol column in the DMSO system ($RR = 0.5$ in Fig. 11.16). This implies that the acetone–chlorobenzene separation is more difficult than the methanol–DMSO separation (compare Fig. 11.15*b* with Fig. 11.15*c*). This may help to explain why the energy demand in the solvent recovery column is higher in the chlorobenzene system.

Notice also that the reflux ratio in the extractive column in the chlorobenzene system ($RR = 1.55$ in Fig. 11.19) is much higher than that in the DMSO extractive column ($RR = 0.842$ shown in Fig. 11.16). This results in a much higher energy consumption in the extractive column (15.97 versus 7.93 MW). These results indicate that the chlorobenzene solvent is less efficient than the DMSO solvent, which is a different conclusion than that reached by Kossack et al.[2]

Figure 11.21 gives temperature and composition profiles for both columns. The large temperature difference in the acetone column will affect the control structure selected.

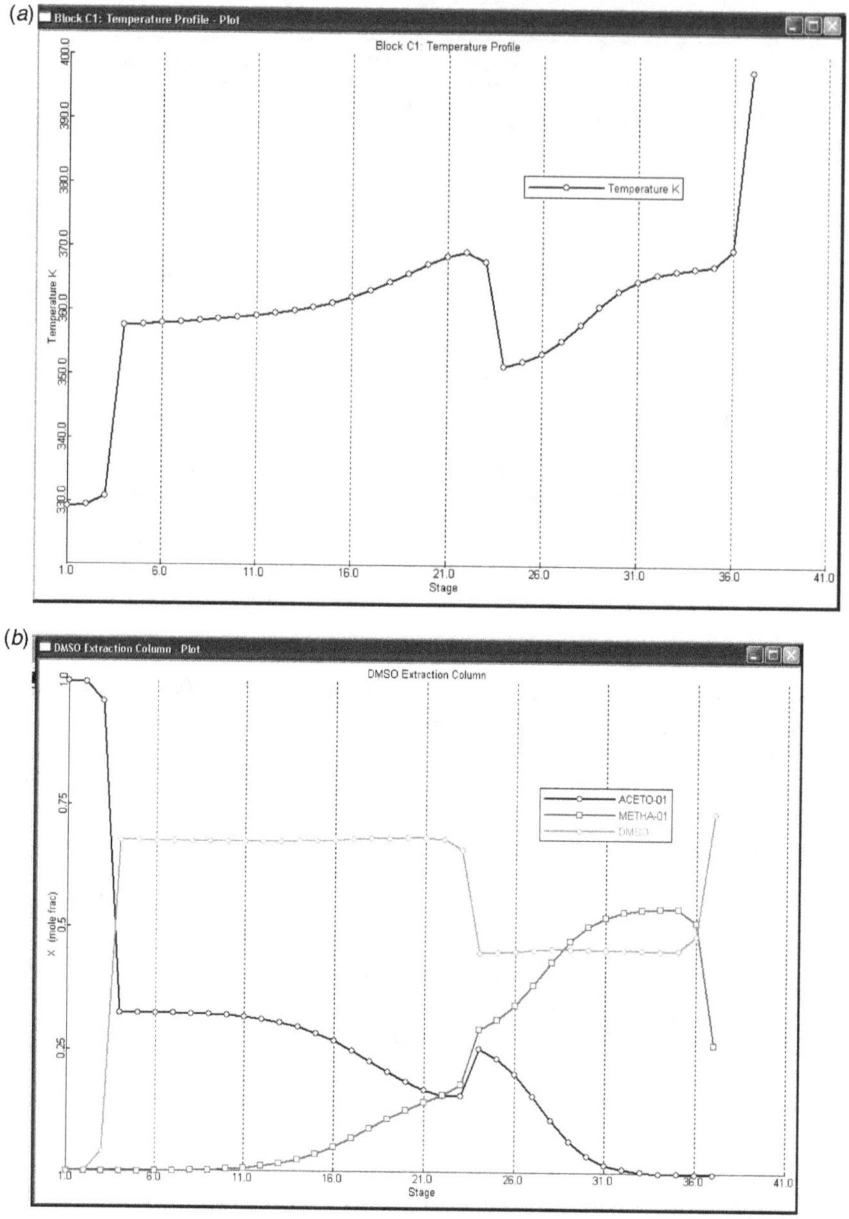


**Figure 11.18** (*a*) Extractive column temperature profile; DMSO solvent. (*b*) Extractive column composition profiles; DMSO solvent.

### 11.5.5 Control System Design

The Aspen Plus files for each of the three extractive distillation systems are exported to Aspen Dynamics after dynamic parameters are specified. Aspen Plus tray sizing is used to fix column diameters. Table 11.4 gives column sizes. Pump heads and valve pressure drops (typically 3 atm) are selected to give reasonable rangeability so that 20% increases

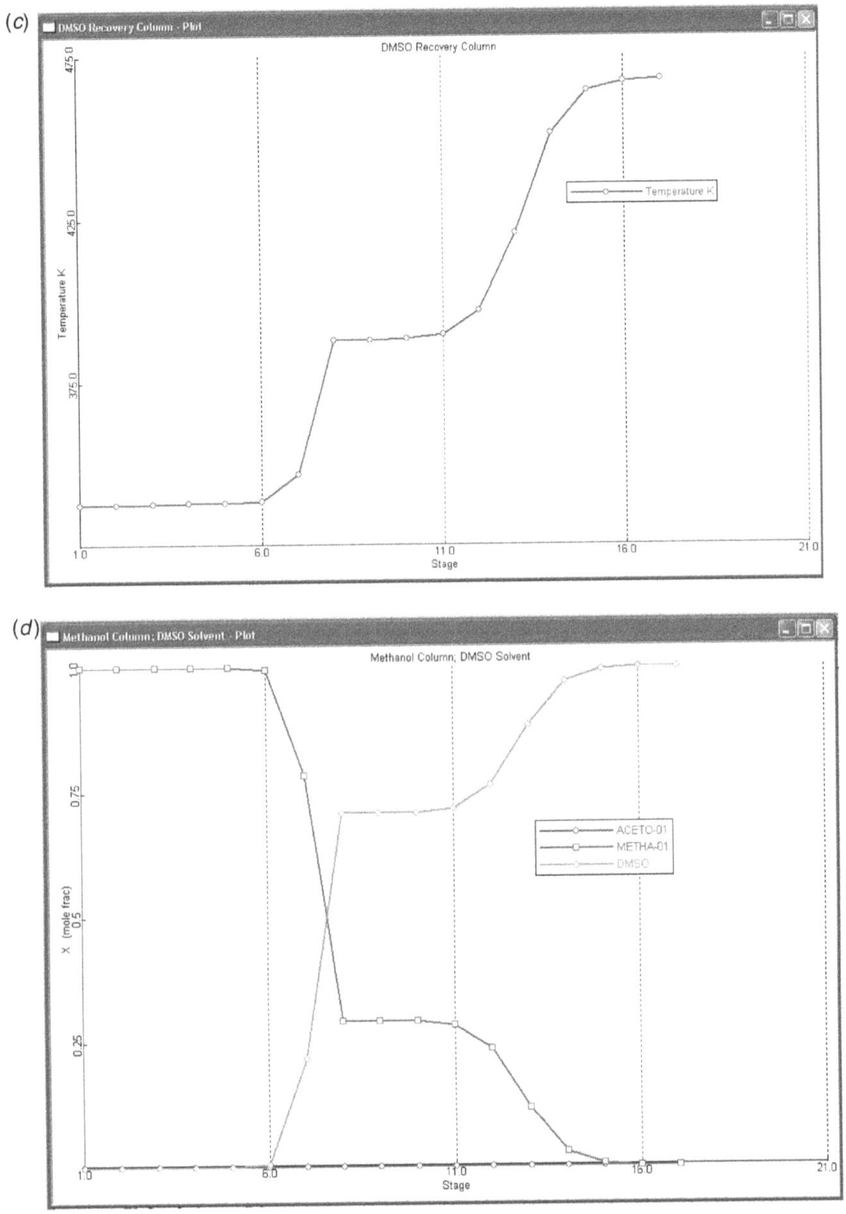


**Figure 11.18** (*c*) Methanol column temperature profile; DMSO solvent. (*d*) Methanol column composition profiles; DMSO solvent.

in throughput can be handled without valve saturation. All reflux drums and column bases are sized to provide 5 min of holdup when 50% full at steady-state conditions. The Aspen Plus files are exported to Aspen Dynamics as pressure-driven simulations.

***Basic Control Structure.*** Figure 11.22 shows the basic control structure used for the water solvent system. All three systems essentially use the same structure with some

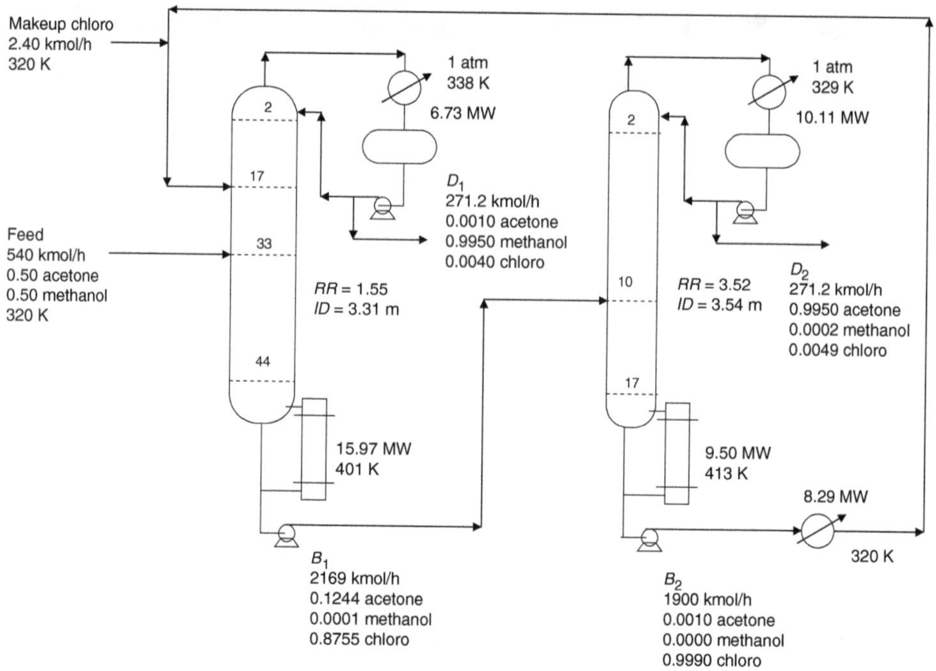


**Figure 11.19** Acetone–methanol extractive distillation with chlorobenzene solvent.

modifications to account for different temperature profiles and sensitivities to disturbances. This structure is suggested by Grassi.[3]

The basic control structure consists of the loops listed below. Conventional PI controllers are used except for level loops, which are proportional with $K_C = 2$.

1. Feed is flow controlled. This is the throughput handle.
2. Solvent flow is ratioed to feed flow.
3. Reflux drum levels are controlled by distillate flows.
4. Base level in the extractive column is controlled by bottoms flow.
5. The reflux ratios are controlled in each column by manipulating reflux flowrate.
6. Base level in the solvent recovery column is controlled by the makeup flow of solvent. The flowrate of this makeup stream is very small compared to the solvent flowrate, so the base level is not controlled tightly. However, the 5 min base holdup is sufficient to handle dynamic disturbances when the solvent flowrate changes, which eventually changes the feed to the solvent-recovery column to bring the base level back to near its steady-state level.
7. Column pressures are controlled by condenser heat removals.
8. The temperature of the solvent entering the extractive column is controlled by the heat removal in the solvent cooler.
9. Tray temperatures are controlled in both columns. A 1 min deadtime is included in the temperature loops. In the water system, a single temperature in the methanol column is controlled. In the DMSO and chlorobenzene systems, an average temperature is controlled because of the sharp temperature profile.

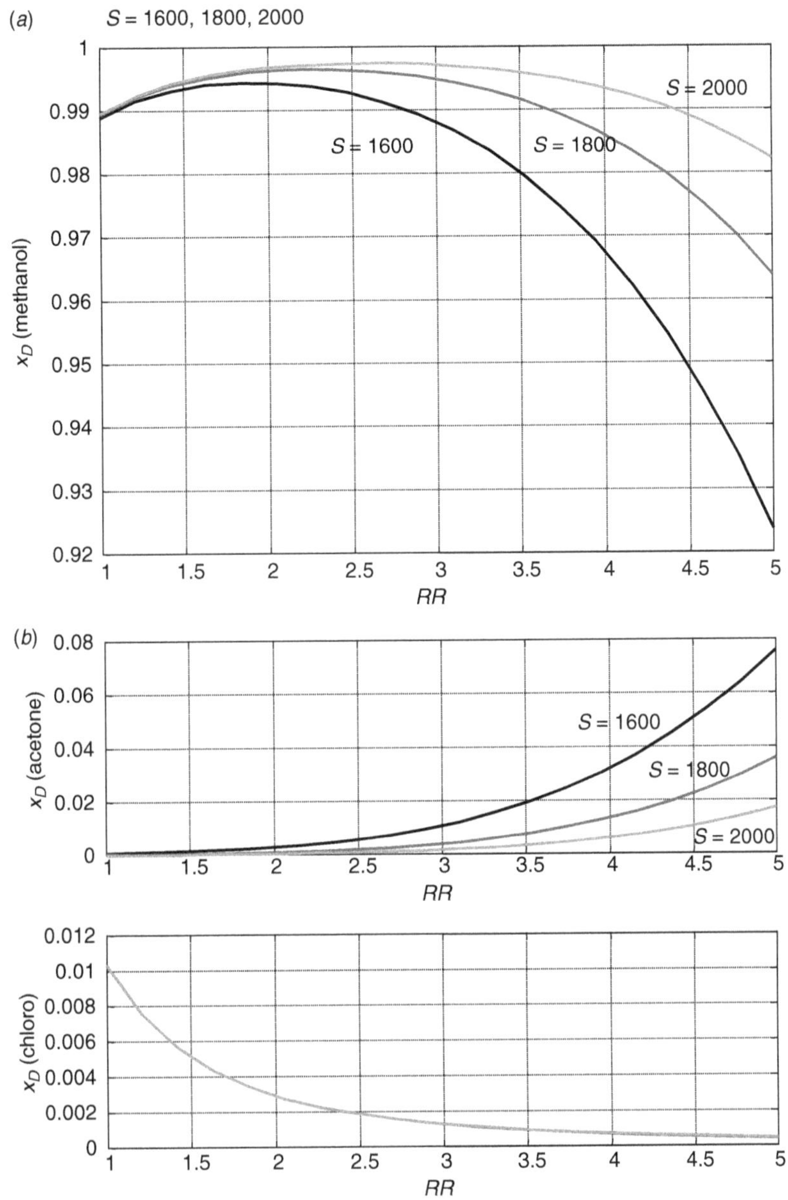


**Figure 11.20** (*a*) Effect of *RR* and solvent on methanol purity with chlorobenzene solvent. (*b*) Effect of *RR* and solvent on impurities in methanol product with chlorobenzene solvent.

Items 2 and 6 are the essential features of this control structure (see block arrows in Fig. 11.22).

Temperature controllers are tuned by running relay-feedback tests and applying Tyreus–Luyben tuning rules. Controller parameters are given in Table 11.6. The dynamic performance is tested by subjecting the system to disturbances in feed flowrate and feed composition.

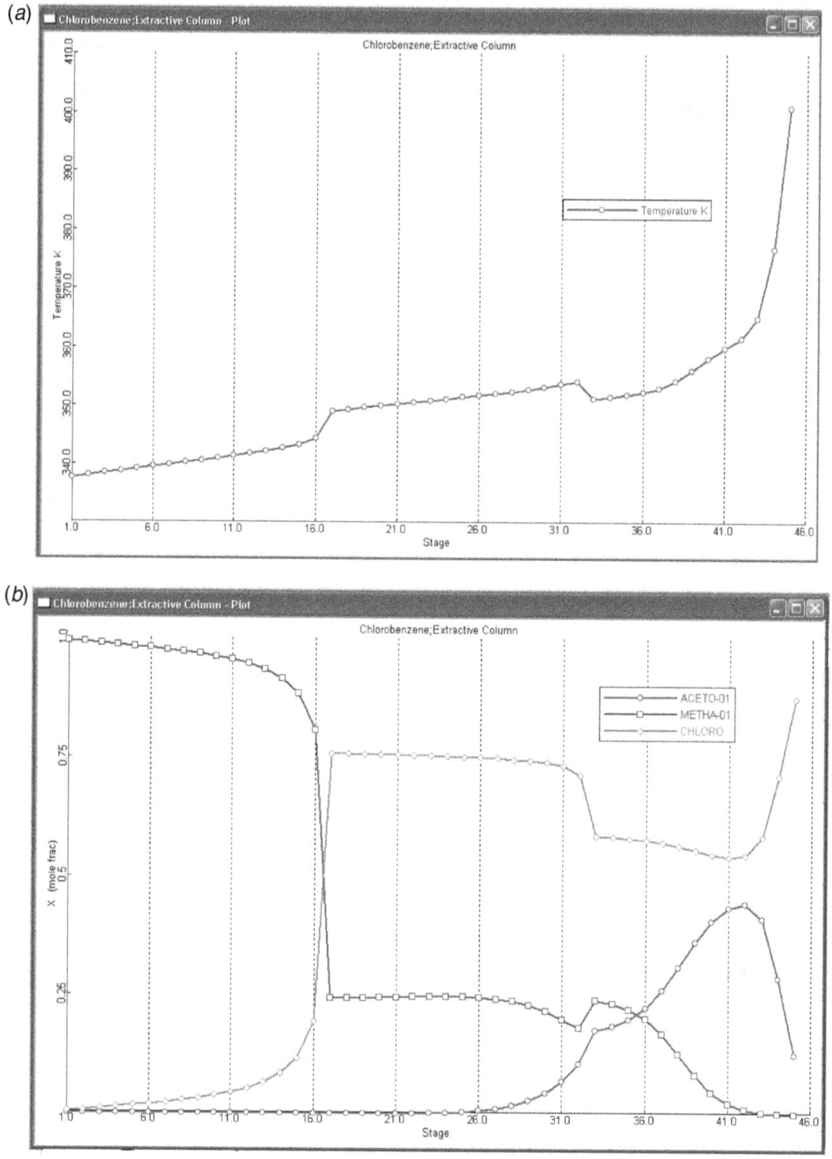


**Figure 11.21** (*a*) Extractive column temperature profile; chlorobenzene solvent. (*b*) Extractive column composition profiles; chlorobenzene solvent.

***Water Solvent Control Structure.*** The temperature control tray location is selected by choosing the tray where there is a large temperature change from stage to stage (Stage 50 at 347.6 K in the extractive column and Stage 22 at 371.7 K in the methanol column). See Figures 11.6*a* and 11.7*a*.

Controller faceplates are shown at the bottom of Figure 11.22. Notice that the controller *FCtot* is on "cascade" because it receives its remote setpoint from a multiplier block "ratio" so that the solvent flowrate is ratioed to the feed flowrate.

(c)

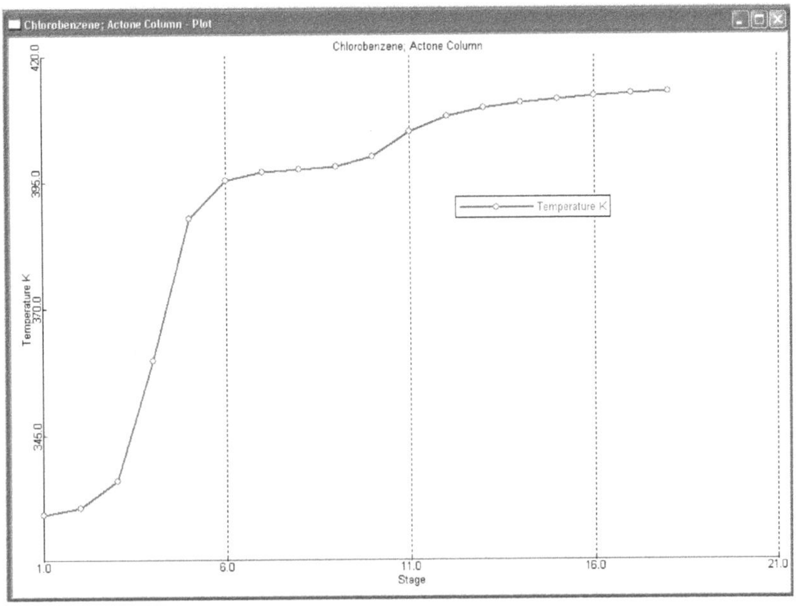


(d)

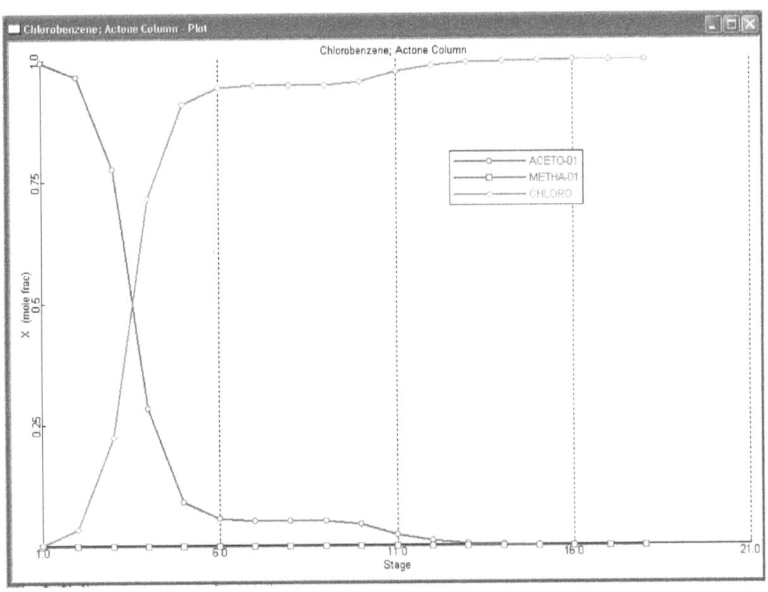


**Figure 11.21** (*c*) Acetone column temperature profile; chlorobenzene solvent. (*d*) Acetone column composition profiles; chlorobenzene solvent.

***DMSO Solvent Control Structure.*** The temperature on Stage 29 in the extractive column (360.9 K) is selected. As shown in Figure 11.18*a*, this is not where the temperature is changing the most rapidly, which is at the very base of the column due to the increase of the DMSO concentration relative to the methanol (see Fig. 11.18*b*). There is a reasonable change in temperature from tray to tray at Stage 29, and this is where the acetone composition

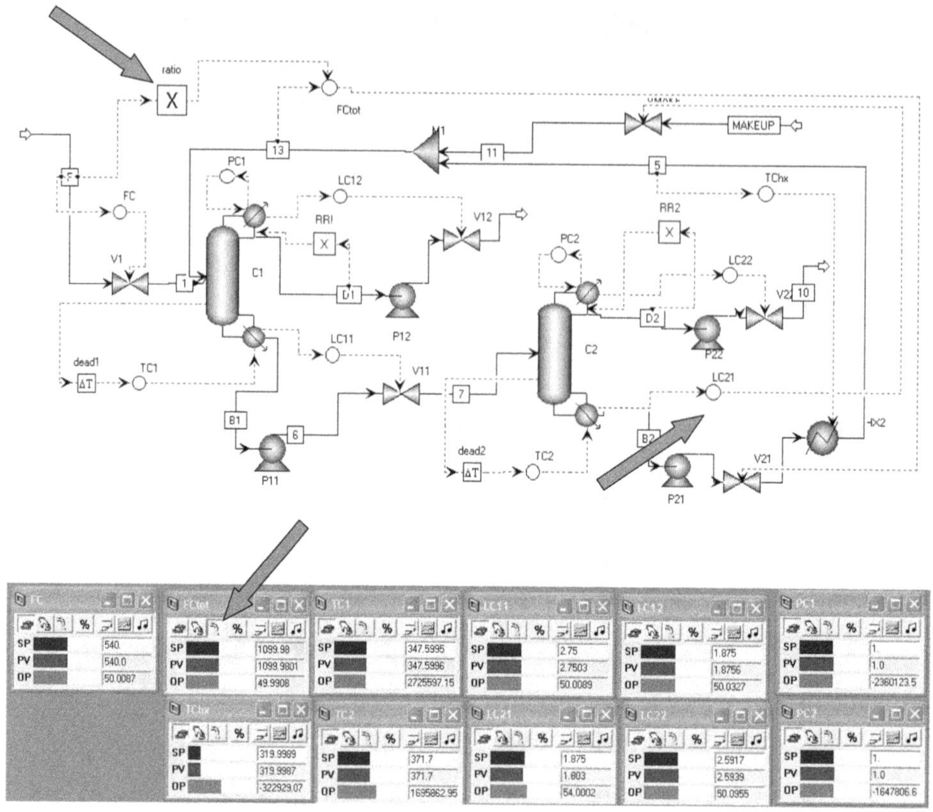


**Figure 11.22** Control structure with water solvent.

is decreasing rapidly. Since we want to keep the acetone from dropping out of the bottom, Stage 29 is a good location to control.

If a single tray temperature is selected in the methanol column, the sharp temperature profile would produce a very small controller gain, which would give poor load performance. Therefore, an average temperature is used by measuring the temperatures on three trays: Stage 12 at 397 K, Stage 13 at 421 K, and Stage 14 at 451 K. Figure 11.23 shows the variables in each of the control devices used to implement this average temperature control in Aspen Dynamics. Notice that Aspen Dynamics uses metric units, so the outputs of the first two summations are in °C. Figure 11.24 shows the structure. The first two summation blocks add the three stage temperatures. The multiplier block "average" multiplies by 0.3333. The last summation block adds the constant "273" to convert the signals back to K from the metric units (°C) used in Aspen Dynamics. The average temperature is 429 K, which is the setpoint of the *TC2* controller shown on the faceplates shown at the bottom of Figure 11.24.

***Chlorobenzene Solvent System.*** Selecting an appropriate tray in the extractive column in the chlorobenzene system is a little less obvious than in the previous cases. As Figure 11.21*a* shows, the sharpest temperature profile is near the base, but this is caused by the rapid increase in chlorobenzene concentration and decrease in the acetone concentration (see Fig. 11.21*b*). Remember that in this system the methanol goes overhead in the

**TABLE 11.6 Controller Tuning Parameters.**

| | Water Solvent | DMSO Solvent | Chlorobenzene Solvent |
|---|---|---|---|
| | | *TC1* | |
| Controlled variable | $T_{1,50} = 347.6$ K | $T_{1,29} = 360.9$ K | $T_{1,39} = 357.9$ K |
| Manipulated variable | $Q_{R1}$ | $Q_{R1}$ | $Q_{R1}$ |
| Transmitter range | 300–400 K | 300–400 K | 300–400 K |
| Controller output range | $0–6 \times 10^6$ cal/s | $0–6.63 \times 10^6$ cal/s | 0–0.2 ratio$^{(1)}$ |
| $K_C$ | 2.23 | 0.72 | 13.7 |
| $\tau_I$ | 9.29 min | 7.0 min | 7.9 min |
| | | *TC2* | |
| Controlled variable | $T_{2,22} = 371.7$ K | $T_{2,\text{average}} = 429$ K | $T_{2,\text{average}} = 381$ K |
| Manipulated variable | $Q_{R2}$ | $Q_{R2}$ | $Q_{R2}$ |
| Transmitter range | 325–425 K | 360–600 K | 350–450 K |
| Controller output range | $0–3.6 \times 10^6$ cal/s | $0–6.63 \times 10^6$ cal/s | 0–0.03 ratio$^{(1)}$ |
| $K_C$ | 1.32 | 0.176 | 0.25 |
| $\tau_I$ | 10.6 min | 11.9 min | 14.5 min |

*Note*: (1) Ratio in GJ/kmol.

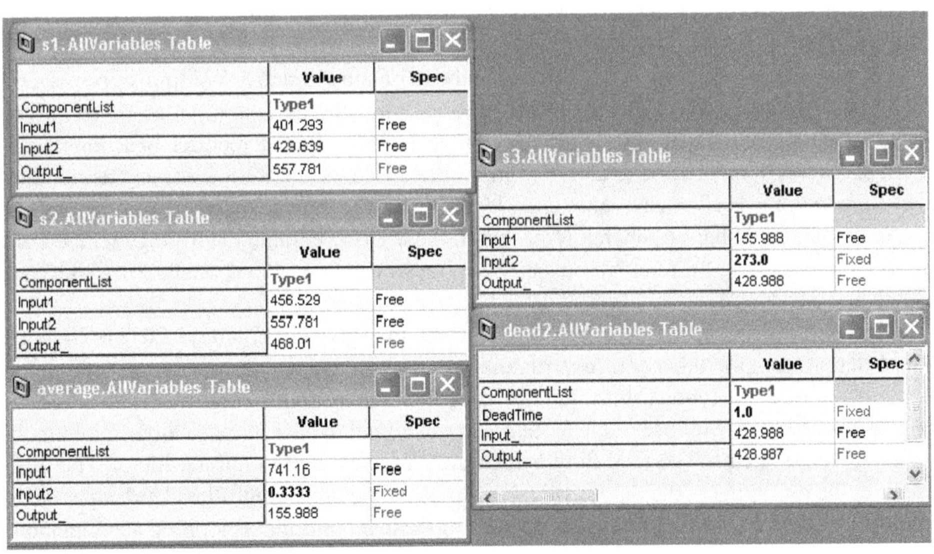


**Figure 11.23** Average temperature control.

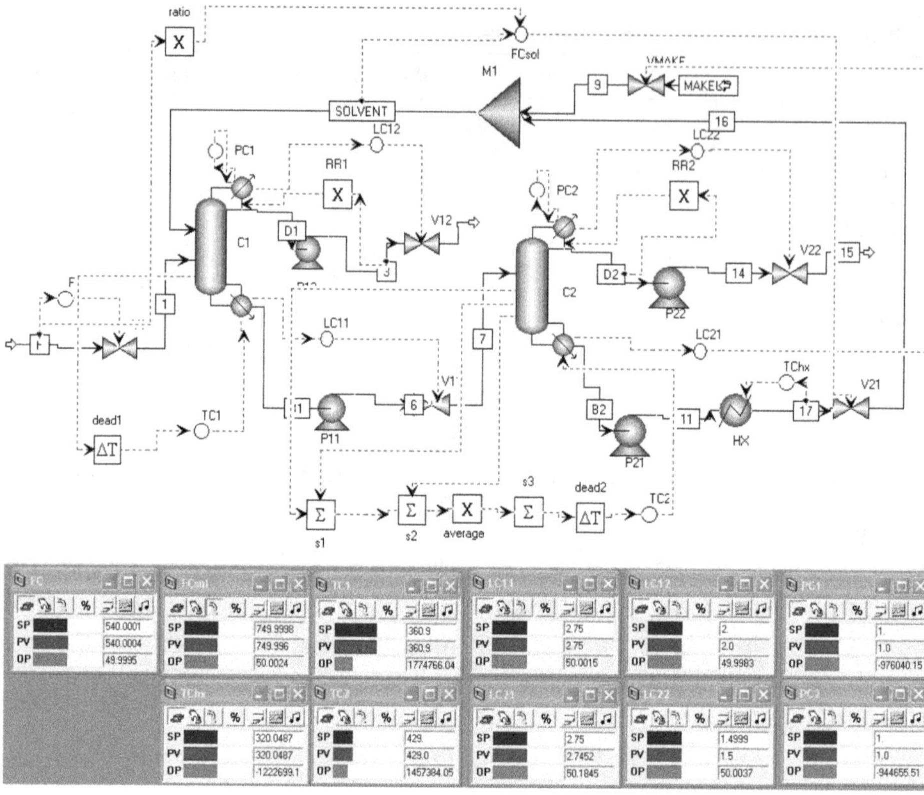

Figure 11.24 Control structure with DMSO solvent.

extractive column, so we must find a tray that will keep the methanol from dropping out the bottom and appearing as an impurity in the acetone product left in the distillate of the acetone column. At the location where the methanol concentration is dropping, the temperature is not changing very rapidly.

To aid in this selection, a sensitivity analysis of the extractive column is performed. Using steady-state Aspen Plus, the reflux ratio and the reboiler heat input are fixed at their steady-state values. Then a small 0.1% change is made in the reboiler heat input. The changes in tray temperatures are plotted in Figure 11.25. Stage 39 is indicated as the most sensitive tray. Its steady-state value is 357.9 K.

Since there is a sharp temperature profile in the acetone column, an average temperature is used by measuring the temperatures on three trays: Stage 4 at 360 K, Stage 5 at 388 K, and Stage 6 at 395 K, giving an average of 381 K.

The control structure for the chlorobenzene system is given in Figure 11.26. Two features are different from the other control structures. They involve the use of steam-to-feed ratios. In both columns, the reboiler heat input is ratioed to the feed to the column. These are added to improve the load response of the system that was found to be inferior to those found in the other solvent systems. The feed flow is measured and sent to a multiplier block. The other input to the multiplier is the output signal from the temperature controller. The output of the multiplier sets the reboiler heat input. So the temperature controller is looking at temperature and outputting a ratio signal. Controller parameters are given in Table 11.6. Notice that the $Q_R/F$ ratios must be in units of GJ per kmol in the Aspen Dynamic convention.

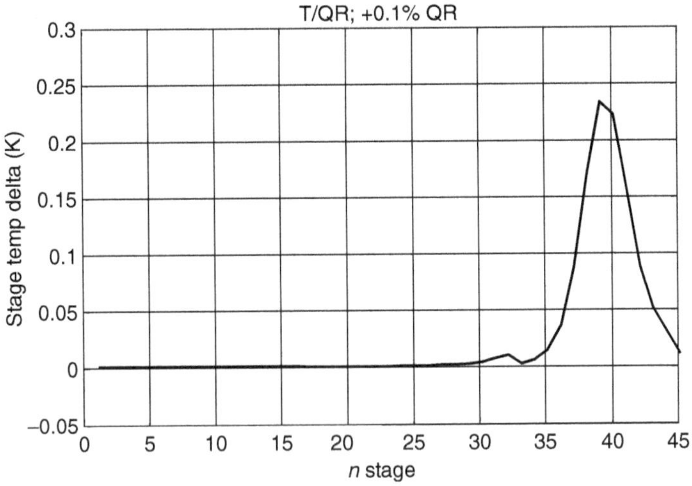


**Figure 11.25** Steady-state gains for extractive column with chlorobenzene solvent.

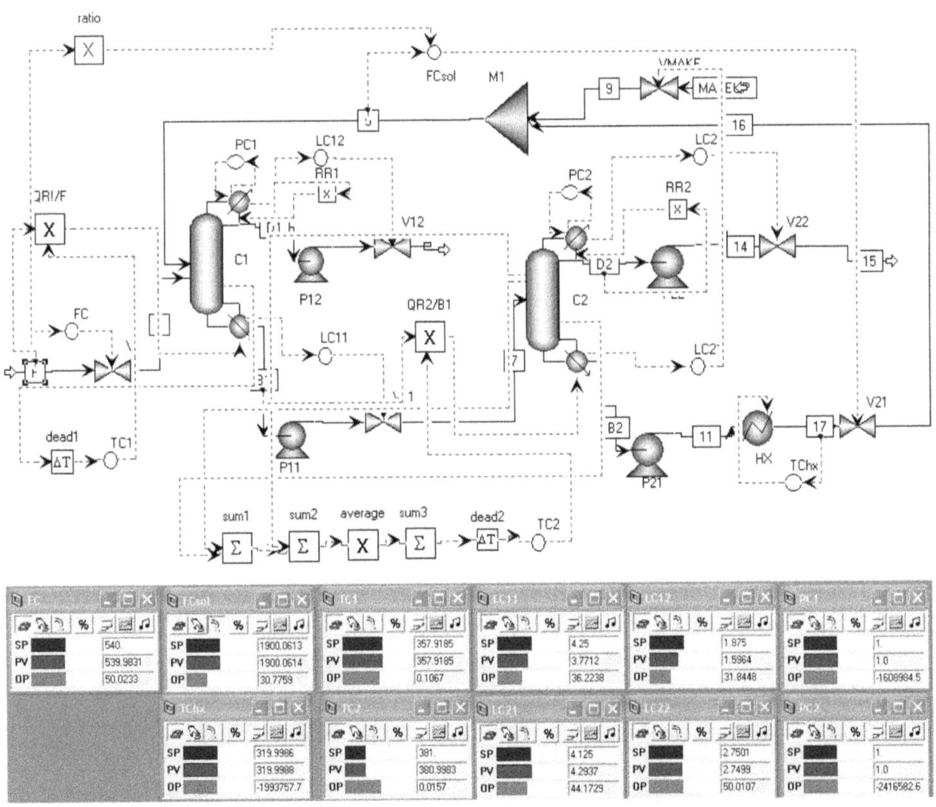


**Figure 11.26** Control structure with chlorobenzene solvent.

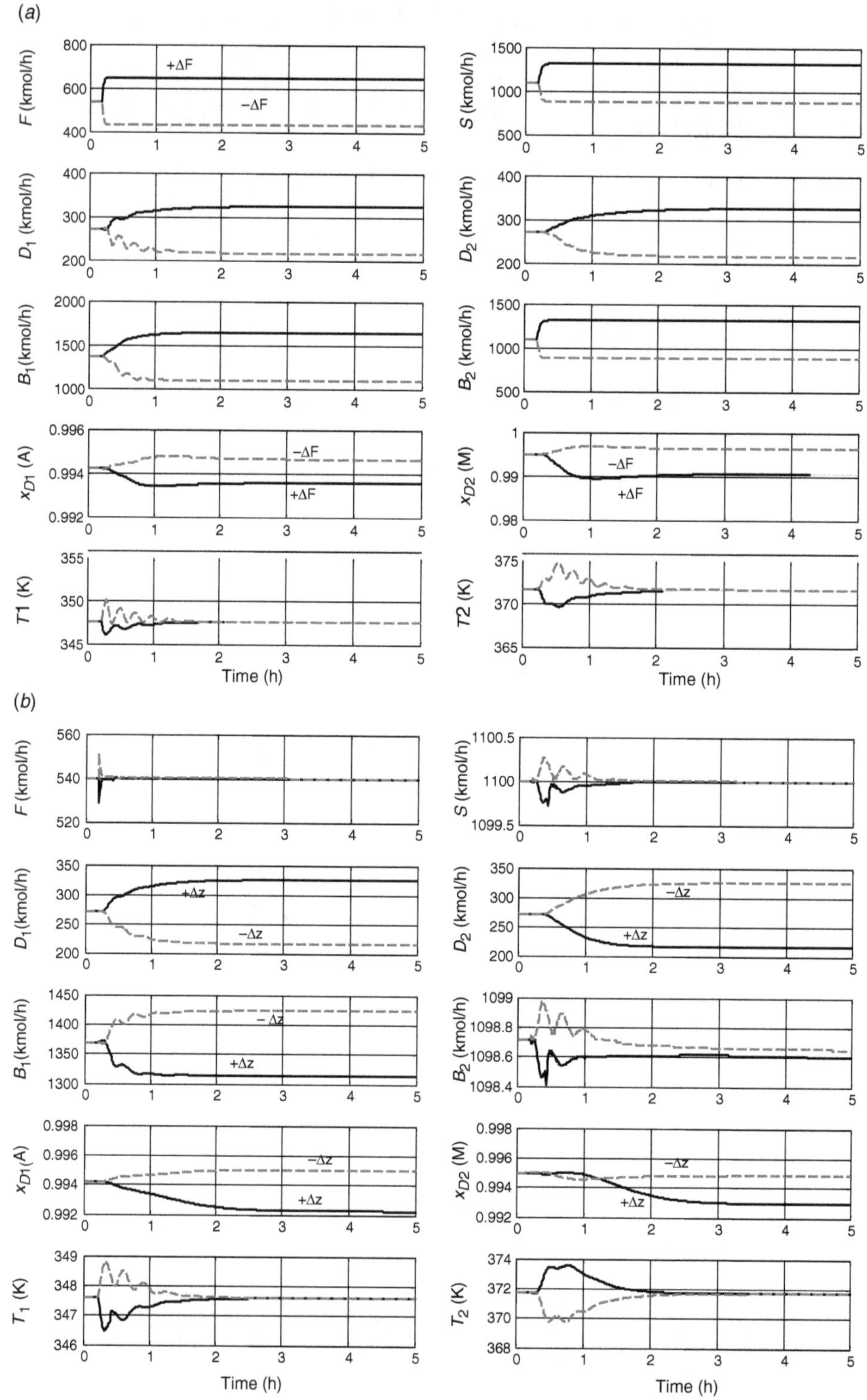


**Figure 11.27** (*a*) 20% feed flowrate disturbances; water solvent. (*b*) Feed composition disturbances; water solvent.

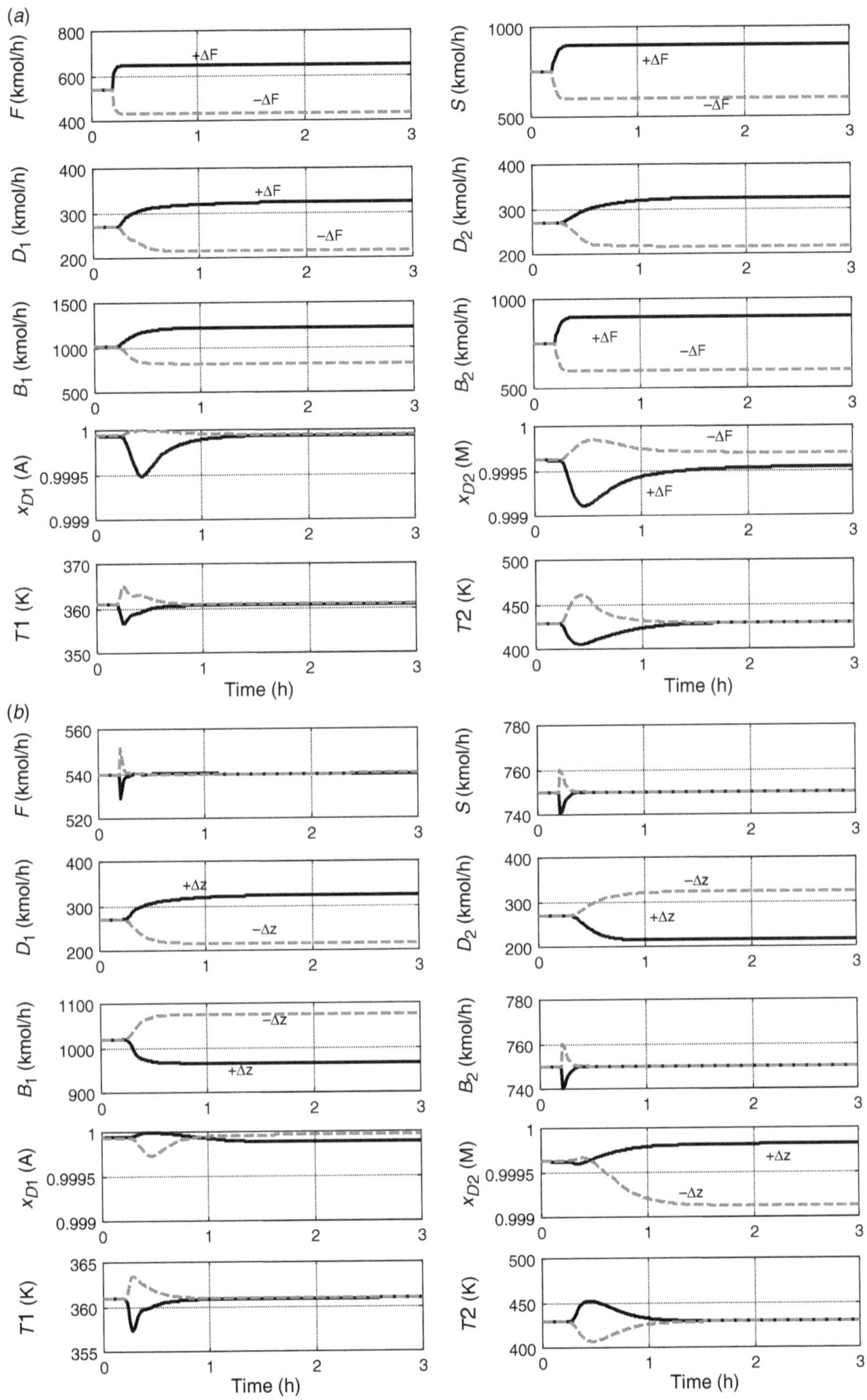


**Figure 11.28** (*a*) 20% feed flowrate disturbances; DMSO solvent. (*b*) Feed composition disturbances; DMSO solvent.

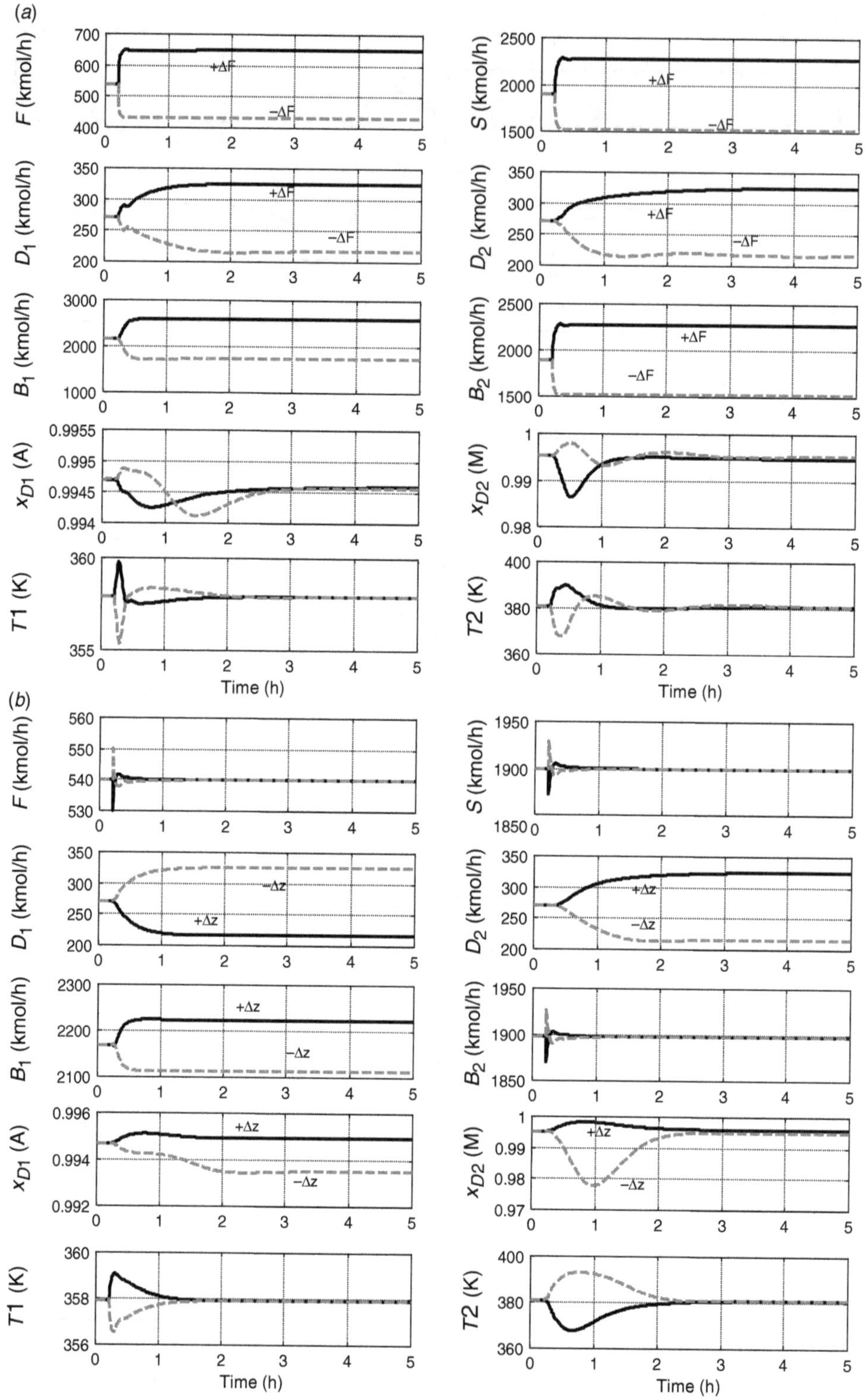


**Figure 11.29** (*a*) 20% feed flowrate disturbances; chlorobenzene solvent. (*b*) Feed composition disturbances; chlorobenzene solvent.

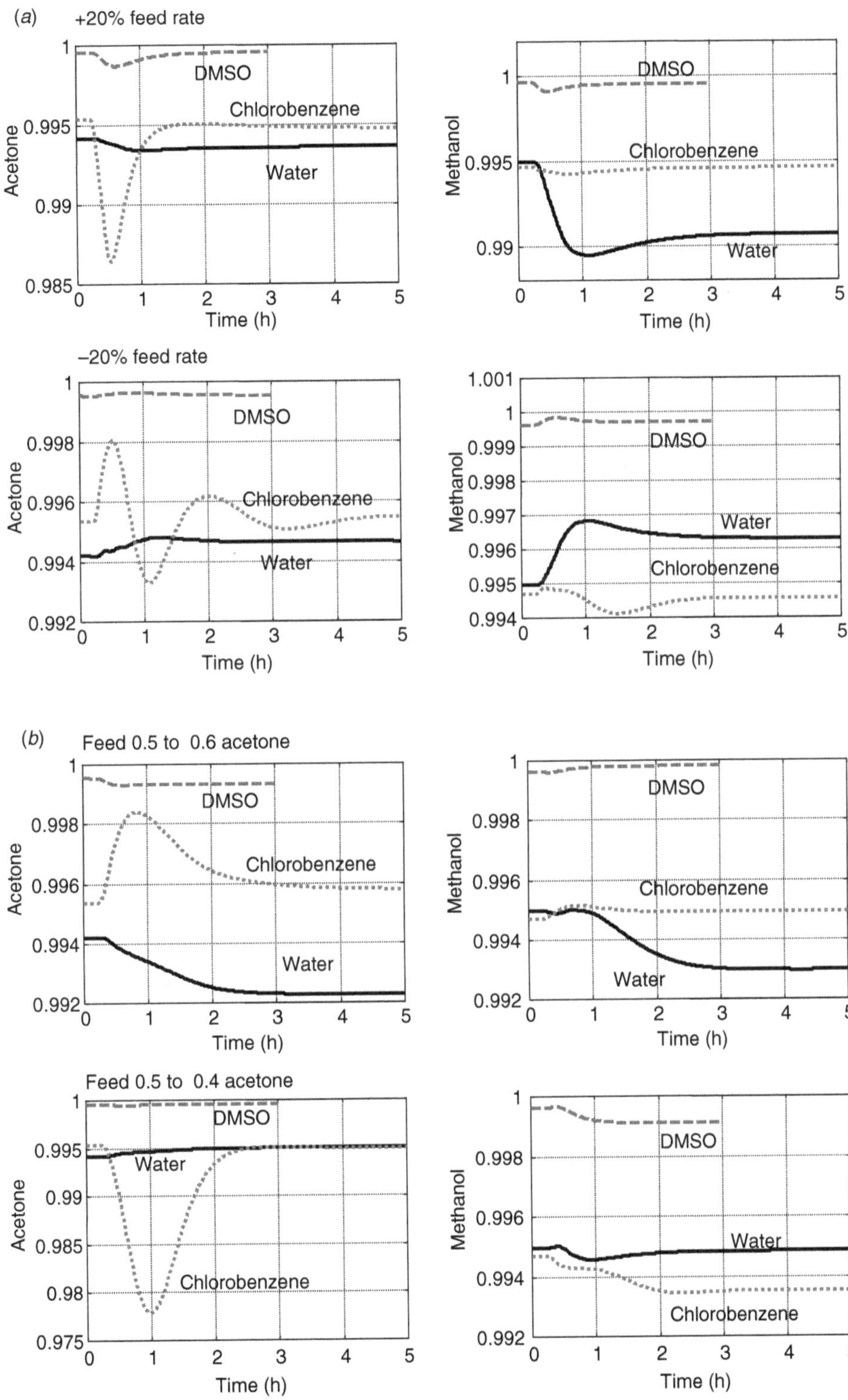


**Figure 11.30** (*a*) Comparison for feed flowrate disturbances. (*b*) Comparison for feed composition disturbances.

### 11.5.6 Dynamic Performance

The performances of these control structures are evaluated by introducing several large disturbances. Stable regulatory control is achieved in all three systems.

The responses of the water system to large 20% increases (solid lines) and decreases (dashed lines) in feed flowrate are shown in Figure 11.27*a*. Responses for feed composition disturbances from 50 to 60 mol% acetone (solid lines) and from 50 to 40 mol% acetone (dashed lines) are shown in Figure 11.27*b*. Product purities ($x_{D1(\mathrm{A})}$ and $x_{D2(\mathrm{M})}$) are held quite close to their specifications. The system takes about 1.5 h to come to a new steady state.

The responses of the DMSO system to large 20% increases and decreases in feed flowrate are shown in Figure 11.28*a*. Responses for feed composition are shown in Figure 11.28*b*. Product purities ($x_{D1(\mathrm{A})}$ and $x_{D2(\mathrm{M})}$) are held quite close to their specifications, which are quite high (99.95 mol%). The system is faster than the water system, taking only about 1 h to come to a new steady state.

The responses of the chlorobenzene system to feed flowrate disturbances are shown in Figure 11.29*a*. Responses for feed composition are shown in Figure 11.29*b*. Product purities ($x_{D1(\mathrm{M})}$ and $x_{D2(\mathrm{A})}$) are held quite close to their specifications for the feed flowrate changes, but the purity of the methanol distillate from the extractive column $x_{D1(\mathrm{M})}$ drifts downward for the disturbance in feed composition ($-\Delta z$) from 0.5 to 0.4 acetone in the feed. More methanol must come out the top of the extractive column in stream $D_1$. Since the solvent flow is unchanged, the increased methanol load produces a drop in purity. The dynamics of the chlorobenzene system are the slowest of the cases, taking over 2 h to come to a new steady state.

At this point it is useful to give a direct comparison of the dynamic performances of the three alternative solvent systems. Figure 11.30 provides this comparison. The solid lines are for the water system, the dashed lines are for the DMSO system, and the dotted lines are for the chlorobenzene system. Control of the DMSO system is excellent. Control of the water system is good. Control of the chlorobenzene system is the poorest of the three. The transient deviations in the purity of the acetone product (the distillate from the solvent-recovery column) are the largest. This is due to the inherent tendency of methanol to drop out the bottom of the extractive column during an upset, and this methanol shows up in the distillate of the acetone column as impurity.

Overall the simulation results demonstrate that all three systems can be controlled by the basic control structures developed. The DMSO solvent is superior in both steady-state economics and dynamic controllability. It also provides the capability of achieving very high purity products, which is not possible using the other two solvents.

### 11.5.7 Conclusions for Controllability

Steady-state and dynamic comparisons have been presented of an extractive distillation process using three different solvents. The numerical example examines the acetone–methanol separation using water, DMSO, or chlorobenzene solvents.

The DMSO solvent gives the best steady-state economics and the best dynamic controllability. This is one of those uncommon situations in which there is no trade-off between controllability and steady-state economics.

## 11.6 CONCLUSION

In this chapter, the use of extractive distillation has been illustrated using the acetone–methanol system as a numerical example. Steady-state and dynamic comparisons have been presented between extractive distillation and a pressure-swing distillation, with and without heat integration. In addition, the effect of solvent selection on dynamic controllability has been investigated.

## REFERENCES


1. Knapp J. P. and M. F. Doherty, Thermal integration of homogeneous azeotropic distillation sequences, *AIChE J.*, **36**, 969–984 (1990).
2. Kossack S., K. Kraemer, R. Gani, and W. Marquardt, A systematic synthesis framework for extractive distillation processes, *Chem. Eng. Res. Des.*, **86**, 781–792 (2008).
3. Grassi V. G., *Practical Distillation Control*, Van Nostrand Reinhold, New York, 1992, pp. 370–404.
4. Luyben W. L., *Distillation Design and Control Using Aspen Simulation*, Wiley, New York, 2006, pp. 209–222.

# CHAPTER 12

# MAXIMUM-BOILING AZEOTROPES

Most azeotropes, whether they are homogeneous or heterogeneous, are minimum boiling, that is, the azeotrope boils at a temperature that is lower than the boiling point of the lighter pure component. Minimum boiling azeotropes occur when there is repulsion between dissimilar molecules. All of the azeotropes considered in other chapters in this book are minimum boiling. However, there are a number of systems that exhibit molecular *attractions*, and these systems can result in maximum-boiling azeotropes. The investigation of the design of this type of azeotropic system is very limited, and studies of their control are almost nonexistent.

In this chapter, we explore the design and control of the maximum-boiling azeotropic acetone–chloroform distillation system as a typical example. Other systems that exhibit maximum-boiling azeotropes include formic acid–water, nitric acid–water, and acetic acid–DMAC (*n,n*-dimethyl acetamide).

A two-column extractive distillation process is used with dimethyl sulfoxide (DMSO) as the solvent. The two components are separated into 99.5 mol% pure products leaving in the distillate streams from two distillation columns. The solvent flowrate that minimizes total energy consumption is determined. A control structure that is capable of handling very large disturbances in throughput and feed composition is developed. The control of two tray temperatures in the extractive column is found to be necessary to handle feed composition disturbances.

## 12.1 ACETONE–CHLOROFORM SYSTEM STEADY-STATE DESIGN

Acetone and chloroform have very similar boiling points (329.4 and 334.3 K), despite having molecular weights that are considerable different (58.08 and 117.4 kg/kmol). These molecules have an attraction for each other, which results in a maximum-boiling

*Design and Control of Distillation Systems for Separating Azeotropes.* By William L. Luyben and I-Lung Chien

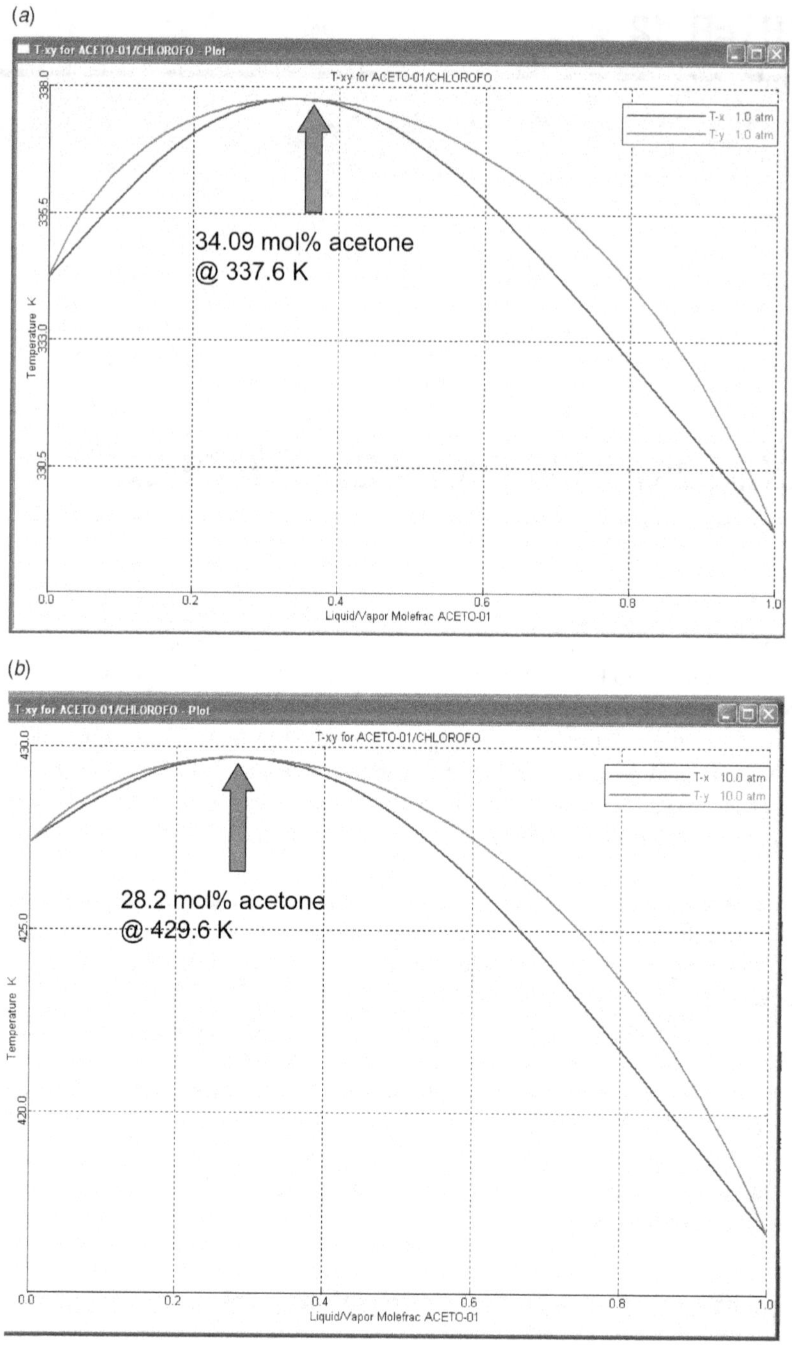


**Figure 12.1** (*a*) *Txy* diagram at 1 atm: acetone–chloroform. (*b*) *Txy* diagram at 10 atm: acetone–chloroform.

azeotrope. Figures 12.1*a* and *b* give *Txy* diagrams for the system at two different pressures. At 1 atm the azeotropic composition is 34.09 mol% acetone and the temperature is 337.6 K. Note that this is higher than the boiling point temperature of either pure component.

### 12.1.1 Pressure-Swing Distillation

Figures 12.1 shows that the effect of pressure on the azeotrope is small, so we would expect pressure-swing distillation to be unattractive. This would be the case for minimum-boiling azeotropes because the circulating streams are the distillates of the two columns that must be vaporized, which translates into high energy consumption, large heat exchangers, and large diameter columns.

However, in a pressure-swing distillation system with a maximum-boiling azeotrope, the circulating streams are the bottoms of the two columns. The distillate streams are the two product streams. The large recycle streams are liquids that do not have to be vaporized; therefore, if the azeotrope is maximum boiling, a pressure-swing system may be feasible despite having relatively small changes in the azeotropic composition with pressure. This inherent difference between minimum-boiling and maximum-boiling systems could be quite significant in terms of flowsheet selection. Investigation of a pressure-swing distillation process for acetone–chloroform indicated that columns with a large number of trays and high reflux ratios are required because the relative volatilities between the two components on both sides of the azeotrope are close to unity. However, extractive distillation was found to be effective for this separation.

### 12.1.2 Extractive Distillation

Figure 12.2 shows the economically optimized flowsheet of the extractive distillation system. A solvent is added near the top of the extractive column. Acetone goes out the top of this column, while the bottoms is fed to the solvent recovery column where chloroform is produced in the distillate stream.

Several solvents were explored using Aspen Plus simulations, including water, chlorobenzene, ethylene glycol, and xylene. UNIQUAC physical properties are used in the Aspen simulations. The only solvent that achieved the desired separation was DMSO. The ternary diagram for the acetone–chloroform–DMSO system is given in Figure 12.3*a*. There is a distillation boundary that separates the diagram into two regions. Notice that the residue curves follow paths that start at either of the light components (acetone or chloroform) and move to the heavy component (DMSO).

In the extractive distillation column, the solvent is essentially pure DMSO and is fed in the upper part of the extractive column. The fresh feed mixture of acetone and chloroform is fed lower in the column. The total feed point shown in Figure 12.3*b* corresponds to a fresh feed flowrate of 100 kmol/h of 50 mol% acetone and 50 mol% chloroform with a solvent flowrate of 164 kmol/h of DMSO. The ternary diagram shows that a distillate of quite pure acetone can be produced from the top of the extractive column, and the bottoms product will contain very small amounts of acetone. The bottoms stream is essentially a binary mixture of chloroform and DMSO that can be easily separated in a recovery column. The chloroform goes overhead in the distillate and the DMSO leaves in the bottoms, which is recycled back to the extractive column.

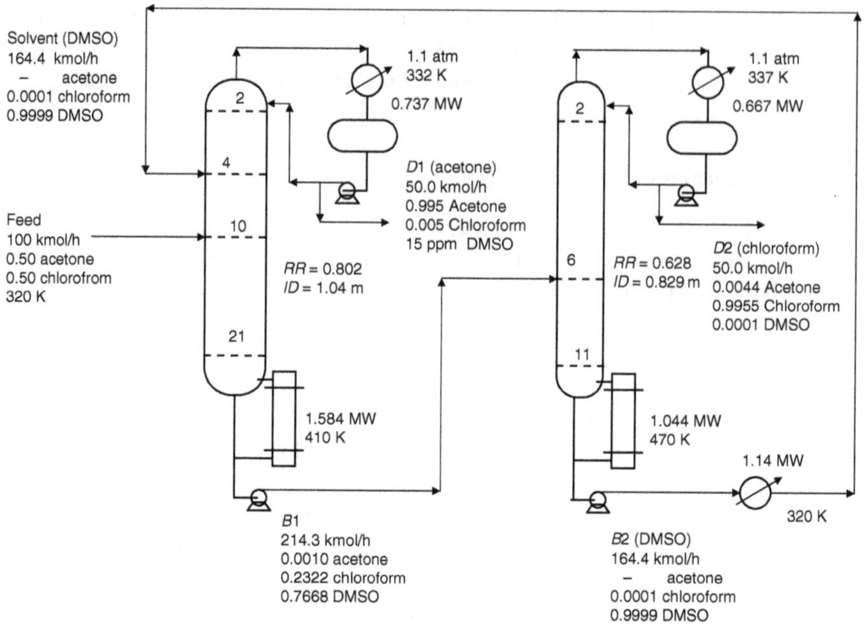


**Figure 12.2** Extractive distillation; acetone–chloroform–DMSO; $S = 164.4$.

It may seem confusing that the feed point of the recovery column (the bottoms $B1$ from the first column) is slightly outside the region in which the distillate $D2$ and bottoms $B2$ points lie in the ternary diagram (Fig. 12.3*b*). This is physically possible because the distillation boundary is curved. Therefore, the feed point can lie outside the region in which the two product streams are located.

The extractive column has several design degrees of freedom. In addition to the number of trays and fresh feed and solvent feed tray locations, the amount of solvent, and the reflux ratio can be varied to achieve the desired purities of the distillate and bottoms streams while minimizing energy consumption. A 22-stage extractive column is used in this study since adding more trays was found to have little effect on the amount of solvent required and the energy consumed. The fresh feed is introduced at Stage 10 and the solvent at Stage 4 (using Aspen tray numbering with the reflux drum Stage 1). A 12-stage solvent recovery column is used that is fed on Stage 6. These feed locations were found empirically by observing their effects on reboiler energy consumptions.

The separation in the extractive column depends on the amount of solvent circulating around the system. Figure 12.4 shows that high solvent flowrates reduce the impurity of chloroform in the distillate acetone product. For each solvent flowrate, there is a nonmonotonic effect of reflux ratio. To achieve the desired distillate purity of 99.5 mol% acetone, the minimum solvent flowrate is 145 kmol/h (solvent-to-feed ratio of 1.45). These results are obtained with the impurity of acetone in the bottoms held at 0.1 mol% acetone using the design spec/vary feature of Aspen Plus and manipulating distillate flowrate.

The optimum solvent flowrate is found by determining the minimum total energy required in the reboilers of the two columns ($Q_{R1}$ and $Q_{R2}$), using four design spec/vary specifications. In the extractive column, the distillate impurity is held at 0.5 mol% chloroform and the bottoms impurity is held at 0.1 mol% acetone by varying distillate flowrate $D_1$ and reflux ratio $RR_1$. In the solvent recovery column, the distillate impurity is held at

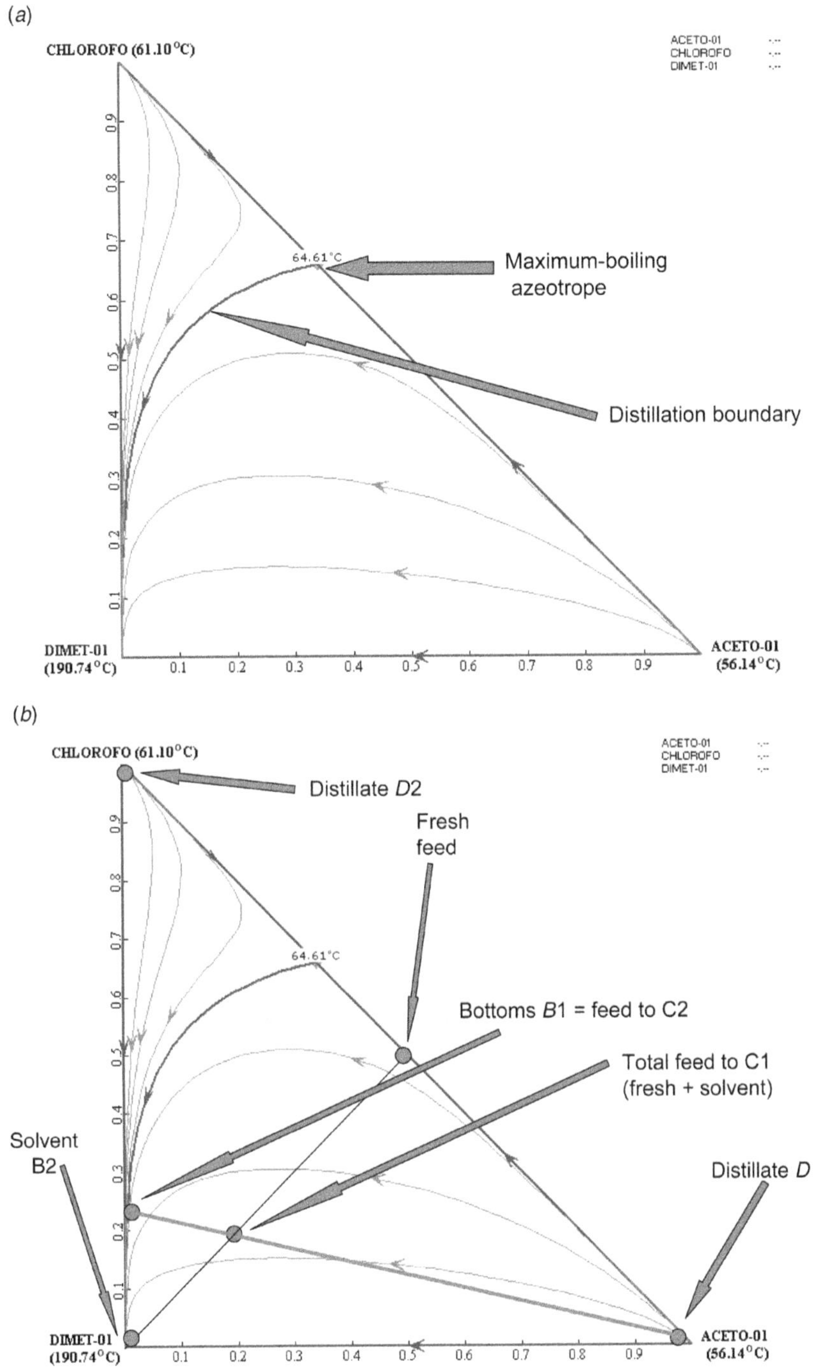


**Figure 12.3** (*a*) Ternary diagram: acetone–chloroform–DMSO. (*b*) Feed, product and solvent points.

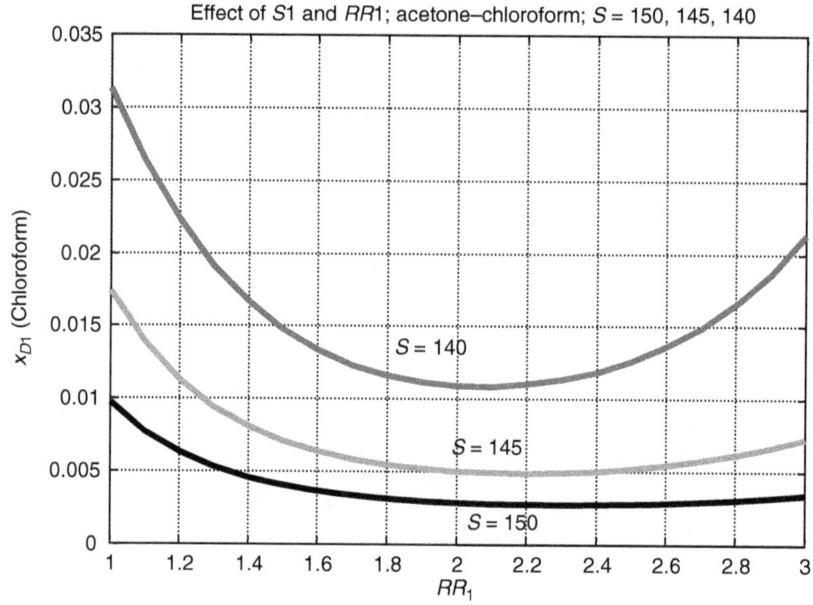


**Figure 12.4** Effect of reflux ratio and solvent on chloroform impurity in acetone product.

0.01 mol% DMSO and the bottoms impurity is held at 0.01 mol% chloroform by varying distillate flowrate $D_2$ and reflux ratio $RR_2$. Figure 12.5 shows that the reboiler duty in the extractive column $Q_{R1}$ decreases as solvent flowrate increases, but the reboiler duty in the solvent recovery column $Q_{R2}$ increases. The solvent flowrate that minimizes total energy consumption is 164 kmol/h.

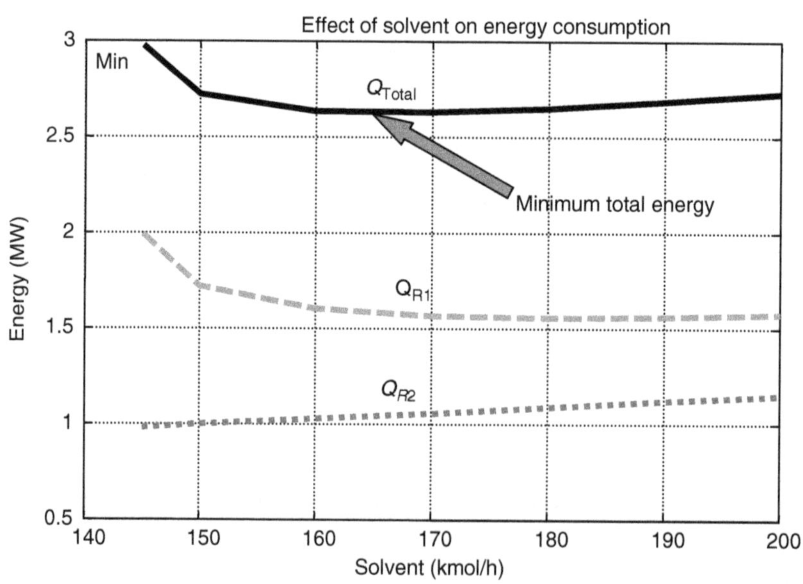


**Figure 12.5** Minimum energy consumption.

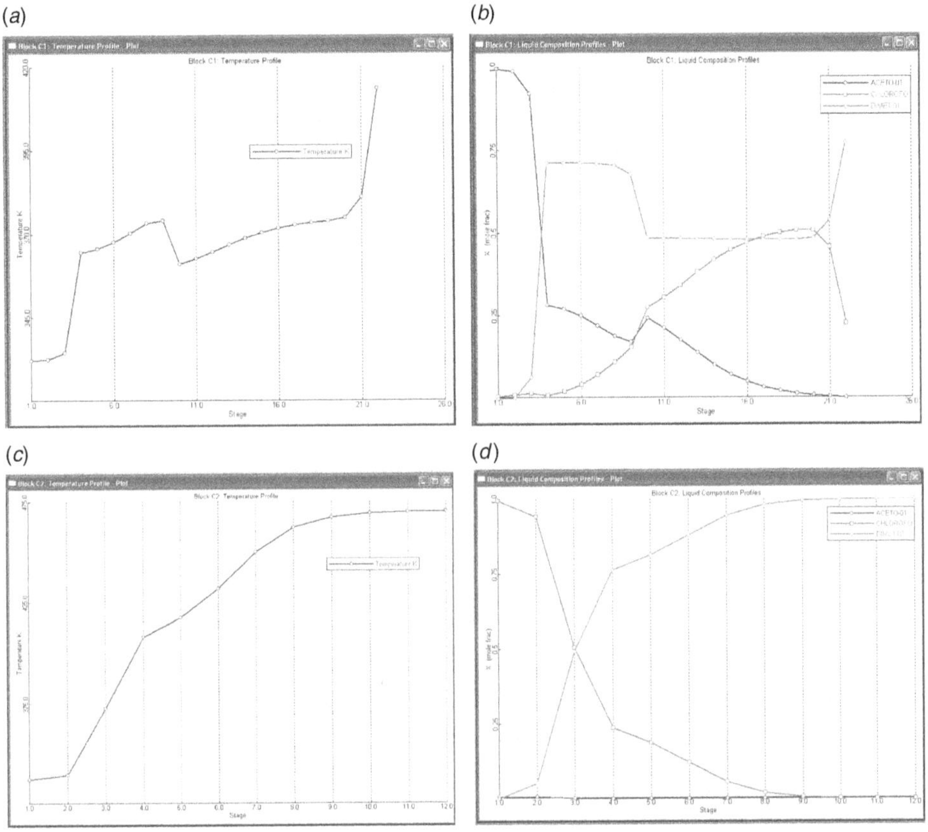

**Figure 12.6** (*a*) Temperature profile in the extractive column. (*b*) Composition profiles in the extractive column. (*c*) Temperature profile in the recovery column. (*d*) Composition profiles in the recovery column.

Figure 12.2 gives the flowsheet with stream conditions, heat duties, reflux ratios, and column diameters. Both columns operate at 1.1 atm. Reflux drum temperatures are 332 and 337 K, which permit the use of cooling water in the condensers. The reflux ratios are fairly small (0.802 and 0.628), which indicates that the separations are not difficult and columns with relatively few trays are required. Notice that a heat exchanger is installed to cool the solvent from the bottom of the solvent recovery column before introducing it into the extractive column. Figure 12.6 gives temperature and composition profiles for the two columns.

## 12.2 DYNAMICS AND CONTROL

Several alternative control structures were investigated. Figure 12.7 shows one control structure CS1, which handled feed rate disturbances fairly well but did not maintain product purities for feed composition changes. A second control structure is developed later in this chapter that handles both types of disturbances effectively.

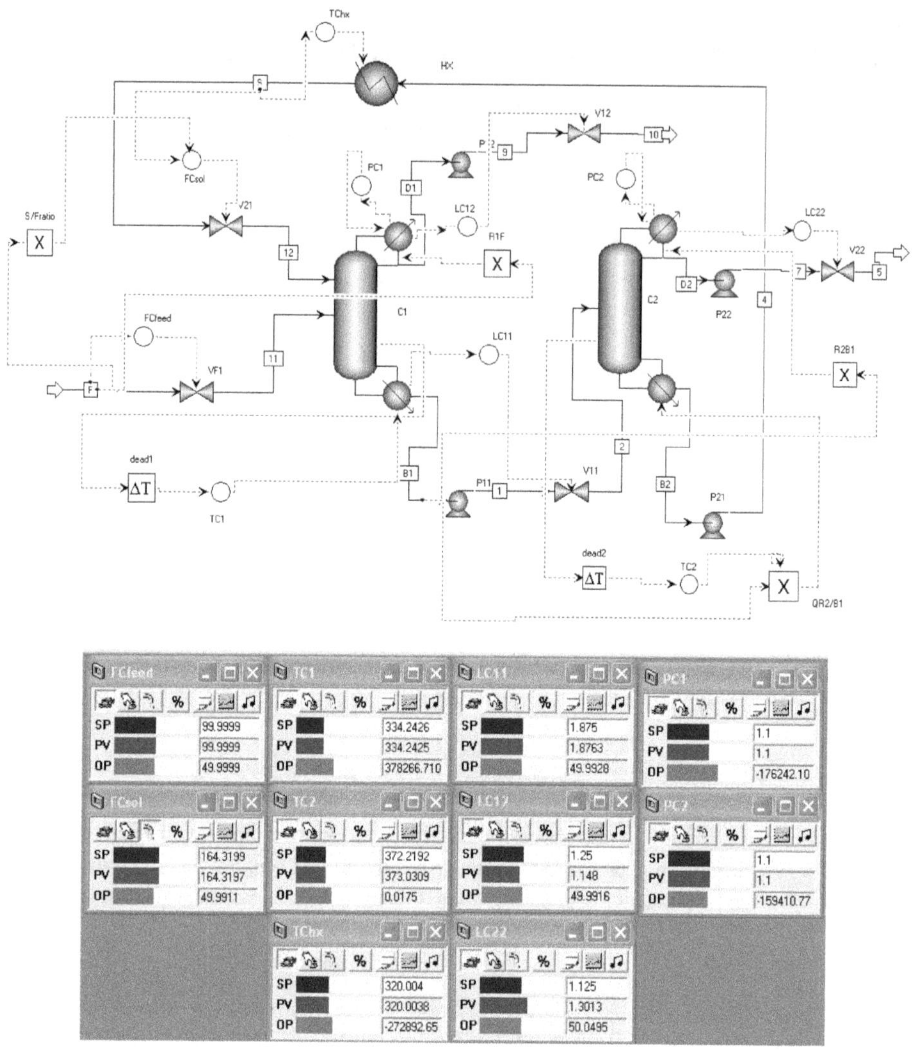


**Figure 12.7** Control structure CS1.

### 12.2.1 CS1 Structure (Single-Temperature Control in each Column)

The features of control structure CS1 shown in Figure 12.7 are listed below. All controllers have conventional PI action except level controllers, which are proportional only.

1. Feed is flow controlled.
2. Solvent is ratioed to feed.
3. Reflux flowrate in each column is ratioed to its feed flowrate. In the extractive column the ratio is $R1/F$. In the solvent recovery column the ratio is $R2/B1$.
4. Reflux drum level in each column is controlled by manipulating distillate.
5. Base level in the extractive column is controlled by manipulating bottoms.
6. Base level in the solvent recovery column is not controlled because the solvent losses are almost negligible. A periodic addition of fresh solvent would be required to make

up for these very small losses, but this would not be part of the basic regulatory control structure.

7. Pressure in each column is controlled by manipulating condenser heat removal.
8. The temperature of the solvent entering the extraction column is controlled by manipulating heat removal in the cooler.
9. A temperature in the extractive column is controlled by manipulating reboiler heat input.
10. Reboiler heat input in the solvent recovery column is ratioed to the feed to the column (stream *B*1).
11. A temperature in the solvent recovery column is controlled by manipulating this heat input-to-feed ratio.

Since both product streams are the distillates from the two columns, conventional process control wisdom suggests that a tray temperature near the top of each column should be selected in order to hold product purities. As shown in Figure 12.6*a*, holding Stage 3 temperature at 334 K in the extractive column should maintain acetone purity $x_{D1(A)}$ of the distillate $D_1$. Figure 12.6*c* shows the holding Stage 3 temperature at 372 K in the solvent recovery column should maintain chloroform purity $x_{D2(C)}$ of the distillate $D_2$. Notice that there is a very large temperature difference between the top and bottom of the solvent recovery column (337 to 470 K). The implications of this on temperature control are considered later in this chapter. Deadtimes of 1 min are inserted in each temperature loop. Relay-feedback tests and Tyreus–Luyben tuning are used to find the temperature controller tuning constants given in Table 12.1.

Notice in the controller faceplates shown at the bottom of Figure 12.7 that the *FCsol* controller is on "cascade" with its setpoint coming from the multiplier *S*/*F ratio*. The

**TABLE 12.1 Controller Tuning Parameters.**

| | CS1 | CS2 |
|---|---|---|
| *Extractive column C1* | | |
| Controlled variable | $T_{1,3} = 334.2$ K | $T_{1,3} = 334.2$ K |
| Manipulated variable | $Q_{R1}$ | $R/F$ |
| Transmitter range | 300–400 K | 300–400 K |
| Controller output range | $0–0.824 \times 10^6$ cal/s | $0–1.5 \times 10^6$ cal/s |
| $K_C$ | 5.4 | 1.35 |
| $\tau_I$ | 7.9 min | 6.6 min |
| Controlled variable | | $T_{1,15} = 369.6$ K |
| Manipulated variable | | $Q_{R1}/F$ |
| Transmitter range | | 300–400 K |
| Controller output range | | 0–0.1 |
| $K_C$ | | 2.0 |
| $\tau_I$ | | 9.2 min |
| *Solvent recovery column C2* | | |
| Controlled variable | $T_{2,3} = 372.6$ K | $T_{2,AVG} = 372.6$ K |
| Manipulated variable | $Q_{R2}/B1$ | $Q_{R2}/B1$ |
| Transmitter range | 300–500 K | 300–500 K |
| Controller output range | 0–0.04 | $0–0.80 \times 10^6$ cal/s |
| $K_C$ | 0.12 | 0.76 |
| $\tau_I$ | 21 min | 21 min |

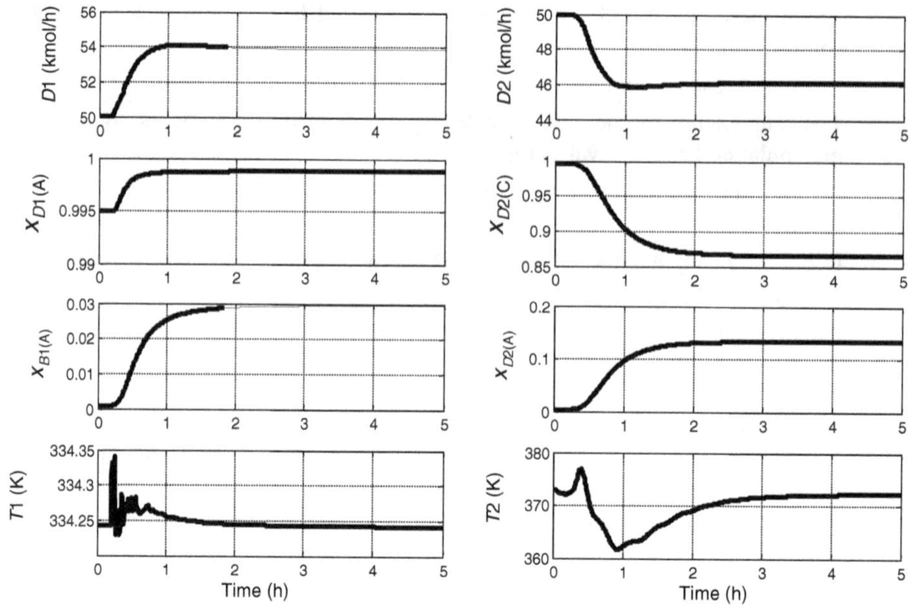

**Figure 12.8** CS1; feed composition 50 to 60 mol% acetone.

output signal from the *TC1* controller is the reboiler heat input in the extractive column. The output signal from the *TC2* controller is the ratio of reboiler heat input in the solvent recovery column to the feed to the column $QR2/B1$.

This control structure handled feed flowrate changes fairly well; however, as Figure 12.8 shows, the response of the system for a feed composition disturbance is very poor. At time equals 0.2 hours, the feed composition is changed from 50/50 molar ratio of acetone to chloroform to a 60/40 ratio. The temperatures at Stage 3 in both columns are controlled. The purity $x_{D1(\mathrm{A})}$ of the acetone product distillate stream from the extractive column increases, but the purity $x_{D2(\mathrm{C})}$ of the chloroform product distillate stream from the solvent recovery column drops drastically from 99.5 to 86 mol% chloroform. This occurs because acetone drops out the bottom of the extractive column. The bottoms composition in the extractive column climbs to $x_{B1(\mathrm{A})} = 3$ mol% acetone, and all of this acetone goes overhead in the solvent recovery column.

Holding a single-temperature constant near the top of the extractive column maintains acetone product purity, but it does not prevent acetone from moving down the column and appearing in the chloroform product from the solvent recovery column. We need to find a control structure that keeps chloroform from leaving out of the top and acetone from leaving out of the bottom. This implies a dual composition or dual temperature control structure.

The alternative of selecting the control tray in the stripping section was evaluated and found to be ineffective because the purity of the distillate was not maintained.

### 12.2.2 Analysis for Dual Temperature Control

In order to understand why the CS1 control structure fails for feed composition disturbances, a series of runs are made using steady-state simulations in Aspen Plus. Feed composition is varied from its design value of 50 mol% acetone. The distillate and bottoms compositions

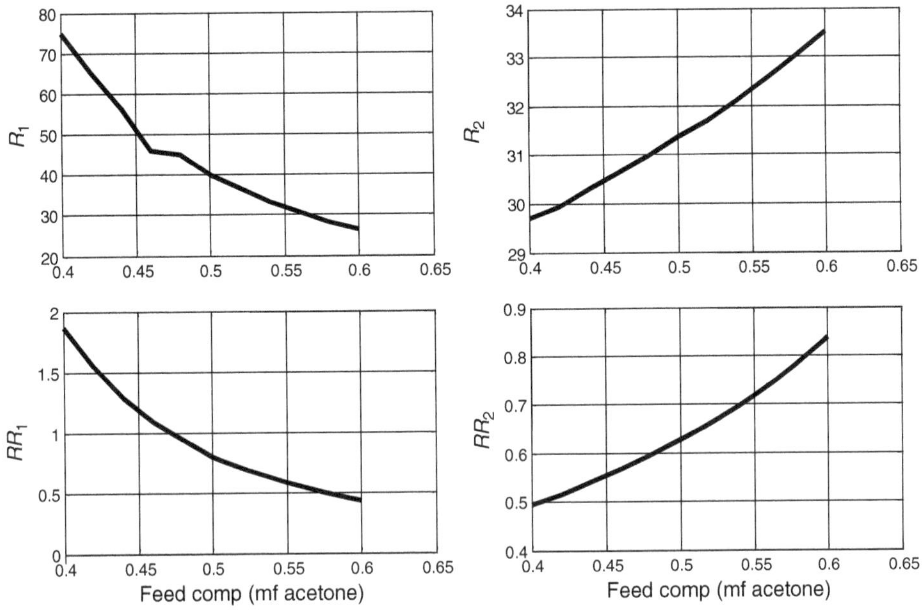


**Figure 12.9** Effect of feed composition on required $R$'s and $RR$'s.

are held constant in both columns at their desired values by using four design spec/vary features. The resulting values of reflux flowrate and reflux ratio for each column are plotted in Figure 12.9. It is clear that all of these variables have to change significantly as feed composition changes in order to maintain product purities. Therefore, a control structure that holds a constant reflux ratio or a constant reflux-to-feed ratio cannot maintain product purities in the face of feed composition changes.

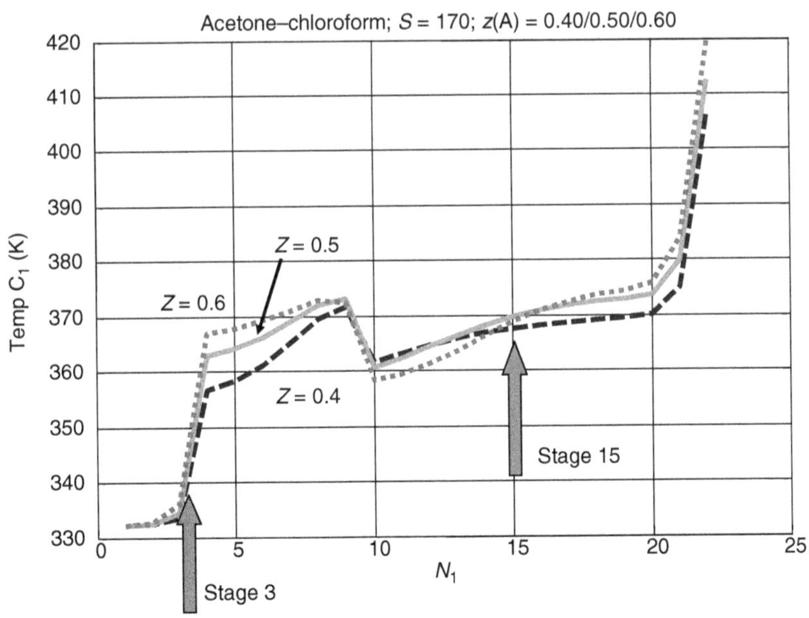


**Figure 12.10** Temperature profiles for feed composition changes.

The required changes in the solvent recovery column are fairly large but not excessive. The changes in the reflux $R2$ are only about 10% over the range of feed compositions from 40 to 60 mol% acetone. This suggests that a reflux-to-feed ratio scheme with one temperature controller should be effective in the solvent recovery column. However, in the extractive column the required changes in both reflux and reflux ratio are very large, changing by a factor of 3 over the range of feed compositions from 40 to 60 mol% acetone. Therefore, the control structure must be able to adjust these variables, which suggests that a dual temperature structure is required in this column.

The next issue is to decide what two temperatures to control. Figure 12.10 illustrates one approach to answering this question. The temperature profiles in the extractive column are plotted for feed compositions of 40, 50, and 60 mol% acetone when the products are all

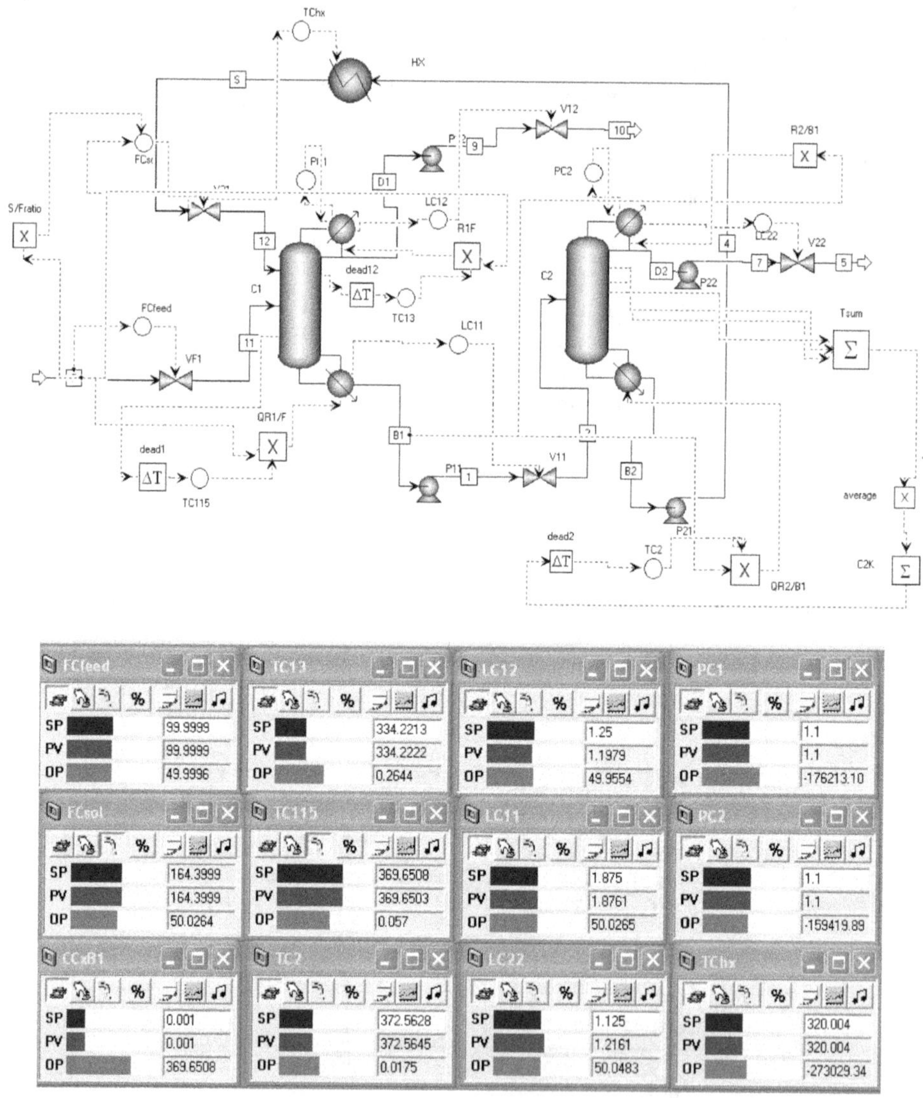


**Figure 12.11** Control structure CS2.

held at their specified values. If there are locations where temperatures do not change much as feed composition changes, these tray locations are candidates for temperature control. Figure 12.10 indicates that Stage 3 and Stage 15 in the extractive column satisfy this criterion.

### 12.2.3 CS2 Control Structure

Figure 12.11 shows the second control structure developed. It differs from the CS1 structure shown in Figure 12.7 in the following ways. The rest of the loops are unchanged.

1. The temperature on Stage 3 in the extractive column is controlled by manipulating the reflux-to-feed ratio.
2. The temperature on Stage 15 in the extractive column is controlled by manipulating the reboiler heat input-to-feed ratio.
3. An average temperature in the solvent recovery column is calculated using Stages 3, 4, and 5. This becomes the process variable in the temperature controller *TC2* that changes the reboiler heat input-to-feed ratio.

Tsum.AllVariables Table

| | Value | Spec |
|---|---|---|
| ComponentList | Type1 | |
| Input_(1) | 371.931 | Free |
| Input_(2) | 339.461 | Free |
| Input_(3) | 406.776 | Free |
| NSignal | 3 | |
| Output_ | 298.718 | Free |

average.AllVariables Table

| | Value | Spec |
|---|---|---|
| ComponentList | Type1 | |
| Input1 | 298.718 | Free |
| Input2 | 0.3333 | Fixed |
| Output_ | 99.5628 | Free |

C2K.AllVariables Table

| | Value | Spec |
|---|---|---|
| ComponentList | Type1 | |
| Input1 | 99.5628 | Free |
| Input2 | 273.0 | Fixed |
| Output_ | 372.563 | Free |

**Figure 12.12** Setting up average temperature control in solvent recovery column.

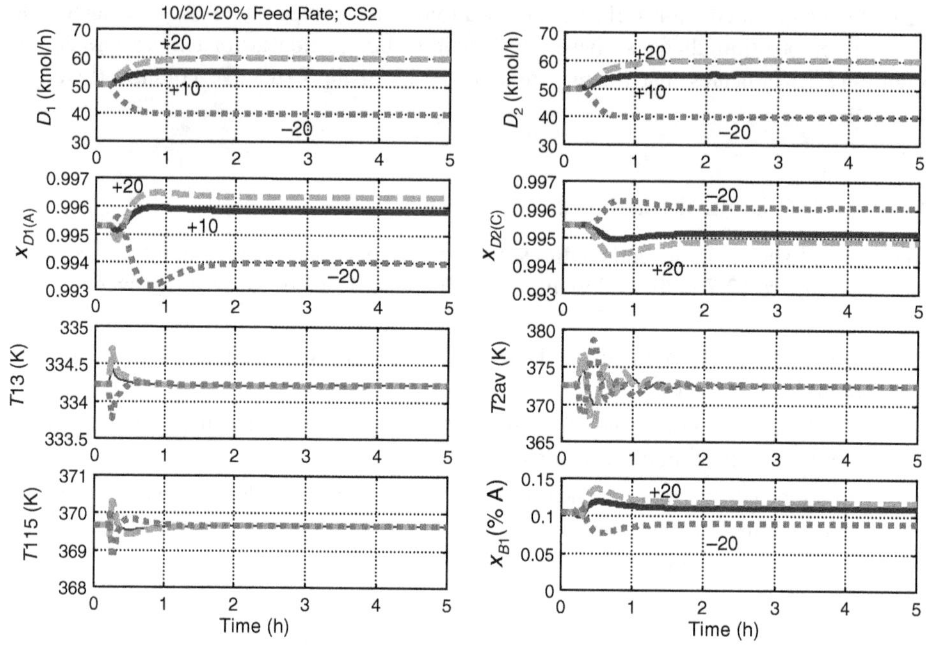


**Figure 12.13** CS2; feed flowrate disturbances.

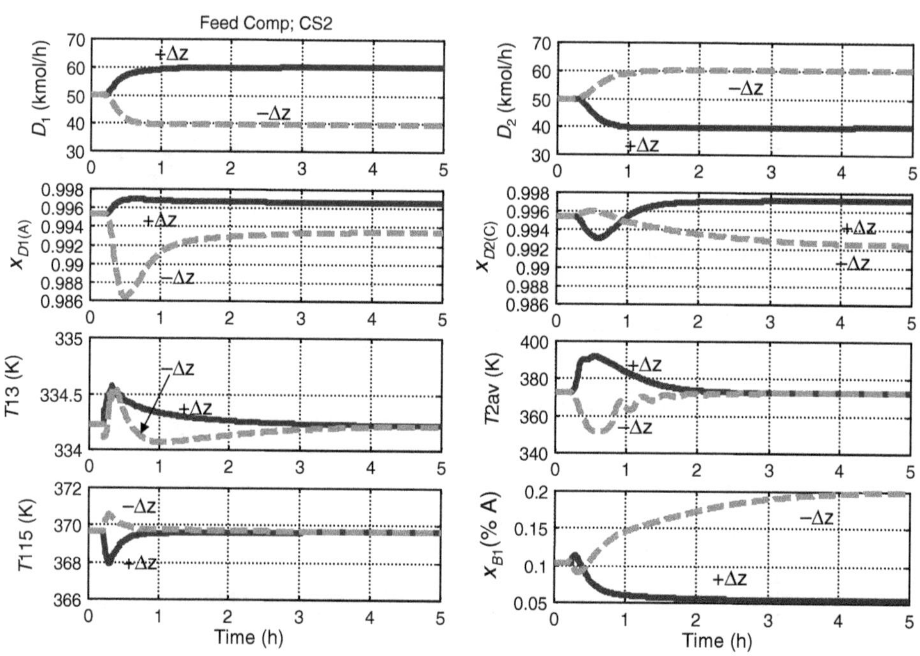


**Figure 12.14** CS2; feed composition disturbances.

An average temperature is used in the solvent recovery column because of the large temperature gradient in this column.[1] Using multiple temperatures is an effective method to achieve profile position control. The temperature profile can be tracked as it moves up and down the column. The process gain is reduced, which permits a larger controller gain. The control structure using multiple temperatures also avoids saturation of the process variable signal, which occurs when a single temperature at only one tray is measured.

Setting up the average temperature in Aspen Dynamics is illustrated in Figure 12.12. The temperatures on the three stages are added together in *Tsum*. The output signal is the sum of the three temperatures in Celsius because Aspen Dynamics uses metric units. This signal is multiplied by 0.3333 in *average* and finally a constant 273 is added to convert the average temperature to Kelvin before feeding into the deadtime element.

Figures 12.13 and 12.14 give responses for feed flowrate and feed composition disturbances. Stable regulatory control is shown with product purities held close to their specifications for all of these large disturbances. Figure 12.13 gives the responses of the system for step changes in the setpoint of the feed flow controller at 0.2 h. The solid lines are for a 10% increase, the dashed lines are for a 20% increase, and the dotted lines are for a 20% decrease. Increasing the feed results in increases in both distillate product streams, as expected. The two temperatures in the extractive column are held at their setpoint values after a short transient period, as is the average temperature in the solvent recovery column. The purities of the two products are held quite close to their desired values. The largest offset in the acetone product purity occurs for the 20% decrease in feed flowrate where it drops to 99.4 mol%. The system comes to a new steady state in less than 2 h.

Figure 12.14 gives the responses of the system for step changes in feed composition at 0.2 h. The solid lines are for an increase from 50 to 60 mol% acetone in the feed with a corresponding reduction in chloroform. The dashed lines are for a decrease from 50 to 40 mol% acetone in the feed. Feeding more acetone and less chloroform produces an increase in $D_1$ and a decrease in $D_2$ as expected. There is a fairly large transient drop in acetone product purity $x_{D1(A)}$ down to 98.6 mol% when less acetone is fed, but the steady-state purity recovers to 99.3 mol%. The chloroform purity is maintained quite close to specification. The system comes to a new steady state in about 3 h.

These results illustrate that the dual-temperature control structure provides effective regulatory control of this maximum-boiling azeotropic system in the face of very large disturbances.

## 12.3 CONCLUSION

The design and control of a maximum-boiling azeotropic system has been studied in this chapter. Extractive distillation is shown to be capable of producing quite pure products. A conventional control structure is developed that provides effective disturbance rejection for both production rate and feed composition changes. Dual temperature control is required in the extractive column in order to handle feed composition disturbances.

## REFERENCE

1. Luyben W. L., *Distillation Design and Control Using Aspen Simulation*, Wiley, New York, 2006, p. 88.

# PART 5

# OTHER WAYS FOR SEPARATING AZEOTROPES

When the production scale is relatively small, it is more beneficial to run the distillation process in batch mode. Batch distillation is commonly used in the fine chemicals, specialty polymer, biochemical, pharmaceutical, and food industries. Only a single column is needed in batch distillation for multi-component separation, while a continuous distillation system requires multiple columns for such separation. In batch distillation, a single column can handle a wide range of feed compositions, various numbers of components, and various degrees of difficulty in the separation, thus providing great flexibility in the operation.

Chapter 13 will illustrate the operations of batch distillation utilizing two previously mentioned separation methods for azeotropic mixtures: extractive distillation and heterogeneous azeotropic distillation. In the first part of Chapter 13, operation of batch extractive distillation is studied for separating acetone and methanol using water as the entrainer and separating IPA and water using DMSO as the entrainer. In the latter part of that chapter, a heteroazeotropic batch distillation system for acetic acid dehydration will be studied.

Besides the methods illustrated so far in this book, there are other ways for separating azeotropes. One way is to react the azeotrope away in a reactive distillation column to form other useful products. The design and control of various reactive distillations have been extensively studied in a recent book by Luyben and Yu.[1] Another way commonly used in ethanol dehydration is to use the hybrid distillation-adsorption process. In this process, distillation is used to purify the mixture to a composition near the ethanol–water azetrope, and then an adsorption unit (e.g., molecular sieves) is used to adsorb the remaining water so that anhydrous ethanol can be obtained.[2] The key technology in this process is the performance of the adsorbent material in removing water from the mixture and is beyond the scope of this book.

*Design and Control of Distillation Systems for Separating Azeotropes.* By William L. Luyben and I-Lung Chien

Another method used in industry, which can be categorized as extractive distillation, is to use a dissolved salt (e.g., calcium chloride) as a separating agent to alter the relative volatility of the two components in the mixture. This behavior, known as salt effect, is due to the preferential salvation of the ions (formed when the salt dissociates in solution) by the less volatile component of the solvent mixture.[3] The bottom product of this saline extractive distillation column is the mixture of the salt and the less volatile component (e.g., water). The separation of this bottom stream may require an evaporative crystallizer and a spray dryer for recovery of the anhydrous salt. A saline extractive distillation method is usually more expensive than conventional extractive distillation.

In the same extractive distillation category, recent research developments have shown that it is possible to use nonvolatile hyperbranched polymers or ionic liquid as entrainers in the extractive distillation process. An example in the literature (Seiler et al.[4]) is for ethanol-water separation using hyperbranched polyglycerol or ionic liquid like $[EMIM]^+[BF_4]^-$ as entrainer. Because the entrainer is nonvolatile, the recovery of entrainer is much easier than saline extractive distillation.

Membranes can also be used to purify a mixture and attain composition beyond the azeotropic composition. The pervaporation process features a liquid feed, a liquid retentate, and a vapor permeate. While gas-phase membrane processes are essentially isothermal, the phase change in the pervaporation process produces a temperature decrease as the retentate flows through the unit. Since flux rates decrease with decreasing temperature, the conventional pervaporation unit consists of several membrane modules in series with interstage heating. The vapor permeate must be condensed for recovery and recycle, and refrigeration is usually required. Hybrid systems of distillation columns and pervaporation units are frequently used in situations where distillation alone is impossible or very expensive. An important application is the removal of water from the ethanol–water azeotrope. Chapter 14 will discuss the details of design and control of such processes.

## REFERENCES

1. Luyben W. L. and C. C. Yu, *Reactive Distillation Design and Control.* Wiley, Hoboken, NJ, 2008.
2. Huang H. J., S. Ramaswamy, U. W. Tschirner, and B. V. Ramarao, A review of separation technologies in current and future biorefineries, *Sep. Pur. Tech.*, **62**, 1–21 (2008).
3. Liano-Restrepo M. and J. Aguilar-Arias, Modeling and simulation of saline extractive distillation columns for the production of absolute ethanol, *Comput. Chem. Engng.*, **27**, 527–549 (2003).
4. Seiler M., C. Jork, A. Kavarnou, W. Arlt, and R. Hirsch, Separation of azeotropic mixtures using hyperbranched polymers or ionic liquids, *AIChE J.*, **50**, 2439–2454 (2004).

# CHAPTER 13

# BATCH DISTILLATION OF AZEOTROPES

Batch distillation is commonly used in the fine chemicals industries, specialty polymer, biochemical, pharmaceutical, and food. In these types of applications, the production scale is usually small, which justifies running the separation process in batch mode. When the mixture contains an azeotrope, the separation methods mentioned in previous chapters can also be operated in batch mode. We will start this chapter by studying the operation of batch extractive distillation for two systems.[1] One is to separate acetone and methanol using water as the entrainer. The other system is to separate IPA and water using DMSO as the entrainer.

## 13.1 BATCH EXTRACTIVE DISTILLATION (ACETONE–METHANOL WITH WATER AS THE ENTRAINER)

The mixture of acetone (with a normal boiling point of 56.14°C) and methanol (with a normal boiling point of 64.53°C) contains an azeotrope with a minimum-boiling temperature of 55.24°C at atmospheric pressure. The mixture cannot be separated into pure components by either a batch or a continuous distillation column. In this chapter, NRTL thermodynamic models are used to find the liquid activity coefficients in the VLE calculations. The Aspen built-in model parameters are used, and the vapor phase was assumed to be ideal. The NTRL model parameters are given in Table 13.1 and show very good agreement with experimental values of the azeotropic composition and the azeotropic temperature. For comparison purposes, the experimental azeotropic composition at atmospheric pressure in Gmehling[2] from several datasets are in the range of 74.94 to 81.60 mol% acetone while the predicted azeotropic composition from our NRTL model is 77.75 mol% acetone. The azeotropic temperatures in Gmehling[2] taken from several datasets are in the range of 55.10 to

*Design and Control of Distillation Systems for Separating Azeotropes.* By William L. Luyben and I-Lung Chien

**TABLE 13.1 NRTL Parameters for an Acetone–Methanol–Water System.**

| Comp. $i$<br>Comp. $j$ | Acetone<br>Methanol | Acetone<br>Water | Methanol<br>Water |
|---|---|---|---|
| $a_{ij}$ | 0 | 6.3981 | −0.693 |
| $a_{ji}$ | 0 | 0.0544 | 2.7322 |
| $b_{ij}$ (K) | 101.89 | −1809.0 | 172.99 |
| $b_{ji}$ (K) | 114.13 | 419.97 | −617.27 |
| $\alpha_{ij}$ | 0.3 | 0.3 | 0.3 |

Aspen Plus NRTL:

$$\ln \gamma_i = \frac{\sum_j x_j \tau_{ji} G_{ji}}{\sum_k x_k G_{ki}} + \sum_j \frac{x_j G_{ij}}{\sum_k x_k G_{kj}} \left[ \tau_{ij} - \frac{\sum_m x_m \tau_{mj} G_{mj}}{\sum_k x_k G_{kj}} \right]$$

where $G_{ij} = \exp(-\alpha_{ij}\tau_{ij})$

$$\tau_{ij} = a_{ij} + \frac{b_{ij}}{T}$$

$$\alpha_{ij} = c_{ij},\ \tau_{ii} = 0,\ G_{ii} = 1$$

56.90°C while the predicted azeotropic temperature from our NRTL model is 55.24°C. The predicted azeotropic composition and temperature are within the range of the experimental data; thus, the prediction should be reliable.

The RCMs and the equivolatility curves of this chemical system can be seen in Figure 13.1, where the numbers in the equivolatility curves denote the relative volatility of acetone versus methanol in the presence of water. The RCM indicates that any mixture of acetone and methanol, even premixed with water, will produce the acetone–methanol azeotrope at the top of the column. However, by continuously adding water (a heavy entrainer) into the column, it can be seen from the equivolatility curves that the acetone is becoming more and more volatile than the methanol in the extractive section. Acetone and methanol can then be separated in the extractive section if the number of trays in this section is sufficient. Acetone will go toward the top of the column while methanol will be carried with the water toward the column bottom. In the rectifying section, owing to the lack of methanol in this section, only the separation of acetone and water is performed. Pure acetone will preferably go to the top of the batch extractive distillation column. After the draw-off of the acetone product and a slop-cut period, where the acetone in the column is completely depleted, the methanol product can be collected at the top of the column. The heavy entrainer (water) can be collected at the column bottom.

### 13.1.1 Batch Distillation Operating Procedure

The operating procedure can be summarized as follows.

***Step 1: Start-up.*** In this step, the acetone and methanol mixture is placed in the bottom of the column together with some entrainer. Heat is added in the reboiler and vapor boil-up moves up the column. At the same time, entrainer water is fed continuously into the middle of the column at a flowrate (to be specified). The column is run under total reflux conditions until the acetone of the top product reaches its purity specification of 95 mol%.

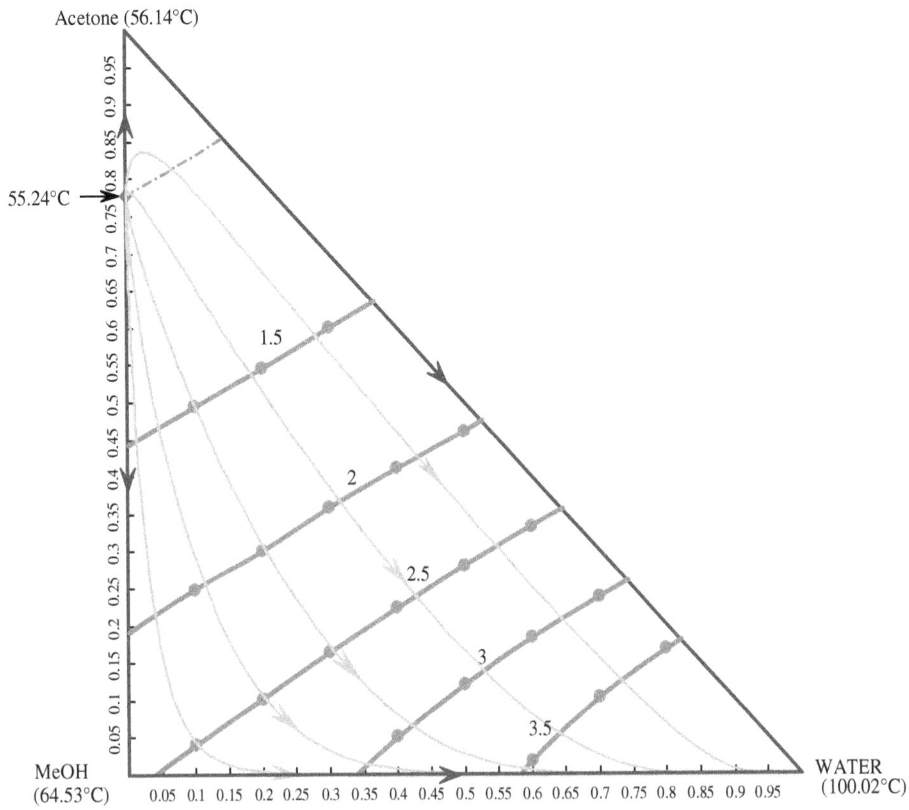


**Figure 13.1** RCM, isovolatility, and equivolatility curves of an acetone–methanol–water system.

***Step 2: Production of Acetone (P1).*** After the acetone reaches its purity specification in the top product, a continuous draw-off of this product into the acetone product tank is conducted with a constant reflux ratio policy (to be specified). In this step, the continuous feeding of the entrainer remains until the end of this step when the acetone purity in the product tank can no longer meet its purity specification. At the end of this step, the continuous feeding of the entrainer is stopped.

***Step 3: Slop-cut (S1) Collection.*** When the acetone purity in the product tank can no longer meet its purity specification, the product draw-off is diverted into another slop-cut tank until the time when the methanol of the top product reaches its purity specification of 92 mol%. Notice that, in this step, the draw-off of the product does not need to use the same reflux ratio as in Step 2.

***Step 4: Production of Methanol (P2).*** After the methanol reaches its purity specification in the top product, continuous draw-off of this product into the methanol product tank is conducted with a constant reflux ratio policy. This step ends when the purity in the methanol product tank can no longer meet the purity specification.

***Step 5: Production of Water (P3) at the Column Bottoms.*** At the end of the Step 4, the water product in the column bottom typically has already reached its purity specification of 99 mol%. The column is shut-down and the bottom product is collected. In some cases another step of collecting another slop-cut (*S*2) from the top of the column may be required until the bottom water product reaches its purity specification. However, in the acetone–methanol case, when the purity of the product in the methanol product tank can no longer meet the purity specification, the bottom product has already reached its purity specification; thus the step for collecting the slop-cut (*S*2) is omitted.

The operation of this batch extractive distillation column can be seen in Figure 13.2. There are four control loops in the figure including a ratio control loop that sets the draw-off rate during the various operating steps, a level control loop for the reflux drum, a pressure control loop to maintain the column pressure, and a flow loop for the continuous feeding of the entrainer. Notice in the above operating procedure, the simple constant reflux ratio policy is used. We will discuss the use of a varying reflux ratio policy in the next example. From the above operating procedure, there are quite a few operating variables that need to be determined. They are the amount of water mixed with the separation mixture, the continuous entrainer feed flowrate, the reflux ratio in Step 2, the reflux ratio in Step 3, and the reflux ratio in Step 4. For the operation of Steps 1 and 2, the objective function presented below is used to determine the operating variables. This objective function can be considered as the efficiency of the entrainer usage.

$$\eta = \frac{\text{Total acetone collected in } P1 \text{ tank (kmol)}}{\text{Total entrainer used (kmol)}} \tag{13.1}$$

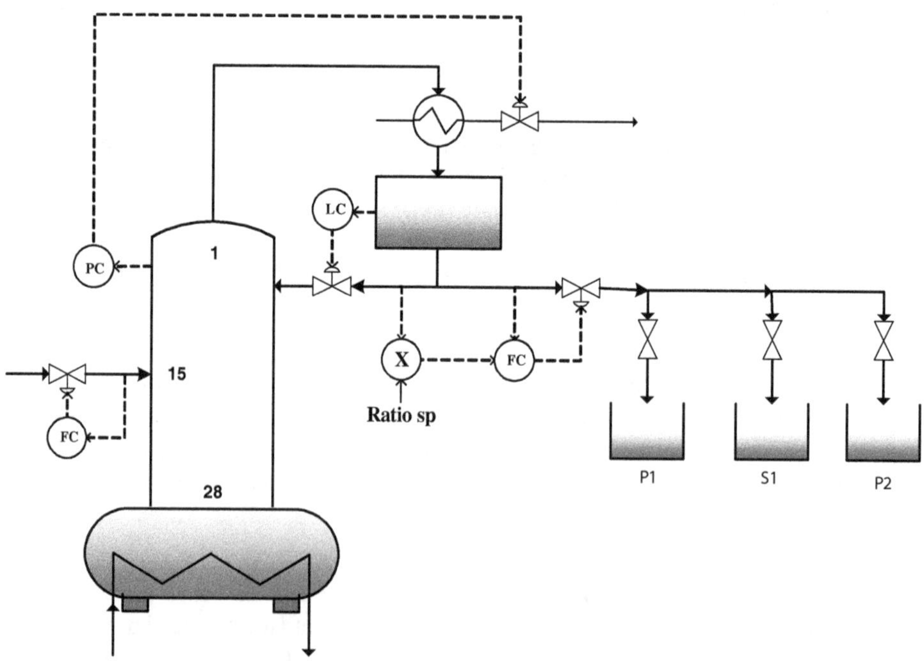


**Figure 13.2** The illustration of a batch extractive distillation column.

### 13.1.2 Simulation Results

The column starts from an empty column prepared by a nitrogen purge. The "Radfrac" module in Aspen was used with rigorous calculation of the total mass balance, component balance, and energy balance of each tray. Pressure-driven simulation of the vapor flow was also carried out with the capability of calculating the pressure drop at each tray. In this Aspen Dynamics simulation, the entrainer is fed on Stage 15 of the 28-stage column (not including the total condenser or the reboiler stages). The reboiler duty is set to be 0.2 GJ/h, which avoids flooding the column. The column diameter is 0.26 m, weir height is 0.05 m, the volumes for the column bottom and reflux drum are 3.14 $m^3$ and 0.1 $m^3$, respectively. The mixture for the separation contains 4 kmol acetone and 4 kmol methanol.

Table 13.2 shows a number of cases with different operating variables in Steps 1 and 2. The two operating variables, the amount of water mixed with the separation mixture, and the continuous entrainer feed flowrate, are fixed at some combinatory values. For each combination of the above two operating variables, operating Steps 1 and 2 are simulated with a prespecified reflux ratio. The objective function in Eq. (13.1) can be calculated for each pre-specified reflux ratio. Figure 13.3 gives a plot of solvent efficiency versus reflux ratio *RR* for the various cases given in Table 13.2. Each case has different fixed values of the amount of water mixed with the separation mixture and the continuous entrainer feed flowrate. From the figure, the best efficiency of the entrainer usage is found by a preloading amount of 10 kmol, a continuous feed rate of 8 kmol/h, and a reflux ratio of 2.5.

Several papers (e.g., Lelkes et al.[3,4] and Steger et al.[5]) have shown that the preloading of the entrainer may not be necessary for obtaining pure acetone in Step 2. However, preloading water with the acetone–methanol mixture does enhance the separation of acetone and methanol in the extractive section (as can be seen in the equivolatility curves in Fig. 13.1). On the other hand, "total entrainer used" is in the denominator of the objective function (Eq. 13.1), which we want to maximize; thus the preloading amount should be as small as possible. With these two conflicting factors, a trade-off results in an optimum preloading amount of 10 kmol.

For Step 3 of the operating procedure, several different values of the reflux ratio have been simulated. The objective is to minimize the batch time until the end of this step because the slop-cut is considered to be waste material that is recycled in the next batch run. Table 13.3

**TABLE 13.2 Summary of Operating Conditions for Steps 1 and 2 of Acetone–Methanol Separation.**

| Operating Condition | Amount of Water Mixed with the Separation Mixture (kmol) | The Continuous Entrainer Feed Flowrate (kmol/h) |
|---|---|---|
| A | 9 | 7 |
| B | 9 | 8 |
| C | 9 | 9 |
| D | 10 | 7 |
| E | 10 | 8 |
| F | 10 | 9 |
| G | 11 | 7 |
| H | 11 | 8 |
| I | 11 | 9 |

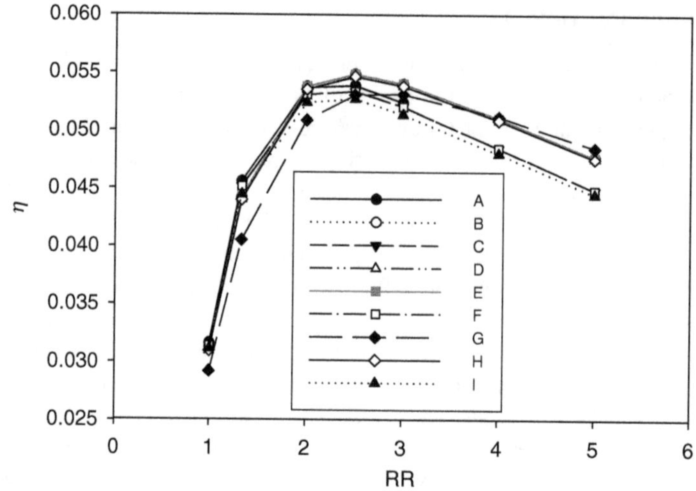


**Figure 13.3** Plot of objective functions at various operating conditions in Steps 1 and 2.

shows the results of six simulation runs. The batch time up to the end of this step is reduced with a smaller reflux ratio value; however, when the reflux ratio is less than 5.0, the purity of the methanol product (*P*2) cannot reach its purity specification.

For Step 4 of the operating procedure, the value of the reflux ratio is determined by maximizing capacity factor (CAP) proposed by Luyben[6] for regular batch distillation. The CAP is defined as:

$$\text{CAP} = \frac{\text{Total methanol collected in } P2 \text{ tank (kmol)}}{0.5 + \text{Batch Time (h)}} \tag{13.2}$$

In Eq. (13.2), 0.5 h is assumed for the time needed for shut-down and preparation time for the next batch. The reason for using this objective function is because Step 4 can be seen as an operation of regular batch distillation. There are only two remaining components (methanol and water) to be separated. Figure 13.4 shows a collection of many simulation runs using various reflux ratios in Step 4. Reflux ratio in this step is selected to be 17.5 to maximize the capacity factor.

**TABLE 13.3 Simulation Results for Step 3 of Acetone–Methanol Separation.**

| Reflux Ratio | Batch Time Until the End of Step 3 (h) |
|---|---|
| 4.5 | Cannot get methanol purity at its specification at the end of this step |
| 5.0 | 8.48 |
| 6.0 | 8.90 |
| 8.0 | 9.78 |
| 10.0 | 10.66 |
| 12.0 | 11.56 |

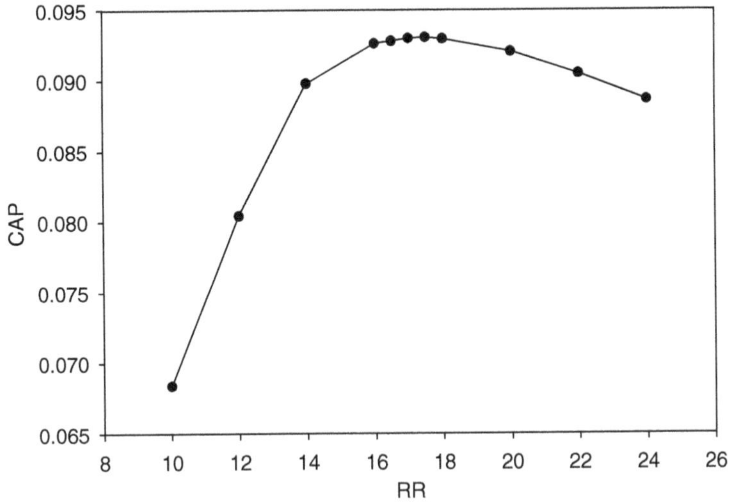


**Figure 13.4** CAP versus reflux ratio in Step 4.

The dynamic simulation run for the liquid compositions in the reflux drum and the column bottoms can be seen in Figure 13.5. In the beginning of the dynamic simulation, 4 kmol of acetone, 4 kmol of methanol, and 10 kmol of water are placed in the column bottom, and then a reboiler duty of 0.2 GJ/h is provided. Also in the dynamic simulation, the column top pressure is controlled at atmospheric pressure, the level of the reflux drum is controlled at 0.4 m, and during Steps 1 and 2 the entrainer feed flowrate is fixed at 8 kmol/h. From the column at an empty state at time = 0 to time = 3.17 h, the acetone purity in the reflux drum continuously increases to the purity specification of 95 mol%. From time = 3.17 to 5.34 h, acetone product is continuously drawn off to *P*1 tank.

At time = 5.34 h, the purity in the *P*1 product tank can no longer be maintained; thus the draw-off is switched to the slop-cut (*S*1) tank until time = 8.48 h. By the end of the slop-cut collection step, the methanol purity in the reflux drum has built up to the purity specification of 92 mol%. From time = 8.48 to 12.26 h, the methanol product is continuously drawn-off

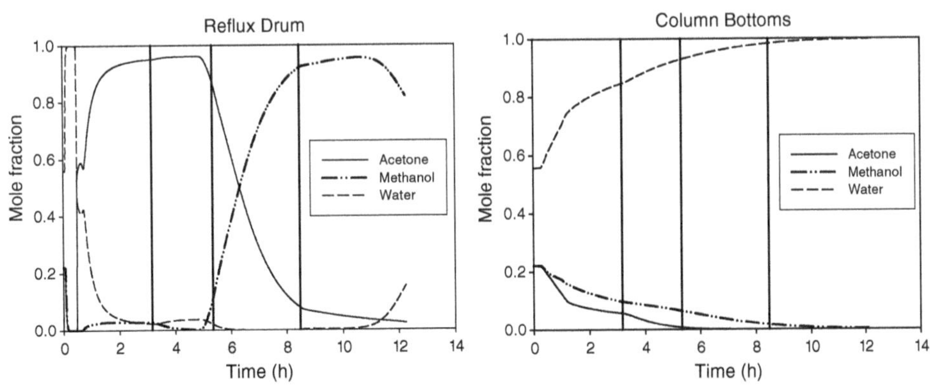


**Figure 13.5** Liquid compositions in the reflux drum and in the column bottoms for the acetone–methanol separation.

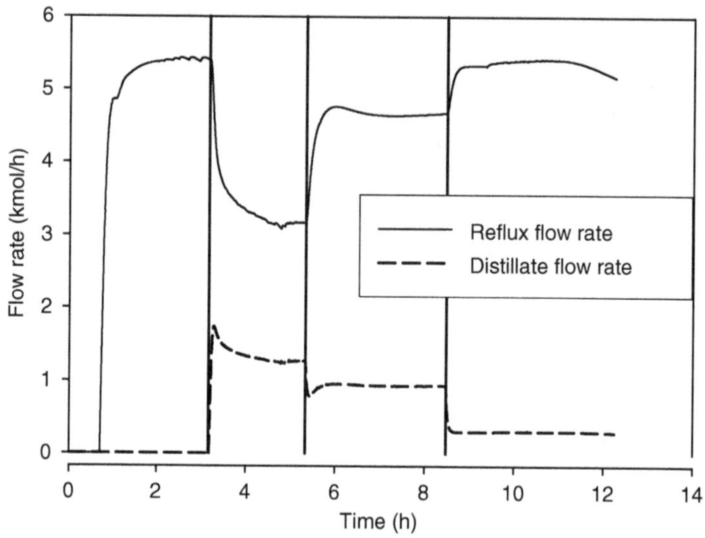


**Figure 13.6** Reflux and distillate flowrates for the acetone–methanol separation.

to the *P*2 tank. At time = 12.26 h, the purity in the *P*2 product tank can no longer be maintained. However, at the same time, the water composition at column bottoms has become very pure (99.8 mol%). The batch run is stopped, and the bottom product is collected.

Figure 13.6 shows the distillate and reflux flowrates during the batch run. From time = 0 to 0.75 h, the liquid holdup builds up gradually in the reflux drum. The column operates under total reflux from time = 0.75 to 3.17 h. After Step 1, the reflux ratio for the following Steps 2 to 4 are set at 2.5, 5, and 17.5, respectively.

The column liquid composition profile at the end of each operating step can be seen in Figure 13.7. Notice that there is no methanol in the upper rectifying section at the end of Step 1 (time = 3.17 h). This is due to the continuous feeding of the entrainer into the column to push the methanol to appear only in the lower extractive section of the column. At the end of Step 2 (time = 5.34 h), the column composition profile has moved from the acetone corner toward the methanol corner. The top liquid composition is close to the methanol corner at the end of Step 3 (time = 8.48 h). At this time, there is almost no acetone inside the column, so the separation is just like regular batch distillation. At the end of Step 4 (time = 12.26 h), the methanol purity in the *P*2 product tank can no longer be maintained at its specification. At the same time, the bottom product has already satisfied the water purity specification. The column is shut down and the bottom product is collected.

### 13.1.3 Comparison of Operating Procedure Given in Earlier Papers

The operating procedures for the acetone–methanol separation system given in the literature (e.g., Lelkes et al.,[4] Lang et al.,[7] and Milani[8]) all included a start-up step with total reflux and no continuous feeding of the entrainer, followed by a run-up step with continuous feeding of the entrainer. Using the same column as shown previously in this section, we can also simulate this operating procedure with the results shown in Figure 13.8.

From time = 0 to 2.21 h, the column is running under total reflux with no continuous feeding of the entrainer. At the end of this step, the top liquid composition approaches the

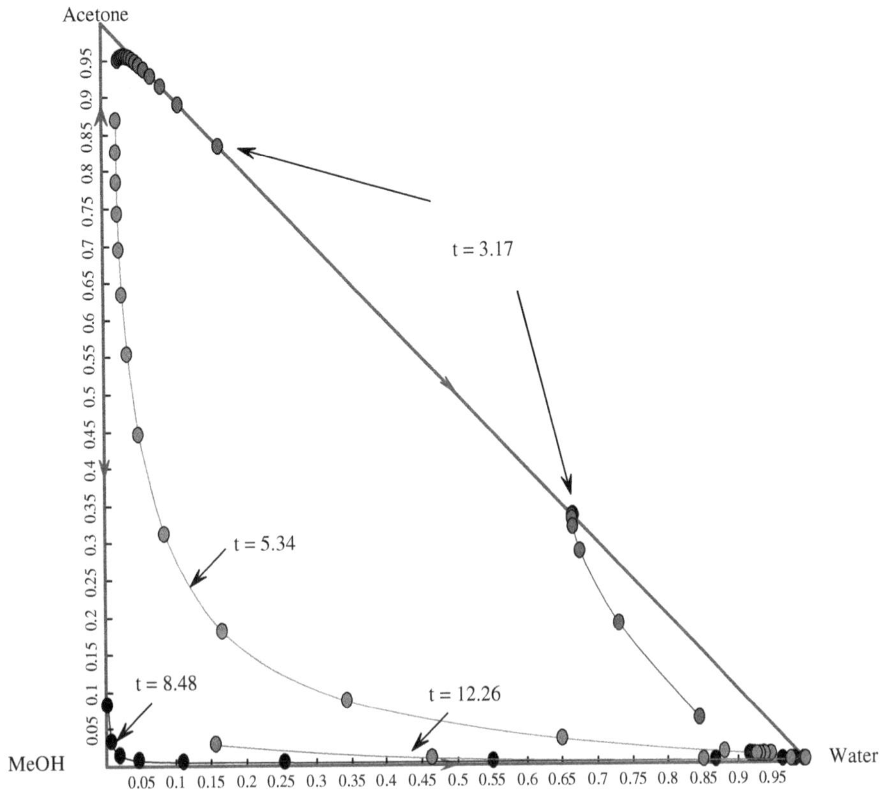


**Figure 13.7** Liquid composition profile at the end of each operating step for the acetone–methanol separation.

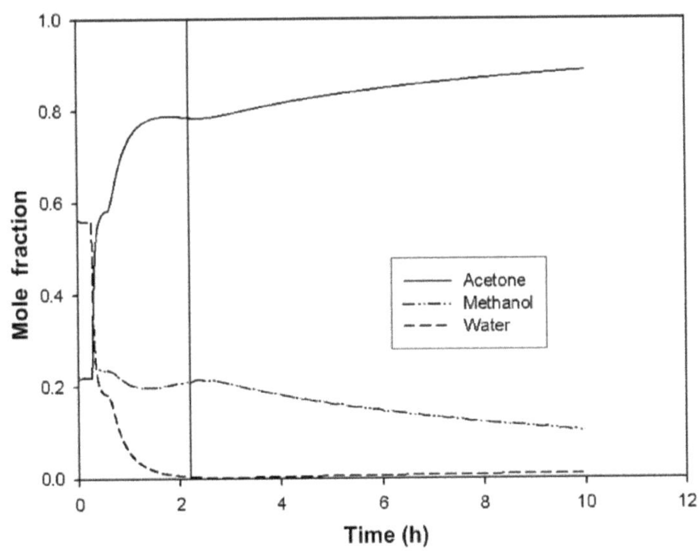


**Figure 13.8** Liquid composition at the top of the column using operating procedure in literature.

azeotropic composition of acetone and methanol even though 10 kmol of the entrainer (water) is already premixed with the acetone–methanol mixture at the beginning of the batch run. This can readily be explained by the RCMs in Figure 13.1. Starting from any compositions inside the ternary plot, even with preloading of the entrainer, the column top of all residue curves arrives at the azeotropic point of acetone–methanol.

With continuous addition of the entrainer at a flowrate of 8 kmol/h after the total reflux step, the top liquid composition gradually becomes purer in acetone. Figure 13.8 shows that the distillate has not reached the desired purity (95 mol% acetone) at the end of 10 h. This should be compared with dynamic response shown on the left-hand side of Figure 13.5. It is clear that the operating procedure of continuously feeding the entrainer at the beginning of the start-up period, as proposed in this chapter, performs better than the procedure given in the literature.

## 13.2 BATCH EXTRACTIVE DISTILLATION (ISOPROPANOL–WATER WITH DMSO AS THE ENTRAINER)

In this section, an IPA dehydration system using batch extractive distillation is investigated. IPA is widely used in the semi-conductor industry as a cleaning agent. Pure IPA is recovered from the waste IPA mixture containing water. The paper by Gmehling and Möllmann[9] showed that DMSO gave the highest value of the separation factor between IPA and water, so it is suitable as an entrainer in an extractive distillation column. In the simulation to follow, the thermodynamic model of NRTL is used to describe the VLE in this column. The Aspen Plus built-in model parameters (see Table 13.4) are used in the dynamic simulation. They show very good agreement between experimental and predicted azeotropic composition and azeotropic temperature. For comparison purposes, the experimental azeotropic composition at atmospheric pressure in Gmehling[2] from several datasets is in the range of 66.45 to 77.41 mol% IPA, while the predicted azeotropic composition is 66.63 mol% IPA. The azeotropic temperature in Gmehling[2] from several datasets is in the range of 80.16 to 80.40°C, while the predicted azeotropic temperature is 80.37°C. The predicted

**TABLE 13.4 NRTL Parameters for IPA–Water–DMSO System.**

| Comp. $i$<br>Comp. $j$ | IPA<br>Water | IPA<br>DMSO | Water<br>DMSO |
|---|---|---|---|
| $a_{ij}$ | −1.3115 | 0 | −1.2449 |
| $a_{ji}$ | 6.8284 | 0 | 1.7524 |
| $b_{ij}$ (K) | 426.40 | 115.28 | 586.80 |
| $b_{ji}$ (K) | −1483.5 | −25.012 | −1130.2 |
| $\alpha_{ij}$ | 0.3 | 0.3 | 0.3 |

Aspen Plus NRTL:

$$\ln \gamma_i = \frac{\sum_j x_j \tau_{ji} G_{ji}}{\sum_k x_k G_{ki}} + \sum_j \frac{x_j G_{ij}}{\sum_k x_k G_{kj}} \left[ \tau_{ij} - \frac{\sum_m x_m \tau_{mj} G_{mj}}{\sum_k x_k G_{kj}} \right]$$

where $G_{ij} = \exp(-\alpha_{ij}\tau_{ij})$

$$\tau_{ij} = a_{ij} + \frac{b_{ij}}{T}$$

$$\alpha_{ij} = c_{ij},\ \tau_{ii} = 0,\ G_{ii} = 1$$

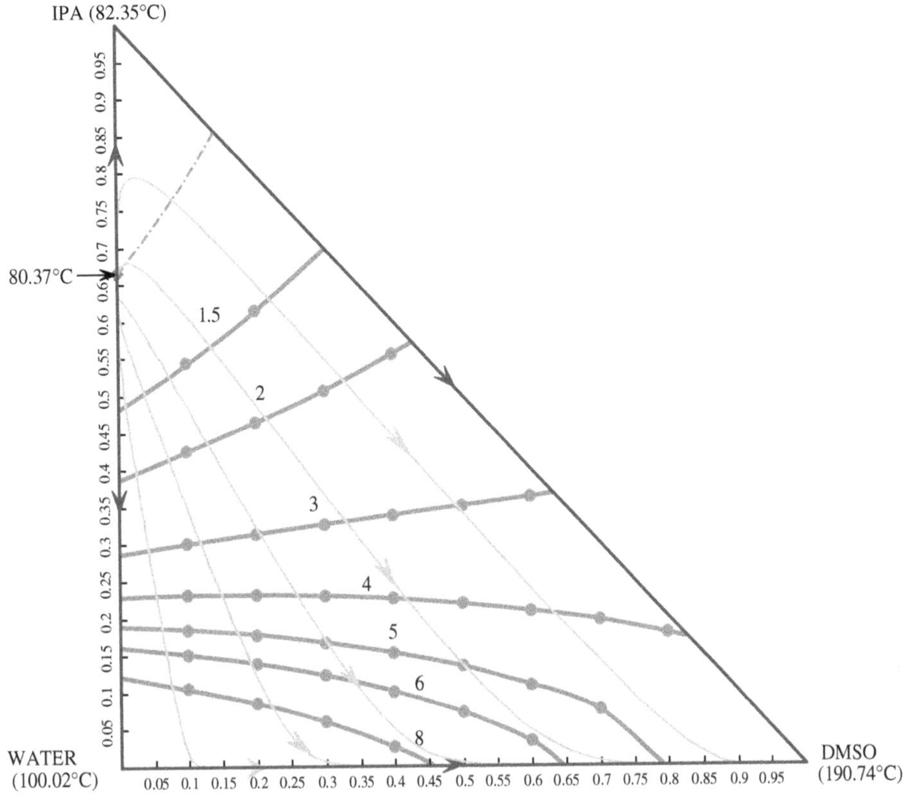


**Figure 13.9** RCM and equivolatility curves of an IPA–water–DMSO system.

azeotropic composition and temperature from our NRTL model are all within the range of the experimental data, so the predictions should be reliable.

The RCMs and the equivolatility curves of this chemical system can be seen in Figure 13.9 where the numbers in the equivolatility curves denote the relative volatility of IPA versus water in the presence of DMSO. Since the thermodynamic nature of this system is similar to the previous acetone–methanol–water system, an optimum operating procedure can be developed the same way. According to Laroche et al.,[10] since the isovolatility curve (relative volatility = 1.0) intersects the IPA–DMSO edge, the first product ($P1$) should be IPA. The second product $P2$ is water. The bottom product is DMSO. The product specifications for this system are set as follows: $P1$(IPA) at 98 mol%, $P2$(water) at 92 mol%, and bottom product (DMSO) at 99 mol%. Also, by comparing the equivolatility curves of this system with DMSO entrainer to those of the previous system with water entrainer shown in Figure 13.1, it should be noticed that the presence of a DMSO entrainer produces more separation enhancement between IPA and water than does a water entrainer between acetone and methanol. This can be seen by comparing the numerical values of the relative volatility achieved in the two cases.

### 13.2.1 Simulation Results with Constant Reflux Ratio and Constant Entrainer Feed Rate

In the simulation results below, the same batch column with a total of 28 stages (not including the total condenser or the reboiler stages) and the entrainer fed on Stage 15 is used. The

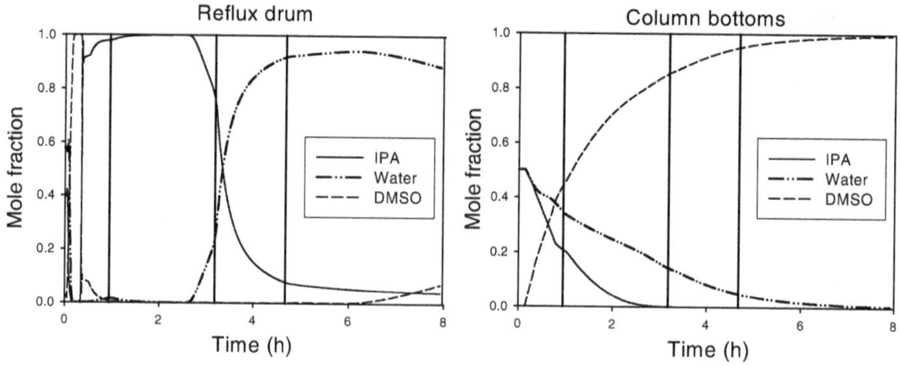


**Figure 13.10** Liquid compositions at the reflux drum and at the column bottoms for the IPA dehydration system.

reboiler duty is again set to be 0.2 GJ/h, suitably within a region to avoid weeping or flooding of the column. The mixture for the separation is 4 kmol of IPA and 4 kmol of water.

Following the same search procedure as before, the values of the operating variables were determined to be no preload DMSO with the IPA and water mixture, a continuous DMSO feed of 6 kmol/h, and reflux ratios in Steps 2, 3, and 4 of 1.0, 2.0, and 6.5, respectively. Notice that there is no need for any preloading amount of the entrainer in the beginning of the run.

The dynamic simulation run for the liquid compositions in the reflux drum and the column bottoms can be seen in Figure 13.10. In the beginning of the dynamic simulation, 4 kmol of IPA and 4 kmol of water are placed in the column bottoms, and the reboiler duty is set at 0.2 GJ/h. Also, in the dynamic simulation, the column top pressure is controlled at atmospheric pressure, and the level of the reflux drum was controlled at 0.4 m.

From time = 0 to time = 0.96 h, the IPA purity in the reflux drum continuously builds up to the purity specification of 98 mol%. From time = 0.96 to 3.19 h, the IPA product is continuously drawn-off to the *P*1 tank. At time = 3.19 h, the purity in the *P*1 product tank can no longer be maintained, so the draw-off is switched to the slop-cut (*S*1) tank until time = 4.68 h. At this time, the water purity in the reflux drum has built-up to the purity specification of 92 mol%. From time = 4.68 to 7.92 h, the water product is continuously drawn-off to the *P*2 tank. At time = 7.92 h, the purity in the *P*2 product tank can no longer be maintained. However, at the same time, the DMSO composition in column bottom has become very pure (99.5 mol%), so the batch run can be stopped and the bottom product collected.

Figure 13.11 shows the distillate and reflux flowrates during the batch run. From time = 0 to 0.56 h, the liquid holdup is gradually built-up in the reflux drum. The column is under total reflux from time = 0.56 to 0.96 h. After Step 1, the reflux ratios for the following Steps 2 to 4 are determined to be 1.0, 2.0, and 6.5, respectively. Notice that for time between 0.96 and 3.19 h, the distillate and reflux flowrates are the same (the two lines coincide in Fig. 13.11). Figure 13.12 shows the dynamic response of the liquid holdup in the column bottom. From time = 0 to 3.19 h (start-up and production of *P*1), the entrainer is continuously fed into the column, which increases the liquid holdup during these two operating steps. After that, the entrainer feeding is stopped, which results in decreasing of the pot liquid holdup as the distillate continues to be removed from the system. The final pot holdup is 18.0 kmol with purity of 99.5 mol%.

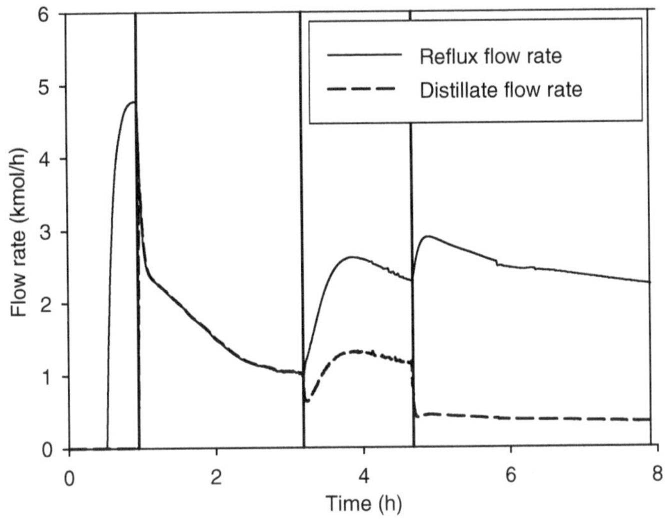


**Figure 13.11** Reflux and distillate flowrates for the IPA dehydration system.

The column liquid composition profile at the end of each operating step can be seen in Figure 13.13. Notice that there is no water in the upper rectifying section at the end of Step 1 (time = 0.96 h). This is due to the continuous feeding of the entrainer into the column to push the water down the column, and it is present only in the lower extractive section of the column. At the end of Step 2 (time = 3.19 h), the column composition profile has moved away from the IPA corner toward the water corner. The top liquid composition has moved close to the water corner at the end of Step 3 (time = 4.68 h). At this time, there is almost no IPA inside the column, so the separation is just a regular binary batch distillation. At the end of Step 4 (time = 7.92 h), the water purity in the $P2$ product tank can no longer be

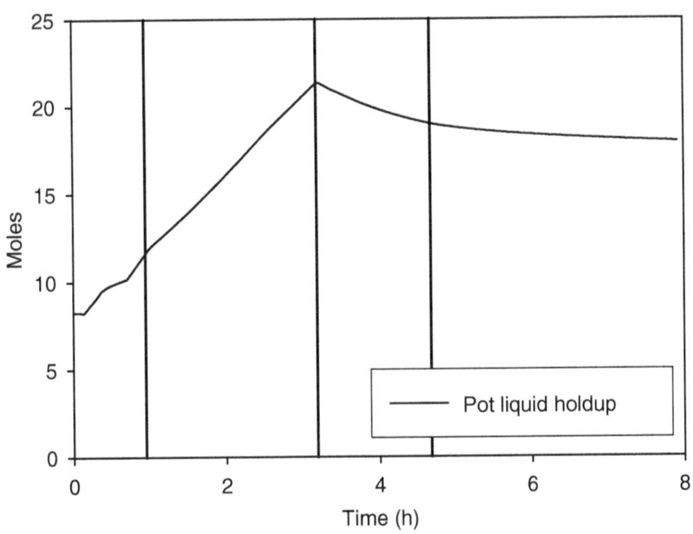


**Figure 13.12** Dynamic response of the liquid holdup in the column bottoms.

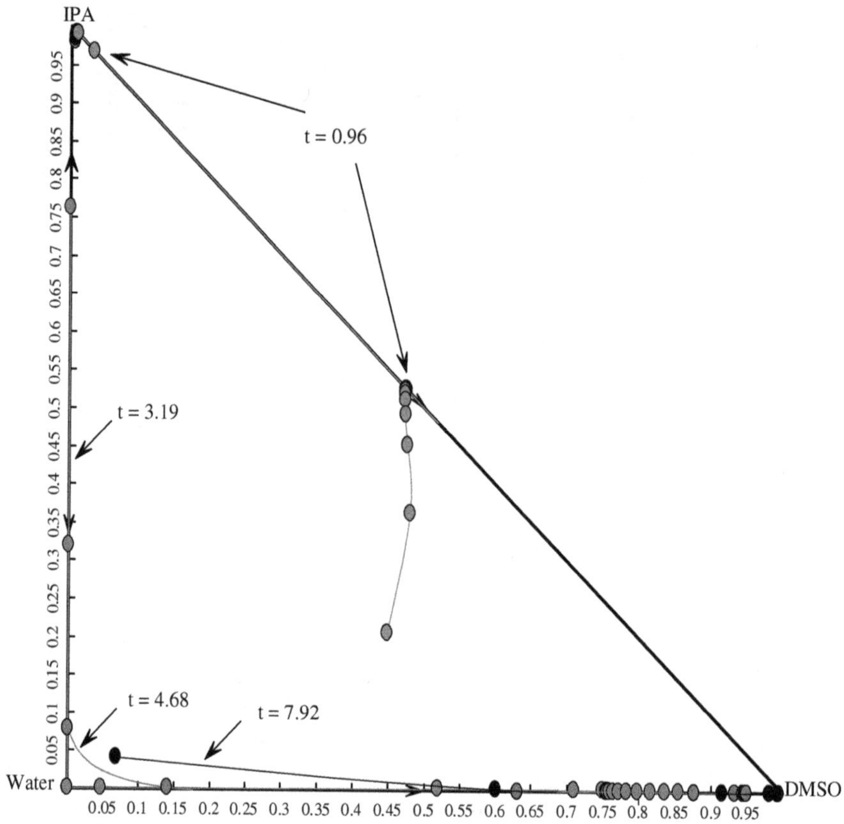


**Figure 13.13** Liquid composition profile at the end of each operating step for the IPA dehydration system.

maintained at its specification. At the same time, the bottom product has already satisfied the DMSO purity specification, so the column is shut-down and the bottom product is collected.

### 13.2.2 Varying the Reflux Ratio and Entrainer Feed Rate During Step 2

It may be possible to reduce the batch time of Step 2 and also reduce the entrainer usage by considering a varying reflux ratio and varying the entrainer feed rate operation. In Step 2, a time-varying reflux ratio (decreasing by increasing the distillate draw-off rate) may reduce the batch time as long as the purity in the *P*1 tank can still be maintained. The way to do this is to add a temperature control loop that changes the reflux ratio. A temperature control strategy was used in Skogestad et al.[11] in multivessel batch distillation systems for mixtures with no azeotrope and was also used in Skouras et al.[13] for mixtures that exhibit a liquid–liquid splitting behavior in a decanter when adding an entrainer. Our system is not classified as either of the above two types of batch distillation systems, so a different temperature control strategy will be used.

The proposed controlled variable is the temperature of the reflux drum, and the strategy is to keep it close to the normal boiling point of IPA. From the dynamic simulation results with a constant reflux ratio policy, it was found that the highest value for the reflux drum

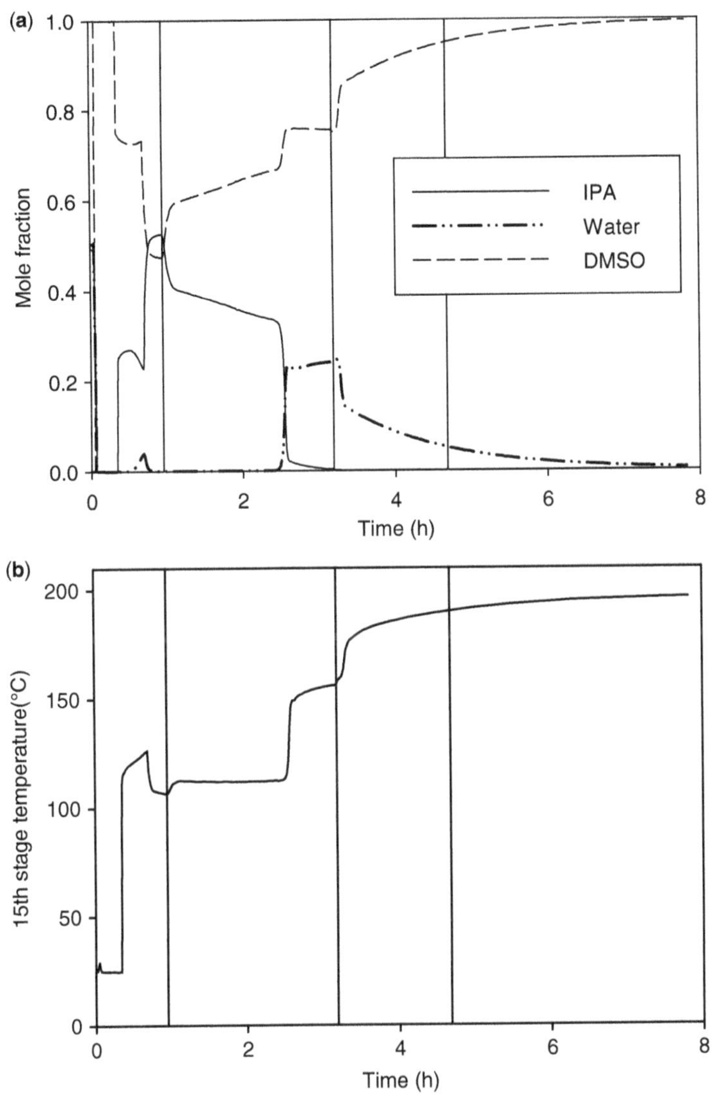


**Figure 13.14** (a) Dynamic response of Stage 15 liquid composition under constant reflux and constant entrainer feed operation. (b) Dynamic response of Stage 15 temperature under constant reflux and constant entrainer feed operation.

temperature was 82.12°C, so this value is used as the setpoint for this temperature control loop. In order to reduce the entrainer usage in Step 2, the idea is to use the temperature at Stage 15 as an indication. Because the purpose of the continuous entrainer feed is to prevent water from getting into the rectifying section, the temperature at Stage 15 should be maintained at a value indicating no, or very little, water in this stage.

Figure 13.14 shows the dynamic responses of the liquid compositions and temperature at Stage 15 observed in the dynamic simulation results when the reflux ratio and entrainer feed rate are constant. Since the IPA is almost completely drawn-off from the top of the column toward the end of this step, the temperature is nearly holding at some constant value until the

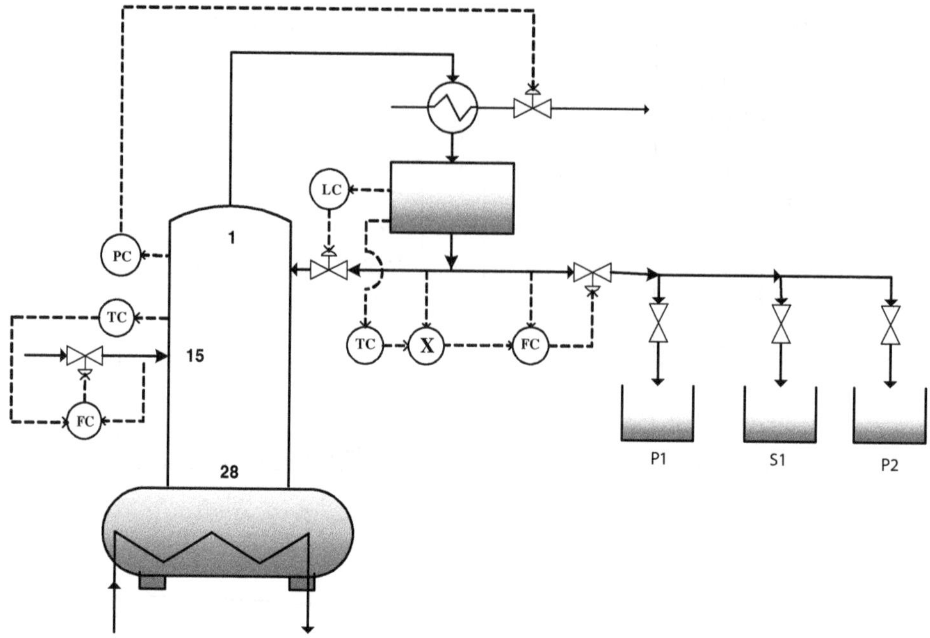


**Figure 13.15** Proposed overall control strategy during Step 2.

time when the water moves up into the rectifying section. The highest temperature at the early stage of Step 2 is used as the setpoint of the second temperature control loop to vary the entrainer feed rate. The value of the setpoint is 112.5°C. If the actual temperature is below this setpoint value, the entrainer feed can be reduced because there is no indication of water entering into the rectifying section. Toward the end of Step 2, when the IPA is almost completely drawn-off from the system, the water composition at Stage 15 increases as seen in Figure 13.14, which causes the temperature to shoot up. Greatly increasing entrainer feed rate at this moment would not prevent water from entering into the rectifying section because the operating step in the rectifying section is in transition from IPA–DMSO separation to water–DMSO separation. For this reason, an upper limit of the entrainer feed rate is set in the dynamic simulation and is set to be the same as in the constant reflux ratio and constant entrainer feed rate operation.

The overall control strategy for Step 2 can be seen in Figure 13.15. The dynamic simulation results for the reflux drum and Stage 15 temperature control loops can be seen in Figures 13.16 and 13.17, respectively. Notice that, in the first part of this step, the reflux ratio is reduced from 1.0 to a lower value because the top liquid composition is purer than the specification, which allows a more rapid draw-off of the IPA product. Towards the end of Step 2, because the IPA is almost completely drawn-off from the system, water composition increases in the reflux drum causing its temperature to drop close to the azeotropic temperature of IPA–water at 80.37°C. This transition is inevitable and further reduction of the reflux ratio cannot reverse this trend.

For the Stage 15 temperature control loop, initially the entrainer feed rate can be reduced without any problem in getting water into the rectifying section. Toward the end of Step 2, because the IPA is almost completely drawn-off from the system, the water composition shoots up on this stage causing the temperature to increase drastically. As explained

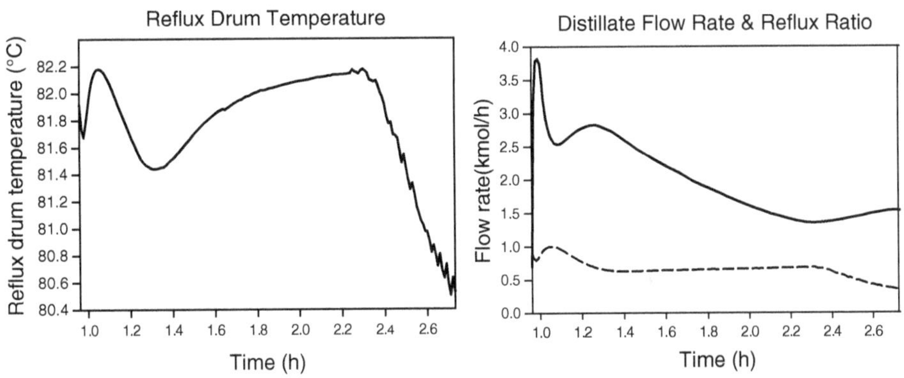


**Figure 13.16** Dynamic response for the reflux drum temperature control loop.

before, further increasing the entrainer feed rate cannot prevent water from entering into the rectifying section, so the entrainer feed rate is maintained at 6 kmol/h. The batch time for Step 2 is reduced from 2.23 h (0.96 to 3.19 h) to 1.77 h (0.96 to 2.73 h) with the same amount of IPA product recovered. The total entrainer usage is also reduced from 19.14 kmol to 16.31 kmol. The objective function $\eta$ [Eq. (13.1)], which indicates the efficiency of the entrainer usage, has been increased from 0.186 to 0.217 (a 16.7% increase) by using a varying reflux ratio and also varying the entrainer feed rate operating policy.

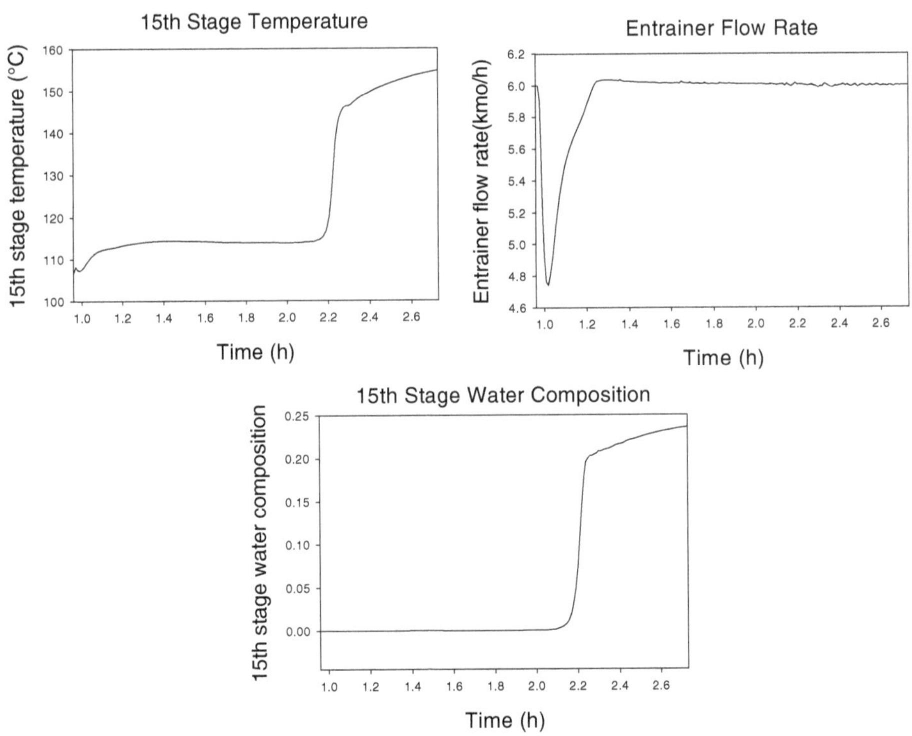


**Figure 13.17** Dynamic response for the Stage 15 temperature control loop.

## 13.3 ACETIC ACID DEHYDRATION VIA HETEROAZEOTROPIC BATCH DISTILLATION

Although acetic acid and water do not form an azeotrope, the use of conventional distillation to separate these two components requires many equilibrium stages because the system has a tangent pinch on the pure water end of the *xy* diagram. Therefore, it is more economical to use a heterogeneous azeotropic distillation system with an entrainer for the separation.

### 13.3.1 Acetic Acid Dehydration via Continuous Process

In Chapter 9, three candidate acetates were studied in detailed process simulations to determine the best entrainer for this system. The three candidate acetates considered were ethyl acetate, isobutyl acetate, and *n*-butyl acetate. Rigorous process simulations were performed to find the optimum design and operating conditions of these three entrainer systems. In each entrainer system, the total annual cost (TAC) was used as the objective function to be minimized. From a comparison of the minimized TAC, the isobutyl acetate was selected as the best entrainer with the lowest TAC and operating utility cost.

The main reason isobutyl acetate is the favorable entrainer can be explained by the RCMs and the LLE of this ternary system given in Figure 13.18. As discussed in Chapter 9, the

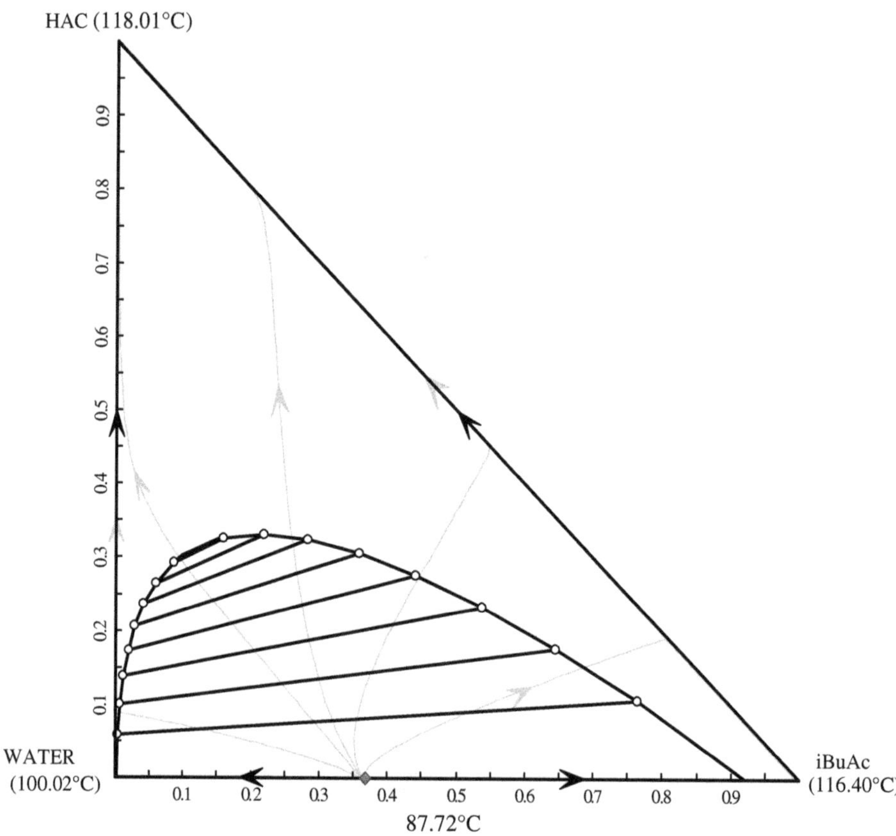


**Figure 13.18** RCMs and LLE using isobutyl acetate as entrainer.

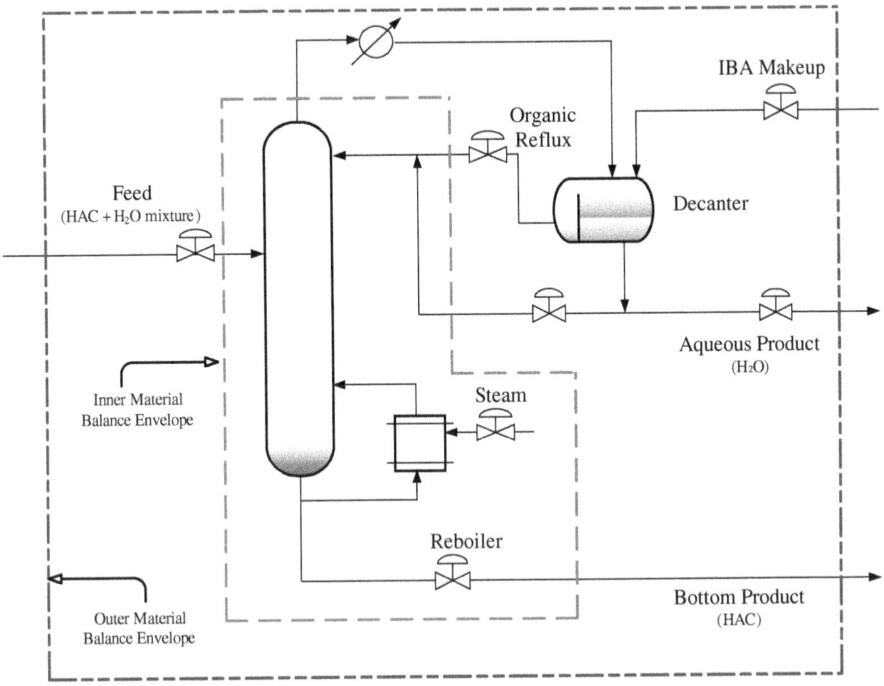


**Figure 13.19** Conceptual design for the separation of acetic acid and water.

azeotropic composition of isobutyl acetate is favorable because this mixture contains more water. This means that this entrainer is more capable of carrying water to the top of the column, so less entrainer is needed inside of the column.

It is better to have a large distance between azeotropic composition point and organic phase composition point because this means that, besides the azeotropic composition containing more water, the organic phase composition contains more entrainer. The locations of these two points give an estimation of the organic reflux flowrate into the heterogeneous column by using the inner material balance envelope shown in Figure 13.19. Less organic reflux flowrate is needed when these two points are further apart as is the case for isobutyl acetate.

Another factor that favors isobutyl acetate to be a suitable entrainer is the aqueous phase composition. The aqueous phase should contain as little entrainer as possible. The aqueous phase stream will be drawn out of the system, so any entrainer loss must be compensated for by the entrainer makeup stream as shown in Figure 13.19. This will contribute to an entrainer cost in the TAC calculations. The makeup flowrate can actually be estimated using the outer material balance envelope shown in Figure 13.19.

The vapor and liquid profiles inside the column for the optimized isobutyl acetate system can be seen in Figure 13.20. Notice that the column behaves as it was designed with the bottom product approaching the pure acetic acid corner and the top vapor approaching the azeotropic composition. The first five stages, counting from the top of the column, have two liquid phases. The combined liquid compositions for these five stages are plotted in this figure. Another thing worth mentioning is that the column composition profiles bypass the corner of pure water, which is the region to be avoided because of the tangent pinch.

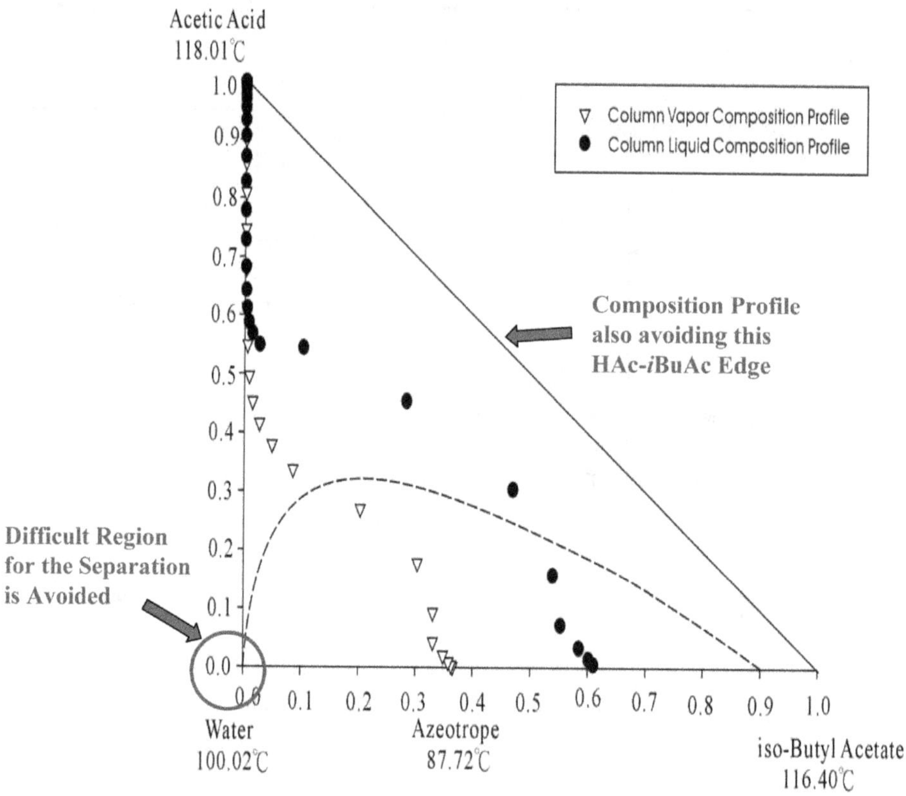


**Figure 13.20** Column vapor and liquid composition profiles for the continuous process.

Notice also that the path of the column composition profile in the continuous process design shown in Figure 13.20 does not go through the edge connecting the two corners of acetic acid and isobutyl acetate. Thus, the closeness of the two normal boiling-point temperatures (acetic acid at 118.01°C and isobutyl acetate at 116.40°C) does not affect the design of the continuous process at all. However, we will demonstrate in the next section that the difficulty caused by the closeness of the normal boiling- point temperatures of these two components will prevent the isobutyl acetate from being a suitable entrainer for acetic acid dehydration when operating in batch mode.[12]

### 13.3.2 Operation and Control of Heteroazeotropic Batch Distillation Column

According to Skouras et al.[13,14] the operation of the heteroazeotropic batch distillation column were defined as the following two modes.

***Mode 1.*** With this operation mode, the mixture is separated by distillation to recover the heteroazeotrope in the decanter. After the distillation step, the heteroazeotrope splits into its two immiscible phases: the entrainer-lean phase and the entrainer-rich phase. Thus, Mode 1 may be viewed as a hybrid process that is a combination of two different separation methods (distillation and liquid–liquid split) operating *sequentially*.

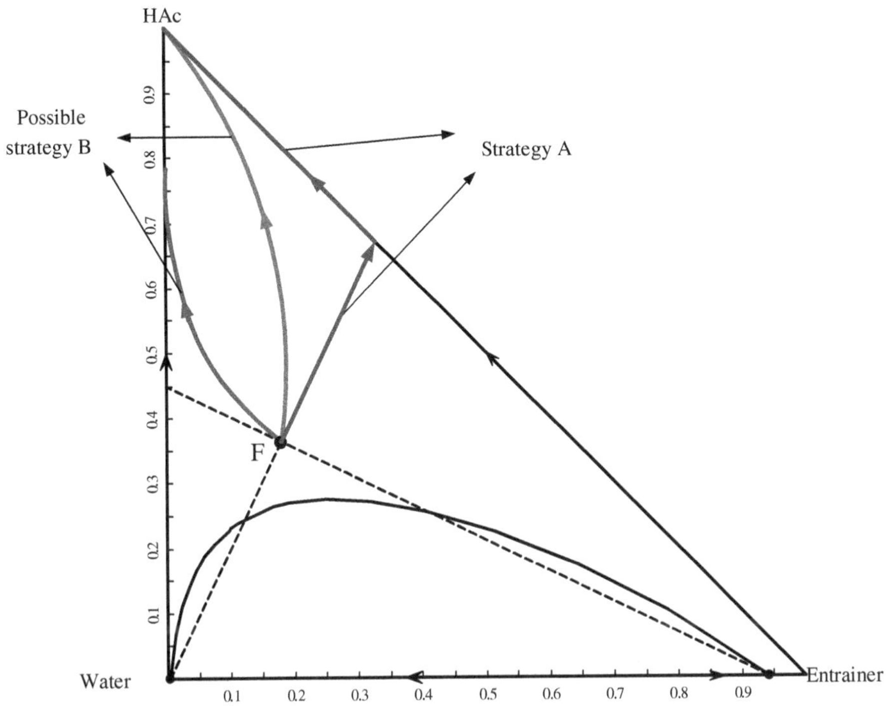


**Figure 13.21** Separation strategies in the batch rectifier with the corresponding path of bottom composition.

***Mode 2.*** With this operation model, the mixture is separated by distillation to obtain some of the heteroazeotrope in the decanter, and the liquid–liquid split takes place during distillation with withdrawal or accumulation of the entrainer-lean phase and reflux of all or part of the entrainer-rich phase. A start-up period is needed for Mode 2 as stated in Skouras et al.,[13,14] where all the mixed phases are refluxed. In our study, this Mode 2 operation will be used to combine two different separation methods (distillation and liquid–liquid splitting) simultaneously. The batch operating sequence was further modified to combine the start-up period with the next batch operating step to save the total batch time.

There are two operating strategies that can be used for the acetic acid dehydration. The two strategies (A or B) were explained in detail in Skouras et al.[13,14] Both operating strategies can be viewed in Figure 13.21.

***Strategy A.*** After a brief period of total reflux of the two liquid phases during the start-up period, the entrainer-lean phase (aqueous phase) is continuously accumulated in the aqueous phase tank, while the entrainer-rich phase is all refluxed back to the column. Since continuous accumulation can be treated the same as continuous withdrawal from the system, the compositions in the bottom are going in the opposite direction than the compositions of the aqueous phase (if total tray holdup is negligible). After the water is completely accumulated in the aqueous phase tank, the second step in the operating sequence can be started with the simple separation of two components (entrainer and acetic acid) with the entrainer going overhead and acetic acid remaining in the bottoms. The path of the bottom composition is shown in Figure 13.21.

***Strategy B.*** After a brief period of total reflux of two liquid phases during the start-up period, the entrainer-lean phase is continuously accumulated in the aqueous phase tank while the entrainer-rich phase is only partially refluxed (not under total reflux). Thus, both the entrainer-rich phase and the entrainer-lean phase are simultaneously accumulated. Since continuous accumulation can be treated the same as continuous withdrawal from the system, the compositions in the bottom are going in the opposite direction as determined by the combination of the two directions by the organic and aqueous phases. Depending on the rate of accumulation of the organic phase product, there are two possible bottom-composition paths, as shown in Figure 13.21.

In order to reduce the total batch time, we will use a strategy similar to Strategy B in the following evaluations of different entrainers. The only difference is that our strategy is further simplified by combing the initial start-up period with the next batch operation sequence. The only drawback of Strategy B compared to Strategy A is that the composition of the entrainer product is limited by the organic phase composition. Thus, pure entrainer cannot be recovered using this operating strategy. However, in the acetic acid dehydration application, there is no requirement of recovering pure entrainer. The entrainer with organic phase composition can be reused in the next batch. We will illustrate this point later.

The control loops used in our dynamic simulation can be seen in Figure 13.22. The total number of stages is chosen to be 31 (including the reboiler). In the figure, the *QC* stands

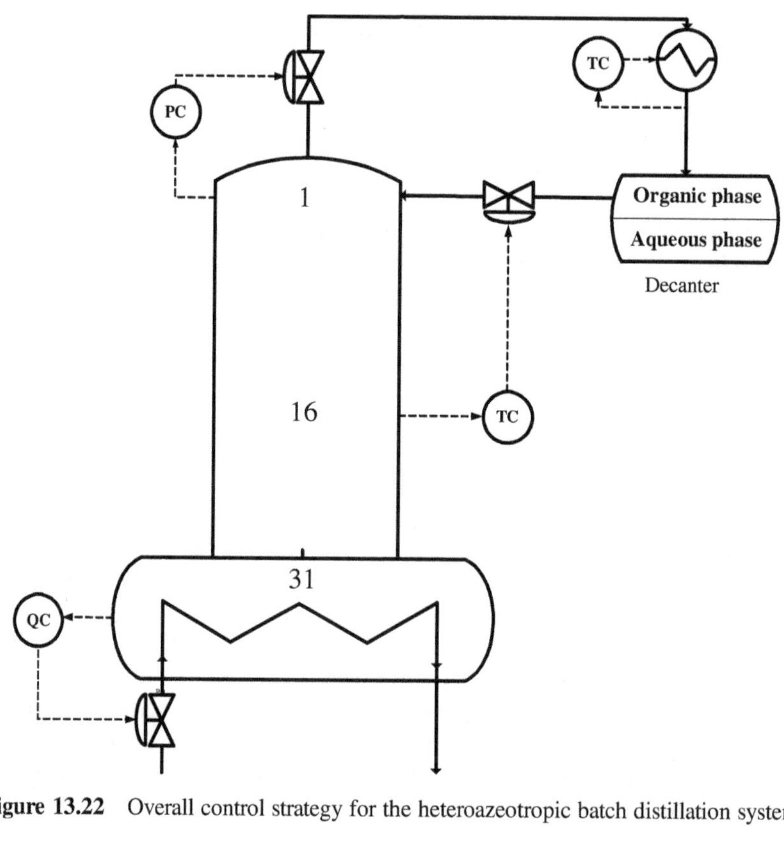


**Figure 13.22** Overall control strategy for the heteroazeotropic batch distillation system.

for control of reboiler heat input ($Q = 0.16$ GJ/h) throughout the whole batch. The top of the column is controlled at 1.2 bar, a little higher than the atmospheric pressure, and the decanter operating temperature is controlled at 40°C.

The most important control loop in this operation is a tray temperature controller at the middle stage of the column. The temperature is controlled at the average temperature of the minimum-boiling azeotropic temperature of the entrainer–water mixture and the normal boiling-point temperature of pure acetic acid. This will ensure that the top vapor of this column will approach the binary azeotrope and the bottoms will approach pure acetic acid. The manipulated variable of this important temperature loop is the organic reflux valve. Thus the accumulation of the organic phase will be automatically adjusted throughout the batch run.

The extremely simple batch operating sequence used in this section is summarized in the following:

*Step 1*: Charge the mixture needed to be separated into the column bottom together with a certain amount of the entrainer. In the simulation results to follow, the feeds include 2.0 kmol of acetic acid, 2.5 kmol of water, and 1.0 kmol of entrainer.

*Step 2*: Start the heat up of the column and put all controllers in automatic mode (middle tray temperature control, top column pressure control, and decanter temperature control).

*Step 3*: When the bottom composition reaches 0.991 (mole fraction of acetic acid product), stop heating and collect bottom, aqueous, and organic products. The reason for using a stop criterion of 0.991 instead of the specification of 0.990 is to compensate for the small amounts of material on the column trays that drain into the bottom.

### 13.3.3 Using Isobutyl Acetate as the Entrainer

Isobutyl acetate was found in Chapter 9 to be the best entrainer for acetic acid dehydration in the continuous process. Several patents have also shown isobutyl acetate to be a favorable entrainer for the separation of acetic acid and water (Costantini et al.[15] and Parten and Ure[16]). However, the closeness of the normal boiling-point temperature of acetic acid (118.01°C) and isobutyl acetate (116.40°C) is definitely troublesome for the second step of operating Strategy A (separation of acetic acid and entrainer) in the batch system. In fact, in the paper by Christensen and Olson[17] and also by Burguet et al.,[18] double azeotropes were found for the mixture of acetic acid and isobutyl acetate in the experimental data because the two normal boiling points are so close together.

The only possibility for isobutyl acetate to work in this batch separation is to follow operating Strategy B, which simultaneously accumulates the entrainer-rich phase and the entrainer-lean phase (e.g., path of bottom composition in Fig. 13.21 that goes to the edge of acetic acid and water first, and then toward the pure acetic acid corner). Simulation results, using the batch operating sequence outlined in previous section, are given in Figures 13.23 to 13.27. In the simulations, the setpoint value of the middle tray temperature control loop is set to be the average value of the pure acetic acid at 118.01°C and the azeotropic temperature of the isobutyl acetate–acetic acid mixture at 87.72°C. The PI controller tuning parameters used in the simulation for the top pressure control loop are: $K_C = 6.0$ and $\tau_I = 0.2$ min. The PI settings for the middle tray temperature, decanter temperature, and the reboiler duty control loops are all with $K_C = 5.0$ and $\tau_I = 5$ min.

Figure 13.23 shows the dynamic response of the middle tray temperature control loop. Controller performance is satisfactory until time = 1.4 h. After that, the temperature of

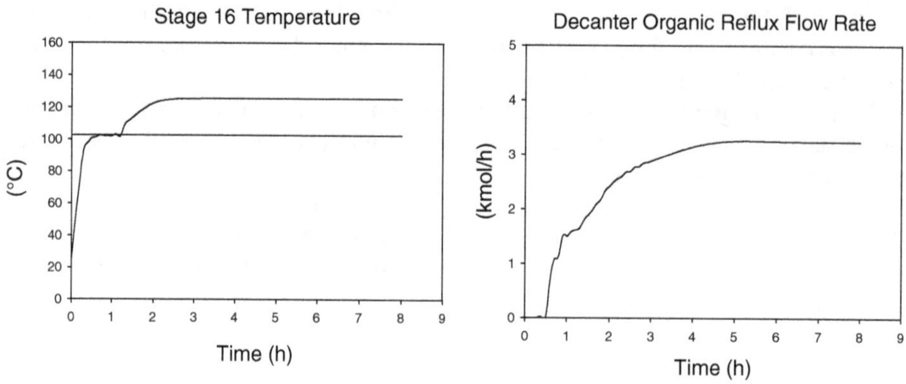


**Figure 13.23** Controlled and manipulated variables for the Stage 16 temperature loop using isobutyl acetate as the entrainer.

middle tray is not held at the setpoint value of 102.87°C. The organic reflux flow continuously increases trying to bring the temperature back to setpoint but is unable to do so. The organic reflux is actually under total reflux condition after time = 1.4 h. This means that all of the organic phase splitting in the decanter is refluxed back to the column. This can be seen in Figure 13.24 showing that the holdup of the decanter organic phase is almost empty after time = 1.4 h.

The compositions of the column bottoms, the decanter organic phase, and the aqueous phase are shown in Figure 13.25. Notice that the separation task has failed with portions of the acetic acid going upward through the column to the top and portions of the isobutyl acetate slipping downward to the column bottoms. Figure 13.26 shows the temperatures on Stages 1, 9, 16, 23, and 30 of the column. There is essentially very little separation for the stages below Stage 9 because all the stage temperatures are close together. Figure 13.27

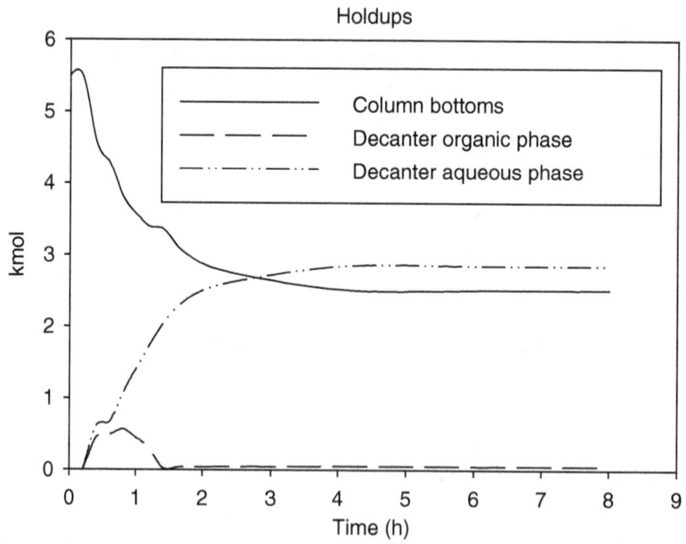


**Figure 13.24** Dynamic responses of three liquid holdups using isobutyl acetate as the entrainer.

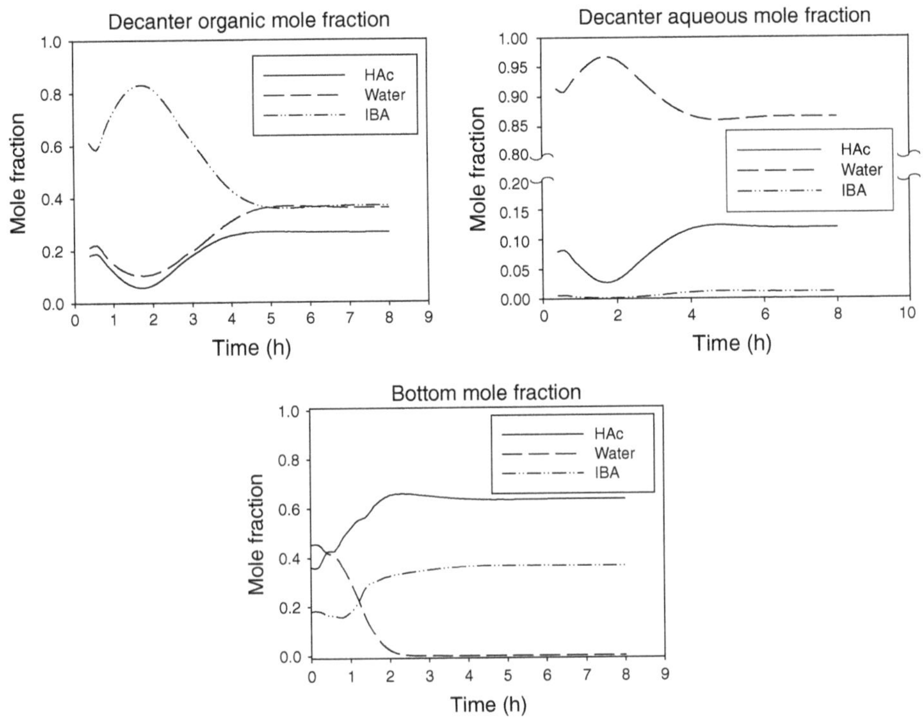


**Figure 13.25** Dynamic responses of three product compositions using isobutyl acetate as the entrainer.

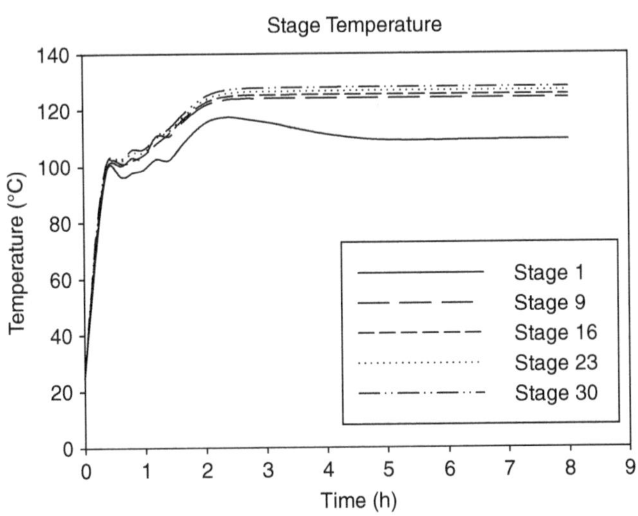


**Figure 13.26** Dynamic temperature responses at selected stages using isobutyl acetate as the entrainer.

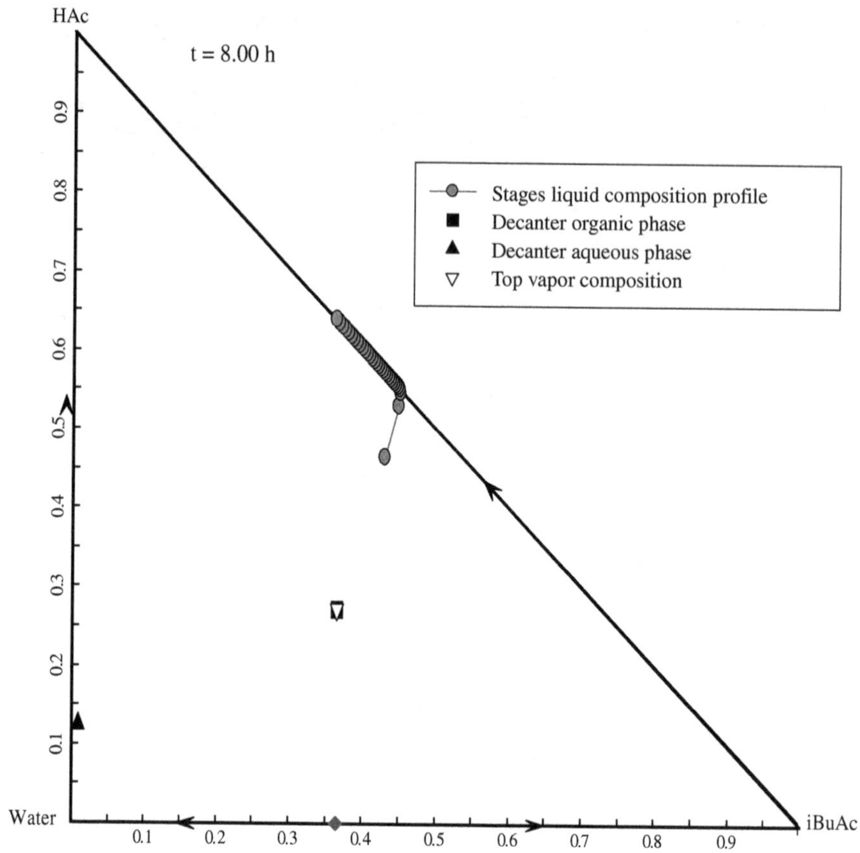


**Figure 13.27** Column liquid composition profile at the end of batch using isobutyl acetate as the entrainer.

reinforces the thought that the separation task has failed as indicated by the liquid compositions of all the stages being very close together on the isobutyl acetate–acetic acid edge.

The main reason for the failure of using isobutyl acetate as the entrainer is because of the poor separation capability between isobutyl acetate and acetic acid. This factor did not have to be considered at all in the continuous process as discussed in Chapter 9. However, it is important in the heteroazeotropic batch distillation.

In the following section, two alternative entrainers considered in the literature will be compared for the purpose of acetic acid dehydration. They are vinyl acetate, which was used in Rodriguez-Donis et al.,[19] and ethyl acetate, which was used in Skouras and Skogestad.[20,21]

### 13.3.4 Using an Alternative Entrainer or no Entrainer

The RCMs and the LLE for the ternary systems using vinyl acetate or ethyl acetate as entrainer are shown in Figures 13.28 and 13.29, respectively. The NRTL parameters used in the simulation are given in Table 13.5. The Hayden–O'Connell[22] second virial coefficient model with association parameters is used to account for the dimerization of acetic acid in

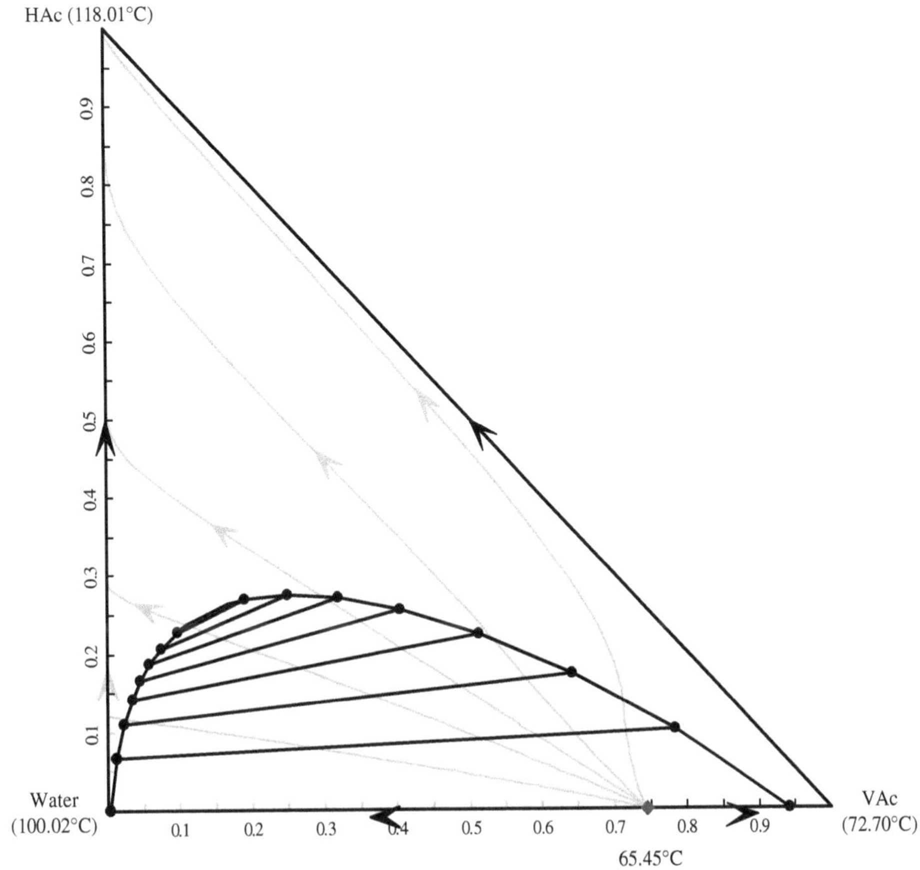


**Figure 13.28** RCMs and LLE using vinyl acetate as entrainer.

the vapor phase. The Aspen built-in association parameters are employed to compute fugacity coefficients.

Neither of these entrainers has an azeotropic composition (entrainer–water) that contains more water than the isobutyl acetate system (compare Figs. 13.28 and 13.29 with Fig. 13.18). However the normal boiling points between the entrainer and the acetic acid are much further apart than in the isobutyl acetate system. The aqueous phase composition of the vinyl acetate system is more favorable than the ethyl acetate system since it has a higher water concentration. The distance between the organic phase composition and the azeotropic composition of the vinyl acetate system is about the same as in the ethyl acetate system. The purity of the entrainer recovered using operating Strategy B is also better for the vinyl acetate system. We will demonstrate these differences in the simulation results using the two alternative entrainers in this subsection.

From the RCMs and the LLE for the ternary system, it appears that vinyl acetate is a better entrainer than ethyl acetate. We will use the same column with the same batch operating sequence outlined in the last subsection. Also the same heat duty will be used to provide a fair comparison of the performance (batch times) of the alternative entrainers.

Figure 13.30 shows the dynamic responses of the middle temperature control loop using vinyl acetate as the entrainer. Very soon after time = 0.5 h, the middle temperature reaches

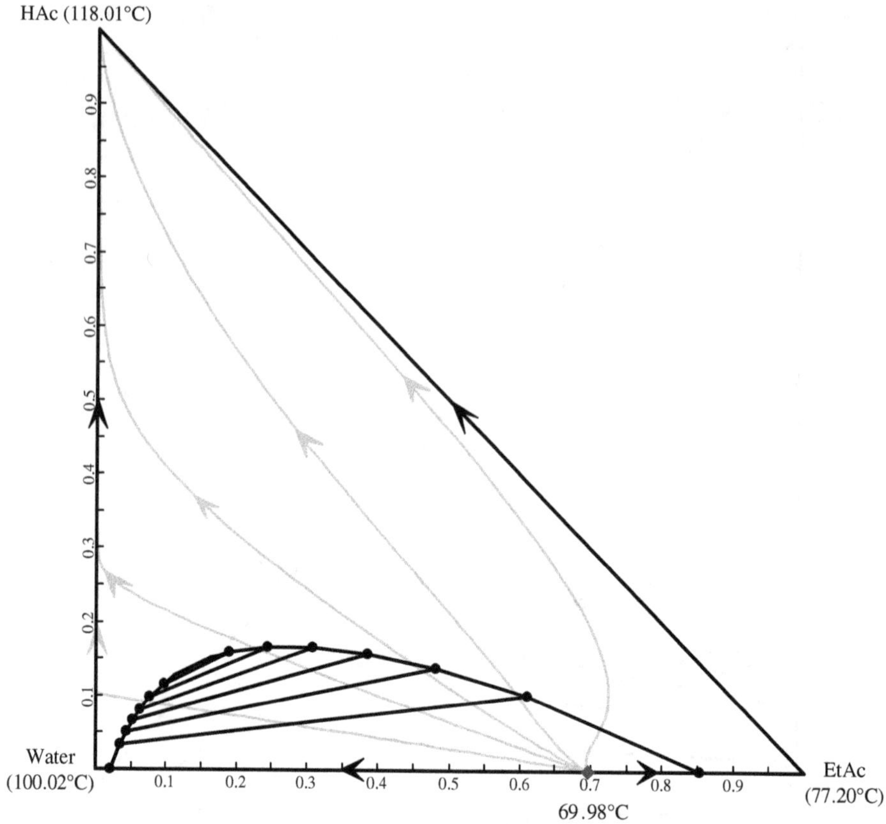


**Figure 13.29** RCMs and LLE using ethyl acetate as entrainer.

**TABLE 13.5 NRTL Parameters for an Acetic Acid–Water–IBA, Vinyl Acetate, or Ethyl Acetate System.**

| Comp. $i$<br>Comp. $j$ | Acetic Acid<br>Water | IBA<br>Water | IBA<br>Acetic Acid | Vinyl Acetate<br>Water | Vinyl Acetate<br>Acetic Acid | Ethyl Acetate<br>Water | Ethyl Acetate<br>Acetic Acid |
|---|---|---|---|---|---|---|---|
| $a_{ij}$ | 0 | 0 | 0 | 0 | 0 | 0 | 0 |
| $a_{ji}$ | 0 | 0 | 0 | 0 | 0 | 0 | 0 |
| $b_{ij}$ (K) | −211.31 | 489.61 | 351.30 | 618.02 | 388.85 | 416.12 | 576.23 |
| $b_{ji}$ (K) | 653.00 | 1809.1 | −68.422 | 1380.0 | −37.901 | 1024.5 | −322.42 |
| $\alpha_{ij}$ | 0.3 | 0.2505 | 0.3 | 0.28 | 0.3 | 0.3067 | 0.3 |

Aspen Plus NRTL:

$$\ln \gamma_i = \frac{\sum_j x_j \tau_{ji} G_{ji}}{\sum_k x_k G_{ki}} + \sum_j \frac{x_j G_{ij}}{\sum_k x_k G_{kj}} \left[ \tau_{ij} - \frac{\sum_m x_m \tau_{mj} G_{mj}}{\sum_k x_k G_{kj}} \right]$$

where $G_{ij} = \exp(-\alpha_{ij} \tau_{ij})$

$$\tau_{ij} = a_{ij} + \frac{b_{ij}}{T}$$

$$\alpha_{ij} = c_{ij},\ \tau_{ii} = 0,\ G_{ii} = 1$$

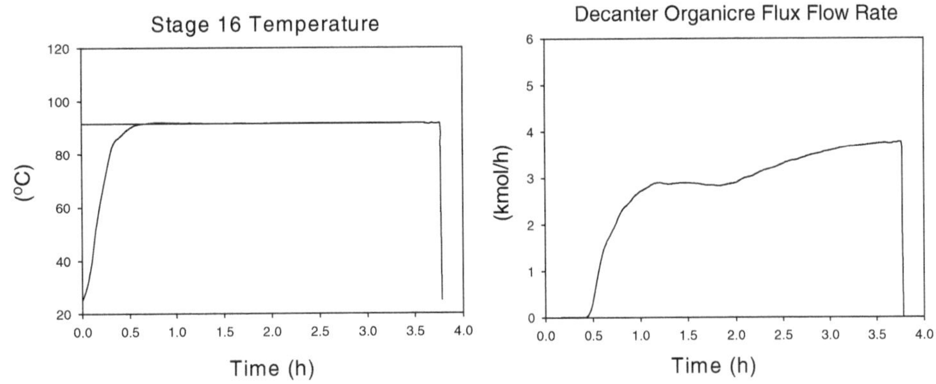


**Figure 13.30** Controlled and manipulated variables for the Stage 16 temperature loop using vinyl acetate as the entrainer.

its setpoint value of 91.73°C (the average of 118.01 and 65.45°C) and stays at this value until the end of the batch. As shown in Figure 13.31, both the organic phase and aqueous phase are accumulated in the decanter according to the plan of operating Strategy B. The liquid compositions in the bottom, the decanter organic phase, and the aqueous phase are shown in Figure 13.32. Note that all three products are produced at high purity. Detailed information of the product compositions, the recovery percentage, and batch time are given in Table 13.6.

Figure 13.33 displays the composition paths taken by the bottom, the decanter organic phase, and the decanter aqueous phase. The bottom composition follows a path that first reaches the acetic acid–water edge and then approaches the pure acetic acid corner. In this figure, time at 0.16 h stands for time when liquids begin to appear in the decanter organic and aqueous phases. Although the bottom composition path is towards the acetic acid–water edge, the column composition profile is near the vinyl acetate–acetic acid edge at the end

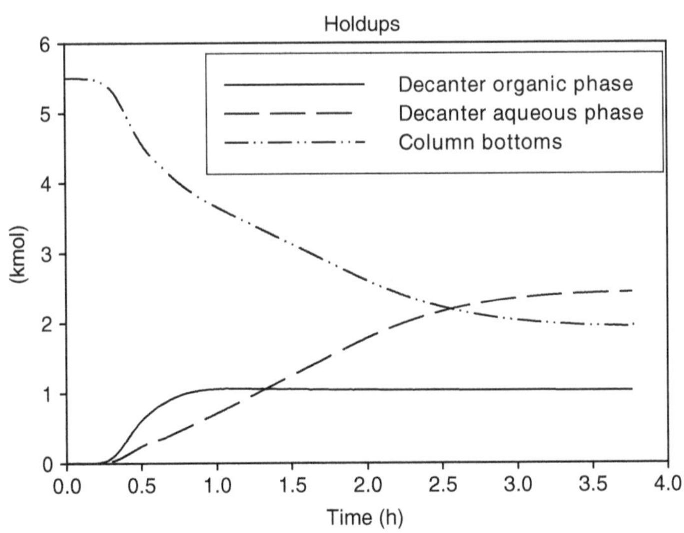


**Figure 13.31** Dynamic responses of three liquid holdups using vinyl acetate as the entrainer.

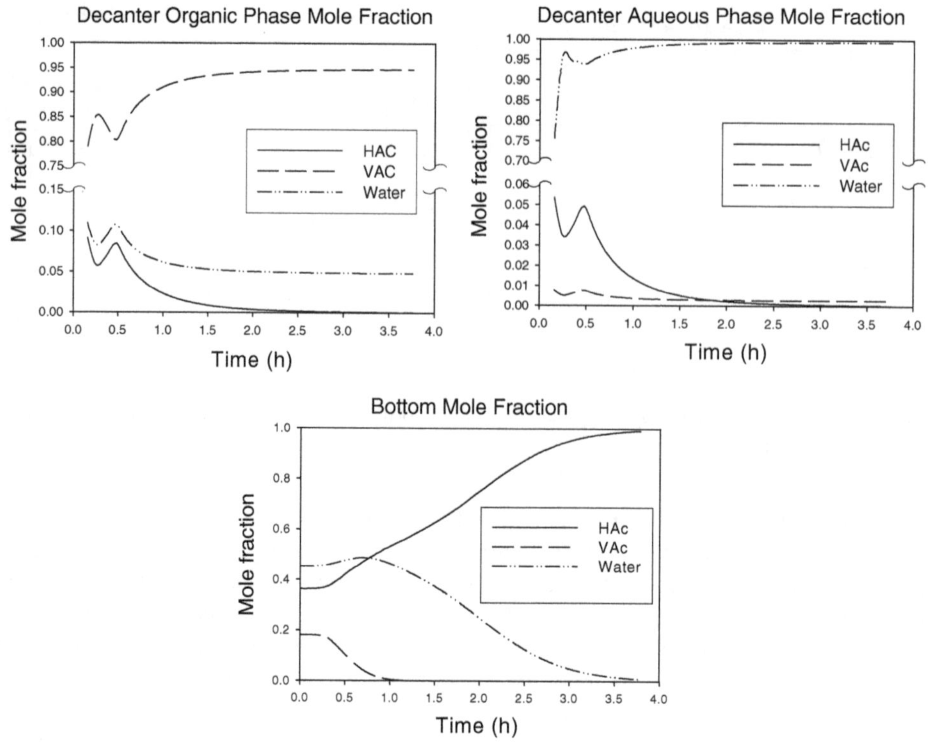


**Figure 13.32** Dynamic responses of three product compositions using vinyl acetate as the entrainer.

of batch because water is almost completely taken out of the column (this can be seen in Fig. 13.34).

Similar dynamic simulations were repeated with the same column and the same batch operating sequence for the case of using ethyl acetate as the entrainer. The only difference is that the setpoint of the middle temperature control loop is changed to 94.00°C (the average of 118.01 and 69.99°C). Detailed simulation results are also given in Table 13.6.

Besides using the two alternative entrainers, the batch distillation of directly separating acetic acid and water without an entrainer was also simulated for comparison purposes. The operation and control strategy as in Skogestad et al.[11] is used in the simulation. The concept of this batch distillation is to collect water in the reflux drum and acetic acid in the bottom. A similar middle tray temperature control loop is used with the setpoint value set at 109.02°C (the average of 118.01 and 100.02°C). The batch was stopped when the product composition at the column bottoms is the same as the case with entrainer (0.990) for direct comparison.

Table 13.6 summarizes all the simulation results using either vinyl acetate or ethyl acetate as the entrainer and the case with no entrainer. Using vinyl acetate as the entrainer requires the least batch time (3.79 h) with a higher recovery percentage of water and the entrainer. The recovered entrainer purity is also much higher (0.953 vinyl acetate) than the case using ethyl acetate as entrainer (only at 0.879). The purity of the water product is also higher (0.997) than the case using ethyl acetate as entrainer (0.989). Compared with the case with no entrainer, a 23% reduction in batch time can be realized with the added benefit of a much higher purity water product (0.997 versus 0.929). From the summary of Table 13.6, by using vinyl acetate

**TABLE 13.6 Comparison of Simulation Results for an Acetic Acid–Water System.**

| Entrainer | Vinyl Acetate | Ethyl Acetate | No Entrainer |
|---|---|---|---|
| Total batch time (h) | 3.79 | 4.04 | 4.92 |
| Total feed charge (kmol) | 5.5 | 5.5 | 4.5 |
| Acetic acid (kmol) | 2.0 | 2.0 | 2.0 |
| Water (kmol) | 2.5 | 2.5 | 2.5 |
| Entrainer (kmol) | 1.0 | 1.0 | 0.0 |
| *Decanter organic phase* | | | |
| Total holdup (kmol) | 1.039 | 1.108 | – |
| Acetic acid (mole fraction) | 0.000 | 0.000 | – |
| Water (mole fraction) | 0.047 | 0.121 | – |
| Entrainer (mole fraction) | 0.953 | 0.879 | – |
| Recovery of entrainer (%) | 99.0 | 97.4 | – |
| *Decanter aqueous phase or reflux drum* | | | |
| Total holdup (kmol) | 2.442 | 2.374 | 2.672 |
| Acetic acid (mole fraction) | 0.000 | 0.000 | 0.071 |
| Water (mole fraction) | 0.997 | 0.989 | 0.929 |
| Entrainer (mole fraction) | 0.003 | 0.011 | – |
| Recovery of water (%) | 97.4 | 93.9 | 99.3 |
| *Column bottoms* | | | |
| Total holdup (kmol) | 2.019 | 2.018 | 1.828 |
| Acetic acid (mole fraction) | 0.990 | 0.990 | 0.990 |
| Water (mole fraction) | 0.009 | 0.009 | 0.010 |
| Entrainer (mole fraction) | 0.001 | 0.001 | – |
| Recovery of acetic acid (%) | 99.9 | 99.9 | 90.5 |

as the entrainer, this simple operating procedure not only produces the three products (acetic acid, water, and vinyl acetate) with high enough purity, but also achieves very high product recovery.

### 13.3.5 Robustness of the Proposed Batch Operation

Vinyl acetate is the best entrainer for acetic acid dehydration using heteroazeotropic batch distillation system. There are only several operating variables needed to be set for this simple batch operating sequence to work. They are the setpoint of the middle tray temperature control loop at 91.73°C, the entrainer preloading amounts of 1.0 kmol, and the reboiler duty fixed at 0.16 GJ/h. In this section, the settings of these three operating variables will be altered to see if the proposed batch operation is robust enough.

We first examine the effect of changing the setpoint of the middle tray temperature control loop. Because the desirable top vapor of this column is the vinyl acetate–water two component azeotrope (at 65.45°C) and the bottom is the pure acetic acid product (n.b.p. at 118.01°C), the setpoint value of this middle temperature control loop was set to be the average of the two temperatures at 91.73°C. Table 13.7 shows five simulation runs with this temperature setpoint altered from 71.73 to 111.73°C in increments of 10°C. This is a very wide range of values for setting this setpoint. However, even with this wide range of

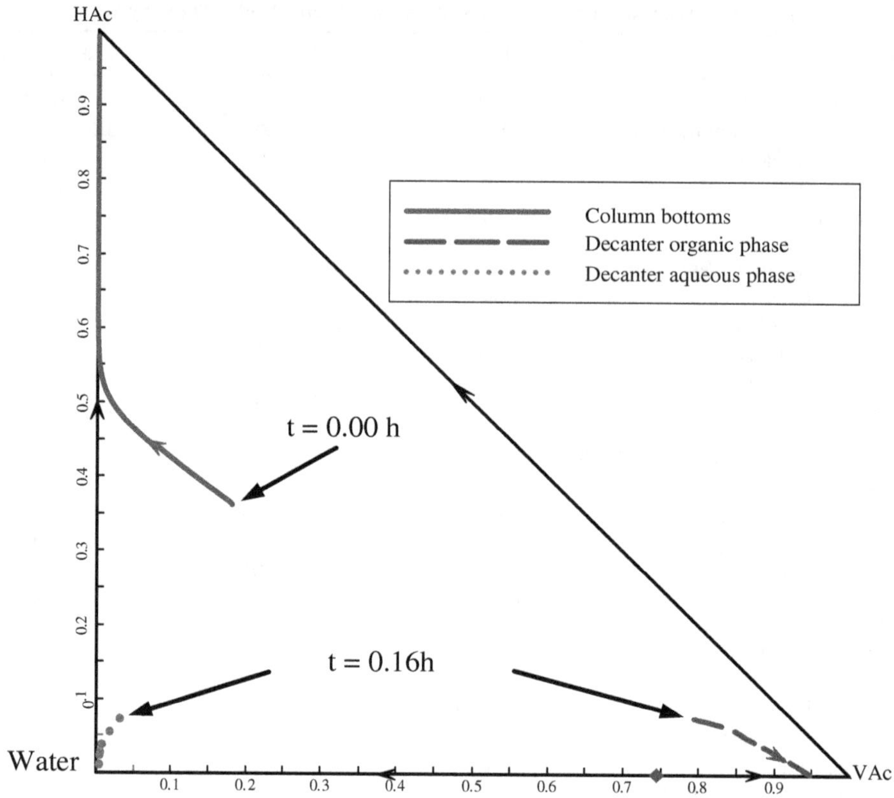


**Figure 13.33** Bottom, organic phase, and aqueous phase composition paths using vinyl acetate as the entrainer.

selection of setpoint values, the performance of this batch heteroazeotropic system did not deteriorate at all.

The total batch time and the recoveries of the three products are pretty much the same with the wide 10°C mismatches of temperature upward or downward (101.73 or 81.73°C). Only when this setpoint is set to be too low at 71.73°C does the batch operation fail with the organic phase totally refluxed but still not able to reach the temperature setpoint. At this low temperature setpoint value causing total refluxing of the organic phase, Strategy A discussed in Section 13.3.2 should be implemented with two sequential steps in the batch sequence (aqueous product recovery step and followed by the entrainer–acetic acid recovery step). However, because our proposed procedure only allows for simultaneous recovery of the three products, the batch operation failed.

The penalty for setting the setpoint value too high is not very severe with only a little increase in the total batch time and an increase in the acetic acid loss toward the top of the column. The reason for the wide tolerance of the setting of the middle tray temperature setpoint is because thermodynamically the bottom product will approach the heavy boiler (pure acetic acid) and the top vapor will approach the lightest boiler (vinyl acetate–water two-component azeotrope). Thus, as long as the middle tray temperature setpoint is set between these two temperatures, and also allowing for enough trays above and below the middle tray, the desirable separation will be achieved.

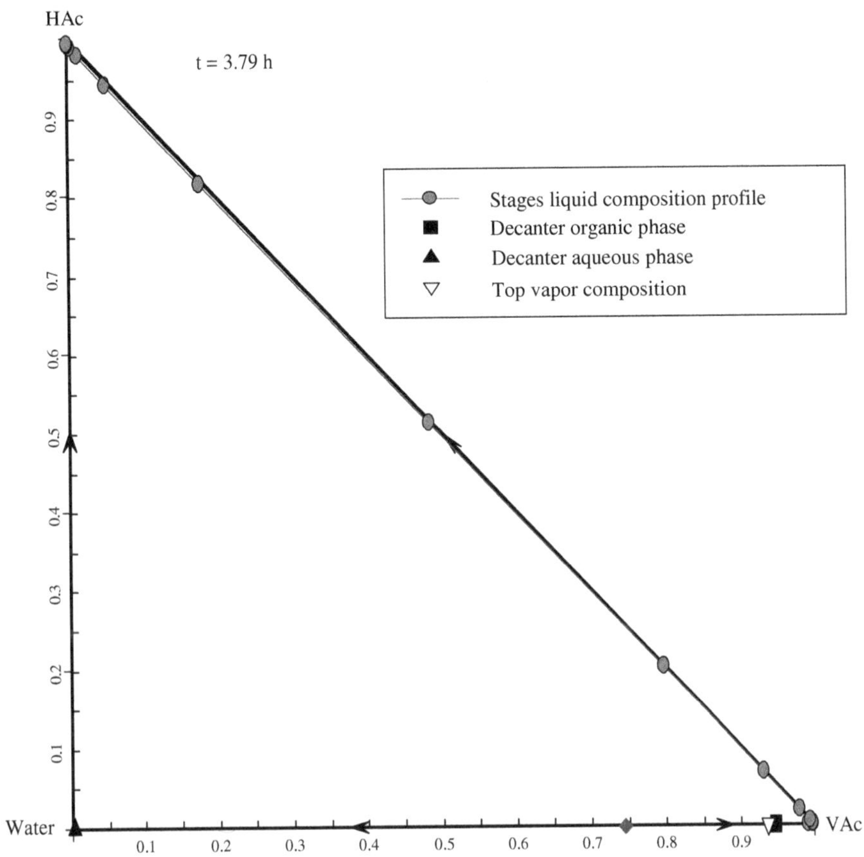


**Figure 13.34** Column liquid composition profile at the end of batch using vinyl acetate as the entrainer.

The amount of entrainer to be preloaded with the mixture to be separated is the next operating variable to check. In the simulation runs in the previous section, 1.0 kmol of vinyl acetate was premixed with the acetic acid–water mixture. Table 13.8 shows the simulation results with this preloading amounts changing from 0.1 to 1.4 kmol. Again with a wide range of this amount, this heteroazeotropic batch distillation system still performs well with a slight reduction in the total batch time for larger preloading amounts. Only when this preloading amount is reduced down to a very small value (0.1 kmol) does the batch operation fail. There is not enough organic reflux to keep the middle tray temperature at its setpoint.

Finally we examine the effect of changing the reboiler duty. Table 13.9 summarizes the six simulation results with the reboiler duty changing from 0.14 to 0.19 GJ/h with 0.16 GJ/h as the base case. The compositions of the decanter organic phase, aqueous phase, and the bottom are all the same with the varying of the reboiler duty. The recovery percentages of the acetic acid, water, and the entrainer (vinyl acetate) are also the same, no matter of what reboiler duty is used. As expected, the total batch time decreases as the reboiler duty increases.

**TABLE 13.7 Effect of Varying the Setpoint Value of the Middle Tray Temperature Control.**

| Setpoint Value (°C) | 71.73 | 81.73 | 91.73 | 101.73 | 111.73 |
|---|---|---|---|---|---|
| Total Batch Time (h) | Not Able to Work | 3.76 | 3.79 | 3.79 | 3.98 |
| *Decanter organic phase* | | | | | |
| Total holdup (kmol) | – | 1.039 | 1.039 | 1.039 | 1.084 |
| Acetic acid (mol. frac.) | – | 0.000 | 0.000 | 0.000 | 0.029 |
| Water (mol. frac.) | – | 0.047 | 0.047 | 0.047 | 0.061 |
| Vinyl acetate (mol. frac.) | – | 0.953 | 0.953 | 0.953 | 0.910 |
| Recovery of vinyl acetate (%) | – | 99.0 | 99.0 | 99.0 | 98.6 |
| *Decanter aqueous phase* | | | | | |
| Total holdup (kmol) | – | 2.442 | 2.442 | 2.442 | 2.464 |
| Acetic acid (mol. frac.) | – | 0.000 | 0.000 | 0.000 | 0.017 |
| Water (mol. frac.) | – | 0.997 | 0.997 | 0.997 | 0.980 |
| Vinyl acetate (mol. frac.) | – | 0.003 | 0.003 | 0.003 | 0.003 |
| Recovery of water (%) | – | 97.4 | 97.4 | 97.4 | 96.6 |
| *Column bottoms* | | | | | |
| Total holdup (kmol) | – | 2.019 | 2.019 | 2.019 | 1.942 |
| Acetic acid (mol. frac.) | – | 0.990 | 0.990 | 0.990 | 0.990 |
| Water (mol. frac.) | – | 0.009 | 0.009 | 0.009 | 0.009 |
| Vinyl acetate (mol. frac.) | – | 0.001 | 0.001 | 0.001 | 0.001 |
| Recovery of acetic acid (%) | – | 99.9 | 99.9 | 99.9 | 96.1 |

**TABLE 13.8 Effect of Varying Pre-mixed Amount of Vinyl Acetate.**

| Pre-mixed Amount (kmol) | 0.1 | 0.2 | 0.4 | 1.0 | 1.2 | 1.4 |
|---|---|---|---|---|---|---|
| Total Batch Time (h) | Not Able to Work | 3.92 | 3.82 | 3.79 | 3.78 | 3.76 |
| *Decanter organic phase* | | | | | | |
| Total holdup (kmol) | – | 0.200 | 0.410 | 1.039 | 1.251 | 1.461 |
| Acetic acid (mol. frac.) | – | 0.000 | 0.000 | 0.000 | 0.000 | 0.000 |
| Water (mol. frac.) | – | 0.047 | 0.047 | 0.047 | 0.047 | 0.047 |
| Vinyl acetate (mol. frac.) | – | 0.953 | 0.953 | 0.953 | 0.953 | 0.953 |
| Recovery of vinyl acetate (%) | – | 95.3 | 98.5 | 99.0 | 99.4 | 99.5 |
| *Decanter aqueous phase* | | | | | | |
| Total holdup (kmol) | – | 2.481 | 2.471 | 2.442 | 2.431 | 2.421 |
| Acetic acid (mol. frac.) | – | 0.000 | 0.000 | 0.000 | 0.000 | 0.000 |
| Water (mol. frac.) | – | 0.997 | 0.997 | 0.997 | 0.997 | 0.997 |
| Vinyl acetate (mol. frac.) | – | 0.003 | 0.003 | 0.003 | 0.003 | 0.003 |
| Recovery of water (%) | – | 98.9 | 98.5 | 97.4 | 96.9 | 96.5 |
| *Column bottoms* | | | | | | |
| Total holdup (kmol) | – | 2.019 | 2.019 | 2.019 | 2.018 | 2.018 |
| Acetic acid (mol. frac.) | – | 0.990 | 0.990 | 0.990 | 0.990 | 0.990 |
| Water (mol. frac.) | – | 0.009 | 0.009 | 0.009 | 0.009 | 0.009 |
| Vinyl acetate (mol. frac.) | – | 0.001 | 0.001 | 0.001 | 0.001 | 0.001 |
| Recovery of acetic acid (%) | – | 99.9 | 99.9 | 99.9 | 99.9 | 99.9 |

**TABLE 13.9 Effect of Varying the Reboiler Duty.**

| Reboiler Duty (GJ/h) | 0.14 | 0.15 | 0.16 | 0.17 | 0.18 | 0.19 Not Able to Work |
|---|---|---|---|---|---|---|
| Total Batch Time (h) | 4.33 | 4.04 | 3.79 | 3.57 | 3.36 | |
| *Decanter organic phase* | | | | | | |
| Total holdup (kmol) | 1.039 | 1.039 | 1.039 | 1.039 | 1.039 | – |
| Acetic acid (mol. frac.) | 0.000 | 0.000 | 0.000 | 0.000 | 0.000 | – |
| Water (mol. frac.) | 0.047 | 0.047 | 0.047 | 0.047 | 0.047 | – |
| Vinyl acetate (mol. frac.) | 0.953 | 0.953 | 0.953 | 0.953 | 0.953 | – |
| Recovery of vinyl acetate (%) | 99.0 | 99.0 | 99.0 | 99.0 | 99.0 | – |
| *Decanter aqueous phase* | | | | | | |
| Total holdup (kmol) | 2.442 | 2.442 | 2.442 | 2.442 | 2.442 | – |
| Acetic acid (mol. frac.) | 0.000 | 0.000 | 0.000 | 0.000 | 0.000 | – |
| Water (mol. frac.) | 0.997 | 0.997 | 0.997 | 0.997 | 0.997 | – |
| Vinyl acetate (mol. frac.) | 0.003 | 0.003 | 0.003 | 0.003 | 0.003 | – |
| Recovery of water (%) | 97.4 | 97.4 | 97.4 | 97.4 | 97.4 | – |
| *Column bottoms* | | | | | | |
| Total holdup (kmol) | 2.019 | 2.019 | 2.019 | 2.019 | 2.019 | – |
| Acetic acid (mol. frac.) | 0.990 | 0.990 | 0.990 | 0.990 | 0.990 | – |
| Water (mol. frac.) | 0.009 | 0.009 | 0.009 | 0.009 | 0.009 | – |
| Vinyl acetate (mol. frac.) | 0.001 | 0.001 | 0.001 | 0.001 | 0.001 | – |
| Recovery of acetic acid (%) | 99.9 | 99.9 | 99.9 | 99.9 | 99.9 | – |

When the reboiler duty is fixed at a high value (0.19 GJ/h), the component acetic acid is starting to boil-up and escape from the top of the column causing the middle tray temperature control loop to fail, even with total organic reflux. This series of dynamic simulation runs show that the reboiler duty is also not too sensitive. The reboiler duty with a range of 0.14 to 0.18 GJ/h all gave acceptable separation performance with trade-off in total batch time and energy consumption.

From the above analysis and many dynamic simulations, it is confirmed that the simple batch operating sequence proposed in the previous section is very robust with respect to the operating variables.

### 13.3.6 Recycle of the Recovered Entrainer for the Next Batch

As explained in Section 13.3.2, the only possible drawback for the proposed batch operation is the limitation of collecting the entrainer in less than pure form because only the composition of the organic phase is collected and not the pure entrainer. However, if the purpose of this batch system is to dehydrate the acetic acid with the addition of entrainer only to help the separation, the recovered entrainer can be reused in the next batch run even though it is not pure. We will illustrate this point in the series of simulation runs to follow.

Table 13.10 summarizes three consecutive batch runs with the next batch run using the recovered entrainer from the organic phase of the previous batch run. These results show that the total batch times of the three runs are pretty much the same. There are small entrainer losses through the aqueous phase and the column bottoms. However, the entrainer loss is very small. The preloaded vinyl acetate component only decreases at a very slow rate

**TABLE 13.10 Three Consecutive Batch Runs Using Recovered Entrainer.**

| Batch Run No. | 1 | 2 (Using Organic Phase Material from the First Run) | 3 (Using Organic Phase Material from the Second Run) |
|---|---|---|---|
| Total batch time (h) | 3.79 | 3.82 | 3.83 |
| Total feed charge (kmol) | 5.5 | 5.539 | 5.530 |
| Acetic acid (kmol) | 2.0 | 2.0 | 2.0 |
| Water (kmol) | 2.5 | 2.5 | 2.5 |
| Vinyl acetate (kmol) | 1.0 | 1.039 | 1.030 |
| *Decanter organic phase* | | | |
| Total holdup (kmol) | 1.039 | 1.030 | 1.023 |
| Acetic acid (mole fraction) | 0.000 | 0.000 | 0.000 |
| Water (mole fraction) | 0.047 | 0.047 | 0.047 |
| Vinyl acetate (mole fraction) | 0.953 | 0.953 | 0.953 |
| Pure vinyl acetate recovered (kmol) | 0.990 | 0.982 | 0.975 |
| *Decanter aqueous phase* | | | |
| Total holdup (kmol) | 2.442 | 2.490 | 2.488 |
| Acetic acid (mole fraction) | 0.000 | 0.000 | 0.000 |
| Water (mole fraction) | 0.997 | 0.997 | 0.997 |
| Vinyl acetate (mole fraction) | 0.003 | 0.003 | 0.003 |
| Pure water recovered (kmol) | 2.435 | 2.483 | 2.481 |
| *Column bottoms* | | | |
| Total holdup (kmol) | 2.019 | 2.019 | 2.019 |
| Acetic acid (mole fraction) | 0.990 | 0.990 | 0.990 |
| Water (mole fraction) | 0.009 | 0.009 | 0.009 |
| Vinyl acetate (mole fraction) | 0.001 | 0.001 | 0.001 |
| Pure acetic acid recovered (kmol) | 1.999 | 1.999 | 1.999 |

from one batch to the next batch. From the simulation results of Table 13.8 showing that the batch operation is still working well with only 0.2 kmol of pure vinyl acetate, many consecutive batch runs can be made with acceptable results using recovered organic phase material from the previous run.

## 13.4 MULTIVESSEL HETEROAZEOTROPIC BATCH DISTILLATION

A multivessel operation to recover water in the decanter aqueous phase, the entrainer (vinyl acetate) in a middle vessel, and acetic acid in the bottom is explored in this section. The most important advantage of this multivessel operation is its ability to recover the entrainer with compositions higher than those obtained by the liquid–liquid split. The operating strategy is to accumulate the aqueous phase as in previous section, to reflux all of the organic phase materials back into the column, and to use the column sections between the middle vessel and the column bottoms to separate the entrainer and acetic acid. The batch operating sequence similar to the works by Skouras and Skogestad[20,21] and by Skouras et al.[13] will be used with some modifications.

The overall control loops used in our dynamic simulation can be seen in Figure 13.35. Note that the column is exactly the same as the one used in Figure 13.22 for fair comparison

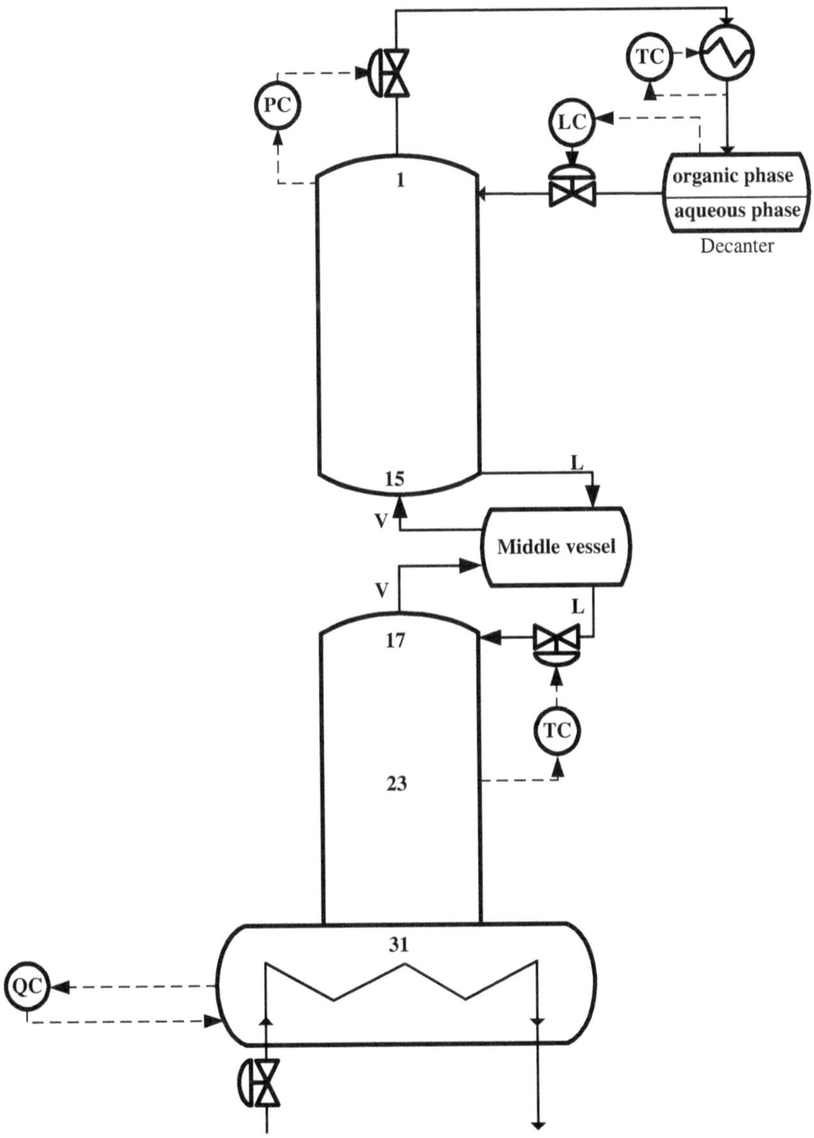


**Figure 13.35** Overall control strategy for the heteroazeotropic batch distillation system with multivessel.

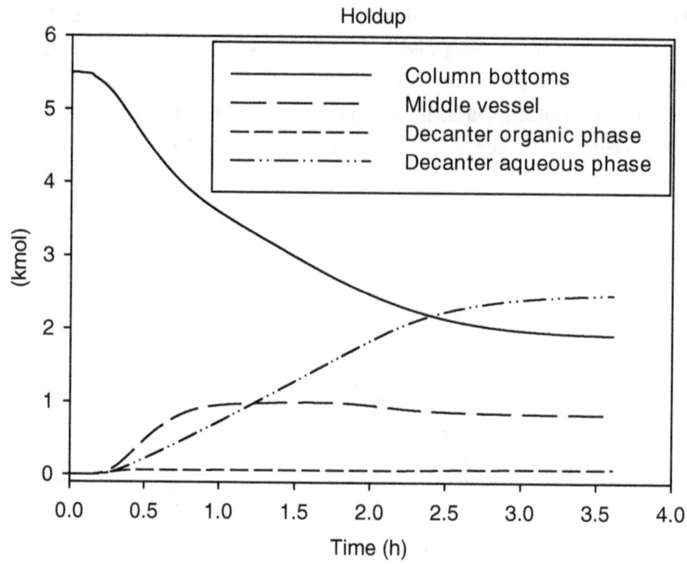


**Figure 13.36** Dynamic responses of four liquid holdups using multivessel operation.

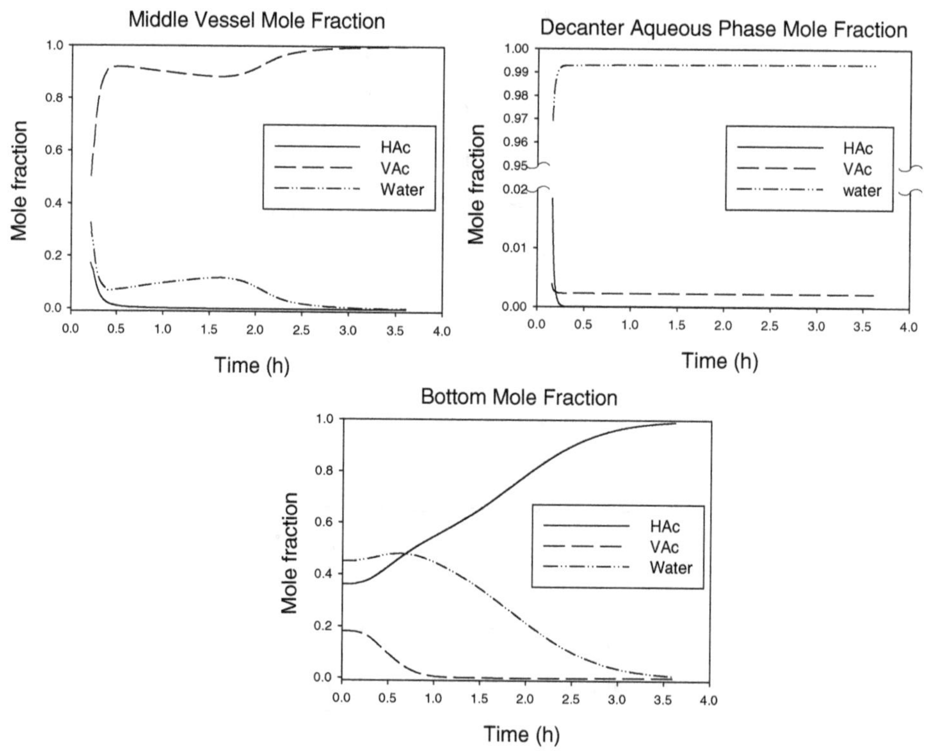


**Figure 13.37** Dynamic responses of three product compositions using multivessel operation.

**TABLE 13.11 Comparison of Results from Batch Rectifier and Multivessel Operations.**

| Operation | Batch Rectifier | Multivessel |
|---|---|---|
| Total batch time (h) | 3.79 | 3.75 |
| Total feed charge (kmol) | 5.5 | 5.5 |
| Acetic acid (kmol) | 2.0 | 2.0 |
| Water (kmol) | 2.5 | 2.5 |
| Vinyl acetate (kmol) | 1.0 | 1.0 |
| *Decanter organic phase* | | *Middle vessel* |
| Total holdup (kmol) | 1.039 | 0.995 |
| Acetic acid (mole fraction) | 0.000 | 0.000 |
| Water (mole fraction) | 0.047 | 0.004 |
| Vinyl acetate (mole fraction) | 0.953 | 0.996 |
| Recovery of vinyl acetate (%) | 99.0 | 99.1 |
| *Decanter aqueous phase* | | |
| Total holdup (kmol) | 2.442 | 2.485 |
| Acetic acid (mole fraction) | 0.000 | 0.000 |
| Water (mole fraction) | 0.997 | 0.997 |
| Vinyl acetate (mole fraction) | 0.003 | 0.003 |
| Recovery of water (%) | 97.4 | 99.1 |
| *Column bottoms* | | |
| Total holdup (kmol) | 2.019 | 2.020 |
| Acetic acid (mole fraction) | 0.990 | 0.990 |
| Water (mole fraction) | 0.009 | 0.009 |
| Vinyl acetate (mole fraction) | 0.001 | 0.001 |
| Recovery of acetic acid (%) | 99.9 | 100 |

of the separation results. The only difference is that an additional middle vessel is used for this operation and an additional level control loop. In practise, the vapor would not pass through the middle vessel. However, for convenience of simulation in Aspen Dynamics, no vapor bypassing around the middle vessel is implemented in the dynamic simulation. According to Skouras and Skogestad,[21] the effect of vapor bypassing to batch-time requirement is found to be negligible.

The column pressure and decanter temperature control loops are the same as in Figure 13.22. The setpoint value of the organic phase level control loop is set to be a very small value so that total refluxing of the organic phase materials back to the column can be achieved. The setpoint value of the Stage 23 temperature control loop is set to be the average of the normal boiling-point temperatures of vinyl acetate (72.70°C) and acetic acid (118.01°C) so that the desirable product in the middle vessel is vinyl acetate and the desirable product in the bottom is acetic acid.

The extremely simple batch operating sequence that is exactly the same as the one used in the previous section is implemented. The batch operating sequence is again summarized below.

*Step 1*: Charge the total mixture to be separated into the bottom of the column together with some entrainer. In the simulation results the feeds include 2.0 kmol of acetic acid, 2.5 kmol of water, and 1.0 kmol of entrainer. Note that no initial distribution of material into the decanter and the middle vessel is used in the dynamic simulations to reduce total batch time.

*Step 2*: Start the heat-up of the column and put all controllers (Stage 23 temperature control, organic phase level control, top column pressure control, and decanter temperature control) on automatic mode. Note that, in this operation, the total reflux period has been eliminated to reduce batch time.

*Step 3*: When the bottom composition reaches 0.991 (mole fraction of acetic acid), stop heating and collect bottom, middle vessel, and aqueous products.

Figure 13.36 shows the liquid holdups in the column bottoms, middle vessel, decanter aqueous phase, and the organic phase. Notice that the holdups of the middle vessel and the decanter aqueous phase are accumulating simultaneously. There is no middle vessel level control loop presetting its level to be approximately equal to the amount of added entrainer. The accumulation of the middle vessel occurs automatically and is dictated by the manipulated variable of the Stage 23 temperature control loop. The liquid compositions of the bottom, middle vessel, and the decanter aqueous phase are shown in Figure 13.37. Notice that all products approach their desired compositions. In particular note that the desired purity of the entrainer vinyl acetate in the middle vessel is achieved.

A comparison between multivessel and conventional batch rectifier operations is given in Table 13.11. The total batch time of this multivessel operation is 3.63 h plus an additional 0.12 h for collecting into the middle vessel the small amount of organic phase on the upper stages of the column and for collecting into the bottoms of the column the small amount of material on the lower stages of the column. This total batch time (3.75 h) should be compared to the total batch time in the previous section (3.79 h). Very little reduction of total batch time is gained. However, additional process, instrumentation, and control equipment (middle vessel, organic phase level transmitter and controller, and additional piping for the middle vessel arrangement) are needed for the multivessel operation. Thus, the batch operation without using a multivessel configuration is recommended for acetic acid dehydration using heterogeneous azeotropic batch distillation.

## 13.5 CONCLUSION

Operation and control strategies of two batch extractive distillation column systems have been investigated. The first system separates acetone and methanol using water as the entrainer, and the second system separates IPA and water using DMSO as the entrainer. Both systems contain a minimum-boiling azeotrope, and the mixture is separated using a heavy entrainer. The operating procedure starting from an empty column to the final shut-down of the column has been dynamically simulated using Aspen Dynamics. Suitable values of the operating variables for collecting the *P*1 product have been determined by maximizing the efficiency of the entrainer usage. An improved operating procedure has been proposed by feeding the entrainer earlier in order to speed up the time required for getting the high-purity *P*1 product. The rest of the operating procedure, including collecting of the *S*1, *P*2, and *P*3 products, has also been investigated.

This chapter also demonstrates that the operation of batch extractive distillation can be improved by allowing the reflux ratio and the continuous entrainer feed rate to be varied with time. The simulation results of the IPA dehydration system demonstrate that batch time and entrainer usage can be reduced by including two temperature control loops to manipulate the reflux ratio and the entrainer feed rate. All dynamic simulations used in this chapter are very rigorous, using the NRTL thermodynamic model, starting up the column from an empty state, using rigorous component and energy balances of each tray, and applying a pressure-driven simulation of the vapor flow. Thus the simulation results should be very reliable.

In the latter part of this chapter, a heteroazeotropic batch distillation system for acetic acid dehydration has been investigated. Unlike the continuous system where isobutyl acetate was found to be a good entrainer for the purpose of acetic acid dehydration, this entrainer is not suitable in batch operation. The main reason is that the column composition profile in batch operation is not fixed but time varying. Therefore, the closeness of the normal boiling-point temperatures between isobutyl acetate and acetic acid cause failure of the batch operation. On the other hand, the column composition profile is fixed in the continuous operation and is designed to avoid this two-component edge.

The performances of the batch operation of two alternative entrainers (vinyl acetate and ethyl acetate) have been compared. The acetic acid dehydration system using vinyl acetate as the entrainer results in less batch time and better recoveries of the water and the entrainer. The proposed batch operation is extremely simple and requires setting of only three operating variables (heat duty, entrainer pre-loading amounts, and setpoint value of the middle temperature control loop). The proposed batch operation is quite insensitive to these settings, and good separation is achieved over a wide range of settings.

One of the potential pitfalls for the proposed batch operation is the limitation of not being able to recover the entrainer in relatively pure form. The organic materials from the liquid–liquid split were collected as the organic product. However, this potential drawback was demonstrated to be not important because the organic materials recovered from one batch run can be reused in the next batch run with little effect on the separations.

The performance of the proposed batch operation has also been compared to a multivessel operation. Although a multivessel operation can recover vinyl acetate in purer form, no significant reduction of the batch time is gained. Because multivessel operation requires additional process, instrumentation, and control equipment, it is not recommended for the acetic acid dehydration system.

## REFERENCES


1. Yao J. Y., S. Y. Lin, and I. L. Chien, Operation and control of batch extractive distillation for the separation of mixtures with minimum-boiling azeotrope, *J. Chin. Inst. Chem. Eng.*, **38**, 371–383 (2007).
2. Gmehling J., *Azeotropic Data*, Second Edition, Wiley-VCH, Weinheim, 2004.
3. Lelkes Z., P. Lang, B. Benadda, and P. Moszkowicz, Feasibility of extractive distillation in a batch rectifier, *AIChE J.*, **44**, 810–822 (1998).
4. Lelkes Z., P. Lang, P. Moszkowicz, B. Benadda, and M. Otterbein, Batch extractive distillation: the process and the operational policies, *Chem. Eng. Sci.*, **53**, 1331–1348 (1998).
5. Steger C., V. Varga, L. Horvath, E. Rev, Z. Fonyo, M. Meyer, and Z. Lelkes, Feasibility of extractive distillation process variants in batch rectifier column, *Chem. Eng. Proc.*, **44**, 1237–1256 (2005).
6. Luyben W. L., Multicomponent batch distillation. 1: Ternary systems with slop recycle, *Ind. Engng. Chem. Res.*, **27**, 642–647 (1988).
7. Lang P., H. Yatim, P. Moszkowicz, and M. Otterbein, Batch extractive distillation under constant reflux ratio, *Comput. Chem. Engng.*, **18**, 1057–1069 (1994).

8. Milani S. M., Optimization of Solvent feed rate for maximum recovery of high purity top product in batch extractive distillation, *Chem. Eng. Res. Des.*, **77**, 469–470 (1999).
9. Gmehling J. and C. Möllmann, Synthesis of distillation processes using thermodynamic models and the Dortmund Data Bank, *Ind. Engng. Chem. Res.*, **37**, 3112–3123 (1998).
10. Laroche L., N. Bekiaris, H. W. Andersen, and M. Morari, Homogeneous azeotropic distillation: comparing entrainers, *Can. J. Chem. Eng.*, **69**, 1302–1319 (1991).
11. Skogestad S., B. Wittgens, R. Litto, and E. Sorensen, Multivessel batch distillation, *AIChE J.*, **43**, 971–978 (1997).
12. Huang, H. J. and I. L. Chien, Choice of suitable entrainer in heteroazeotropic batch distillation system for acetic acid dehydration, *J. Chin. Inst. Chem. Eng.*, **39**, 503–517 (2008).
13. Skouras S., S. Skogestad, and V. Kiva, Analysis and control of heteroazeotropic batch distillation, *AIChE J.*, **51**, 1144–1157 (2005).
14. Skouras S., V. Kiva, and S. Skogestad, Feasible separations and entrainer selection rules for heteroazeotropic batch distillation, *Chem. Eng. Sci.*, **60**, 2895–2909 (2005).
15. Costantini G., M. Serafini, and P. Paoli, Process for the recovery of the solvent and of the by-produced methylacetate in the synthesis of terephthalic acid, U.S. Patent 4,250,330 (1981).
16. Parten W. D. and A. M. Ure, Dehydration of acetic acid by azeotropic distillation in the production of an aromatic acid, U.S. Patent 5,980,696 (1999).
17. Christensen S. P. and J. D. Olson, Phase equilibria and multiple azeotrope of the acetic acid–isobutyl acetate system, *Fluid Phase Equilibria*, **79**, 187–199 (1992).
18. Burguet M. C., J. B. Montón, R. Muñoz, and J. Wisniak, Polyazeotropy in associating systems: the 2-methylpropyl ethanoate + ethanoic acid system, *J. Chem. Engng. Data*, **41**, 1191–1195 (1996).
19. Rodriguez-Donis I., E. Pardillo-Fontdevila, V. Gerbaud, and X. Joulia, Synthesis, experiments and simulation of heterogeneous batch distillation processes, *Comput. Chem. Engng.*, **25**, 799–806 (2001).
20. Skouras S. and S. Skogestad, Separation of ternary heteroazeotropic mixtures in a closed multivessel batch distillation-decanter hybrid, *Chem. Eng. Proc.*, **43**, 291–304 (2004).
21. Skouras S. and S. Skogestad, Time requirements for heteroazeotropic distillation in batch columns, *Comput. Chem. Engng.*, **28**, 1689–1700 (2004).
22. Hayden J. G. and J. P. A. O'Connell, Generalized method for predicting second virial coefficients, *Ind. Eng. Chem. Process Des. Dev.*, **14**, 209–216 (1975).

# CHAPTER 14

# HYBRID DISTILLATION–PERVAPORATION SYSTEMS

The combination of distillation and membrane separation has been explored in the literature for many years. One of the applications is to separate azeotropic mixtures. A distillation column is used to make a preliminary separation with one of the product streams being one of the pure components. The other product stream has a composition fairly close to the azeotropic composition. This stream is fed to a membrane separation unit that achieves the separation not on the basis of VLE, but based on differences in diffusion rates through a solid membrane. The fluid passing through the membrane is called the permeate. The fluid that does not pass through the membrane is called the retentate. Composition differences are the driving forces for mass transfer.

Despite the many papers dealing with pervaporation, the issue of dynamic control seems to be almost completely unexplored. That is the subject of this chapter. A hybrid column–pervaporation process is studied that is designed to produce 99.77 wt% ethanol from a feed stream of an ethanol–water mixture with composition near the azeotrope. The control objective is to maintain the purity of the ethanol product retentate stream in the face of disturbances in feed flowrate and feed composition. There are two possible manipulated variables: permeate pressure and retentate temperature. Permeate flux is increased by decreasing permeate pressure or increasing retentate temperature. A simplified dynamic pervaporation model is developed in Aspen Custom Modeler (ACM), which captures the essential features of the process using energy and component balances along with overall pervaporation performance relationships. Dynamic simulations are used to demonstrate the effectiveness of a control structure that uses a cascade composition–temperature structure. A modification to the basic process is demonstrated to give improved dynamic control.

*Design and Control of Distillation Systems for Separating Azeotropes.* By William L. Luyben and I-Lung Chien

## 14.1 INTRODUCTION

There is a vast amount of literature dealing with membranes and an extensive amount of literature dealing with the special membrane process called pervaporation. This process features a liquid feed and retentate and a vapor permeate. While the gas-phase membrane processes are essentially isothermal, the phase change in the pervaporation process produces a temperature decrease as the retentate flows through the unit. Since flux rates decrease with decreasing temperature, the conventional pervaporation unit consists of several membrane modules in series with interstage heating. The vapor permeate must be condensed for recovery and recycle, and refrigeration is usually required. Hybrid systems of distillation columns and pervaporation units are frequently used in situations where distillation alone is impossible or very expensive. An important application is the removal of water from the ethanol–water azeotrope.

Membrane processes are becoming more important in this era of escalating energy costs because they typically require much less energy than alternative separation methods such as distillation. Gas permeation using membranes is widely used for separating mixtures such as oxygen–nitrogen, hydrogen–carbon dioxide, methane–nitrogen, and so on. These gaseous membrane systems operate essentially isothermally because there are no phase changes.

There are several types of liquid membrane systems. One of the most important is pervaporation in which the feed and retentate streams are liquid on the high-pressure side of the membrane, but the permeate stream is removed as a vapor on the low-pressure side of the membrane. The phase change produces a drop in temperature in adiabatic operation owing to the heat of vaporization. Since flux rates through the membrane decrease with decreasing temperature, the conventional pervaporation unit consists of several membrane stages in series with interstage heating. The vapor permeate must be condensed, which often requires expensive refrigeration.

Pervaporation has been the subject of many papers for several decades. The literature is quite extensive with many papers and several books on the subject. The 1991 book edited by Huang[1] gives 262 citations in the first chapter. The book by Rautenbach and Albrect,[2] written in 1989, has a chapter on pervaporation. There are many papers that discuss the fundamentals of diffusion, adsorption, evaporation, and membrane characteristics. Some papers develop detailed mass-transfer models. A number of papers discuss design considerations and present comparative economic studies of pervaporation versus alternatives such as distillation.

Industrial applications began in the 1970s. This was made possible by the development of highly selective polyvinyl alcohol composite membranes. The process is clearly discussed by Sander and Soukup.[3] They provide engineering performance curves for removing water from an ethanol–water mixture. The curves show how the temperature and ethanol concentration of the feed affect the flux of water through the membrane and the composition of the permeate stream. These performance curves are used in this chapter to provide a simple pervaporation model that captures the behavior of the process with sufficient fidelity for use in control studies.

Many papers point out the economic advantages of pervaporation over conventional processes. A quantitative economic comparison between azeotropic distillation and hybrid column–pervaporation is presented by Guerreri,[4] whose results show higher capital cost but much lower energy cost for the ethanol–water separation. A review of industrial applications of pervaporation, coupled with either distillation columns or chemical reactors, is given by Lipnizki et al.[5] They show a variety of flowsheets, list 22 chemical systems, and provide 72 references. Another review has been given by Feng and Huang[6] who cite over 100 references. Daviou et al.[7] study the dehydration of isopropanol using a hybrid column–pervaporator process. They develop a process design using the HYSYS commercial simulator. The model of the pervaporator was developed in gProms and integrated into HYSYS using a user extension module. They found the optimum economic design using two design optimization parameters: number of modules and amount of interstage heating. Van Hoof et al.[8] presented an economic comparison between azeotropic distillation and different hybrid systems combining distillation with pervaporation for the dehydration of isopropanol.

No papers have been found that specifically deal with the issue of dynamic control in any quantitative way. Sander and Soukup[3] make the statement that "the product concentration is simply controlled by adjusting the alcohol feed rate." In most chemical plants, throughput is set by product demand and therefore cannot be used as a manipulated variable in a process control structure.

The purpose of this chapter is to study the dynamic control of a pervaporator system coupled with a distillation column. The example system is the important ethanol–water separation. The commercial simulation tools of Aspen Technology are used in this study.

## 14.2 PERVAPORATION MODEL

The study of the dynamic control of any process requires a dynamic model that is capable of capturing the important dynamic features of the process. The macroscopic effects of changing manipulated variables on the variables to be controlled must be capture with sufficient fidelity to give realistic dynamic and steady-state responses. The model used for control does not need to be as detailed as one used for gaining fundamental understanding of the microscopic details of mass and energy transfer and phase equilibria in these complex, multi-phase systems.

### 14.2.1 Diffusivities

In this spirit, an approximate dynamic model was developed that is based on pervaporation performance characteristic curves. Sander and Soukup[3] give curves for the ethanol–water system showing how the flux of water through a pervaporation membrane and the composition of the permeate vapor vary with the concentration of ethanol in the liquid feed. Figures 14.1*a* and *b* reproduce these curves. Note that these curves show concentrations in wt% and mass flowrates.

The model used in this work assumes that the diffusion through the membrane depends on the concentration difference between the retentate and permeate sides of the membrane for each component.

$$\mathrm{Flux}_j = D_j(C_{Rj} - C_{Pj}) \tag{14.1}$$

where

$\mathrm{Flux}_j$ = molar flux of component $j$ (kmol h$^{-1}$ m$^{-2}$)
$D_j$ = diffusivity of component $j$ (m/h)
$C_{Rj}$ = concentration of component $j$ in retentate liquid (kmol/m$^3$)
$C_{Pj}$ = concentration of component $j$ in permeate vapor (kmol/m$^3$)

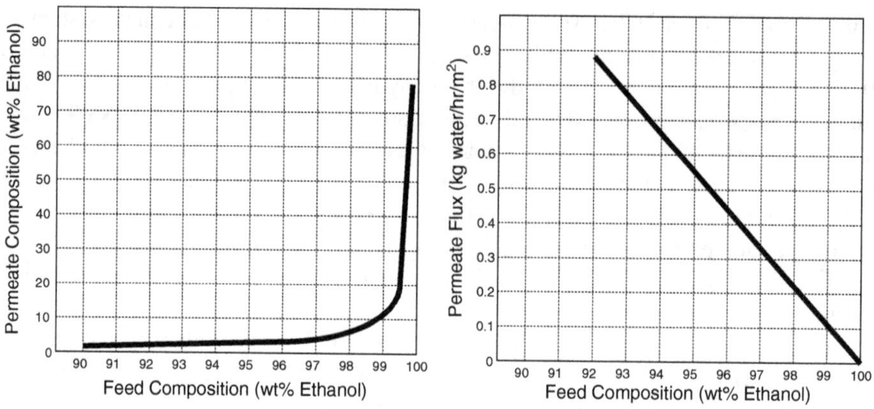


**Figure 14.1** (*a*) Permeate composition. (*b*) Permeate water flux at 373 K.

The density of the incompressible liquid permeate (731 kg/m$^3$) is assumed to be independent of temperature and composition over the small ranges present in the pervaporation unit. Therefore, the liquid molar concentrations can be calculated from the weight fractions $zw_{Rj}$ of the retentate liquid.

$$C_{Rj} = \frac{731\ \text{kg/m}^3}{MW_j}(zw_{Rj}) \tag{14.2}$$

where

$MW_j$ = molecular weight of component $j$ (kg/kmol)
$zw_{Rj}$ = weight fraction of component $j$ in the liquid retentate

The molar concentrations in the permeate vapor can be calculated assuming ideal gases.

$$C_{Pj} = \frac{(P_{\text{perm}})(z_{Pj})}{(0.08206)(T_P)} \tag{14.3}$$

where

$P_{\text{perm}}$ = pressure on the permeate side of membrane (atm)
$z_{Pj}$ = mole fraction of component $j$ in the permeate vapor
$T_P$ = temperature of the permeate (K)

The permeate pressure for the data given in Figure 14.1 is assumed to be 0.015 atm. The temperatures of the retentate and permeate are assumed to be the same and decrease as mass transfer occurs in the cell because of the phase change.

The data given in Figures 14.1*a* and *b* are used to calculate the diffusivities of both ethanol and water. A point on the abscissa (wt% ethanol in the feed, which is assumed to be the retentate in each cell) is selected. The corresponding composition of the permeate vapor is read from Figure 14.1*a* and the corresponding flux of water is read from Figure 14.1*b*. The water compositions of both the retentate and the permeate are converted to mole fractions, so $z_{P,\text{water}}$ and $z_{R,\text{water}}$ are known. Since the system is binary, the mole fractions of ethanol in both the retentate and permeate can be calculated ($z_{P,\text{ethanol}}$ and $z_{R,\text{ethanol}}$). The liquid retentate concentrations $C_{R,\text{water}}$ and $C_{R,\text{ethanol}}$ can then be calculated using the liquid density of 731 kg/m$^3$ and Eq. (14.2).

The flux of water is converted to molar units, so $\text{Flux}_{\text{water}}$ is known. The molar flux of ethanol is calculated by multiplying the molar water flux by the ratio of the molar composition of ethanol to water in the permeate.

$$\text{Flux}_{\text{ethanol}} = \text{Flux}_{\text{water}}\left(\frac{z_{P,\text{ethanol}}}{z_{P,\text{water}}}\right) = \text{Flux}_{\text{water}}\left(\frac{1 - z_{P,\text{water}}}{z_{P,\text{water}}}\right) \tag{14.4}$$

The concentrations of both components in the vapor permeate are calculated using Eq. (14.3). Finally, the diffusivities of both components are calculated using Eq. (14.1) since all fluxes and concentrations are known.

Diffusivities were calculated over a range of feed compositions given in Figures 14.1*a* and 14.1*b*. The calculated diffusivities are almost constant over the range from 92 to 99.5 wt% ethanol. Average values of $D_{\text{water}} = 0.0137$ m/h and $D_{\text{ethanol}} = 2.4 \times 10^{-5}$ m/h are used. These values are at 373 K. The temperature dependence of the diffusivities on temperature is reported[3] to be of the Arrhenius type, with a doubling for every 10 K. Therefore, the temperature dependent diffusivities used in the model are given in Eq. (14.5).

$$\begin{aligned} D_{\text{water}} &= 1.158 \times 10^{9} e^{-9385/T} \\ D_{\text{ethanol}} &= 2.028 \times 10^{6} e^{-9385/T} \end{aligned} \tag{14.5}$$

### 14.2.2 Dynamic Component and Energy Balances

A lumped model of the pervaporation unit is used to describe the dynamic behavior. Three cells are assumed in each pervaporation module. Since the liquid retentate provides most of the capacitance of the pervaporation unit, the ordinary differential equations describing the dynamic changes compositions and temperature of the retentate in each cell are used. The molar holdup in each cell $\text{M}_\text{R}$ is assumed constant, so the total molar balance is algebraic.

$$\frac{\text{dM}_\text{R}}{\text{d}t} = 0 = F_{R,n-1} - F_{R,n} - F_{P,n} \tag{14.6}$$

$$\text{M}_\text{R}\frac{\text{d}h_{R,n}}{\text{d}t} = F_{R,n-1}h_{R,n-1} - F_{R,n}h_{R,n} - F_{P,n}H_{P,n} \tag{14.7}$$

$$\text{M}_\text{R}\frac{\text{d}z_{R,n,j}}{\text{d}t} = F_{R,n-1}z_{R,n-1,j} - F_{R,n}z_{R,n,j} - F_{P,n}z_{P,n,j} \tag{14.8}$$

where

$F_{R,n}$ = molar flowrate of liquid retentate from cell $n$ (kmol/h)
$F_{P,n}$ = molar flowrate of vapor permeate from cell $n$ (kmol/h)
$h_{R,n}$ = molar enthalpy of liquid retentate in cell $n$ (GJ/kmol)
$H_{P,n}$ = molar enthalpy of vapor permeate leaving cell $n$ (GJ/kmol)

TABLE 14.1 ACM Pervaporator Model.

```
Model pervap
// Pervaporation Module with composition driving forces
// Nonisothermal; ethanol and water
// parameters and variables
  NCells                              as        IntegerParameter (Description:"Number of cross flow cells");
  A                                   as Area            (Description:"Total membrane area");  // (m2)
  ACell                               as Area            (Description:"Area per cross flow cell (m2)");// (m2)
  CRet(ComponentList,[1:NCells])  as  Notype         (Description: "retentate composition of each component (kmol/m3)");
  CPerm(ComponentList,[1:NCells])  as  Notype         (Description: "permeate composition of each component (kmol/m3)");
  cvvalve                   as Notype     (Description: "flow resistance");
  Flux(ComponentList,[1:NCells])  as Flow_Mol        (Description:"Molar flux from cell");  // (kmol/h)
  FRet([0:NCells])                as Flow_Mol        (Description:"Retentate mole flow from cell"); // (kmol/h)
  FPerm([1:NCells])               as Flow_Mol        (Description:"Permeate mole flow from cell");  // (kmol/h)
  Diffusion(ComponentList,[1:NCells])      as Notype       (Description:"diffusion coefficients (m3/hr)");
  MRet           as Notype     (Description: "molar holdup in each retentate cell");  //(kmol)
  Pin                              as Pressure         (Description:"Pressure feed"); // (bar)
  Pout                              as Pressure         (Description:"Pressure retentate discharge"); // (bar)
  PPerm                               as Pressure        (Description:"Permeate pressure"); // (bar)
  RetHoldup(ComponentList,[1:NCells])  as Notype
         (Description:"Component molar holdup of retentate cell");  // (kmol)
  RetTotalHoldup([1:NCells])      as Notype            (Description:"Total molar holdup in cell");  // (kmol)
  TRet([0:NCells])      as Temperature      (Description:"Temperature of retentate in cell");  // (C)
  TPerm     ([1:NCells])  as Temperature (Description:"Temperature of permeate from each cell");  // (C)
  RetVol        as Volume           (Description:"Total Retentate Volume  Membrane"); // (m3)
  RhoRet                              as hidden Dens_Mol;  // Retentate molar density (kmol/m3)
  RhoPerm                             as hidden Dens_Mol;  // Permeate molar density (kmol/m3)
  VCell          as Volume     (Description:"Volume in each cell, both retentate and permeate"); // (m3)
  ZPerm(ComponentList,[1:NCells]) as Molefraction
        (Description:"Permeate mole fraction from each cell");
  ZRet(ComponentList,[0:NCells])  as Molefraction     (Description:"Retentate mole fraction in cell");
  ZRetwt(ComponentList,[0:NCells])  as Notype     (Description:"Retentate wt. fraction in cell");
//Additional variables for Aspen properties
  vaporenthalpy_P_ASPEN([1:NCells])     as Enth_Mol_Vap
       (Description: "Aspen liquid enthalpy of Permeate from each cell"); // (GJ/kmol)
  liquidenthalpy_R_ASPEN([0:NCells])     as Enth_Mol_Liq
       (Description: "Aspen liquid enthalpy of Retentate in each cell"); // (GJ/kmol)
  MPerm           as Notype     (Description: "molar holdup in each permeate cell");  //(kmol)
  // ports
```

```
  Inlet     as Input  MaterialPort;
  Retentate as Output  MaterialPort;
  Permeate  as Output  MaterialPort;
// equations
// Retentate inlet conditions
  TRet(0)=Inlet.T;
  Pin=Inlet.P;
For comp in ComponentList Do
    ZRet(comp,0) = Inlet.z(comp);  EndFor
// Pressure driven flow of retentate through cell
Pout=Retentate.P;
FRet(0)=cvvalve*(Pin-Pout);
PPerm=Permeate.P;
 Inlet.F=FRet(0);
//*************************************************************************************
// Calculate temperature and feed composition dependent Diffusivities
     For k in [1:NCells] Do
              Tret(k):initial;
          Diffusion("Ethanol",k)=2.028e6*exp(-9385/(273+TRet(k)));
          Diffusion("Water",k)=1.158e9*exp(-9385/(273+TRet(k)));     EndFor
// Calculate wt. fractions of retentate
   For k in [1:NCells] Do
        ZRetwt("Ethanol",k)=ZRet("Ethanol",k)*46.1 /(ZRet("Ethanol",k)*46.1 + ZRet("Water",k)*18);
     ZRetwt("Water",k)=ZRet("Water",k)*18 /(ZRet("Ethanol",k)*46.1 + Ret("Water",k)*18);EndFor;
// Calculate retentate concentrations (liquid density = 800 kg/m3)
     For k in [1:NCells] Do
          CRet("Ethanol",k)=731*ZRetwt("Ethanol",k)/46.1;
          CRet("Water",k)=731*ZRetwt("Water",k)/18;EndFor;
// Calculate permeate concentrations in vapor phase at TPerm and PPerm and ZPerm
     For k in [1:NCells] Do
          CPerm("Ethanol",k)=PPerm*ZPerm("Ethanol",k)/(TPerm(k)+273)/0.08206;
          CPerm("Water",k)=PPerm*ZPerm("Water",k)/(TPerm(k)+273)/0.08206;EndFor;
// Molar Flux of water and ethanol; vapor permeate flow from each cell (FPerm)
     ACell = A/NCells;
       For k in [1:NCells] Do
              For comp in ComponentList Do
                Flux(comp,k)  = ACell * Diffusion(comp,k) * (CRet(comp,k) - CPerm(comp,k));EndFor
          FPerm(k)=sigma(for each (comp in componentlist) Flux(comp,k));     EndFor
```

*(Continued)*

**TABLE 14.1** *Continued*

```
// Liquid retentate flowrates
     For k in [1:NCells] Do
     FRet(k)=FRet(k-1) - FPerm(k);EndFor
// *******************************************
// Dynamic component balance equation for retentate compositions in each cell
    Vcell=RetVol/NCells;
    MRet=Vcell*731/46.1;
    For k in [1:NCells] Do
         For comp in ComponentList Do
           $ZRet(comp,k)=(FRet(k-1)*ZRet(comp,k-1)-Flux(comp,k)- FRet(k)*ZRet(comp,k))/(MRet); EndFor
    EndFor
  // Calculate Liquid retentate enthalpy and Vapor permeate Enthalpies in each cell and liquid feed
     For k in [1:NCells] Do
       Call(liquidenthalpy_R_ASPEN(k)) = pEnth_mol_liq(TRet(k),Retentate.P,ZRet(ComponentList,k)) ComponentList;
Call(vaporenthalpy_P_ASPEN(k)) = pEnth_mol_vap(TPerm(k),Permeate.P,ZPerm(ComponentList,k)) ComponentList;
        EndFor
        Call(liquidenthalpy_R_ASPEN(0)) = pEnth_mol_liq(TRet(0),Retentate.P,ZRet(ComponentList,0)) ComponentList;
 //*******************************************
/// Dynamic Energy balance for Retentate in each cell
     For k in [1:NCells] Do
      $liquidenthalpy_R_ASPEN(k)=(FRet(k-1)*liquidenthalpy_R_ASPEN(k-1) - FRet(k)*liquidenthalpy_R_ASPEN(k) -
FPerm(k)*vaporenthalpy_P_ASPEN(k))/(MRet); EndFor
// ************************************
// Retentate total flow and composition from last cell in module
       Retentate.F = FRet(NCells);
     Retentate.T = TRet(NCells);
     For comp in ComponentList Do
         Retentate.z(comp) = ZRet(comp,NCells);  EndFor
// Permeate total flow and composition
       Permeate.F = Sigma(Foreach (cell in [1:NCells]) FPerm(cell));
```

```
 For comp in ComponentList Do
    Permeate.z(comp) = Sigma(Foreach (cell in [1:NCells]) FPerm(cell)*ZPerm(comp,cell))  /Permeate.F;  EndFor
    For k in [1:NCells] Do
    For comp in ComponentList Do
    zPerm(comp,k)=Flux(comp,k)/Fperm(k);EndFor
    EndFor
//***************************************
// Thermal equilibirum constraint in each cell
    For k in [1:NCells] Do
    TPerm(k)=TRet(k);     EndFor
// Defining new variables
       CpPerm([1:NCells]) as cp_mol_vap;
       Permeate_Cp as cp_mol_vap;
// Calculating temperature of total permeate from heat capacities of permeate from each cell
       For k in [1:NCells] Do
       Call(CpPerm(k)) = pCp_Mol_Vap(TPerm(k),Permeate.P,ZPerm(ComponentList,k)) ComponentList; EndFor
       Call(Permeate_Cp) = pCp_Mol_Vap(Permeate.T,Permeate.P,Permeate.z);
       Permeate.T * Permeate_Cp * Sigma(FPerm([1:NCells])) = Sigma (Foreach (cell in [1:NCells])
FPerm(cell)*CpPerm(cell)*TPerm(cell));
       Call(Permeate.h)=pEnth_Mol_Vap(Permeate.T, Permeate.P, Permeate.z);
 // Call (<output argument list>)=<procedure name>(<input argument list>);
 Call(Retentate.h)=pEnth_Mol_Liq(Retentate.T, Retentate.P, Retentate.z);
  Call(RhoRet)=pDens_Mol_Liq(Retentate.T, Retentate.P, Retentate.z);
 Call(RhoPerm)=pDens_Mol_Vap(Permeate.T,  Permeate.P,  Permeate.z);
 Retentate.v=1/RhoRet;
 Permeate.v=1/RhoPerm;
End
```

The flowrate of permeate is the sum of the two component fluxes times the membrane area ($A_{\text{mem}}$), and the composition of the permeate is determined by these two fluxes.

$$
\begin{aligned}
F_{P,n} &= A_{\text{mem}}(\text{Flux}_{n,\text{ethanol}} + \text{Flux}_{n,\text{water}}) \\
z_{P,n,j} &= A_{\text{mem}}\text{Flux}_{n,j}/F_{P,n}
\end{aligned}
\tag{14.9}
$$

The flux of each component in each cell is calculated using Eq. (14.1) and Eq. (14.5) since all compositions and temperatures are known. The temperature of the retentate is calculated from the known liquid enthalpy $h_{R,n}$ and the known retentate composition $z_{R,n}$ using physical property relationships. The UNIQUAC relationships are used in both the pervaporation unit and the distillation column.

The differential equations [Eq. (14.7) and Eq. (14.8)] for each cell and each module are incorporated in an Aspen Custom Modeler program, which is given in Table 14.1. The invaluable help of Jay Yun of Aspentech is gratefully acknowledged in overcoming the significant complexities of translating the model into a workable Aspen Custom Modeling file.

## 14.3 PERVAPORATION-COLUMN SYSTEM

The system studied is based on a process discussed by Ryan and Doherty[9] that produces a 85 mol% ethanol distillate from a 22-stage distillation column with a feed flowrate of 700 kmol/h, feed composition of 7 mol% ethanol, and 93 mol% water. The bottoms product from the column has a composition of 99.9 mol% water. Figure 14.2 gives stream conditions,

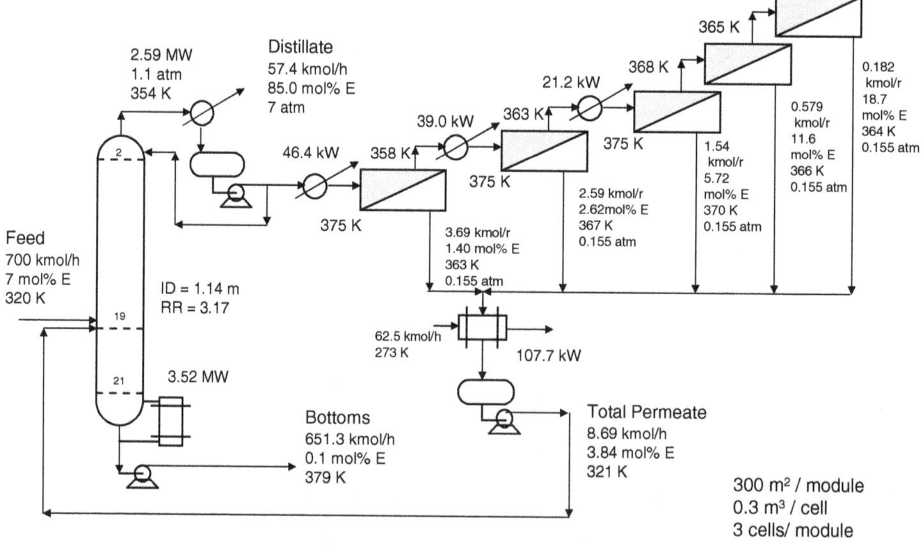


**Figure 14.2** Column–pervaporation flowsheet.

heat duties, and equipment sizes. The column operates at 1.1 atm, giving a reflux drum temperature of 354 K. The column feed is introduced at Stage 19, which is the location that minimizes reboiler heat input.

The distillate is pumped up to a pressure of 7 atm to ensure that the retentate remains liquid. The distillate is heated to 375 K and fed to the first of five pervaporation modules in series. Each module has three cells. The retentate decreases in temperature from 375 to 358 K as it flows through the first adiabatic module because of the phase change. The temperature of the permeate from each cell is assumed to be equal to the temperature of the retentate in that cell.

The retentate streams leaving the first and second modules are heated back up to 375 K. The drop in retentate temperature in the last three modules is quite small (7 K in module 3 and 3 K in module 4) because the flux rates are small owing to the decrease in retentate water composition. Therefore, it seems logical to save the capital investment in additional heat exchanger area and use no heating on these retentate streams. We will return to this question of how many modules should use reheating later in this chapter and show that it may be attractive to modify the design of the process to improve dynamic controllability by using reheating on all modules.

The permeate pressure is 0.155 atm, and the vapor permeate streams from all of the modules flow to a single refrigerated condenser (with area 2.74 $m^2$ and overall heat transfer coefficient of 730 kcal $h^{-1}$ $m^{-2}$ $K^{-1}$), which uses 62.5 kmol/h of 273 K refrigerant to condense the permeate vapor. Condenser heat duty is 108 kW. The total permeate flowrate is 8.69 kmol/h with a composition of 3.84 mol% ethanol, and this stream is fed back into the distillation column along with the fresh feed at Stage 19 (numbering from the top with the reflux drum as Stage 1).

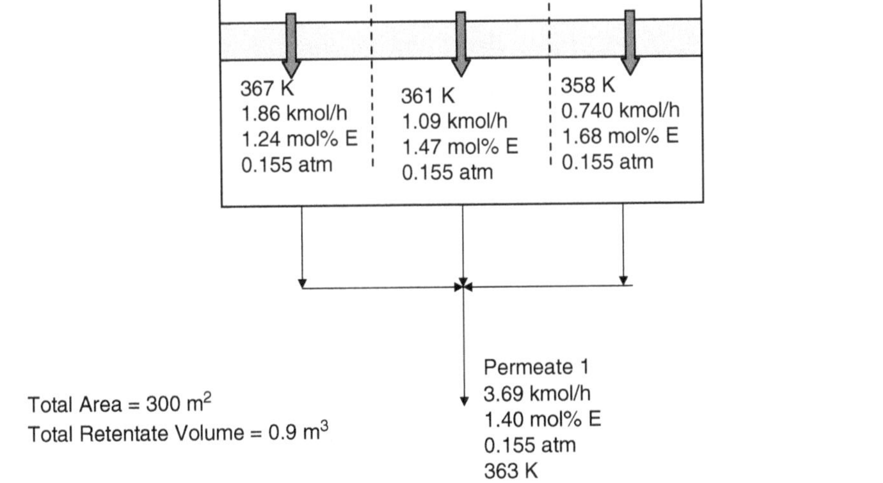


**Figure 14.3** Module 1.

Figure 14.3 illustrates what happens inside the first three-cell module. Temperatures decrease from cell to cell. Permeate ethanol composition increases and permeate flowrate decreases from cell to cell. The fluxes from each of the three cells (with different temperatures and compositions) are mixed to give the total vapor permeate from the module. Each module has a total area of 300 m$^2$ or 100 m$^2$ per cell.

## 14.4 DYNAMICS AND CONTROL

The control objective is to maintain the purity of the retentate ethanol product stream leaving the last stage of the pervaporation unit. The desired purity is 99.42 mol% ethanol (99.77 wt%). Process control wisdom says that it is more effective to control the impurity level than the purity level in a high-purity stream. Therefore, we choose to control the water impurity at 0.58 mol% water. A 5 min deadtime is assumed for the composition measurement.

An important dynamic parameter is the holdup in the pervaporation unit. Since the mass of the liquid retentate phase is much larger than that of the vapor permeate phase, it will dominate the dynamic response. The holdup depends on the volume. In this study, the data given in Geankoplis[10] is used (328 m$^2$/m$^3$). Thus the 300 m$^2$ module has a volume of 0.9 m$^3$ that is split among the three cells in the module.

There are two potential variables that can be manipulated to control the product composition: retentate temperature or permeate pressure. To see what the dynamic response of product impurity is to changes in these two input variables, openloop step changes were made. Results are shown in Figure 14.4. The solid line is for a step change in the setpoint

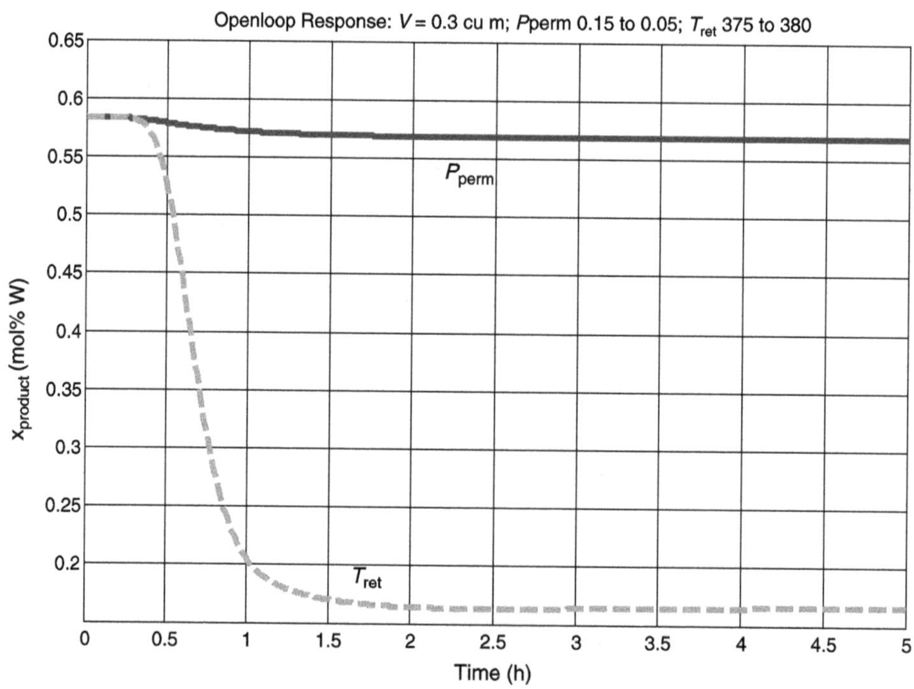


**Figure 14.4** Openloop responses.

(a)

(b)

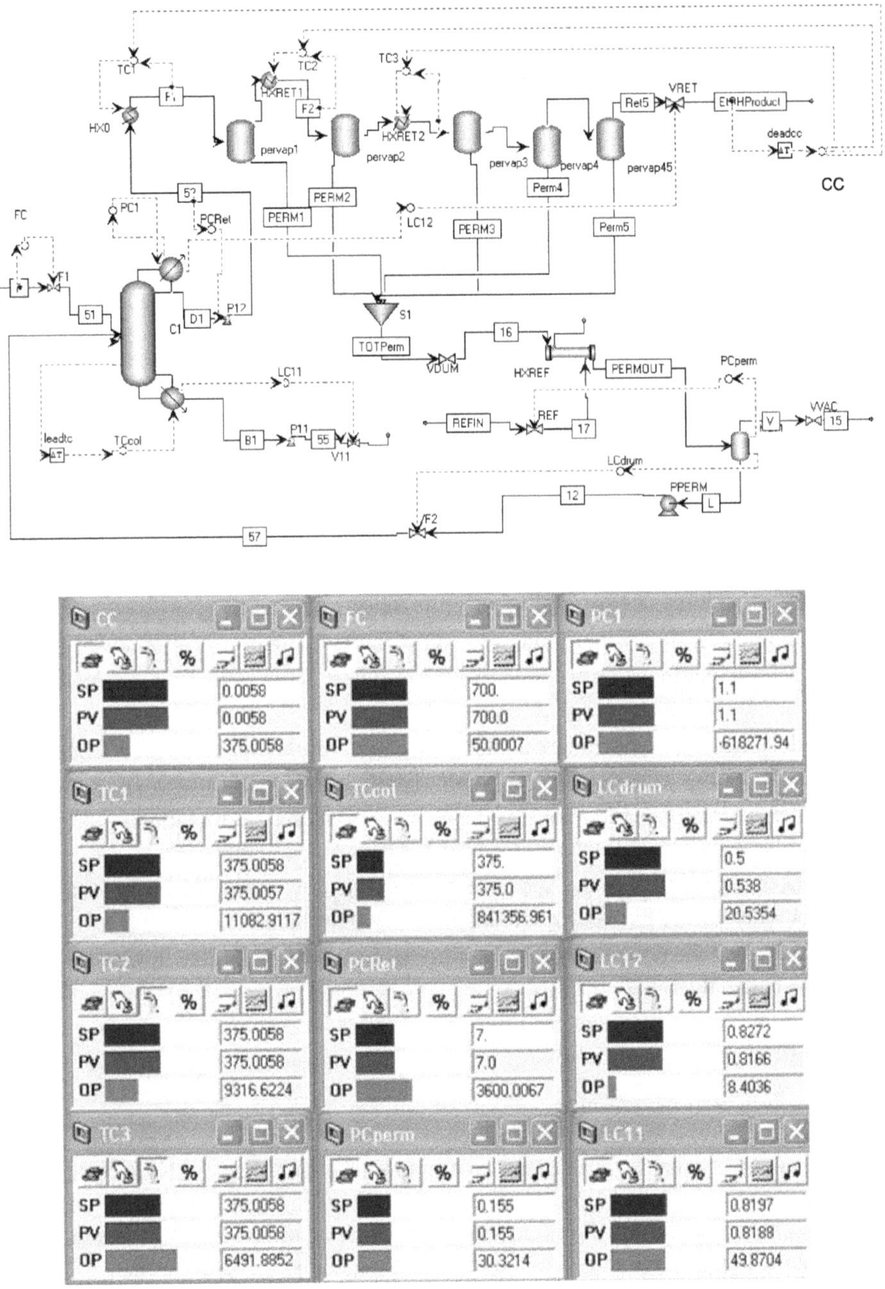


**Figure 14.5** (*a*) Aspen Dynamics PFD; control structure using $T_{ret}$. (*b*) Controller faceplates using $T_{ret}$.

of the permeate pressure controller from 0.15 to 0.05 atm made at time = 0.2 h. The response of product impurity $x_{product,water}$ is slow, but more importantly there is very little change. Thus the process gain between product impurity and permeate pressure is very small, so using this manipulated variable would only be able to handle very small disturbances.

The dash line in Figure 14.4 is for a step change in the setpoints of the three temperature controllers that manipulate heat inputs to the three heaters on the retentate feed streams to the first, second, and third pervaporation modules. At time = 0.2 h, these three setpoints are raise from 375 to 380 K. There is a very significant change in product impurity. Therefore, it appears that a composition/temperature cascade control structure should provide effective control.

The Aspen Dynamics process flow diagram is shown in Figure 14.5*a*. The control structure is summarized below. All controllers have conventional PI action except level controllers, which are proportional only.

1. Feed to the column is flow controlled.
2. Reflux is flow controlled.
3. Base level is controlled by manipulating bottoms flowrate.
4. Reflux drum level is controlled by the valve on the retentate stream leaving the last module.
5. Retentate pressure is controlled by manipulating the speed of the distillate pump.
6. The temperature at Stage 19 is controlled by manipulating reboiler heat input. A 1 min deadtime is inserted in this temperature loop.
7. Column pressure is controlled by manipulating condenser heat removal.
8. Inlet feed temperature to the pervaporation unit is controlled by manipulating heat input to the heater HX0 (TC1 controller).
9. Inlet retentate temperature to the second pervaporation module is controlled by manipulating heat input to the heater HXRET1 (TC2 controller).
10. Inlet retentate temperature to the third pervaporation module is controlled by manipulating heat input to the heater HXRET2 (TC3 controller).
11. The composition of the retentate product stream is controlled by changing the setpoints of the three retentate temperature controllers (CC controller).
12. Pressure in the permeate drum is controlled by manipulating the flow of refrigerant to the permeate condenser.
13. Liquid level in the permeate drum is controlled by manipulating the flow of recycle permeate back to the column.

The product composition, the permeate pressure, and the column temperature loops are tuned by running relay-feedback tests and applying either Ziegler–Nichols or Tyreus–Luyben tuning rules. Table 14.2 gives controller parameters. Figure 14.5*b* shows the controller faceplates. Notice that the three retentate temperature controllers are on cascade, receiving their setpoint signals from the output signal from the composition controller.

Figure 14.6 shows the responses of the system for a 20% increase in feed flowrate to the column. The upper two graphs show what happens when product composition is controlled by permeate pressure. The refrigerant control valve goes wide open and product impurity cannot be controlled at the desired 0.58 mol% water. This poor result is due to the very small process gain between product impurity and permeate pressure.

The lower graphs in Figure 14.6 give results when retentate temperature is used. Product impurity is well controlled with the steady-state retentate temperature increased from 375 to 378 K. It should be remembered that the performance of this loop could be improved if the

**TABLE 14.2 Controller Tuning Parameters.**

| Manipulated Variable | Retentate Temperature | Permeate Pressure | Refrigerant Flowrate | Column Reboiler Heat Input |
|---|---|---|---|---|
| Controlled variable | Retentate impurity | Retentate impurity | Permeate pressure (atm) | Column Stage 19 temperature (K) |
| SP | 0.0058 (mf water) | 0.0058 (mf water) | 0.15 atm | 375 K |
| Transmitter span (mf water) | 0–0.01 | 0–0.01 | 0–0.5 | 350–450 K |
| OP range | 370–390 K | 0–0.5 atm | 0–100% | 0–6.63 × $10^6$ cal/s |
| Ultimate gain | 0.76 | 31 | 35 | 1.3 |
| Ultimate period (min) | 71 | 28 | 3 | 13 |
| $K_C$ | 0.34 (ZN) | 23 (ZN) | 11 (TL) | 0.4 (TL) |
| $\tau_I$ (min) | 60 (ZN) | 44 (ZN) | 6.6 (TL) | 6.6 (TL) |
| Controller action | Direct | Reverse | Direct | Reverse |

composition deadtime can be reduced or if a feedforward structure were used to increase retentate temperature if feed flowrate increases.

Figure 14.7 gives the responses using the retentate temperature as the manipulated variable for changes in either feed flowrate or feed composition to the distillation column. The feed flowrate changes are 20% increases (solid lines) and 20% decreases (dashed lines) at 0.2 h. The feed composition disturbances are changes from 7 to 8 mol% ethanol (solid

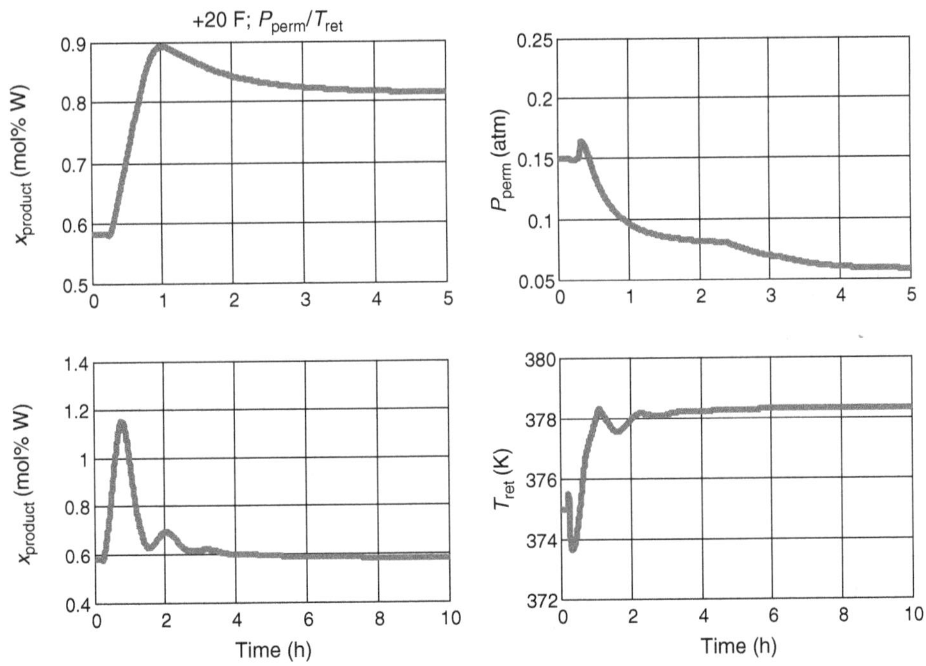


**Figure 14.6** 20% Increase in feed flowrate using $P_{perm}$ or $T_{ret}$.

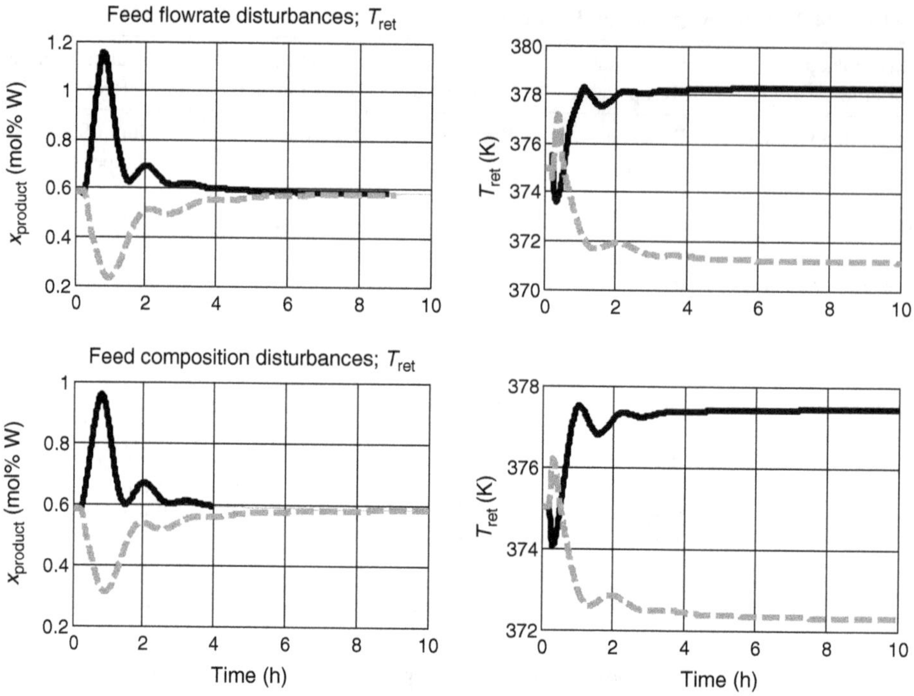


**Figure 14.7** Feed disturbances using $T_{ret}$.

lines) and changes from 7 to 6 mol% ethanol in the column feed at 0.2 h. Good regulatory performance is demonstrated; however, the closed-loop dynamics are fairly slow, and it takes about 3 h to return to the desired purity. In the following section, we discuss a modification of the process to improve dynamics.

## 14.5 PROCESS MODIFICATION

The results shown in Figure 14.7 reveal fairly large closed-loop time constants. This occurs because of the large holdup of retentate in the five modules relative to the small flowrate of retentate. The total retentate holdup in the five modules is 4.5 $m^3$. The flowrate of distillate to the pervaporation unit is 57.4 kmol/h, which corresponds to 54 L/min. Thus the residence time of retentate is about 80 min.

In addition, the corrective action is being taken in the front end of the pervaporation unit, using the heat input to the three heat exchangers on the inlet streams to the first three modules. Thus there is a significant lag between the variable to be controlled (product impurity) and the variable to be manipulated (heat input). This suggests that it may be possible to improve control by modifying the process. Installing heat exchangers on the feed streams to the last two modules would reduce the lag in the loop.

Figure 14.8 shows the resulting revised flowsheet. The composition of the product stream (retentate leaving module 5) is kept the same as in the previous design. The retentate feed temperatures to the five modules are reduced from 375 K in the previous design with three heat exchangers to 371 K with five heat exchangers. The required temperature is

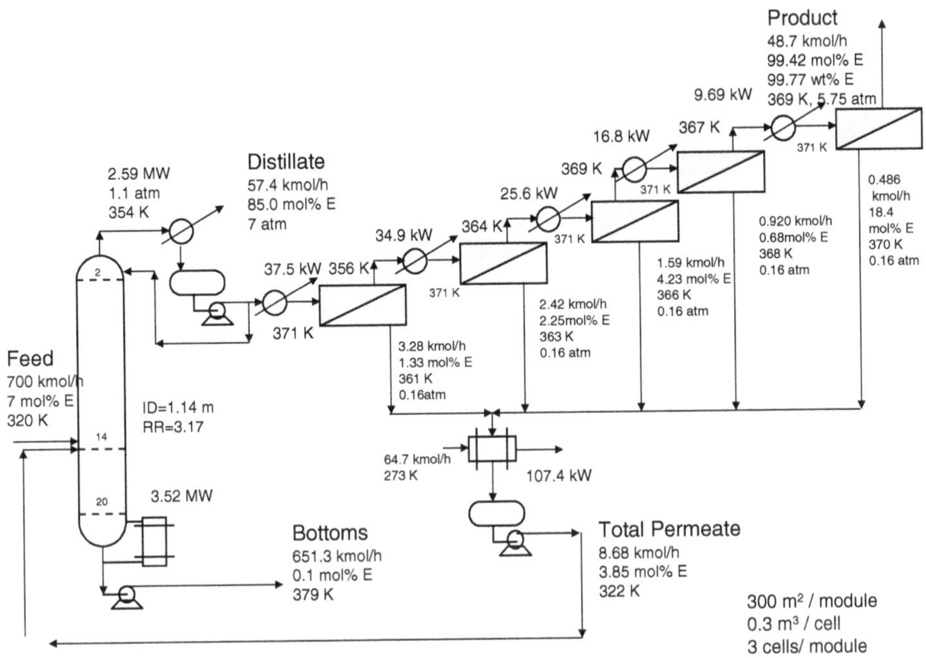


**Figure 14.8** Flowsheet with five heat exchangers.

lower because the fluxes in the last two modules have been increased because of their higher inlet temperatures. In the three heat exchanger designs, these temperatures are 368 and 365 K. In the five heat exchanger designs, they are 371 K. The heat duties in the two new heat exchangers are smaller than those in the first three.

From a control point of view, the best control would be to use only the heat input to the last module. However, it is quite small (9.69 kW) compared to the heat inputs in the upstream heaters (37.5, 34.9, 25.6, and 16.8 kW) and would not be able to handle large disturbances. A decrease in throughput would drive the heat input in this last heat exchanger to zero, but the composition setpoint would not be maintained because the temperature would be too high (control would be lost). In order to handle a 20% decrease in feed and not encounter a zero heat input constraint, the heat inputs in the last three heaters must be adjusted. Thus, with the modified process, the output signal from the composition controller changes the setpoints of the feed temperature controllers to the last three modules. The temperature controllers on the feeds to the first two modules have a fixed setpoint of 371 K.

With this new control structure and new process, the composition controller was retuned. Table 14.3 compares controller tuning parameters in the three heat-exchanger process with those in the five heat-exchanger process. Note that the ultimate period is smaller in the new design (49 versus 71 min). Figure 14.9 compares the responses of the two designs for a 20% increase in feed flowrate. The solid lines are for the original three heat-exchanger process. The dashed lines are for the modified five heat-exchanger process. The control performance is improved in terms of both settling time and maximum transient deviation. Notice that the initial steady-state temperature is lower in the five heat-exchanger process.

**TABLE 14.3 Controller Tuning Parameters for Three and Five Heat-Exchanger Processes.**

| Manipulated Variable | Retentate Temperature into Modules 1, 2, and 3 | Retentate Temperature into Modules 3, 4, and 5 |
|---|---|---|
| Controlled variable | Retentate impurity | Retentate impurity |
| SP | 0.0058 (mf water) | 0.0058 (mf water) |
| Transmitter span (mf water) | 0–0.01 | 0–0.01 |
| OP range | 370–390 K | 370–390 K |
| Ultimate gain | 0.76 | 0.77 |
| Ultimate period (min) | 71 | 49 |
| $K_C$ | 0.34 (ZN) | 0.35 (ZN) |
| $\tau_I$ (min) | 60 (ZN) | 40 (ZN) |
| Controller action | Direct | Direct |

Figures 14.10 and 14.11 show how a number of variables change for disturbances in feed flowrate and feed composition. The output signal from the composition controller (OPCC) changes the setpoints of the temperature controllers on the last three heaters (TC3, TC4, and TC5) and the corresponding heat duties change (Q2, Q3, and Q4). The setpoints of the temperature controllers on the first two heaters (TC1 and TC2) are constant at 371 K, but their heat duties change as the load changes. Figure 14.10 shows that none of the last three heat exchangers saturate at zero for the 20% decrease in feed. If only the last two

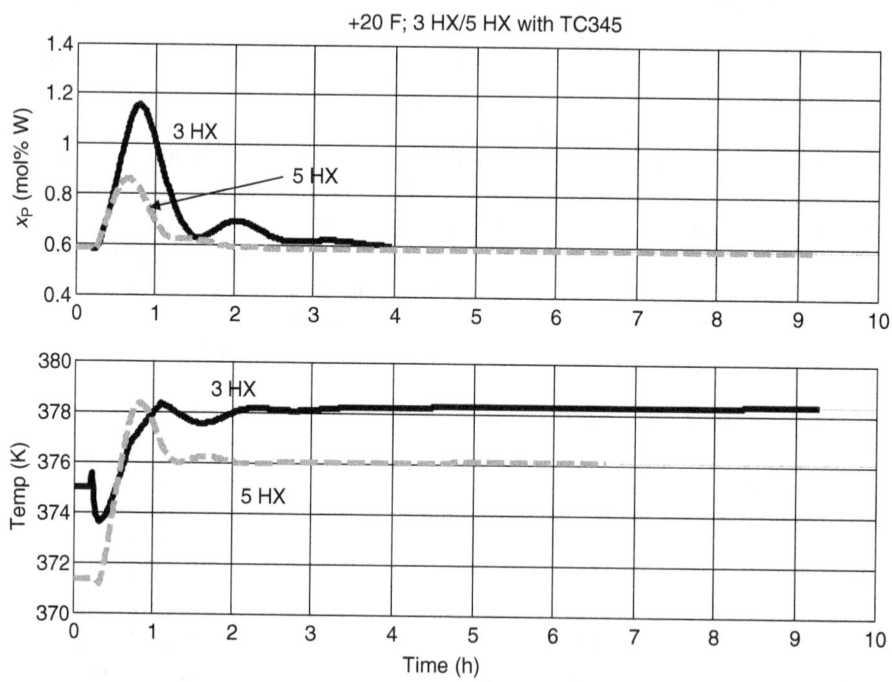


**Figure 14.9** Comparison of three and five heat-exchanger processes; +20% feed flowrate disturbance.

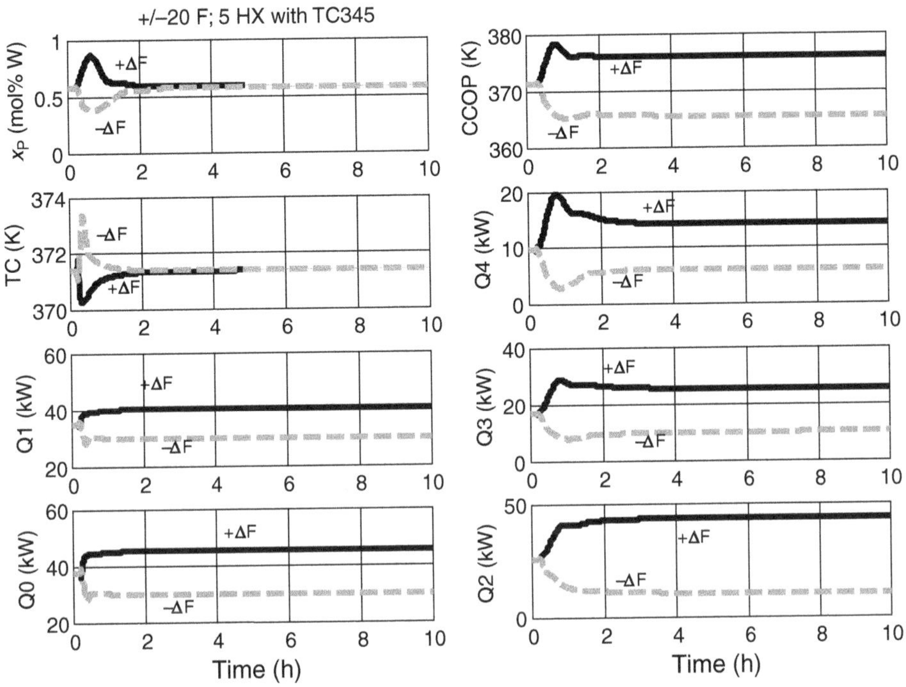


**Figure 14.10** Feed flowrate disturbances with five heat-exchanger process.

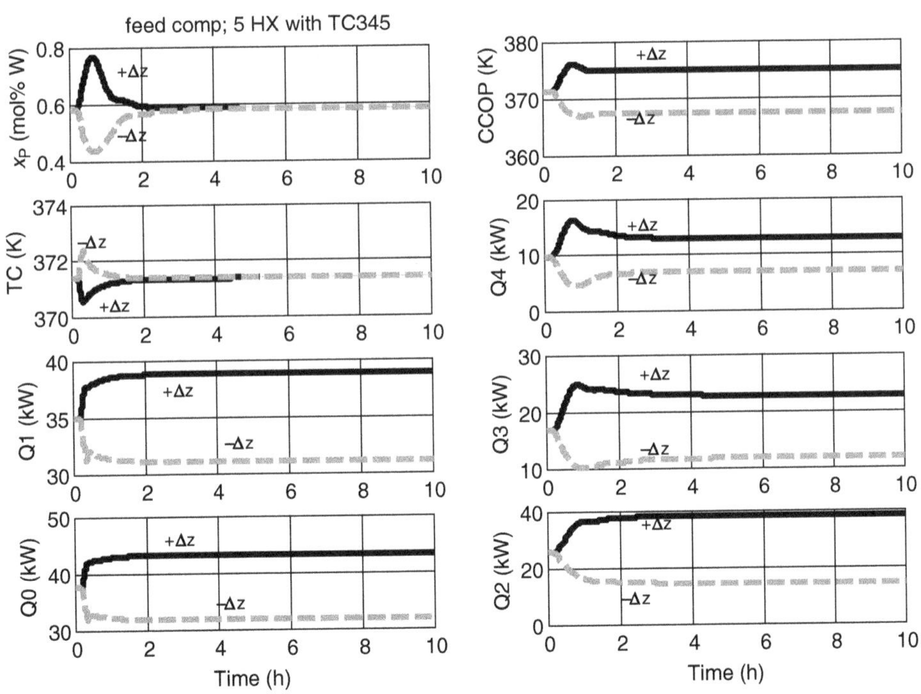


**Figure 14.11** Feed composition disturbances with five heat-exchanger process.

heat exchangers were used, saturation would occur for this size drop in feed. Figure 14.11 shows that feed composition disturbances are well handled.

Another type of disturbance would be to change the setpoint of the composition controller. Figure 14.12 gives results when the desired product impurity is reduced from the original 0.58 mol% water. At time = 0.2 h, the composition setpoint is reduced to 0.4 mol% water. The setpoints of the last three heaters are increased from 371 to 374 K to achieve the higher purity. The setpoints of the first two heaters remain at 371 K. At time = 5 h, the composition setpoint is further reduced to 0.2 mol% water. The setpoints of the last three heaters are increased again up to 378 K to achieve the still higher purity. The heat duties in all the heat exchangers increase, but the heat duty in the fifth heat exchanger changes very little after going through some transient responses.

It is interesting to explore what would happen if a pervaporation unit with less area, and therefore less holdup, were used. Would the dynamic control improve because of the smaller residence time of the retentate? Cutting the area in half (300 to 150 $m^2$ in each module), cutting the volume in half (0.45 $m^3$ in each module), and maintaining the same product composition requires that the feed temperatures of a five heat-exchanger process be increased from 371 to 381 K. Tuning the composition controller (using all five heat exchangers) gives an ultimate period of 53 min. This should be compared to the 75 min ultimate period for the larger area unit with the same control structure. Figure 14.13 shows that dynamic control improves in terms of settling time. The disturbance is a 20% increase in feed flowrate. The maximum deviation is not reduced but is actually slightly larger. This is due to the disturbance affecting the process more quickly when there is less capacitance to attenuate disturbances.

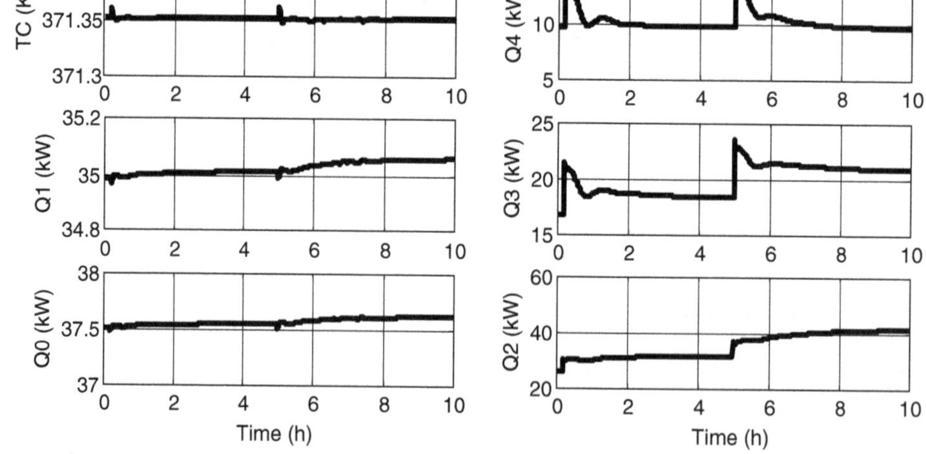


**Figure 14.12** Product purity setpoint changes.

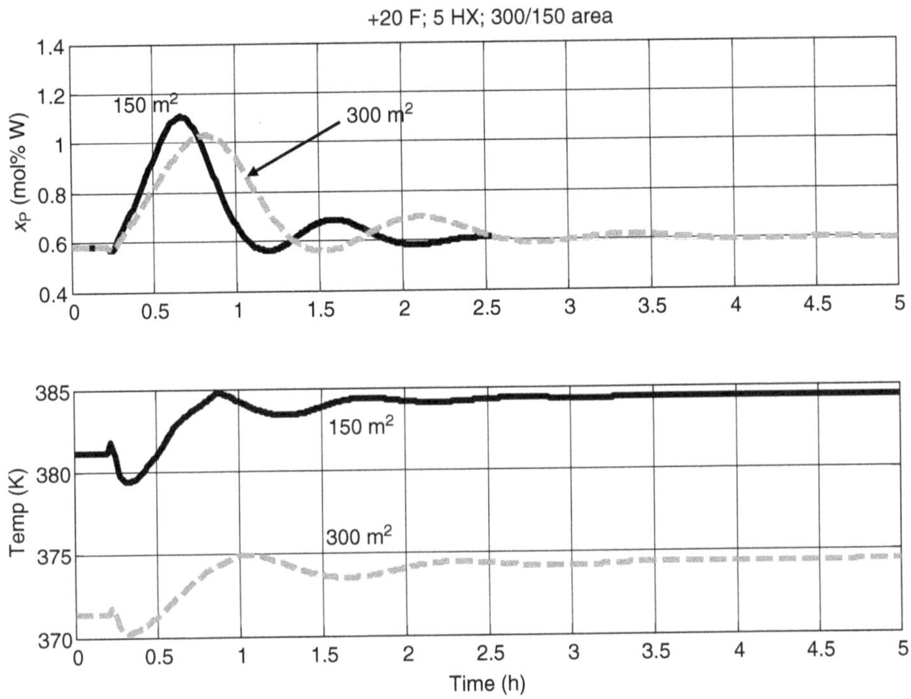


**Figure 14.13** Smaller membrane; +20% feed flowrate.

## 14.6 CONCLUSION

This chapter has studied the control of a column–pervaporation process for producing high-purity ethanol to overcome the azeotropic limitation encountered in distillation. A conventional control structure is developed that provides effective disturbance rejection for both production rate and feed composition changes. A simple pervaporation model is developed in Aspen Custom Modeler that captures the important dynamic features of the process. The model uses pervaporation characteristic performance curves to determine diffusivities. Component fluxes depend upon composition driving forces between the retentate and permeate sides of the membrane. The dynamics of the pervaporation cells are assumed to be dominated by composition and energy capacitance of the liquid retentate.

A composition cascade to retentate temperature control structure is demonstrated to provide stable regulatory control. The use of permeate pressure as a manipulated variable is shown to provide poor control because of the very small process gain between product purity and permeate pressure.

The dynamic controllabilities of two different process designs are compared. As expected, moving the manipulated variable (heat input to feed preheaters) closer to the controlled variable (retentate product composition from the last module) improved control. Smaller pervaporation units give faster dynamics.

## REFERENCES

1. Huang R. Y. M., Ed., *Pervaporation Membrane Separation Processes*, Elsevier, Amsterdam, 1991.

2. Rautenbach R. and R. Albrecht, *Membrane Processes*, Wiley, Chichester, 1989.
3. Sander U. and P. Soukup, Design and operation of a pervaporation plant for ethanol dehydration, *J. Membrane Sci.*, **36**, 463–475 (1988).
4. Guerreri G., Membrane alcohol separation process–integrated pervaporation and fractional distillation, *Trans. IChemE*, **70**(A), 501–508 (1992).
5. Lipnizki F., R. W. Field, and P. Ten, Pervaporation-based hybrid process: a review of process design, applications and economics, *J. Membrane Sci.*, **153**, 183–210 (1999).
6. Feng X. and R. Y. M. Huang, Liquid separation by membrane pervaporation, *Ind. Engng. Chem. Res.*, **36**, 1048–1066 (1997).
7. Daviou M. C., P. M. Hoch, and A. M. Eliceche, Design of membrane modules used in hybrid distillation/pervaporation systems, *Ind. Engng. Chem. Res.*, **43**, 3403–3412 (2004).
8. Van Hoof V., L. Van den Abeele, A. Buekenhoudt, C. Dotremont, and R. Leysen, Economic comparison between azeotropic distillation and different hybrid systems combining distillation with pervaporation for the dehydration of isopropanol, *Separation and Purification Tech.*, **37**, 33–49 (2004).
9. Ryan P. J. and M. F. Doherty, Design/optimization of ternary heterogeneous azeotropic distillation systems, *AIChE J.*, **35**, 1592–1601 (1989).
10. Geankoplis C. J., *Transport Processes and Unit Operations*, Third Edition, Prentice-Hall, Upper Saddle River, NJ, 1993, p. 761.

# INDEX



*Design and Control of Distillation Systems for Separating Azeotropes*. By William L. Luyben and I-Lung Chien

www.ingramcontent.com/pod-product-compliance
Lightning Source LLC
LaVergne TN
LVHW062108200926
848484LV00007B/1617
* 9 7 8 0 4 7 0 4 4 8 6 2 5 *